Modern Lens Design

Optical and Electro-Optical Engineering Series
Robert E. Fischer and Warren J. Smith, *Series Editors*

Published

ALLARD • *Fiber Optics Handbook*
HECHT • *The Laser Guidebook*
NISHIHARA, HARUNA, SUHARA • *Optical Integrated Circuits*
RANCOURT • *Optical Thin Films Users' Handbook*
SIBLEY • *Optical Communications*
SMITH • *Modern Optical Engineering*
STOVER • *Optical Scattering*
WYATT • *Electro-Optical System Design*

Other Published Books of Interest

CSELT • *Fiber Optic Communications Handbook*
KAO • *Optical Fiber Systems*
KEISER • *Optical Fiber Communications*
MACLEOD • *Thin Film Optical Filters*
OPTICAL SOCIETY OF AMERICA • *Handbook of Optics*

Modern Lens Design

A Resource Manual

Warren J. Smith
Chief Scientist
Kaiser Electro-Optics, Inc.
Carlsbad, California

Genesee Optics Software, Inc.
Rochester, New York

McGraw-Hill, Inc.
New York St. Louis San Francisco Auckland Bogotá
Caracas Lisbon London Madrid Mexico Milan
Montreal New Delhi Paris San Juan São Paulo
Singapore Sydney Tokyo Toronto

Library of Congress Cataloging-in-Publication Data

Smith, Warren J.
Modern lens design : a resource manual / Warren J. Smith and Genesee Optics Software, Inc.
p. cm.—(Optical and electro-optical engineering series)
Includes index.
ISBN 0-07-059178-4
1. Lenses—Design and construction—Handbooks, manuals, etc.
I. Genesee Optics Software, Inc. II. Title. III. Series.
QC385.2.D47S65 1992
681'.423—dc20 92-20038
CIP

1 2 3 4 5 6 7 8 9 0 DOC/DOC 9 8 7 6 5 4 3 2

ISBN 0-07-059178-4

The sponsoring editor for this book was Daniel A. Gonneau, the editing supervisor was David E. Fogarty, and the production supervisor was Suzanne W. Babeuf. It was set in Century Schoolbook by McGraw-Hill's Professional Book Group composition unit.

Printed and bound by R. R. Donnelley & Sons Company.

Contents

Preface

This book had its inception in the early 1980s, when Bob Fischer and I, as coeditors of the then Macmillan, now McGraw-Hill, Series on Optical and Electro-Optical Engineering, were planning the sort of books we wanted in the series. The concept was outlined initially in 1982, and an extensive proposal was submitted to, and accepted by, Macmillan in 1986. At this point my proposed collaborators elected to pursue other interests, and the project was put on the shelf until it was revived by the present set of authors.

My coauthor is Genesee Optics Software, Inc. Obviously the book is the product of the work of real people, i.e., myself and the staff of Genesee. In alphabetical order, the Genesee personnel who have been involved are Charles Dubois, Henry Gintner, Robert MacIntyre, David Pixley, Lynn VanOrden, and Scott Weller. They have been responsible for the computerized lens data tables, lens drawings, and aberration plots which illustrate each lens design.

Many of the lens designs included in this book are from OPTICS TOOLBOX® (a software product of Genesee Optics Software), which was originally authored by Robert E. Hopkins and Scott W. Weller. OPTICS TOOLBOX is a collection of lens designs and design commentary within an expert-system, artificial-intelligence, relational data base.

This author's optical design experience has spanned almost five decades. In that period lens design has undergone many radical changes. It has progressed from what was a semi-intuitive art practiced by a very small number of extremely patient and dedicated lovers of detail and precision. These designers used a very limited amount of laborious computation, combined with great understanding of lens design principles and dogged perseverance to produce what are now the classic lens design forms. Most of these design forms are still the best, and as such are the basis of many modern optical systems. However, the manner in which lenses are designed today is almost completely different in both technique and philosophy. This change is, of course, the result of the vastly increased computational speed now available to the lens designer.

In essence, much modern lens design consists of the selection of a starting lens form and its subsequent optimization by an automatic lens design program, which may or may not be guided or adjusted along the way by the lens designer. Since the function of the lens design program is to drive the design form to the nearest local optimum (as defined by a *merit function*), it is obvious that the starting design form and the merit function together uniquely define which local optimum design will be the result of this process.

Thus it is apparent that, in addition to a knowledge of the principles of optical design, a knowledge of appropriate starting-point designs and of techniques for guiding the design program have become essential elements of modern lens design. The lens designs in this book have been chosen to provide a good selection of starting-point designs and to illustrate important design principles. The design techniques described are those which the author has found to be useful in designing with an optimization program. Many of the techniques have been developed or refined in the course of teaching lens design and optical system design; indeed, a few of them were initially suggested or inspired by my students.

In order to maximize their usefulness, the lens designs in this book are presented in three parts: the lens prescription, a drawing of the lens which includes a marginal ray and a full-field principal ray, and a plot of the aberrations. The inclusion of these two rays allows the user to determine the approximate path of any other ray of interest. For easy comparison, all lenses are shown at a focal length of approximately 100, regardless of their application. The performance data is shown as aberration plots; we chose this in preference to MTF plots because the MTF is valid only for the focal length for which it was calculated, and because the MTF cannot be scaled. The aberration plots *can* be scaled, and in addition they indicate what aberrations are present and show which aberrations limit the performance of the lens. We have expanded on the usual longitudinal presentation of spherical aberration and curvature of field by adding ray intercept plots in three colors for the axial, 0.7 zonal, and full-field positions. We feel that this presentation gives a much more complete, informative, and useful picture of the characteristics of a lens design.

This book is intended to build on some knowledge of both geometrical optics and the basic elements of lens design. It is thus, in a sense, a companion volume to the author's *Modern Optical Engineering*, which covers such material at some length. Presumably the user of this text will already have at least a reasonable familiarity with this material.

There are really only a few well-understood and widely utilized principles of optical design. If one can master a thorough understanding of these principles, their effects, and their mechanisms, it is easy to recognize them in existing designs and also easy to apply them to one's own design work. It is our intent to promote such understanding by presenting both expositions and annotated design examples of these principles.

Readers are free to use the designs contained in this book as starting points for their own design efforts, or in any other way they see fit. Most of the designs presented have, as noted, been patented; such designs may or may not be currently subject to legal protection, although there may, of course, be differences of opinion as to the effectiveness of such protection. The reader must accept full responsibility for meeting whatever limitations are imposed on the use of these designs by any patent or copyright coverage (whether indicated herein or not).

Warren J. Smith

Modern Lens Design

Chapter 1

Introduction

Modern Lens Design is intended as an aid to lens designers who work with the many commercially available lens design computer programs. We assume that the reader understands basic optical principles and may, in fact, have a command of the fundamentals of classical optical design methods. For those who want or need information in these areas, the following books should prove helpful. This author's *Modern Optical Engineering: The Design of Optical Systems,* 2d ed., McGraw-Hill, 1990, is a comprehensive coverage of optical system design; it includes two full chapters which deal specifically with lens design in considerable detail. Rudolf Kingslake's *Optical System Design* (1983), *Fundamentals of Lens Design* (1978), and *A History of the Photographic Lens* (1989), all by Academic Press, are complete, authoritative, and very well written.

Authoritative books on lens design are rare, especially in English; there are only a few others available. The Kingslake series *Applied Optics and Optical Engineering,* Academic Press, contains several chapters of special interest to lens designers. Volume 3 (1965) has chapters on lens design, photographic objectives, and eyepieces. Volume 8 (1980) has chapters on camera lenses, aspherics, automatic design, and image quality. Volume 10 (1987) contains an extensive chapter on afocal systems. Milton Laikin's *Lens Design,* Marcel Dekker, 1991, is a volume similar to this one, with prescriptions and lens drawings. Its format differs in that no aberration plots are included; instead, modulation transfer function (MTF) data for a specific focal length and *f* number are given. Now out of print, Arthur Cox's *A System of Optical Design,* Focal Press, 1964, contains a complete, if unique, approach to lens design, plus prescriptions and (longitudinal) aberration plots for many lens design patents.

This book has several primary aims. It is intended as a source book for a variety of designed lens types which can serve as suitable starting points for a lens designer's efforts. A study of the comparative characteristics of the annotated designs contained herein should also illustrate the application of many of the classic lens design principles. It is also intended as a handy, if abridged, reference to many of the equations and relationships which find frequent use in lens design. Most of these are contained in the Formulary at the end of the book. And last, but not least, the text contains extensive discussions of design techniques which are appropriate to modern optical design with an automatic lens design computer program.

The book begins with a discussion of automatic lens design programs and how to use them. The merit function, optimization, variables, and the various techniques which are useful in connection with a program are covered. Chapter 3 details many specific improvement strategies which may be applied to an existing design to improve its performance. The evaluation of a design is discussed from the standpoint of ray and wave aberrations, and integrated with such standard measures as MTF and Strehl ratio. The sample lens designs follow. Each presents the prescription data, a drawing of the lens with marginal and chief rays, and an aberration analysis consisting of ray intercept plots for three field angles, longitudinal plots of spherical aberration and field curvature, and a plot of distortion. A discussion of the salient features of each design accompanies the sample designs, and comments (in some cases quite extensive) regarding the design approach are given for each class of lens. The Formulary, intended as a convenient reference, concludes the book.

The design of the telescope objective is covered in Chap. 6, beginning with the classic forms and continuing with several possible modifications which can be used to improve the aberration correction. These are treated in considerable detail because they represent techniques which are generally applicable to all types of designs. For similar reasons, Chap. 8 deals with the basic principles of airspaced anastigmats in a rather extended treatment. The complexities of the interrelationships involved in the Cooke triplet anastigmat are important to understand, as are the (almost universal) relationships between the vertex length of an ordinary anastigmat lens and its capabilities as regards speed and angular coverage.

Chapter

2

Automatic Lens Design: Managing the Lens Design Program

2.1 The Merit Function

What is usually referred to as *automatic lens design* is, of course, nothing of the sort. The computer programs which are so described are actually optimization programs which drive an optical design to a local optimum, as defined by a *merit function* (which is not a true merit function, but actually a defect function). In spite of the preceding disclaimers, we will use these commonly accepted terms in the discussions which follow.

Broadly speaking, the merit function can be described as a combination or function of calculated characteristics, which is intended to completely describe, with a single number, the value or quality of a given lens design. This is obviously an exceedingly difficult thing to do. The typical merit function is the sum of the squares of many image defects; usually these image defects are evaluated for three locations in the field of view (unless the system covers a very large or a very small angular field). The squares of the defects are used so that a negative value of one defect does not offset a positive value of some other defect.

The defects may be of many different kinds; usually most are related to the quality of the image. However, any characteristic which can be calculated may be assigned a target value and its departure from that target regarded as a defect. Some less elaborate programs utilize the third-order (Seidel) aberrations; these provide a rapid and efficient way of adjusting a design. These cannot be regarded as optimizing the image quality, but they do work well in correcting ordinary lenses. Another type of merit function traces a large number of rays from an object point. The radial distance of the image plane intersection of the ray from the centroid of all the ray intersections is then the image defect. Thus the merit function is effectively the sum of the root-mean-square (rms) spot sizes for several field angles. This type of merit function, while inefficient in that it requires many rays to be traced, has the advantage that it is both versatile and in some ways relatively foolproof. Some merit functions calculate the values of the classical aberrations, and convert (or weight) them into their equivalent wavefront deformations. (See Formulary Sec. F-12 for the conversion factors for several common aberrations.) This approach is very efficient as regards computing time, but requires careful design of the merit function. Still another type of merit function uses the variance of the wavefront to define the defect items. The merit function used in the various David Grey programs is of this type, and is certainly one of the best of the commercially available merit functions in producing a good balance of the aberrations.

Characteristics which do not relate to image quality can also be controlled by the lens design program. Specific construction parameters, such as radii, thicknesses, spaces, and the like, as well as focal length, working distance, magnification, numerical aperture, required clear apertures, etc., can be controlled. Some programs include such items in the merit function along with the image defects. There are two drawbacks which somewhat offset the neat simplicity of this approach. One is that if the first-order characteristics which are targeted are not initially close to the target values, the program may correct the image aberrations without controlling these first-order characteristics; the result may be, for example, a well-corrected lens with the wrong focal length or numerical aperture. The program often finds this to be a local optimum and is unable to move away from it. The other drawback is that the inclusion of these items in the merit function has the effect of slowing the process of improving the image quality. An alternative approach is to use a system of constraints outside the merit function. Note also that many of these items can be controlled by features which are included in almost all programs, namely *angle-solves* and *height-solves*. These algebraically solve for a radius or space to produce a desired ray slope or height.

In any case, the merit function is a summation of suitably weighted defect items which, it is hoped, describes in a single number the worth of the system. The smaller the value of the merit function, the better the lens. The numerical value of the merit function depends on the construction of the optical system; it is a function of the construction parameters which are designated as variables. Without getting into the details of the mathematics involved, we can realize that the merit function is an n-dimensional space, where n is the number of the variable constructional parameters in the optical system. The task of the design program is to find a location in this space (i.e., a lens prescription or a solution vector) which minimizes the size of the merit function. In general, for a lens of reasonable complexity there will be many such locations in a typical merit function space. The automatic design program will simply drive the lens design to the nearest minimum in the merit function.

2.2 Optimization

The lens design program typically operates this way: Each variable parameter is changed (one at a time) by a small increment whose size is chosen as a compromise between a large value (to get good numerical accuracy) and a small value (to get the *local* differential). The change produced in every item in the merit function is calculated. The result is a matrix of the partial derivatives of the defect items with respect to the parameters. Since there are usually many more defect items than variable parameters, the solution is a classical least-squares solution. It is based on the assumption that the relationships between the defect items and the variable parameters are linear. Since this is usually a false assumption, an ordinary least-squares solution will often produce an unrealizable lens or one which may in fact be worse than the starting design. The *damped least-squares* solution, in effect, adds the weighted squares of the parameter changes to the merit function, heavily penalizing any large changes and thus limiting the size of the changes in the solution. The mathematics of this process are described in Spencer, "A Flexible Automatic Lens Correction Program," *Applied Optics,* vol. 2, 1963, pp. 1257–1264, and by Smith in W. Driscoll (ed.), *Handbook of Optics,* McGraw-Hill, New York, 1978.

If the changes are small, the nonlinearity will not ruin the process, and the solution, although an approximate one, will be an improvement on the starting design. Continued repetition of the process will eventually drive the design to the nearest local optimum.

One can visualize the situation by assuming that there are only two variable parameters. Then the merit function space can be compared to a landscape where latitude and longitude correspond to the variables and the elevation represents the value of the merit function. Thus the starting lens design is represented by a particular location in the landscape and the optimization routine will move the lens design downhill until a minimum elevation is found. Since there may be many depressions in the terrain of the landscape, this optimum may not be the best there is; it is a local optimum and there can be no assurance (except in very simple systems) that we have found a global optimum in the merit function. This simple topological analogy helps to understand the dominant limitations of the optimization process: the program finds the nearest minimum in the merit function, and that minimum is uniquely determined by the design coordinates at which the process is begun. The landscape analogy is easy for the human mind to comprehend; when it is extended to a 10- or 20-dimension space, one can realize only that it is apt to be an extremely complex neighborhood.

2.3 Local Minima

Figure 2.1 shows a contour map of a hypothetical two-variable merit function, with three significant local minima at points *A*, *B*, and *C*; there are also three other minima at *D*, *E*, and *F*. It is immediately apparent that if we begin an optimization at point *Z*, the minimum at point *B* is the only one which the routine can find. A start at *Y* on the ridge at the lower left will go to the minimum at *C*. However, a start

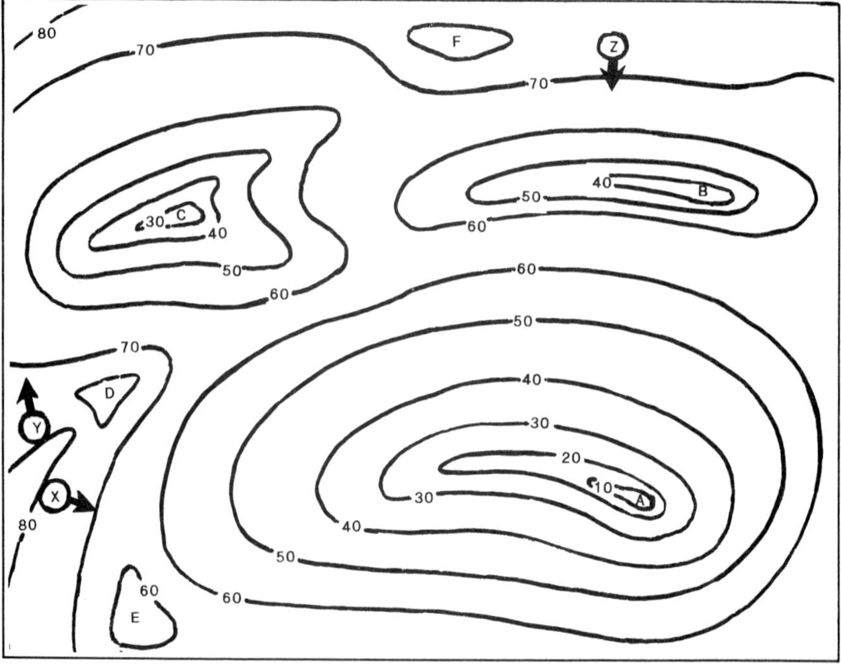

Figure 2.1 Topography of a hypothetical two-variable merit function, with three significant minima (*A*, *B*, *C*) and three trivial minima (*D*, *E*, *F*). The minimum to which a design program will go depends on the point at which the optimization process is started. Starting points *X*, *Y*, and *Z* each lead to a different design minimum; other starting points can lead to one of the trivial minima.

at X, which is only a short distance away from Y, will find the best minimum of the three, at point A. If we had even a vague knowledge of the topography of the merit function, we could easily choose a starting point in the lower right quadrant of the map which would guarantee finding point A. Note also that a modest change in any of the three starting points could cause the program to stagnate in one of the trivial minima at D, E, or F. It is this sort of minimum from which one can escape by "jolting" the design, as described below.

The fact that the automatic design program is severely limited and can find only the nearest optimum emphasizes the need for a knowledge of lens design, in order that one can select a starting design form which is close to a good optimum. This is the only way that an automatic program can systematically find a good design. If the program is started out near a poor local optimum, the result is a poor design.

The mathematics of the damped least-squares solution involves the inversion of a matrix. In spite of the damping action, the process can be slowed or aborted by either of the following conditions: (1) A variable which does not change (or which produces only a very small change in) the merit function items. (2) Two variables which have the same, nearly the same, or scaled effects on the items of the merit function. Fortunately, these conditions are rarely met exactly, and they can be easily avoided.

If the program settles into an unsatisfactory optimum (such as those at D, E, and F in Fig. 2.1) it can often be jolted out of it by manually introducing a significant change in one or more parameters. The trick is to make a change which is in the direction of a better design form. (Again, a knowledge of lens designs is virtually a necessity.) Sometimes simply freezing a variable to a desirable form can be sufficient to force a move into a better neighborhood. The difficulty is that too big a change may cause rays to miss surfaces or to encounter total internal reflection, and the optimization process may break down. Conversely, too small a change may not be sufficient to allow the design to escape from a poor local optimum. Also, one should remember that if the program is one which adjusts (optimizes) the damping factor, the factor is usually made quite small near an optimum, because the program is taking small steps and the situation looks quite linear; after the system is jolted, it is probably in a highly nonlinear region and a big damping factor may be needed to prevent a breakdown. A manual increase of the damping factor can often avoid this problem.

Another often-encountered problem is a design which persists in moving to an obviously undesirable form (when you *know* that there is a much better, very different one—the one that you want). Freezing the form of one part of the lens for a few cycles of optimization will often allow the rest of the lens to settle into the neighborhood of the desired optimum. For example, if one were to try to convert a Cooke triplet into a split front crown form, the process might produce either a form which is like the original triplet with a narrow airspaced crack in the front crown, or a form with rather wild meniscus elements. A technique which will usually avoid these unfortunate local optima in this case is to freeze the front element to a plano-convex form by fixing the second surface to a plane for a few cycles of optimization. Again, one must know which lens forms are the good ones.

2.4 Types of Merit Functions

Many programs allow the user to define the merit function. This can be a valuable feature because it is almost impossible to design a truly *universal* merit function. As an example, consider the design of a simple Fraunhofer telescope objective: a merit function which controls the spherical and chromatic aberrations of the axial marginal ray and the coma of the oblique ray bundle (plus the focal length) is all that is necessary. If the design complexity is increased by allowing the airspace to vary and/or adding another element, the merit function may then profitably include entries which will control zonal spherical, spherochromatism, and/or fifth-order coma. But as long as the lens is thin and in contact with the aperture stop, it would be foolish to include in the merit function entries to control field curvature and astigmatism. There is simply no way that a thin stop-in-contact lens can have any control over the inherent large negative astigmatism; the presence of a target for this aberration in the merit function will simply slow down the solution process. It would be ridiculous to use a merit function of the type required for a photographic objective to design an ordinary telescope objective. (Indeed, an attempt to correct the field curvature may lead to a compromise design with a severely undercorrected axial spherical aberration which, in combination with coma, may fool the computer program into thinking that it has found a useful optimum.)

There are many design tasks in this category, where the requirements are effectively limited in number and a simple, equally limited merit function is clearly the best choice. In such cases, it is usually obvious that some specific state of correction will yield the best results; there is no need to *balance* the correction of one aberration against another.

More often, however, the situation is not so simple; compromises and balances are required and a more complex, suitably weighted merit function is necessary. This can be a delicate and somewhat tricky matter. For example, in the design of a lens with a significant aperture and field, there is almost always a (poor) local optimum in which (1) the spherical aberration is left quite undercorrected, (2) a compromise focus is chosen well inside the paraxial focus, (3) the Petzval field is made inward-curving, and (4) overcorrected oblique spherical aberration is introduced to "balance" the design. A program which relies on the rms spot radius for its merit function is very likely to fall into this trap. A better design usually results if the spherical (both axial and oblique) aberrations are corrected, the Petzval curvature is reduced, and a small amount of overcorrected astigmatism is introduced. When one recognizes this sort of situation, it is a simple matter to adjust the weighting of the appropriate targets in the merit function to force the design into a form with the type of aberration balance which is desired. Another way to avoid this problem is to force the system to be evaluated/designed at the paraxial focus rather than at a compromise focus, i.e., to not allow defocusing. As can be seen, the design of a general-purpose merit function which will optimally balance a wide variety of applications is not a simple matter.

Although it is not always necessary, there are occasions when it is helpful to begin the design process by controlling only the first-order properties (image size, image location, spatial limitations, etc.). Then one proceeds to control the chromatic and perhaps the Petzval aberrations. (Things may even go better if the first-order and the chromatic are fairly completely worked out by hand before submitting the system to an automatic design process.) The next step in the sequence is to correct the primary aberrations (spherical, coma, astigmatism, and distortion), either directly or by using the Seidel coefficients, and finally proceed to balancing and correcting the higher-order residuals. This sort of ordered approach is sometimes useful (or even necessary) when one is exploring terra incognita, and, of course, it requires a user-defined merit function if it is to be implemented.

2.5 Stagnation

Sometimes the automatic design process will stagnate and the convergence toward a solution will become so slow as to be imperceptible. This can result from being in a very flat and broad optimum in the merit function. It can also result from an ill-designed merit function. Often first-order properties which are specified in the merit function are the cause of the problem. It is only too easy to require contradictory or redundant characteristics. This is especially true for zoom lenses or multiconfiguration systems, which can be confusingly complex. When stagnation occurs, or convergence is slower than you know it should be, it is wise to stop and examine the merit function for problems. Look critically at every item in the merit function and consider what it is intended to be doing and what it *actually* does. Eliminate redundancies and try to make each entry in the merit function explicitly control its intended characteristic. Stagnation may also result from a starting design which is so far from a solution that differential changes to the variables have a negligible effect.

2.6 Generalized Simulated Annealing

The discussions above have centered on the standard damped least-squares program, or its equivalent. There have been several versions of *random search* programs proposed in the past. The most recent of these is quite sophisticated and is called *generalized simulated annealing.* In this, the computer randomly selects the lens dimensions (within a limited range and according to some probability distribution) and evaluates the resulting lens prescription. If the new version is better than the old, it is unconditionally accepted. If it is worse, it *may* be accepted, on the basis of random chance, weighted by a probability function which reduces the chance of acceptance in proportion to the amount that the lens is worse than the original form. This sort of approach obviously allows the program an easy escape from the local minima described with Fig. 2.1, but it equally obviously requires a very large number of trials before a random chance can find a good combination of dimensions for the lens. Nonetheless, it does work, but not rapidly. Perhaps as computers increase in speed, a program of this sort will displace or supplement the damped least-squares as the routine of choice for automatic lens design.

2.7 Considerations about Variables for Optimization

The potential variables for use in optimization include: the surface curvatures, conic constants, and asphericities; the surface spacings; and the refractive characteristics of the materials involved. Occasionally tilts and decentrations are also included as variables.

Materials

Although the material characteristics are not continuous variables, for optical glasses at least, the index and dispersion (or V value) can be varied within the boundaries of the glass map (Fig. 2.2) as if they were. The real glass nearest the optimized values can be substituted for the optimized glass to achieve nearly the same resultant design after another cycle or two of optimization with the real glass. Note that this is *not* true for partial dispersions, since there are relatively few glasses with partial dispersions unusual enough to be useful in the

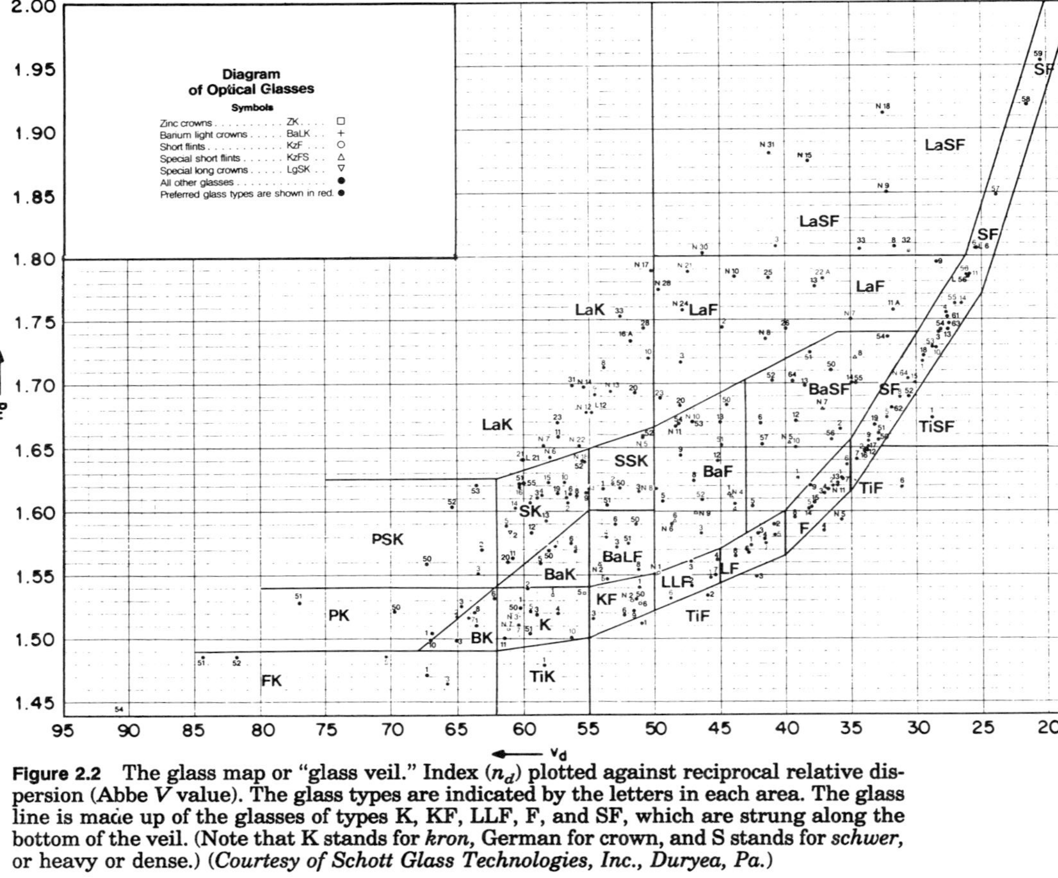

Figure 2.2 The glass map or "glass veil." Index (n_d) plotted against reciprocal relative dispersion (Abbe V value). The glass types are indicated by the letters in each area. The glass line is made up of the glasses of types K, KF, LLF, F, and SF, which are strung along the bottom of the veil. (Note that K stands for *kron*, German for crown, and S stands for *schwer*, or heavy or dense.) (*Courtesy of Schott Glass Technologies, Inc., Duryea, Pa.*)

correction of secondary spectrum. Obviously, for applications outside the spectral regions where optical glass is usable, one cannot treat the refractive characteristics as variables, since the available materials tend to be few and far between.

In many of the simpler types of designs it is essential to allow the glass characteristics to vary. In the Cooke triplet for example, the relationship between the *V* values of the crown and flint elements determines the overall length of the lens. As described in Sec. 8.3, the length of a triplet (and that of most anastigmats) determines the amount of higher-order spherical aberration and astigmatism; these in turn determine the aperture and field coverage capabilities of the lens. If these types of lenses are to be optimized to suit the application at hand, the glass characteristics *must* be allowed to vary.

Some optimization programs have difficulty with the bounds of the glass map; if this is a problem, the optimization process is often facilitated by starting the variable glass well away from the boundary, so that it can find its best value before encountering the boundary problem.

It is often better to vary the flint glasses than to vary the crowns. This is because the crowns usually tend to go to the upper left corner (high index, high *V* value) of the glass map. Flints head for the lower right corner, and are then, of course, constrained to lie on the glass line. The glasses along the glass line are numerous, inexpensive, and almost universally well behaved. On the other hand, the crown glasses in the upper left corner include in their number many which are expensive and/or easily attacked by the environment. Thus one might be willing to accept the computer's choice of a glass along the glass line, but would prefer to make a more discriminating selection from among the others.

Curvatures

In general, one would expect to want to make use of every available variable. This is almost always true regarding the curvatures, all of which, unless there is a reason to constrain the shape of an element, are usually allowed to vary.

Airspaces

Ordinarily, airspaces may be regarded in the same light as curvatures, since they are continuously variable and are very effective variables.

Defocusing

Although the distance by which the design image plane departs from the paraxial focus is usually an airspace, and can be regarded as a variable, its effects can be insidious. If the image surface is allowed to depart from the paraxial focus from the beginning of the optimization process, an unfortunate lens may result. In some lenses, and with some optimization merit functions, the tendency is to produce a lens with:

1. The image plane well inside the paraxial focus
2. A large undercorrected spherical aberration
3. A strongly inward-curving field
4. A heavy overcorrecting oblique spherical

Although this combination occupies a local optimum in the merit function, this is usually not the best state of correction. It fools the optimization program because the undercorrected spherical causes the best axial focus to lie to the left of the paraxial focus and the overcorrected oblique spherical causes the best off-axis focus to lie to the right of the inward-curving field; the net result is that, to the program, the field seems flat. One can usually avoid this pitfall by not allowing any defocus in the early stages of the optimization and/or putting a heavy penalty on the defocusing. Note that for *some* lenses, such as non-diffraction-limited systems used with detectors, the correction described above may in fact be a good one. See the comments on aberration balance in Sec. 3.8.

Thickness

Element thicknesses must be regarded quite differently than airspaces. They must of course be bounded by the necessity for a practical edge thickness for the positive elements and a reasonable center thickness for the negative elements. In many designs, element thickness is an insignificant and ineffective variable (and one whose effects are easily duplicated by an adjacent airspace). In this circumstance one can arbitrarily select a thickness on the basis of economy or ease of fabrication. The elements in such designs are typically quite thin; see, for example, a telescope objective or an ordinary Cooke triplet.

There are, however, many systems in which the element thickness is not only an effective variable, but one which is essential to the success of the design type. The older meniscus lenses (Protar, Dagor, etc.) and the double-Gauss forms depend on the separation of the concave and convex surfaces of their thick meniscus components to control the Petzval curvature and, in many instances, the higher-order aberrations. In lenses of this type it is absolutely essential that these glass thicknesses be allowed to vary.

One must be wary of and skeptical toward a thickness variable which is very weakly effective. Occasionally an optimization program will produce a design with an overly thick element, where the large thickness produces only a very small improvement (which is not worth the added cost of producing the thick element). This occurs because the optimization routine will seek out any improvement that it can get, no matter how small, and without concern as to the cost. It is wise to test the value of a thick element if there is any doubt about its utility. This is readily accomplished with another optimization run which fixes the thickness in question to a smaller value. Very often the performance of the thin version will not be noticeably different from that of the "optimum" thicker version. Although most significant with respect to lens thickness, this same rationale obviously applies to airspaces as well.

Aspheric surfaces

Surface asphericity can be an extremely effective (if sometimes expensive) variable, but it is one that often requires a bit of finesse. On occasion, one may be ill-advised to begin an optimization with the conic constant and all the aspheric deformation coefficients used simultaneously as variables. The conic constant and the fourth-order deformation coefficient both affect the third-order aberrations in exactly the same way. Thus they are at least partially redundant, but more significantly, identical variables have an undesirable effect on the mathematics of the optimization process. It is often advisable to vary one or the other, but not both. A safe practice is to vary only the conic constant (or the fourth-order term) at first, and then add the higher-order terms (sixth, eighth, tenth) one at a time, as necessary. The tenth-order term is, in many systems, totally unnecessary, adding little or nothing to the quality of the system; in fact, the eighth-order term is often something that can be done without.

A surface defined by a tenth-order polynomial can cause the spherical aberration to be corrected exactly to zero at four ray heights. If there are only four axial rays in the merit function, their ray intercept errors may all be brought to zero; the danger is that, between these rays, the residual aberration may be unacceptably large. A tenth-order surface can be a rather extreme shape. Thus the use of an aspheric surface sometimes calls for more rays in the merit function than one might otherwise expect to need. With a program which allows wavefront deformation or optical path difference (OPD) targets in the merit function, the severity of this problem can be lessened.

2.8 How to Increase the Speed or Field of a System and Avoid Ray Failure Problems

Very often, the lens designer is faced with the necessity of increasing the speed (i.e., relative aperture, numerical aperture, etc.) and/or the field of view of an existing optical system. There are two common reasons why this may be desirable. One may want to adapt an existing design (such as those in this book) to an application which requires a larger aperture or wider field than that for which the original lens has been configured. The other common situation is simply the creation of an entirely new system with a relatively large aperture and/or field. In either case, the difficulty which can arise is that the rays which are needed to design the system may not be able to get through the initial lens prescription.

There are two reasons that a ray may not be able to get through. One reason is that the height of the ray at a surface may be greater than the radius of the surface; the ray simply misses the surface entirely and its path obviously cannot be calculated any further. The second reason is that the ray may encounter total internal reflection (TIR) in passing from a higher index to a lower; again, the ray path cannot be calculated. Each of these conditions represents a boundary which, if crossed, causes failure of the ray trace. Note also that, as these boundaries are approached, the situation rapidly becomes very unstable. This is because, close to the boundary, the angle of incidence (for the case of the ray approaching the value of the radius) or the angle of refraction (for the case of the ray approaching TIR) is very near to 90°. Near this angle, Snell's law of refraction ($n \sin I = n' \sin I'$) becomes *very* nonlinear, producing a highly unstable situation which often explodes as the lens construction parameters are incremented in the course of the optimization.

A good way of dealing with this situation is simply to back off from the aperture and/or field requirement that is causing the problem. Many design programs have the capability to easily adjust or scale the aperture and field angle. If a change is made to smaller values of aperture or field, the rays will no longer be so near to the failure boundary. If the lens is now optimized, the program is very likely to adjust the lens parameters so as to reduce the angles of incidence, because this is usually a factor which causes the aberrations to be reduced. The optimizing changes can thus be expected to pull the problem situation in the system further away from the ray failure boundary, *if this is possible.*

After the optimization has relaxed the problem, the field and/or the aperture can usually be adjusted (scaled) to a moderately larger value without again encountering the failure boundary. Depending on just how sensitive the system is to the problem, an increase of about 10 to 50 percent may be appropriate. Now another cycle of optimization will strongly tend to again reduce the troublesome angles of incidence.

This process of adjusting (scaling) the field or the aperture to larger values and then optimizing is continued until the desired aperture or field is attained without ray failures. This works well, *provided* that this desired result is possible for the lens configuration which is under study. It may be necessary to choose another configuration, usually one with more elements. When this is necessary, a drawing of the lens and rays (of the last design form which has successfully passed all the rays) will usually indicate which rays and which surfaces are causing the problem. One simply looks for angles of incidence or refraction which are large (and often near 90°). Then the offending element can be split into two (or more) elements which are shaped to reduce these angles. The scale-and-optimize process can now be repeated with a much improved chance of success. Note that, in general, the examination of the critical ray paths (typically those of the marginal rays) for large angles of incidence or refraction is a technique which will often indicate the source of a design problem.

2.9 Test Plate Fits, Melt Fits, and Thickness Fits

When the deleterious effects of fabrication tolerances become too large to bear, a technique commonly used to reduce these effects is to fit the lens design to the known values for the radius tooling and/or to the measured glass indices. The former is called a *test plate* (or *test glass* or *tooling*) fit; the latter is called a *melt fit*.

The *test plate fit* is begun by first obtaining a list of the available test plate radii from the shop which is scheduled to fabricate the lens. It is wise to ascertain that the radius values of the list are not just nominal values, but are based on accurate and recent measurements of the test plates, since there is often a significant difference between the two.

The fit is carried out as follows: A surface is selected at which to begin the process. This selection is based on one of the following criteria: (1) the surface most sensitive to change,* (2) the surface with the shortest radius, (3) the strongest surface [i.e., with the largest surface power $(n' - n)/r$], or (4) the surface which shows the largest curvature difference from the nearest available test plate radius. Very often all four of these criteria will indicate the same surface; if not, the choice of which one to use is almost a matter of taste. The nearest radius on the test plate list is substituted for the selected surface and the lens is reoptimized, allowing all the variable parameters to change

*Note that the relative sensitivity of any dimension of the system can be determined quite easily by making an incremental change in the dimension and noting the effect it produces on the merit function. This, of course, assumes a merit function which accurately represents the quality of the image.

(except of course, the radius which has been set to the test plate value).

Another surface is then chosen and fixed to the nearest test plate radius; the lens is reoptimized again. This process is repeated until all the surfaces have been fitted to test plates. It is usually wise to avoid fitting both surfaces of a singlet (or all surfaces of a component) until all components have at least one fitted surface each. This allows the unfitted surface to vary and adjust for power, chromatic, and Petzval aberrations and the like for as long as possible.

With a reasonably complete test plate list, all radii can usually be fitted without significantly degrading the image quality (i.e., the merit function). In fact, the merit function is often slightly improved by this process, because of the additional cycles of optimization which have been performed. If the test plate list is limited or has a gap in it, it may not be possible to fit all the design radii to test plates without degrading the performance. In such a case, one must either fabricate new tooling and test plates, or seek a new vendor for the parts.

A *melt fit* reoptimizes the lens design by using measured data for the material indices instead of the nominal values from the glass catalog. This measured data comes in the form of what is called a *melt sheet* provided by the glass manufacturer or supplier. For noncritical applications, this data is usually sufficient. A worthwhile elaboration of the process is to determine the difference between the measured index values and the catalog values. These differences are then plotted against wavelength. This plot should be a smooth, relatively level line. A data point which does not plot smoothly is suspect; the measured data may well be in error. Next, a smoothly drawn curve through the points is used to determine improved values for the index differences. These differences are then applied to the catalog values to arrive at better values for the measured melt data, and the improved values are used in the melt fit reoptimization. The smoothed curve of differences can also be used to determine the index for wavelengths which are not included in the melt sheet.

An ordinary melt sheet will list the indices for the wavelengths of d, e, C, F, and g light. These are usually not individually measured. Instead, the index is measured for d light and the index difference between C and F light is measured. These two measurements are fed into a computer program which uses the known characteristics of the glass type to calculate the indices for e, C, F, and g light. This, while not ideal, is adequate for many, even most, applications. However, for some critical applications, and for all applications in which an attempt has been made to reduce or correct the secondary spectrum, it is usually quite unsatisfactory. For an additional charge, the glass manufacturer can provide what is usually called a *precision* melt sheet, for which the indices have been measured individually, and to a greater precision. Index values for wavelengths specific to the application can also be measured. It is, of course, wise to subject even this data to the difference test and smoothing process described in the preceding paragraph.

We called the last of these fits a *thickness fit* in the heading for this section. This is a process which is carried out during the final assembly of the lens. In sum, one reoptimizes the lens by varying the airspaces of the system, using accurately measured data for the radii, thicknesses, and indices of the fabricated elements. If the melt data and the test plate data are available, this just amounts to adding the measured element thicknesses and reoptimizing.

As can be seen, all of the above procedures are designed to almost completely eliminate the effects of the fabrication tolerances on the performance of the system. What is left, instead of the tolerances, is the uncertainty or inaccuracy of the various measurements on which the fits are based. The effect is usually quite modest and, therefore, acceptable. However, these uncertainties are, in fact, the exact equivalent of tolerances in determining the performance of the fabricated system.

2.10 Spectral Weighting

In any ray-tracing process, the index of refraction of the material is, of necessity, that corresponding to a single specific wavelength. Most lens *design* programs allow the use of only three wavelengths to represent the spectral bandpass for which the system is to be designed. Several *analysis* programs allow the use of five or ten suitably weighted wavelengths in calculating such things as MTF, point spread functions, radial energy distributions, and the like.

In either case, the question becomes, "What wavelengths should I use, and how should they be weighted?" For visual systems and just three wavelengths, the classical answer is to use d (or e), C, and F light, with weightings typically set at 1.0, 0.5, and 0.5 respectively. For other applications, this approximately corresponds to using the central wavelength and the wavelengths 25 percent from the extreme edges of the passband. If the results of an image analysis need not be especially precise, three wavelengths may be sufficient. For calculations done in the midst of the design process, this is often good enough to enable a judgment as to the relative merit of, or the rate of improvement between, two stages in the design process.

However, what should one do *in general?* To immediately dispose of the obvious, it is apparent that the more wavelengths that are used, the more accurate the results can be. That said, how do we select the wavelengths to be used? There are three obvious choices. In order to assess the full effects of the chromatic aberrations, one would like to include the extreme long and short wavelengths. One would also probably like to consider dividing the spectral weighting function into increments of equal power or response, so that each wavelength would represent an equally weighted sector. But if this is done, the extremes of the spectral passband are not included. Another possible option is to choose wavelengths which are evenly distributed across the spectral band, and to weight them according to the spectral response function. One might choose an even distribution on a wavelength scale, or what might be a bit better, an even distribution on a wave number (reciprocal wavelength) scale.

From this it should be apparent that most wavelength and weighting choices are based on some sort of compromise. Often the outer two wavelengths are chosen to be fairly close to the ends of the passband and the intermediate wavelengths and weights are a compromise between an even power distribution and an even wavelength spacing. If there are peaks or bumps in the spectral response function, wavelengths are often selected to be at or near the peaks. Note that, to a limited extent, one's choices will partially control the design process: heavier weighting at the ends of the spectrum will obviously emphasize both primary and secondary chromatic aberration, spherochromatism, and the like; heavy weights on the central wavelengths will emphasize the monochromatic aberrations at the expense of the chromatic.

2.11 How to Get Started

Experienced designers are often asked questions such as "How do you know where to start?" or "How did you decide that a (name of a design type) could meet the specs?" or "Why did you shift the components around?"

The answer to these questions is almost always "Experience," which probably means that the full answer is different for every problem. It would be foolhardy to pretend to be able to give a complete, definitive answer or set of answers to such questions, but there are a few guidelines which should be reasonably dependable.

Figure 2.3 is, in effect, a compilation of some of that experience. If the system to be designed roughly corresponds to a photographic objective or to one of the other types indicated, the figure can be used as an easy guide to the selection of an appropriate form. In this plot the areas corresponding to various combinations of field and aperture are labeled to indicate the type of system which is commonly used there. Obviously the boundary of each area is fuzzy and ill-defined. In a pre-

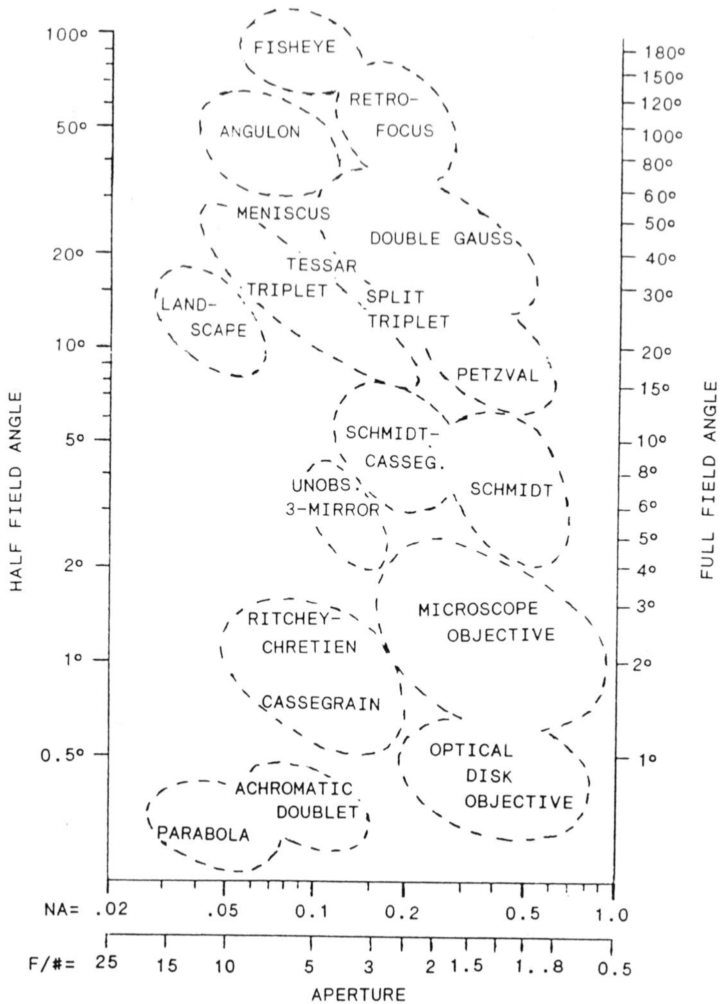

Figure 2.3 Map showing the design types which are commonly used for various combinations of aperture and field of view.

sentation of this type there is also an implied level of performance within each area, which is presumably typical of each particular lens type. In general, the performance (image quality, resolution, or whatever) is better when a given lens type is optimized for a smaller field and/or smaller aperture, that is, for a combination closer to the origin of the plot. Thus one can select the design type from Fig 2.3 corresponding to the field-aperture combination required with reasonable assurance that it is an appropriate choice. But should the performance prove inadequate for the application at hand, one can move up to the form which is above and to the right on the plot, which should have a greater performance capability. Of course, if the required level of performance is known in advance to be relatively high, one would select the more capable type to begin with. The same selection approach can be applied to the components of a more complex system. There are many design types and applications which are not represented on this necessarily abbreviated chart. However, many of these are presented in the lens designs included in this book; the reader may wish to mark up Fig. 2.3 to add the types which are of particular interest.

For a complex system, the approach must start on a more fundamental level. The first step is to collect and tabulate the requirements to be met; these may include such things as:

Resolution or performance (versus diffraction limit)

Wavelength

Fields of view

Image size

Aperture, numerical aperture (NA), and *f* number

Vignetting or illumination uniformity

Focal lengths or magnifications

Space limitations

The next step is to make a first-order layout of the system which will satisfy the requirements. The first-order layout is simply an arrangement of component powers (or focal lengths) and spacings which will produce an image in the required location, in the required orientation, and of the required size. At this stage, no consideration is given to the design type of each component; one is concerned only with its power, aperture, and field as first-order, i.e., paraxial, characteristics. For systems (or portions of systems) which consist of two components, the equations of Sec. F.7 can be extremely useful. For more complex systems, the component by component ray-tracing equations of Sec. F.6 may be used. A general approach is to trace rays which will define the required characteristics. The ray-trace results (ray heights, ray slopes, intersection lengths, etc.) can be expressed as equations with the component powers and spacings as unknowns to be solved for. An important facet of this stage is that one should try to find a layout which minimizes the component powers, or minimizes (or equalizes) the "work" (ray height times component power). Doing this will almost always produce a system with less aberration residuals, one which is less expensive to fabricate and less sensitive to fabrication and alignment errors.

Sometimes one can leap directly from the first-order layout to choosing the component design types, and thence to the optimization stage. However, if this is not the case, the next step is usually to analyze and/or correct the chromatic aberrations. Equations F.9.6 and F.10.7 can be used for the whole system, or the components can be individually achromatized. To this end, the element powers for a thin achromatic doublet are given by

$$\phi_A = \frac{V_A}{(V_A - V_B)F} \tag{2.1}$$

$$\phi_B = \frac{V_B}{(V_B - V_A)F} = \frac{1}{F} - \phi_A \tag{2.2}$$

and the element surface curvatures are determined from

$$C_1 - C_2 = \frac{\phi}{n-1} = \frac{1}{r_1} - \frac{1}{r_2} \tag{2.3}$$

At this point a sketch of the system is often helpful. Simply make a scale drawing, showing each element as either a plano-convex or an equi-convex form (or plano-concave or equi-concave for negative elements). If the elements look too fat, they should be split into two or more elements. Be sure that the element diameters are sized properly for the rays that they must pass. At this stage the system should begin to look like a good lens.

The next step is to give some consideration to the Petzval curvature. Choosing suitable anastigmat types for the components is one way to handle this. Another is to use a field flattener in an appropriate location (i.e., near an image) in the system. And, of course, the usual device of configuring the system or component with separated positive and negative elements or surfaces to reduce the Petzval sum can always be utilized if a new design must be created from scratch.

Often many of these steps can be handled conveniently and expeditiously by the automatic design program. The first-order layout can be done with zero-thickness plano-convex or plano-concave elements, allowing the spaces and the curvatures of the curved surfaces (but not the plano surfaces) to vary. The merit function is a simple one, configured to define the required first-order characteristics. Note well the comments in Sec. 2.5 regarding stagnation and contradictory or redundant first-order entries in the merit function.

Since the chromatic aberrations and the Petzval curvature depend on element power and not on element shape, the lens design program can also be used to find a layout which is a preliminary solution with the chromatic and Petzval adjusted to desired, reasonable values, which typically should be small and negative.

Sometimes it is useful as a next step to allow the elements to bend, and to correct the third-order aberrations. This can be done by putting an angle-solve on the second surface of each element so that the axial ray slope is maintained. More often than not, however, the next step will skip over the third-order and go directly to a full-dress thick lens optimization run.

And, of course, in the best of all worlds, one simply sets up what seems to be a likely layout and proceeds directly to the automatic lens design program, which promptly turns out an excellent design. "Experience!" Lots of luck!

Chapter

3

Improving a Design

3.1 Standard Improvement Techniques

There are several classic design modification techniques which can be reliably used to improve an existing lens design. They are:

1. Split an element into two (or more) elements
2. Compound a singlet into a doublet (or triplet)
3. Raise the index of the positive singlets
4. Lower the index of the negative singlets
5. Raise the index of the elements in general
6. Aspherize a surface (or surfaces)
7. Split a cemented doublet
8. Use unusual partial dispersion glasses to reduce secondary spectrum (see Chap. 6, "Telescope Objectives")

The simple, straightforward application of these techniques is no guarantee of improvement in a lens, in that they do not automatically correct the defects that they are intended to address. In general, these changes tend to reduce the aberration contributions of the modified components; in order to take full advantage of this, the aberrations of the balance of the system must be reduced as well. The operative principle is this: if large amounts of aberrations are corrected or balanced by equally large amounts of aberrations of opposite sign, then the residual aberrations also tend to be large. Conversely, if the balancing aberrations are both small, then the residuals tend to be correspondingly small.

3.2 Glass Changes: Index and *V* Value

The refractive characteristics of the materials used in a lens are obviously significant and important to the design. In general, for a positive element, the higher the index the better. The higher index reduces the inward Petzval curvature which plagues most lenses. It also tends to reduce most of the other aberrations as well. As an example, see Fig. 3.1, which clearly indicates the effect of higher index in reducing the spherical aberration of a single element. This sort of reduction is primarily a result of the fact that the surface curvature required to produce a given element power is inversely proportional to

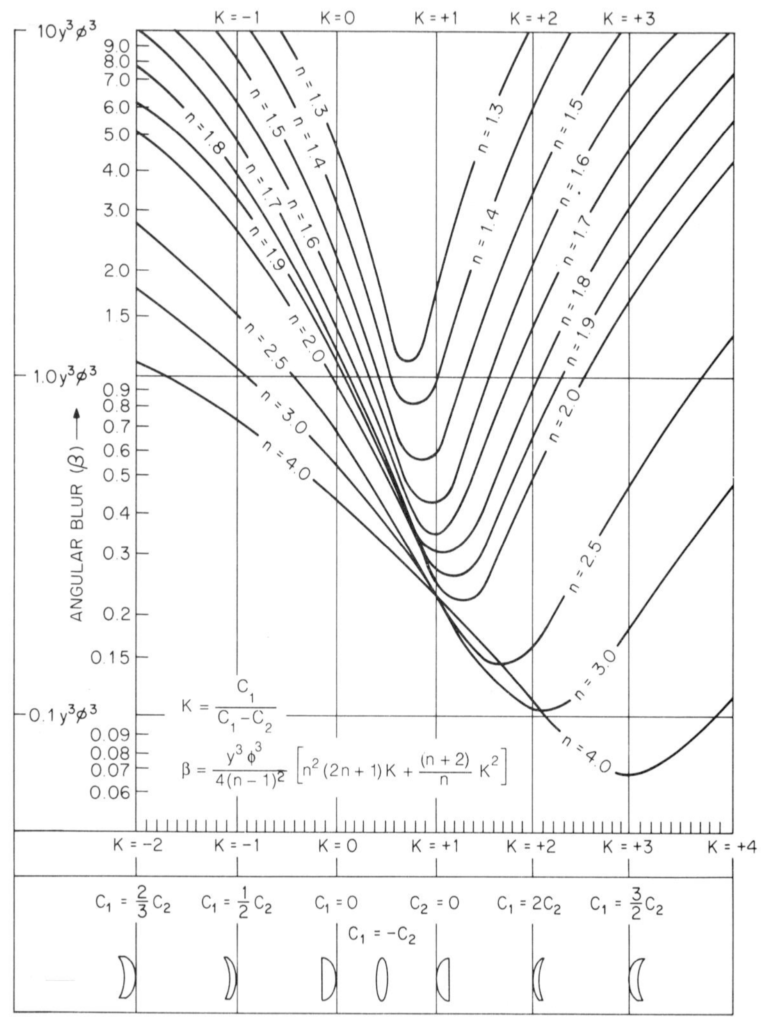

Figure 3.1 The angular spherical aberration blur of a single lens element as a function of lens shape, for various values of the index of refraction; ϕ is the element power and y is the semiaperture. The angular blur can be converted to longitudinal spherical aberration by LA = $2B/y\phi^2$, or to transverse aberration by TA = $-2B/\phi$. (The object is at infinity.)

$(n - 1)$. The improvement also results from the reduction of the angles of incidence at the surfaces of the element.

In a negative element, the situation is less clear. From the standpoint of the Petzval correction, a low index would increase the overcorrecting contribution of a negative element. This can help to offset the (inward) undercorrection which is a major problem in most lenses. On the other hand, a higher index would reduce the surface curvatures and have a generally desirable effect on the overall state of correction. The situation is usually resolved with the negative elements made from a glass along the glass line boundary of the glass map (Fig. 2.2).

A high V value for the positive element and a low V value for the negative element of an achromatic doublet reduce the element powers; this is ordinarily desirable. In lenses (such as the Cooke triplet) where the relative V values of separated elements control the element spacing or the system length, this desideratum may be overridden by other concerns.

Note that, as usual, when you are dealing with components of negative focal length, many of the considerations outlined above are reversed. In a negative achromatic doublet, the negative element is often made of crown glass and the positive is made of flint. Here a high-index (flint) positive element will reduce the inward Petzval curvature, as will a low-index (crown) negative element.

3.3 Splitting Elements

Splitting an element into two (or more) approximately equal parts whose total power is equal to the power of the original element can reduce the aberration contribution by a significant factor. The reason that this reduces aberrations is that it allows the angles of incidence to be reduced; the nonlinearity of Snell's law means that smaller angles introduce less aberration than do large ones. This technique is often used in high-speed lenses to reduce the zonal spherical residual and in wide-angle lenses to control astigmatism, distortion, and coma.

Figure 3.2 shows the thin lens third-order spherical aberration for spherical-surfaced positive elements which are shaped (or bent) to minimize the undercorrected spherical. The upper plot shows the spherical as a function of the index of refraction for a single element with a distant object. The curve labeled $i = 2$ shows the spherical for two elements whose total power is equal to that of the single element. The best split is 50-50—i.e., the split elements have equal power; this minimizes the spherical. (The same is true for a split into more than two elements, i.e., three, four, etc., as shown in the curves labeled $i = 3$ and $i = 4$.) The improvement produced by splitting an element in

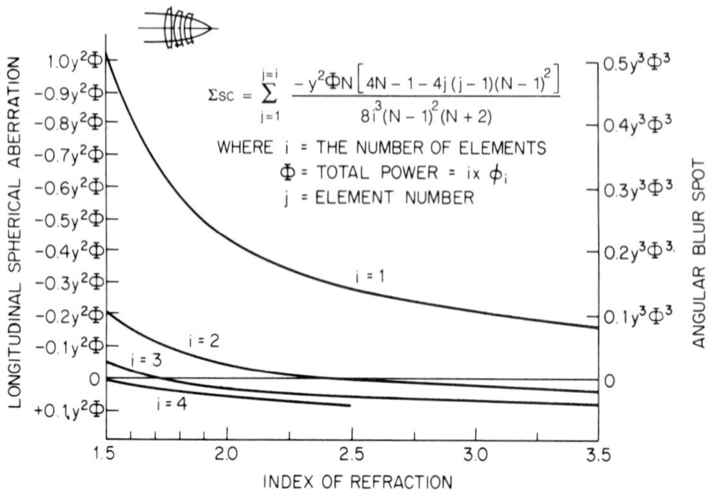

Figure 3.2 The spherical aberration of one, two, three, and four thin positive elements, each bent for minimum spherical aberration, plotted as a function of the index of refraction, and showing the reduction in the amount of aberration produced by splitting a single element into two or more elements (of the same total power). Each plot is labeled with *i*, the number of elements in the set. (The object is at infinity.)

two can be seen to be a factor of about 5 for lenses of index equal to 1.5. The higher the index, the greater the reduction; for an index of 1.8, the factor is about 7. At an index of 2.5 or higher, the spherical can be brought to zero or even overcorrected with just two positive elements. Most other aberrations are similarly affected by splitting, although it should be obvious that neither Petzval nor chromatic is changed by splitting.

In high-speed lenses this technique is frequently used to reduce the residual zonal spherical; the positive elements are split. This illustrates the basic idea. If the residual zonal spherical is negative (undercorrected), one splits a positive element; in the rare event that the zonal is positive (overcorrected), one would split a negative element. A similar philosophy can be applied for troublesome residuals of the other aberrations as well.

The choice of which element to split is often less apparent. The logical candidate would obviously seem to be the element which contributes most heavily to the problem aberration. (An examination of the third- and fifth-order surface contributions can often locate the source of the aberration.) However, other considerations often become significant. For example, in the Cooke triplet, the rear element is the prime candidate for the split, and such a split is quite effective in reducing the zonal spherical, as Figs. 14.1 and 14.2 will attest. But the better choice for the split is the front element, not because it does a better job of reducing the zonal spherical, but because the resulting lens is better corrected for the other aberrations. Figure 14.3 shows the simple split-front triplet. This is the ancestor of the Ernostar family of lenses; Figs. 14.10 through 14.15 illustrate designs which can be considered as descendants from the split-front triplet. Although they have been largely superseded by the more powerful double-Gauss form, they are nonetheless excellent design types.

Many retrofocus and wide-angle lenses which use strong outer meniscus negative elements illustrate the use of this technique for the control of coma, astigmatism, and distortion by splitting these negative elements.

The implementation of this technique with an automatic design program is often far from easy. For example, if one decides to split one of the crowns of a Cooke triplet and simply replaces one crown with two, after the computer optimization has run its course, the resultant lens may look like an ordinary triplet with a narrow cracklike airspace in the split element (a *cracked crown triplet*). The performance of the lens is the same as the original triplet; the split has not improved a thing. This is because the original lens was in a local optimum of the merit function. Aberrations other than the zonal spherical dominated the design; this caused the program to return the lens to its original design configuration. What is necessary in this situation is to force the split elements into a configuration which will accomplish the desired result.

Consider the split-front triplet. There are two ways to get to a design like Fig. 14.3. One approach is to make the lens so fast that the zonal spherical is by far the single dominant aberration in the merit function. Then the program will probably choose a form which reduces the zonal spherical; the lens shapes in Fig. 14.3 are a likely result. A difficulty with this approach may be that you aren't interested in a very fast lens, or if you are, the rays may miss the surfaces of the initial design completely, or encounter TIR. The alternative approach is to constrain the front elements to a configuration in which the spherical is minimized. Simply fixing the first element to a plano-convex form (by not allowing the plano surface curvature to vary) or holding the second to an aplanatic meniscus shape is usually sufficient to obtain a stable design which is enough different from the cracked crown triplet. When this has been accomplished the constraint can be released and the automatic design routine allowed to find what is (one hopes) a new and better local optimum. The problem here is that this approach presupposes a knowledge of the configuration which will produce a good result. Obviously a knowledge of both aberration theory and of successful design forms is a useful tool to the designer.

3.4 Separating a Cemented Doublet

Airspacing a cemented doublet can provide two additional degrees of freedom: two bendings instead of one, plus an airspace. While this technique does not have the inherent aberration reduction capability that many other modifications possess, the extra variables may indirectly make a design improvement possible. A difficulty in implementing this is that the refraction at the cemented surface is apt to become much more abrupt when it is split into two glass-air interfaces than when it was a cemented surface; in fact rays may encounter TIR if a simple split is attempted without a concomitant reduction of the angle of incidence. Manual intervention in the form of adjusting the radii to reduce the angle of incidence is often necessary.

3.5 Compounding an Element

Compounding a singlet to a doublet can be viewed in two different ways:

1. As a way of simulating a desirable but nonexistent glass type
2. As a way of introducing a cemented interface into the element in order to control the ray paths

Note that in almost all examples of Tessar-type lenses (and other types which utilize compounded elements), the doublets have positive elements with high index and high V values, while the negative element of the doublet has both a lower index and V value. See Chap. 12 for examples.

The longitudinal axial chromatic of a singlet is given by $LA_{ch} = -f/V$. Thus a fully achromatized lens (with $LA_{ch} = 0.0$) has the chromatic characteristic of a lens made from a material with a V value of infinity; a partially achromatized doublet acts like a singlet with a very high V value.

The Petzval radius of a singlet is given by $\rho = -nf$, where n is its index. An *old achromat* with a low-index crown and a higher-index flint has a shorter Petzval radius than a singlet of the crown glass. For example, an achromat of BK7 (517:649) and SF1 (717:295) has a Petzval radius $\rho = -1.37f$; in other words, in regard to Petzval field curvature, it behaves like a singlet with an index of 1.37. A *new achromat* has a high-index crown and a lower-index flint. A new achromat of SSKN5 (658:509) and LF5 (581:409) has a Petzval radius $\rho = -2.19f$ and a Petzval curvature which is characteristic of a singlet with an index of 2.19.

Thus an achromatized (or partially achromatized) doublet with a high-index positive element and a low-index negative element has many of the characteristics of a lens made of a high-index, high-V-value crown glass. (Note that for a negative focal length doublet, the reverse is true.) Both conditions are usually to be desired, in order to flatten the Petzval field and to achieve achromatism.

Figure 3.3 shows a singlet, an old achromat, and a new achromat, each with the same focal length. The equivalent V value of each achromat is, of course, equal to infinity. The Petzval radius for each is given in the figure caption.

The cemented interface of the doublet can be used for specific control of specific rays. In a lens such as the Tessar, where the doublet is located well away from the aperture stop, the upper and lower rim rays of the oblique fan have very different angles of incidence at the cemented surface. In Fig. 3.4 it can be seen that the angle of incidence at this surface is much larger for the upper ray than for the lower. In this type of lens the cemented surface is typically a convergent one, and the (trigonometric) nonlinear characteristic of Snell's law means that the upper ray is, in this case, refracted downward more than it would be were the refraction linear with angle. Thus the upper ray is deviated in such a way as to reduce any positive coma of this ray. This

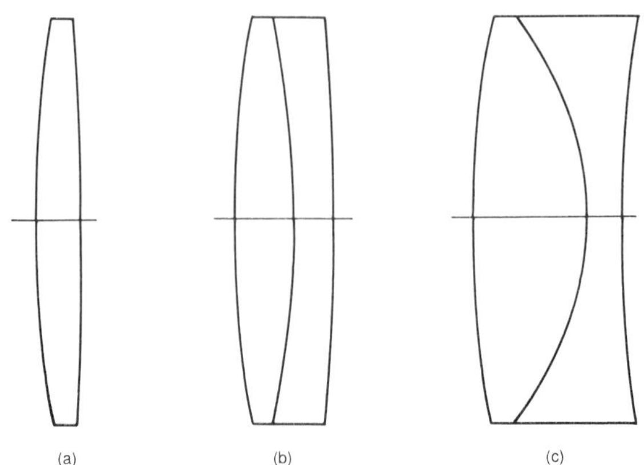

Figure 3.3 Three lenses, each with the same focal length f. (*a*) A singlet of BK7 (517:642) glass; Petzval radius equals $-1.52f$. (*b*) An old achromat of BK7 (517:642) and SF1 (717:295) glasses; Petzval radius equals $-1.37f$. (*c*) A new achromat of SSKN5 (658:509) and LF5 (581:409); Petzval radius is $-2.19f$.

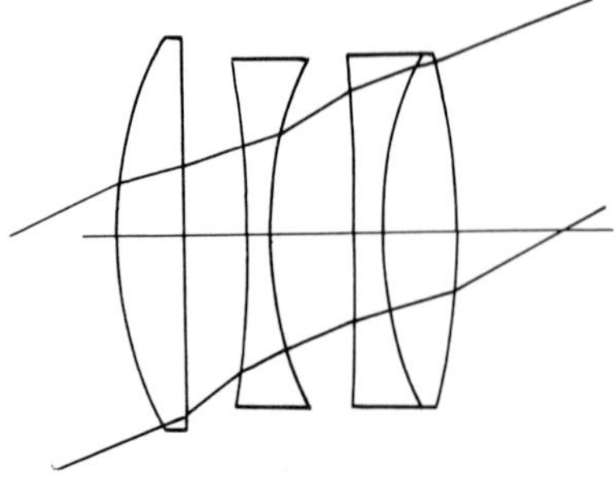

Figure 3.4 The upper and lower rim rays have significantly different angles of incidence at the cemented interface in the rear doublet of this Tessar design. Properly handled, this difference can be used to modify the correction of the coma-type aberrations.

illustrates the manner in which a cemented surface can be used for an asymmetrical effect on an oblique beam.

The Merté surface

A strongly curved, collective cemented surface with a small index break (to the order of 0.06) has an effect which can be used to reduce the undercorrected zonal spherical aberration. The central doublet of the Hektor lens shown in Fig. 3.5 illustrates this principle. The ce-

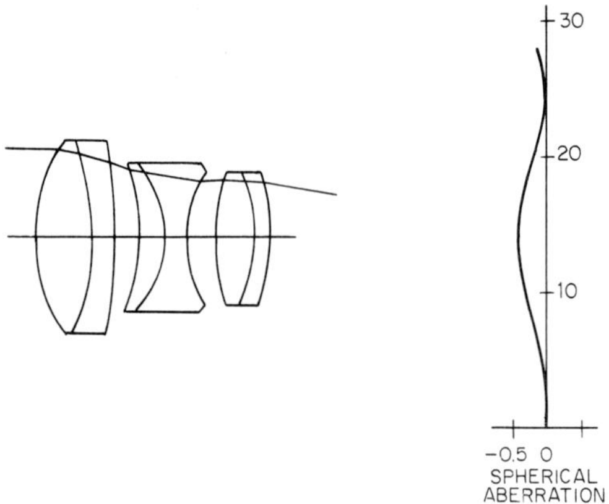

Figure 3.5 The cemented surface in the center doublet of this Hektor lens is what is called a Merté surface. The index break $(n' - n)$ across the surface is small, but at the margin of the aperture the angle of incidence for the axial ray becomes quite large. This combination produces an undercorrecting seventh-order spherical aberration which, as the plot shows, dominates the spherical aberration at the margin of the aperture, causing the marginal spherical to be negative rather than the usual positive value.

mented surface is a collective one (in that $[n' - n]/r$ is positive) and contributes undercorrected spherical aberration. For rays near the axis, the spherical aberration contribution of the surface is modest. However, when the ray intersection height increases and the angle of incidence becomes large, as shown in Fig. 3.5, the trigonometric nonlinearity of Snell's law causes the amount of ray deviation to be disproportionately increased. This causes the undercorrection from this surface to dominate the spherical aberration. The result is a spherical aberration characteristic like that shown in Fig. 3.5. The spherical in the central part of the aperture appears quite typical: the undercorrected third-order dominates close to the axis and the overcorrecting fifth-order causes the plot to curve back as the ray height increases. However, toward the edge of the aperture the undercorrection of the Merté surface becomes dominant and the aberration plot reverses direction again. The net result is the equivalent of a reduced zonal spherical aberration.

It is rare to see as extreme an example of the Merté surface as that illustrated in the Hektor of Fig. 3.5. Such a surface is very sensitive to fabrication errors and is thus expensive to make. It is also often best used close to the aperture stop because, if it is located away from the stop, the asymmetrical effects described two paragraphs above can become quite undesirable. However, it is well worth noting that an ordinary collective cemented surface has a tendency to behave as a mild Merté surface and to reduce the spherical zonal, at least somewhat.

3.6 Vignetting and Its Uses

Vignetting, which is simply the mechanical limitation or obstruction of an oblique beam, is usually regarded primarily as something which reduces the off-axis illumination in the image. However, vignetting often plays an essential role in determining the off-axis image quality as well as the illumination. Of course there are many applications for which vignetting cannot be tolerated; the illumination must be as uniform as possible across the entire field of view. The complexity of the lens design, therefore, must be sufficient to produce the required image quality at full aperture over the full field.

But for many applications, vignetting is, in fact, quite tolerable. In commercial applications the clear apertures may well be established so as to be just sufficient to pass the full aperture rays for the axial image. It is not at all unusual for vignetting to exceed 50 percent at the edge of the field. For a camera lens, this vignetting will, of course, completely disappear when the iris of the lens is stopped down to an aperture below the vignetting level. Since camera lenses are most often used at less than full aperture, the vignetting is not as significant as it is in a lens which is always used at full aperture, such as a microscope or projection lens.

The *benefit* of vignetting is that it cuts off the upper and/or lower rim rays of the oblique tangential fan. Since these are ordinarily the most poorly behaved rays, the image quality may well be improved by their elimination. Most lenses which cover a significant field are afflicted with oblique spherical aberration, a fifth-order aberration which looks like third-order spherical aberration, but which varies as the square of the field angle. And since its magnitude is different for sagittal than for tangential rays, it can be seen to have characteristics of both astigmatism and spherical aberration. Oblique spherical aberration usually causes the rays at the edge of the oblique bundle to show strongly overcorrected spherical aberration; vignetting is a simple way to block these aberrant rays from the image.

Another factor favoring the use of vignetting is that it results from lens elements with small diameters. In general, one can count on a smaller-diameter lens being less costly to fabricate.

For a camera lens, one must be certain that the iris diaphragm is located centrally in the oblique beam so that, when the iris is closed down, the central rays of the beam are the ones which are passed. These are usually the best-corrected rays of the oblique beam. Also this location assures that the vignetting will be eliminated at the largest possible aperture.

3.7 Eliminating a Weak Element; the Concentric Problem

Occasionally an automatic design program will produce a design with an element of very low power. Frequently this means that the element can be removed from the design without adversely affecting the quality of the design. Often a straightforward removal will not work; the design process may simply "blow up." An approach which usually works (if anything will) is to add the thickness and the surface curvatures of the element to the merit function with target values of zero, allowing them to continue as variable parameters. Sometimes targeting the difference between the two curvatures is also useful. Usually, if the element isn't necessary to the design, a few cycles of optimization, possibly with gradually increasing weights on these targets, will change the element to a very thin, nearly plane parallel plate, which can then be removed without severe trauma to the design. If your design program will not accept curvatures and thicknesses as targets, an alternative technique is to remove the curvatures and the thickness as variables and to gradually weaken the curvatures and reduce the thickness (by hand) while continuing to optimize with the other variables.

An unfortunate form of the "weak" element is a fairly strongly bent meniscus, which the computer uses for a relatively important design function, such as the correction of spherical aberration or the reduction of the Petzval curvature. It is rarely possible to eliminate such an element because it is an integral part of the design. The unfortunate aspect of this situation is that, if the surfaces of the meniscus element are concentric or nearly so, the customary centering process used in optical manufacture is impossible or impractical, and the element is costly to fabricate. This situation can be ameliorated by forcing the centers of curvature of the surfaces apart by a distance sufficient to allow the use of ordinary centering techniques. Again, including the required center-to-center spacing in the merit function and reoptimizing will usually modify the offending element to a more manufacturable form without any significant damage to the system performance.

3.8 Balancing Aberrations

The optimum balance of the aberrations is not always the same in every case; the best balance varies with the application and depends on the size of the residual aberrations. In general, for well-corrected lenses, the aberrations should be balanced so as to minimize the OPD, i.e., the wavefront variance, but there are significant exceptions.

Spherical aberration

If a lens is well-corrected and the high-order residual spherical aberration is small, so that the OPD is to the order of a half-wave or less, then the best correction is almost always that with the marginal spherical corrected to zero, as illustrated in Fig. 3.6*b*. However, when the zonal spherical is large, there are two situations where one may want to depart from complete correction of the marginal spherical.

If the lens will always be used at full aperture (as a projection lens, for example), and if the spherical aberration residual is large (say to the order of a wave or so), the diffraction effects will be small when compared to the aberration blur; then the spherical aberration should be corrected to minimize the size of the blur spot rather than to minimize the OPD. This will produce the best contrast for an image with relatively coarse details, i.e., for a resolution well below the diffraction limit. As an example, at a speed of *f*/1.6, a 16-mm projection lens has a diffraction cutoff frequency of about 1100 line pairs per millimeter (lpm). But its performance is considered quite good if it resolves 100

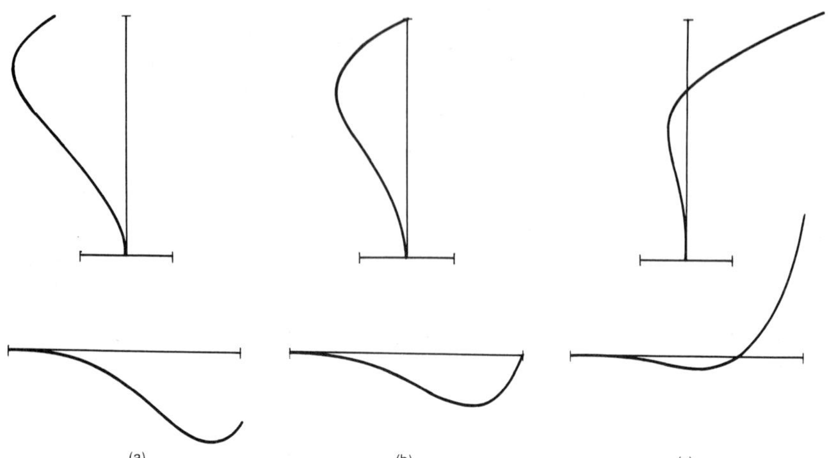

Figure 3.6 Three states of correction of spherical aberration are shown. Each has the same amount of fifth-order spherical, but different amounts of third-order. (*a*) Spherical aberration balanced to give the smallest possible size blur spot. This correction may be optimum when the aberration is large and the required level of resolution is low compared to the diffraction limit. (*b*) Spherical aberration balanced for minimum OPD. This is optimum when the system is diffraction-limited. (*c*) Spherical aberration balanced to minimize the focus shift as the lens aperture is stopped down. This correction is used in camera lenses when the residual spherical is large. The upper row is longitudinal spherical versus ray height; the lower row is transverse ray intercept plots.

lpm, an order of magnitude less than the diffraction limit. Such a lens can advantageously be corrected for the minimum diameter geometrical blur spot. This state of correction occurs (for third- and fifth-order spherical) when $LA_z = 1.5LA_m$, or $TA_z = 1.05TA_m$; the result is a high-contrast, but low-resolution, image. This correction is illustrated in Fig. 3.6*a*. See also the comments on defocusing in Secs. 2.4 and 2.7.

For a lens which is used at varying apertures, as is a typical camera lens, it is important that the best focus position not shift as the size of the aperture stop is changed. If the spherical aberration is corrected at the margin of the aperture, or corrected as described in the paragraph above, the position of the best focus will shift as the aperture is changed. The best focus will move toward the paraxial focus as the aperture is reduced. The state of correction which is often used in such a case is overcorrection of the marginal spherical, as shown on Fig. 3.6*c* (assuming an undercorrected zonal residual). The result is a design in which the focus is quite stable as the lens is stopped down. The *resolution* is better than it would be otherwise, but, at full aperture, the *contrast* in the image is quite low. This works out reasonably well in a high-speed camera lens because camera lenses are only infrequently used at full aperture. Typically, photographs are taken with the lens stopped down well below the full aperture, and, when the camera is stopped down, this state of correction yields a much better photograph.

The three correction states shown in Fig. 3.6 also indicate the manner in which the spherical aberration is changed when the third-order aberration is changed. This is a typical situation often encountered in lens design: the fifth- (and higher-) order aberrations are relatively stable and difficult to change, but the third order is easily modified (by bending an element, for example). In the figure, all three illustrations have exactly the same amount of fifth-order spherical; the difference is solely in the amount of third-order. Note that, in the (upper) longitudinal plots, the change from one illustration to the next varies as y^2, whereas in the (lower) transverse plots the differences vary as y^3.

Chromatic aberration and spherochromatism

Here the question is how to balance the spherochromatism, which typically causes the spherical aberration at short (blue) wavelengths to be overcorrected and that at the long (red) wavelengths to be undercorrected. If the aberration is small (diffraction-limited), the best correction is probably with the chromatic aberration corrected at the 0.7 zone of the aperture. This means that the central half of the aperture area is undercorrected for color and the outer half of the aperture is overcorrected, as shown in Fig. 3.7*a*. But if the amount of the aberration is large, the spherical overcorrection of the blue marginal ray causes a blue flare and a low contrast in the image. In these circumstances the correction zone can advantageously be moved to (or toward) the marginal zone, as shown in Fig. 3.7*b*. This will probably reduce the resolution somewhat, because it increases the size of the core of the image blur, but it improves the contrast significantly and yields a more pleasing image, free of the blue flare and haze. This state of correction is accomplished by increasing the undercorrection of the chromatic aberration of the paraxial rays.

Astigmatism and Petzval field curvature

In a typical anastigmat lens the fifth-order astigmatism tends to become significantly undercorrected (i.e., negative) as the field angle is increased. In order to minimize the astigmatism over the full field, the third-order astigmatism is made enough overcorrected to balance the undercorrected fifth-order astigmatism. The result is the typical field curvature correction with the sagittal focal surface located inside the tangential focal surface in the central part of the field because of the overcorrected third-order astigmatism, and the reverse arrangement

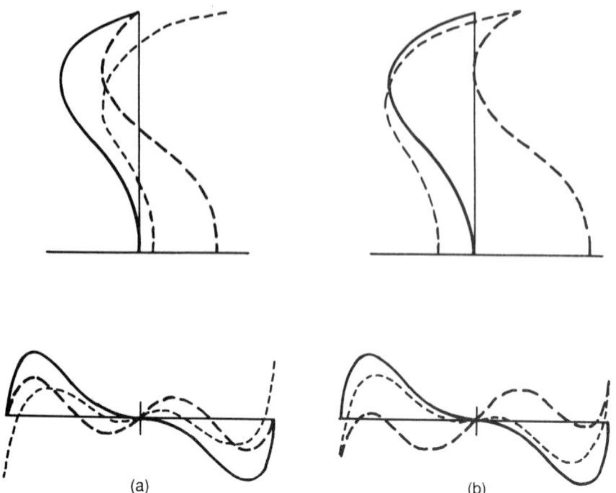

Figure 3.7 Spherochromatism. (*a*) Chromatic aberration balanced so that the outer half of the aperture is overcorrected and the inner half is undercorrected. This may be best if the amount of aberration is small. (*b*) Chromatic aberration balanced so that it is corrected at the margin. If the aberration is large, this correction eliminates the blue flare which can result from the type of correction in (*a*). Note that the state of correction is more easily perceived in the upper, longitudinal aberration plots, whereas the effect on the blur spot size and flare is much more apparent in the lower transverse ray intercept plots.

in the extreme outer portions. The field angle at which the *s* and *t* fields cross (i.e., where the astigmatism is zero) is called the *node*. Usually the two fields separate very rapidly outside the node, and the image quality quickly deteriorates, often suddenly. The Petzval curvature is usually made somewhat negative, so that both fields are slightly inward-curving and the effective field is as flat as possible. A typical state of correction is shown in Fig. 3.8.

Note that a field correction with the *s* and *t* focal surfaces spaced equally on either side of the focal plane (so that the compromise "smallest circle of confusion" focal surface is flat) is definitely *not* the best state of correction. In considering the correction of the field curvature, one should bear in mind that the oblique spherical aberration (a fifth-order aberration which varies as the cube of the aperture and the square of the field angle) typically goes overcorrected with increasing field angle. In addition, the oblique spherical is usually more significant for the tangential fan of rays than for the sagittal. Thus the effective field curvature for the full fan of rays is usually more backward-curving than the X_s and X_t field curves indicate. These curves indicate the imagery of a very small bundle of rays close to the principal ray, and do not take the oblique spherical of the full aperture

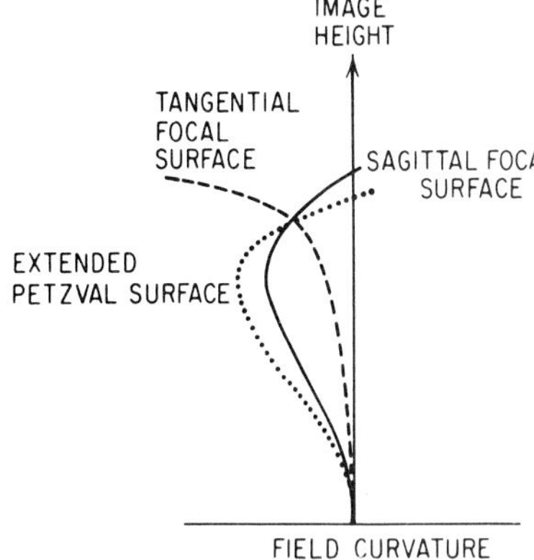

Figure 3.8 This is the typical balance of astigmatism and Petzval curvature in the presence of undercorrecting fifth-order astigmatism and overcorrecting high-order Petzval curvature. This is achieved by leaving the third-order Petzval slightly inward-curving and overcorrecting the third-order astigmatism by a small amount. This is the usual aberration balance for most anastigmats.

into account. Thus, for most designs, the astigmatism and field curvature are usually arranged somewhere between the state at which the *s* and *t* curves are superimposed (i.e., zero astigmatism) and that at which the *t* field is approximately flat. Often a through-focus MTF plot which includes both on-axis and off-axis plots will indicate quite clearly the effective field curvature, which is, of course, more informative than the X_s and X_t curves.

Note well that these discussions have assumed the type of higher-order residual aberrations which one ordinarily finds: overcorrected fifth-order spherical aberration, undercorrected fifth-order astigmatism, and overcorrected spherochromatism for the shorter wavelengths. Although rare, the reverse is sometimes encountered. In such circumstances the obvious move is to apply the above advice in reverse.

As an additional consideration, note that the undercorrection of either the chromatic aberration or the Petzval curvature has the usually desirable side effect of reducing the power of the elements of the lens system. Thus a secondary benefit of this undercorrection is the reduction of residual aberrations in general, because a lower-power element produces less aberration, which means that there is less higher-order residual aberration left when the aberrations are balanced out.

3.9 The Symmetrical Principle

When an optical system has mirror symmetry about the aperture stop (or a pupil), as shown in Fig. 3.9, the system is free of coma, distortion, and lateral color. This results from the fact that the components on one side of the stop have aberrations which exactly cancel the aberrations from the components on the other side of the stop. Obviously, to have mirror symmetry, the system must work at unit magnification,

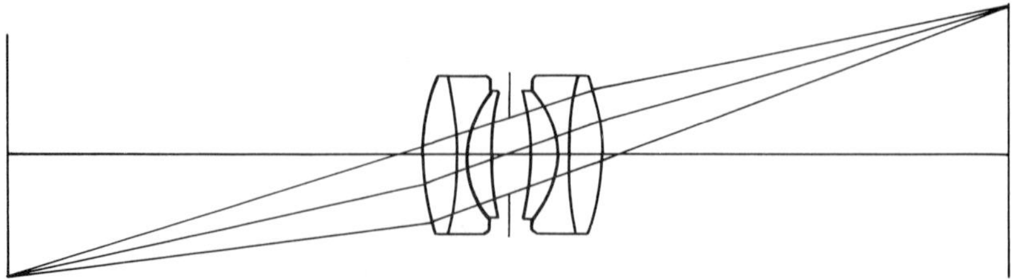

Figure 3.9 A fully (left to right) symmetrical system is completely free of coma, distortion, and lateral color, because the aberration in one half of the system exactly cancels out the aberration in the other half.

with equal object and image distances. For the symmetry to be *absolutely* complete, the object and image surfaces must be identical in shape; this then would imply separately curved sagittal and tangential surfaces at both object and image. However, the third-order coma, distortion, and lateral color are completely removed by symmetry, even with flat object and image surfaces.

Of course, most systems do not operate at unit magnification, and therefore a symmetrical construction of the lens will not completely eliminate these aberrations. However, even for a lens with an infinitely distant object, these aberrations are markedly reduced by symmetry, or even by an approximately symmetrical construction. This is why so many optical systems which cover a significant angular field display a rough symmetry of construction. Consider the Cooke triplet: it has outer crown elements which are similar, but not identical in shape, and the center flint, while not equi-concave, is bi-concave, and, except for slow-speed triplets, the airspaces are quite similar in size. The benefit of this is that the higher-order residuals of coma, distortion, and lateral color are markedly reduced by this symmetry. This is especially true for wide-angle lenses when good distortion correction is important.

3.10 Aspheric Surfaces

Many designs can be improved by the use of one or more aspheric surfaces. Except for the case of a molded or diamond-turned element, an aspheric surface is many times more expensive to fabricate than a spherical surface. A conic aspheric is easier to test than a general aspheric and is therefore somewhat less costly. For many systems, e.g., mirror objectives, aspheric surfaces are essential to the design and cannot be avoided.

One technique for introducing an aspheric into an optical system is to first vary only the conic constant. (Note that the conic constant and the fourth-order deformation term have exactly the same effect on the third-order aberrations. Thus, allowing both to vary in an automatic design program may cause a slowing of the convergence or, in extreme cases, a failure of the process. Occasionally the difference between the effect of the conic and the fourth-order term on the fifth- and higher-order aberrations may be useful in a design, but more often than not the two are redundant.) If the effect of varying the conic constant alone is inadequate, one can then allow the sixth-order term to vary, then the eighth-order, etc. Some designs have aspherics specified to the tenth-order term when just the sixth or eighth would suffice. It is a good idea to calculate the surface deformation caused by the highest-order term used; if it is a fraction of a wave at the edge of the surface aperture, its utility may well be totally imaginary.

Occasionally one encounters a design specification or print in which the aspheric is specified by a tabulation of sagittal heights instead of an equation. The optimization program can be used to fit the constants of the standard aspheric surface equation to the tabulated data. The specification table is entered in the merit function as the sag of the intersections of (collimated) rays at the appropriate heights. The surface coefficients are allowed to vary, and the result is a least-squares fit to the sag table.

The equations of Sec. F.11 indicate the effects of a conic or a fourth-order aspheric term on the third-order aberrations. Several points are worthy of note. The conic has no effect on the Petzval curvature or on axial or lateral chromatic. Further, if the conic is located at the aperture stop or at a pupil, then the principal ray height, y_p, is zero and the conic has no effect on third-order coma, astigmatism, or distortion; it can only affect third-order spherical. In the Schmidt camera the aspheric surface is located at the stop because the coma and astigmatism are already zero, because the stop is at the center of curvature of the spherical mirror; the purpose of the aspheric is to change *only* the spherical aberration. Conversely, if the purpose of an aspheric is to affect the coma, astigmatism, or distortion, then it must be located a significant distance from the aperture stop.

It is also worth noting that the primary effect of the conic, or fourth-order, deformation term is on the third-order aberrations. The primary effect of the sixth-order deformation term is on the fifth-order aberrations, etc., etc.

Chapter

4

Evaluation: How Good Is This Design?

4.1 The Uses of a Preliminary Evaluation

At some point in the process of designing an optical system, the designer must decide whether the design is good enough for the application at hand. With modern computing power, it is not a difficult matter to calculate the MTF or the point spread function (PSF), and to accurately include the effects of diffraction in the calculations. The process does consume a finite amount of time, however (which, on a slow computer, may be a significant amount), and it is useful to be able to make a reasonable estimate of the system performance from a more limited amount of data. A good estimate can avoid wasting time and computer paper in evaluating a clearly deficient design, or it can signal an appropriate point at which to conduct a full-dress evaluation.

4.2 OPD versus Measures of Performance

The distribution of illumination in the point spread function, particularly in the diffraction pattern of a reasonably well-corrected lens, is often used as a measure of image quality. The *Strehl ratio* (or *Strehl definition*) is the ratio of the illumination at the center of an (aberrated) point image to the illumination at the center of the point image formed by an aberration-free system. Figure 4.1 illustrates the concept. Another measure of image quality uses the percentage of the total energy in the point image which is contained within the diameter of the Airy disk. This diameter remains relatively constant in size for small amounts of aberrations. The table of Fig. 4.2 gives the relationships between the wavefront deformation (or OPD), the Strehl ratio, and the energy distribution.

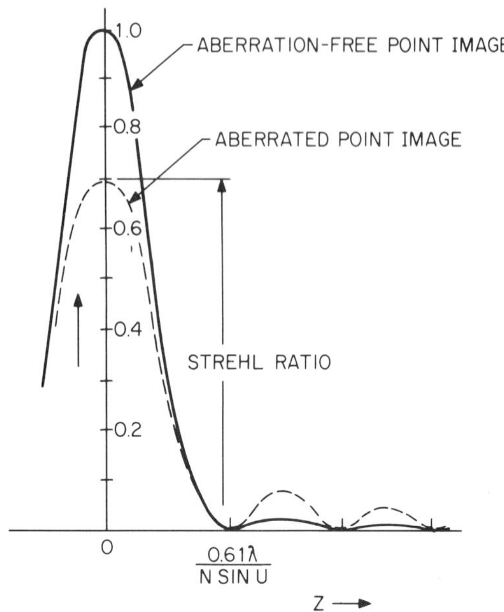

Figure 4.1 The Strehl ratio is the illumination at the center of the diffraction pattern of an aberrated image, relative to that of an aberration-free image.

Relation of Image Quality Measures to OPD

P-V OPD	RMS OPD	Strehl Ratio	% energy in Airy Disk	% energy in Rings
0.0	0.0	1.00	84	16
0.25RL = λ/16	0.018λ	0.99	83	17
0.5RL = λ/8	0.036λ	0.95	80	20
1.0RL = λ/4	0.07λ	0.80	68	32
2.0RL = λ/2	0.14λ	0.4*	40	60
3.0RL = 0.75λ	0.21λ	0.1*	20	80
4.0RL = λ	0.29λ	0.0*	10	90

*The smaller values of the Strehl ratio do not correlate well with image quality.

Figure 4.2 Tabulation of the Strehl ratio and the energy distribution as a function of the wavefront deformation. RL means the Rayleigh limit of one-quarter-wavelength peak-to-valley OPD.

Another commonly utilized measure of performance is the modulation transfer function, which describes the image modulation or contrast as a function of the spatial frequency of the object or image. The MTF of a perfect, aberration-free system is given by

$$\text{MTF}(v) = \frac{2}{\pi}(\phi - \cos\phi \sin\phi) \tag{4.1}$$

where

$$\phi = \arccos\left(\frac{\lambda v}{2\text{NA}}\right) \tag{4.2}$$

This is plotted as curve *A* in Figs. 4.3 and 4.4. Figure 4.3 shows the effect on the MTF of defocusing an otherwise aberration-free lens. The spatial frequency in these plots is normalized to the cutoff frequency

$$v_0 = \frac{2\text{NA}}{\lambda} = \frac{1}{\lambda(f\text{ number})} \tag{4.3}$$

Figure 4.4 shows the effect of simple third-order spherical aberration on the MTF. Note that, although the curves of Figs. 4.3 and 4.4 are not identical, they *are* quite similar. This similarity of effect is the basis for the common rule of thumb that a given amount of OPD will de-

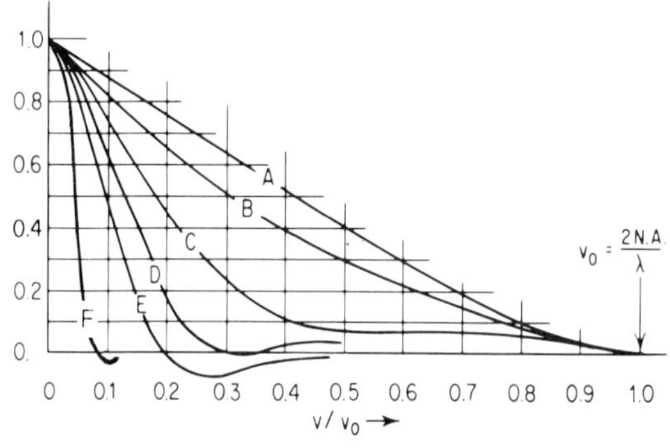

Figure 4.3 The effect of defocusing on the modulation transfer function of an aberration-free system.
(*A*) In focus OPD = zero
(*B*) Defocus = $\lambda/2n \sin^2 U$ OPD = $\lambda/4$
(*C*) Defocus = $\lambda/n \sin^2 U$ OPD = $\lambda/2$
(*D*) Defocus = $3\lambda/2n \sin^2 U$ OPD = $3\lambda/4$
(*E*) Defocus = $2\lambda/n \sin^2 U$ OPD = λ
(*F*) Defocus = $4\lambda/n \sin^2 U$ OPD = 2λ
(Curves are based on diffraction effects—not on a geometric calculation.)

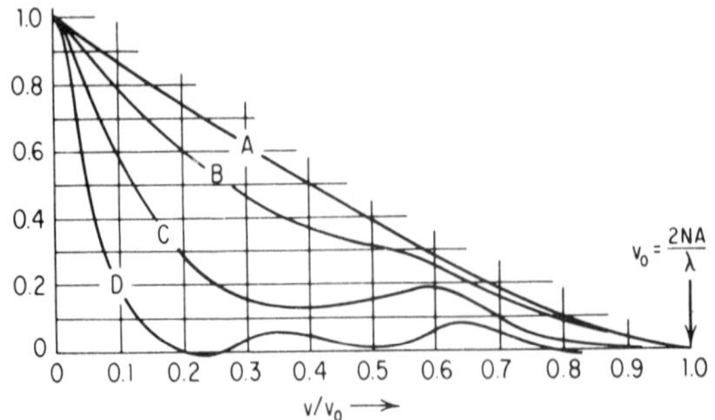

Figure 4.4 The effect of third-order spherical aberration on the modulation transfer function.

(*A*) LA_M = zero	OPD = 0
(*B*) $LA_M = 4\lambda/n \sin^2 U$	OPD = $\lambda/4$
(*C*) $LA_M = 8\lambda/n \sin^2 U$	OPD = $\lambda/2$
(*D*) $LA_M = 16\lambda/n \sin^2 U$	OPD = λ

grade the image by the same amount regardless of what type of aberration produced the OPD.

When the OPD is large (say more than one or two waves), the following geometrical approximation (derived from the geometric defocusing expression) can be used to calculate the MTF with reasonable accuracy:

$$\text{MTF}(v) = \frac{J_1[8\pi n\text{OPD}(v/v_0)]}{4\pi n\text{OPD}(v/v_0)} \tag{4.4}$$

where v_0 is the cutoff frequency (Eq. 4.3), n is the index of the image medium, OPD is the peak-to-valley wavefront deformation in waves, and

$$J_1[x] = \frac{x}{2} - \frac{(x/2)^3}{1^2 2} + \frac{(x/2)^5}{1^2 2^2 3} - \cdots$$

The relationships between the basic aberrations and the OPD are given in Sec. F.12, as are the relationships between rms OPD and peak-to-valley OPD and between rms OPD and the Strehl ratio.

A convenient relationship to remember is that a quarter-wave of OPD corresponds to a transverse spherical aberration (either marginal or zonal) of about

$$\text{TA} = \frac{4\lambda}{\text{NA}} \tag{4.5}$$

This is a useful way to make a quick and dirty evaluation from just the ray intercept plots.

4.3 Blur Spot Size versus Certain Aberrations

Many times, the system characteristic of interest is the size of the blur produced as the image of a point source. There are a few simple relationships which are useful in this regard:
Third-order spherical at best focus (three/fourths of the way from paraxial to marginal focus):

$$B = 0.5\ \mathrm{LA}_m \tan U_m = 0.5\mathrm{TA}_m \tag{4.6}$$

Third- and fifth-order spherical (with the marginal spherical corrected, focused at 0.42 LA_z from the paraxial focus):

$$B = 0.84\ \mathrm{LA}_z \tan U_m = 0.59\ \mathrm{TA}_z \tag{4.7}$$

Third- and fifth-order spherical (corrected so that $\mathrm{LA}_z = 1.5\ \mathrm{LA}_m$, or $\mathrm{TA}_m = 1.06\ \mathrm{TA}_z$, and focused at 0.83 LA_z from the paraxial focus; this correction yields the smallest-diameter blur spot for a given amount of fifth-order spherical):

$$\begin{aligned} \mathrm{B} &= 0.5\ \mathrm{LA}_m \tan U_m = 0.5\mathrm{TA}_m \\ &= 0.33\ \mathrm{LA}_z \tan U_m = 0.47\mathrm{TA}_z \end{aligned} \tag{4.8}$$

Note that the above are based on the idea of the smallest spot containing 100 percent of the energy in the image of a point. For many applications this concept is valid and useful, but for best image quality there is usually another focus or correction at which the image has a smaller, brighter core and a larger flare; this is usually judged to be better for definition and pictorial purposes.

The effect of a large amount of defocusing on a well-corrected image is to produce a uniformly illuminated blur disk with a diameter of

$$B = 2(\text{defocus}) \tan U_m \approx \frac{\text{defocus}}{f\,\text{number}} \tag{4.9}$$

Astigmatism and field curvature can be evaluated by applying Eq. 4.9 separately in the sagittal and tangential meridians.

Although ordinary axial chromatic is also defocusing, the blur it produces is not uniformly illuminated, but has the energy centrally concentrated. At the midway focus point, the diameter of the blur containing 100 percent of the energy is

$$B = \mathrm{LA}_{ch} \tan U_m = \mathrm{TA}_{ch} \tag{4.10}$$

However, the central concentration leads to a situation where 75 to 90 percent of the energy is in a spot only half this size, and 40 to 60 percent is in a spot one-quarter as large. (The smaller percentages apply for a uniform spectral response distribution and the larger for a triangular spectral distribution.)

For third-order coma, the blur is the typical comet shape, and has a height equal to the tangential coma and a width (in the sagittal direction) two-thirds this size. Note, however, that some 50 to 60 percent of the energy in the coma patch is in the point of the figure, whose size equals one-third the tangential coma.

4.4 MTF—The Modulation Transfer Function

The interpretation of an MTF plot is often problematical; it is not the easiest thing in the world to decide how good an image is on the basis of an examination of its MTF plot.

The limiting resolution is easily determined if the system sensor can be characterized by an aerial image modulation (AIM) curve. This is a plot of the threshold, or minimum, modulation required in the image for the sensor to produce a response. When plotted against spatial frequency, the intersection of the AIM curve and the MTF plot clearly indicates the limiting resolution, as shown in Fig. 4.5.

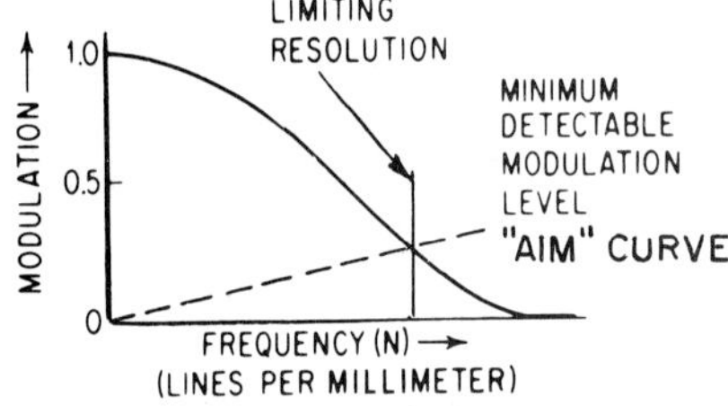

Figure 4.5 The intersection of the AIM curve and the MTF curve indicates the limiting resolution of the system.

A criterion for *excellent* performance (one which is often used as a design goal for top-of-the-line professional motion picture camera lenses) is to look for a 50 percent MTF at 50 lpm. Another criterion which has been presented for commercial 35-mm camera lenses is 20 percent MTF at 30 lpm over 90 percent of the field. Both criteria are applied at full aperture. These will give some idea of the range of the MTF values which are more or less standard for this type of work.

Chapter 5

Lens Design Data

5.1 About the Sample Lenses

One of the features of this volume is a fairly large set of lens designs, their prescriptions, and their aberration plots. This set is not intended to be a complete or extensive *collection* of all or even most of the published lens designs. We happily leave that to others. The set is intended to be a *selection* of lens designs which will serve two primary purposes. These are to serve as a set of suitable starting designs and to serve as a set of designs which illustrate to the reader the principles and techniques of successful lens designs.

The designs in this book were drawn from many sources. A significant number are from the original version of OPTICS TOOLBOX.* (All of the designs in this text, plus many others which were considered but not chosen for inclusion, have been incorporated in the *Warren J. Smith Lens Library*.†) Many are derived from the patent literature, or books which include patent references. Some of the designs are from the technical literature, such as journals, proceedings, or other books about lenses. Some have never been published previously.

In most cases the published designs have been modified to some extent. For the majority of the designs, we have specified the optical glass as one of those from the Schott (Schott Glass Technologies, Inc.) catalog. We have chosen what we feel is the nearest Schott glass to that indicated in our source for the lens data. Occasionally this may constitute a significant change, but we have attempted to stay as close to the original data as possible. In a few designs, non-Schott glasses have been used.

---

*OPTICS TOOLBOX is a product of Genesee Optics Software, Inc.

†Warren J. Smith Lens Library is a trademark of Genesee Optics Software, Inc., and is incorporated in their optical design software products.

The aperture and field which are indicated for any given lens design are more a matter of taste than anything else. What constitutes an acceptable level of aberration depends mostly on the application to which the lens is to be put. Thus the values for field and aperture which accompany each design in this book have often been selected somewhat arbitrarily to yield a level of correction which we thought reasonable.

The choice of the clear apertures for the lens elements is equally arbitrary. Obviously, the clear aperture of an element cannot be so large that the edge thickness at that diameter becomes negative or impractically thin. We have selected what seemed to be reasonable values for the clear apertures, based on both edge thickness considerations and the choice of a vignetting factor which allows a reasonably sized oblique beam through the lens and also trims the oblique beam to eliminate the worst-behaved rays.

5.2 Lens Prescriptions, Drawings, and Aberration Plots

The lens design data and the associated graphics for this book have been produced by computer. While data input errors and other glitches are always possible, by producing the lens data table, the lens drawing, and the aberration plots all from the same lens data file, we hope to prevent most of the errors which have afflicted some other efforts of this type. The design examples have all been scaled to focal lengths which are within a few percent of 100 units in order that comparisons can be easily made. So that the details of the aberration correction will be readily apparent, the computer was programmed to select the scale of each aberration plot to make the plot fill the available space. This does, of course, have the disadvantage that each lens and its aberrations may be plotted to a different scale. In order to minimize this disadvantage we have limited the scales used to decimal factors of 2, 5, and 10.

Lens prescription

A sample is shown in Fig. 5.1. The lens construction data are tabulated in a quite straightforward way. The columns are headed *radius, thickness, mat'l, index, V-no,* and *sa* (for semiaperture); the meanings should be apparent. The radius value follows the usual sign convention that a positive radius has its center of curvature to the right of the surface. Plano surfaces (i.e., with infinite radius) are indicated by a blank entry in the radius column. The thickness and material following a surface are presented on the same line as the surface radius, and have the same number.

F/4.5 25.2deg TRIPLET US 1,987,878/1935 SCHNEIDER

radius	thickness	mat'l	index	V-no	sa
26.160	4.916	LAK12	1.678	55.2	11.7
1201.700	3.988	air			11.7
-83.460	1.038	SF2	1.648	33.8	10.2
25.670	4.000	air			10.2
	6.925	air			9.2
302.610	2.567	LAK22	1.651	55.9	10.3
-54.790	81.433	air			10.3

EFL	= 98.56	= EFFECTIVE FOCAL LENGTH
BFL	= 81.43	= BACK FOCAL LENGTH
NA	= -0.1127 (F/4.4)	= NUMERICAL APERTURE (F-NUMBER)
GIH	= 46.33 (HFOV=25.17)	= IMAGE HEIGHT (HALF FIELD IN DEGREES)
PTZ/F	= -2.831	= (PETZVAL RADIUS)/EFL
VL	= 23.43	= VERTEX LENGTH
OD	infinite conjugate	= OBJECT DISTANCE

Figure 5.1 Sample lens prescription.

With few exceptions, the material names are those of Schott Glass Technologies, Inc. The index and V number values correspond to the wavelengths given with the ray intercept plot (e.g., see Fig. 5.3); for most lenses we have used the d, F, and C lines. The location of the aperture stop is indicated by a blank in the radius column with air on both sides of the surface. Aspheric surfaces are specified by the conic constant kappa and/or the aspheric deformation coefficients. The equation for the surface is

$$x = \frac{cy^2}{1 + [1 - (1 + \kappa)c^2y^2]^{1/2}} + ADy^4 + AEy^6 + AFy^8 + AGy^{10} \tag{5.1}$$

The data below the prescription tabulation has the following meanings:

EFL	Effective focal length
BFL	Back focal length (the distance from the last surface to the paraxial focal point)
NA	Numerical aperture (the corresponding f number is in parentheses)
GIH	Gaussian (paraxial) image height (half-field in degrees is in parentheses)
PTZ/F	Petzval radius as a fraction of EFL
VL	Vertex length from first to last surface
OD	Object distance

Lens drawing

A sample lens drawing is shown in Fig. 5.2. The scale of the lens drawing is indicated by the dimensioned length of the line immediately below the lens sketch. The two rays in the sketch are the marginal and principal rays corresponding to the aperture and field angle which are

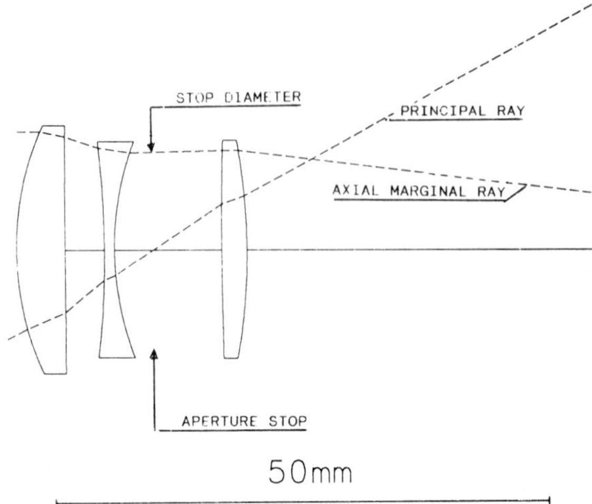

Figure 5.2 Sample lens drawing.

tabulated with the lens data. The aperture stop location is indicated by the point at which the principal ray crosses the optical axis. The lens elements are drawn to the clear apertures given in the prescription table as *sa*, the semiaperture.

Additional rays can easily be added to the lens drawing if desired, by using a technique which is exact only for paraxial rays, but which is often accurate enough for use in estimating or drawing ray paths. A ray may be scaled by simply multiplying the heights at which it strikes the surfaces by a scaling constant. Also, rays may be added by adding their intersection heights together. In each case the result is a reasonable approximation to the path of another ray. Obviously, two rays can be scaled and then added. Thus any desired third ray can be drawn by determining its intersection heights from

$$Y_3 = AY_1 + BY_2 \tag{5.2}$$

where A and B are scaling factors and Y_1 and Y_2 are the ray heights of the rays in the lens drawing. If one defines the desired third ray by its intersection with any two surfaces (which may include the object or image surface), then a simultaneous solution for A and B may be found from the two equations which result when the appropriate values of Y_1, Y_2, and Y_3 are substituted into the equation above.

Aberration plots

A sample aberration plot is shown in Fig. 5.3. The aberration plots include both tangential and sagittal ray intercept plots (sometimes

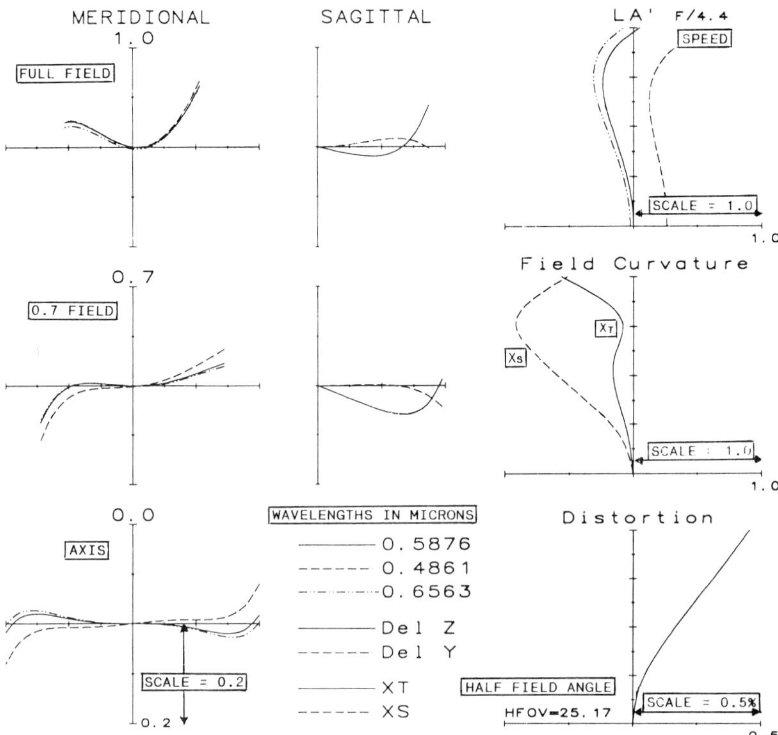

Figure 5.3 Sample aberration plot.

called *H*-tan *U* curves) for the axis, 0.7 field, and full field. The ray displacements are plotted vertically, as a function of the position of the ray in the aperture. The vertical scale is given at the lower end of the vertical bar for the axial plot; the number given is the half-length (i.e., from the origin to the end) of the vertical line in the plot. The horizontal scale is proportional to the tangent of the ray slope angle. Following the usual convention, the upper ray of the ray fan is plotted to the right. In the sagittal plots, the solid line is the transverse aberration in the *z*, or sagittal, direction and the dashed line is the ray displacement in the *y* direction (which is sagittal coma).

In addition to the ray intercept plots (which are, in general, probably the most broadly useful presentation of the aberration characteristics of a design), two aberrations are also presented as longitudinal plots. The longitudinal representations of spherical aberration and field curvature have been the classical, conventional presentation for decades, despite the fact that they give a very incomplete picture of the state of correction of the lens. However, a longitudinal plot of the spherical aberration in three wavelengths does allow a much clearer understanding of the spherochromatism, as well as the secondary spectrum. The scale factor for this plot is the number given at the right end of the horizontal axis; the number is the half length of the horizontal line. The vertical dimension of the plot is the height of the ray at the pupil; the *f* number is given at the top of the plot. This is the *f* number of the imaging cone and is equal to 1/2NA. The longitudinal field curvature plots yield an excellent picture of the correction of the Petzval curvature and the astigmatism. The scale for X_s and X_t is given at the right end of the horizontal axis; again the number is the half length of the horizontal line. The solid line is X_t and the dashed line is X_s. The vertical scale is the fraction of the gaussian image height (GIH); the half field angle is given at the left side of the distortion plot. The scale for distortion is in percent, and the number is the half-length of the horizontal line.

5.3 Estimating the Potential of a Design

It is relatively easy to estimate the effects of a modest redesign on the aberration plots of an existing design by applying a knowledge of third-order aberration theory. This is because the third-order aberrations of a lens are easily adjusted by changing the spaces or the shapes of the elements, whereas the amount of higher-order aberration tends to be quite stable and resistant to change.

Equations 5.3 and 5.4 are a power series expansion of the relationships between the ray intersection with the image plane (y', z') as a function of the object height h and the ray position in the pupil (defined in polar coordinates s and θ), as shown in Fig. 5.4.

$$\begin{aligned} y' = {} & A_1 s \cos\theta + A_2 h + B_1 s^3 \cos\theta + B_2 s^2 h(2 + \cos 2\theta) \\ & + (3B_3 + B_4) s h^2 \cos\theta + B_5 h^3 + C_1 s^5 \cos\theta + (C_2 + C_3 \cos 2\theta) s^4 h \\ & + (C_4 + C_6 \cos^2\theta) s^3 h^2 \cos\theta + (C_7 + C_8 \cos 2\theta) s^2 h^3 + C_{10} s h^4 \cos\theta \\ & + C_{12} h^5 + D_1 s^7 \cos\theta + \cdots \end{aligned} \tag{5.3}$$

$$\begin{aligned} z' = {} & A_1 s \sin\theta + B_1 s^3 \sin\theta + B_2 s^2 h \sin 2\theta \\ & + (B_3 + B_4) s h^2 \sin\theta + C_1 s^5 \sin\theta + C_3 s^4 h \sin 2\theta \\ & + (C_5 + C_6 \cos^2\theta) s^3 h^2 \sin\theta + C_9 s^2 h^3 \sin 2\theta + C_{11} s h^4 \sin\theta \\ & + D_1 s^7 \sin\theta + \cdots \end{aligned} \tag{5.4}$$

Notice that, in the A terms, the exponents of s and h are unity. In the B terms, the exponents total 3, as in s^3, s^2h, sh^2, and h^3. In the C

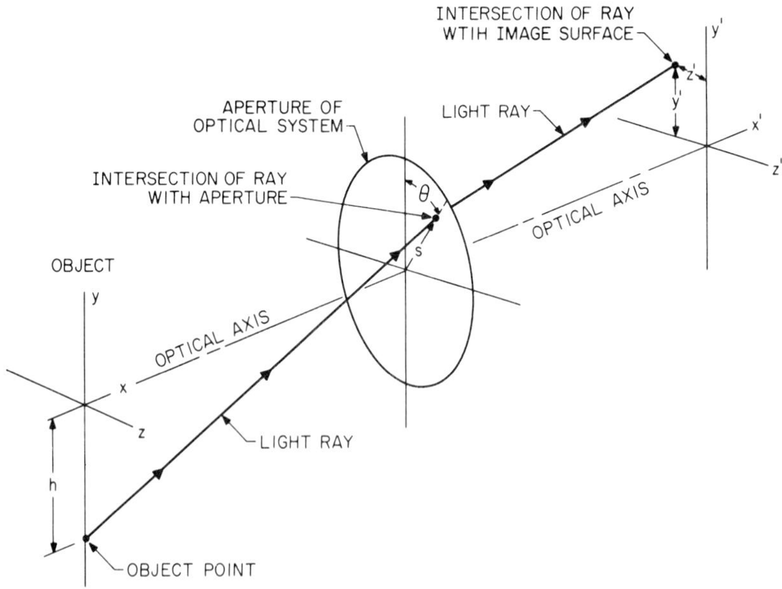

Figure 5.4 A ray from the point $y = h$, $z = 0.0$ in the object passes through the aperture of the optical system at a point defined by its polar coordinates (s, θ), and intersects the image surface at point y', z'.

terms, the exponents total 5, and in the D terms, 7. These are referred to as the first-order, third-order, and fifth-order terms, etc. There are 2 first-order terms, 5 third-order, 9 fifth-order, and $[(n + 3)(n + 5)/8 - 1]$ nth-order terms. In an axially symmetrical system there are no even-order terms; only odd-order terms may exist (unless we depart from symmetry as, for example, by tilting a surface or introducing a toroidal or other nonsymmetrical surface).

It is apparent that the A terms relate to the paraxial (or first-order) imagery. A_2 is simply the magnification (h'/h), and A_1 is a measure of the distance from the paraxial focus to our "image plane." All the other terms in Eqs. 5.3 and 5.4 are called *transverse aberrations*. They represent the distance by which the ray misses the ideal image point as described by the paraxial imaging equations.

The B terms are called the *third-order,* or *Seidel,* or *primary* aberrations. B_1 is spherical aberration, B_2 is coma, B_3 is astigmatism, B_4 is Petzval, and B_5 is distortion. Similarly, the C terms are called the *fifth-order* or *secondary* aberrations. C_1 is fifth-order spherical aberration; C_2 and C_3 are linear coma; C_4, C_5, and C_6 are oblique spherical aberration; C_7, C_8, and C_9 are elliptical coma; C_{10} and C_{11} are Petzval and astigmatism; and C_{12} is distortion.

The 14 terms in D are the seventh-order or tertiary aberrations; D_1 is the seventh-order spherical aberration. A similar expression for OPD, the wavefront deformation, is given in Sec. 5.4 (Eqs. 5.5 and 5.6).

The terms with the B coefficients are the transverse third-order aberrations. The longitudinal aberrations are equal to the transverse aberrations divided by $-u_k'$; since u_k' is a direct function of the ray height s, one can convert these into longitudinal aberrations by reducing the exponent of s by 1 (and adjusting the coefficients).

Thus one can assume that if the spherical aberration is adjusted so as to correct it at a given aperture, the change in the longitudinal aberration will be proportional to the square of the ray heights. For example, if the spherical at ray height Y_1 is to be changed by $d\mathrm{LA}_1$, then at ray height Y_2 it will change by $d\mathrm{LA}_2 = d\mathrm{LA}_1(Y_2/Y_1)^2$. The change of the transverse spherical will vary as a cubic function, so that the transverse change $d\mathrm{TA}_2 = d\mathrm{TA}_1(Y_2/Y_1)^3$. Figure 3.6 shows the effect of changing the third-order spherical (assuming a constant fifth-order spherical).

If the axial chromatic is changed, the ray intercept plots for the different colors will simply rotate with respect to each other. The secondary spectrum and spherochromatism will change very little, so that one can readily estimate the ray intercept plots which will result from a simple change in the axial chromatic. Figure 3.7 shows two different balances of axial chromatic and spherochromatism. Similarly, a lateral chromatic change will change the relative heights of the different color ray plots, and the amount of the height change will be proportioned to the image height.

The change in the longitudinal astigmatism and field curvature is proportional to the square of the image height or field angle. Thus if the field curvature is changed by dX_1 at image height H_1, the change at H_2 is $dX_2 = dX_1(H_2/H_1)^2$. The change in the tangential field curvature X_t is three times the change in the sagittal field curvature X_s if the change is produced by changing the amount of the astigmatism. However, if the change results from a change in the Petzval curvature, both X_s and X_t are shifted by the same distance. Note that the slope of the ray intercept plot ($dH'/d \tan U$) at the principal ray is equal to the tangential field curvature (X_t) (or X_s for the sagittal plot).

Changes in the third-order coma produce a parabolic-shaped change in the ray intercept plot. If the plot is raised by an amount dH at the ends of the plot, it will be raised by $(0.7)^2\, dH = 0.5\, dH$ at the 0.7 zones of the aperture. The amount of the coma change for other field angles will vary directly with the field angle or image height.

The change in percent distortion will vary with the square of the field angle. The change in the lateral color varies directly with the field angle.

Thus one can look at the aberration plots for a given design and, by applying the techniques outlined above, easily visualize what they will look like after an adjustment has been made to fit the design to the application at hand.

5.4 Scaling a Design, Its Aberrations, and Its MTF

A lens prescription can be scaled to any desired focal length simply by multiplying all of its dimensions by the same constant. All of the *linear* aberration measures will then be scaled by the same factor. Note however, that percent distortion, chromatic difference of magnification (CDM), the numerical aperture or f number, aberrations expressed as angular aberrations, and any other *angular* characteristics remain completely unchanged by scaling.

The exact *diffraction* MTF cannot be scaled with the lens data. The diffraction MTF, since it includes diffraction effects which depend on wavelength, will not scale because the wavelength is not (ordinarily) scaled with the lens. A *geometric* MTF can be scaled by dividing the spatial frequency ordinate of the MTF plot by the scaling factor. Of course, because it neglects diffraction, the geometric MTF is quite inaccurate unless the aberrations are very large (and the MTF is correspondingly poor).

A diffraction MTF can be scaled *very* approximately as follows: Determine the OPD which corresponds to the MTF value of the lens for several spatial frequencies. This can be done by comparing the MTF plot for the lens to Figs. 4.3 and 4.4, which relate the MTF to OPD. Then multiply the OPD by the scaling factor and, again using Figs 4.3 and 4.4, determine the MTF corresponding to these scaled OPD values. Obviously the accuracy of this procedure depends on how well the simple relationships of Figs. 4.3 and 4.4 represent the usually complex mix of aberrations in a real lens.

In the event that a proposed change of aperture or field is expected to produce a change in the amount of the aberrations, one can attempt to scale the MTF as affected by aberration. This is done by determining the type of aberration which most severely limits the MTF, then scaling the OPD according to the way that this aberration scales with aperture or field, in a manner analogous to that described in Sec. 5.3. In general, OPD as a function of aperture varies as one higher exponent of the aperture than does the corresponding transverse aberration. For example, the OPD for third-order transverse spherical (which varies as Y^3) varies as the fourth power of the ray height. In a form analogous to Eqs. 5.3 and 5.4, which indicate a power series expansion of the transverse aberrations as a function of aperture and field, Eq. 5.5 gives the relationship for OPD. As in Sec. 5.3, the terms of the equation refer to Fig. 5.4.

$$\begin{aligned}\mathrm{OPD} = {} & A'_1s^2 + A'_2sh\cos\theta + B'_1s^4 + B'_2s^3h\cos\theta \\ & + B'_3s^2h^2\cos^2\theta + B'_4s^2h^2 + B'_5sh^3\cos\theta + C'_1s^6 + C'_2s^5h\cos\theta \\ & + C'_4s^4h^2 + C'_5s^4h^2\cos^2\theta + C'_7s^3h^3\cos\theta + C'_8s^3h^3\cos^3\theta \\ & + C'_{10}s^2h^4 + C'_{11}s^2h^4\cos^2\theta + C'_{12}sh^5\cos\theta + D'_1s^8 + \cdots \end{aligned} \tag{5.5}$$

Note that although the constants here correspond to those in Eqs. 5.3 and 5.4, they are not numerically the same. However, the expressions are related by

$$y' = \mathrm{TA}_y = \frac{l}{N}\frac{\partial \mathrm{OPD}}{\partial y} \quad \text{and} \quad z' = \mathrm{TA}_z = \frac{l}{N}\frac{\partial \mathrm{OPD}}{\partial z} \tag{5.6}$$

where l is the pupil-to-image distance and N is the image space index. Note that the exponent of the semiaperture term s is larger by 1 in the wavefront expression than in the ray-intercept equations.

5.5 Notes on the Interpretation of Ray Intercept Plots

When the image plane intersection heights of a fan of meridional rays are plotted against the slope of the rays as they emerge from the lens, the resultant curve is called a ray intercept curve, an $H' - \tan U'$ curve, or sometimes (erroneously) a rim ray curve. The shape of the intercept curve not only indicates the amount of spreading or blurring of the image directly, but also can serve to indicate which aberrations are present.

In Fig. 5.5 an oblique fan of rays from a distant object point is brought to a perfect focus at point P. If the reference plane passes through P, it is apparent that the $H' - \tan U'$ curve will be a straight horizontal line. However, if the reference plane is behind P (as shown) then the ray intercept curve becomes a tilted straight line since the height, H', decreases as $\tan U'$ decreases. Thus it is apparent that shifting the reference plane (or focusing the system) is equivalent to a rotation of the $H' - \tan U'$ coordinates. A valuable feature of this type of aberration representation is that one can immediately assess the effects of refocusing the optical system by a simple rotation of the abscissa of the figure. Notice that the slope of the line ($\Delta H'/\Delta \tan U'$) is equal to the distance δ from the reference plane to the point of focus, so that for an oblique ray fan the tangential field curvature is equal to the slope of the ray intercept curve.

Figure 5.6 shows a number of intercept curves, each labeled with

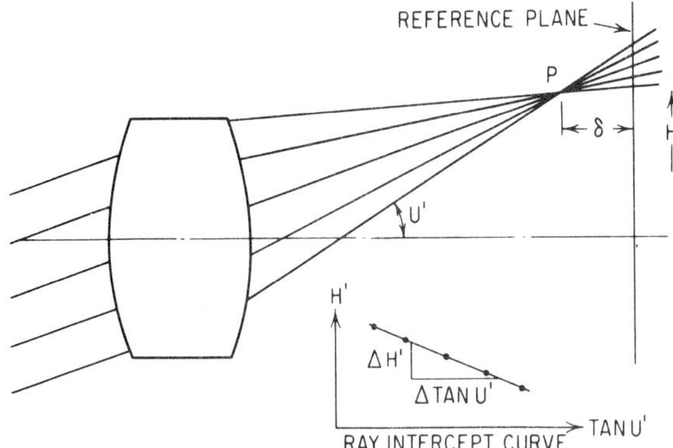

Figure 5.5 The ray intercept curve (H' versus $-\tan U'$) of an image point which does not lie in the reference plane is a tilted straight line. The slope of the line ($dH'/d \tan U'$) is mathematically identical to the distance from the reference plane to the point P. Note that this distance is equal to X_t, the tangential field curvature (if the reference plane is the paraxial focal plane).

the aberration represented. The generation of these curves can be readily understood by sketching the ray paths for each aberration and then plotting the intersection height and slope angle for each ray as a point of the curve. Distortion is not shown in Fig. 5.6; it would be represented as a vertical displacement of the curve from the paraxial image height h'. Lateral color would be represented by curves for two colors which were vertically displaced from each other. The ray intercept curves of Fig. 5.6 are generated by tracing a fan of meridional or tangential rays from an object point and plotting their intersection heights versus their slopes. The imagery in the other meridian can be examined by tracing a fan of rays in the sagittal plane (normal to the meridional plane) and plotting their z-coordinate intersection points against their slopes in the sagittal plane (i.e., the ray slope relative to the principal ray lying in the meridional plane).

It is apparent that the ray intercept curves which are "odd" functions, that is, the curves which have a rotational or point symmetry about the origin, can be represented mathematically by an equation of the form

$$y = a + bx + cx^3 + dx^5 + \cdots$$

or

$$H' = a + b \tan U' + c \tan^3 U' + d \tan^5 U' + \cdots \tag{5.7}$$

All the ray intercept curves for *axial* image points are of this type. Since the curve for an axial image must have $H' = 0$ when $\tan U' = 0$,

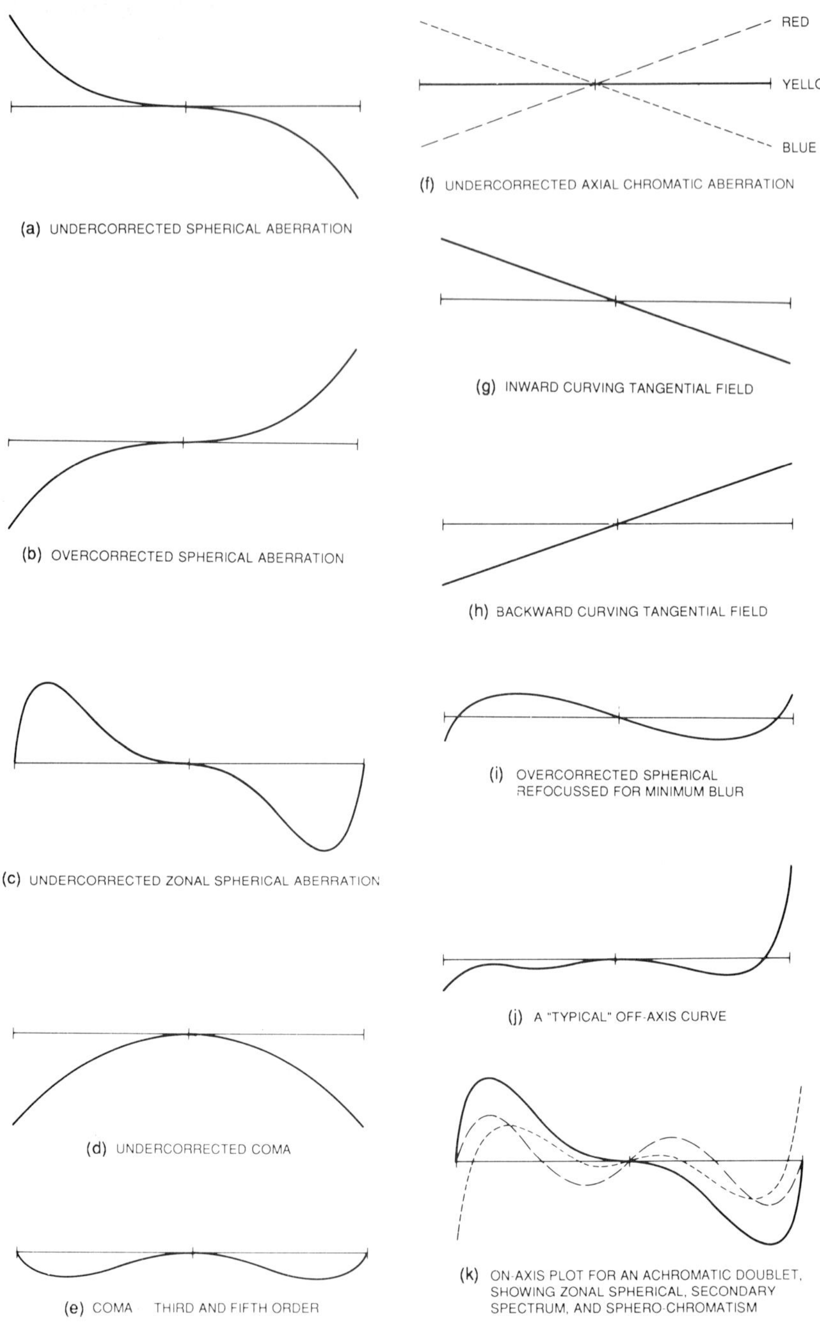

Figure 5.6 Sample ray intercept plots for various aberrations. The ordinate for each curve is the height at which the ray intersects the (paraxial) image plane; usually H is plotted relative to the principal ray height, which is set to zero. The abscissa is tan U, the final slope of the ray with respect to the optical axis. Note that, regardless of the sign convention for the ray slope, it is conventional to plot the ray through the top of the lens at the right of the figure, and that the curves for image points above the axis are usually shown. Observance of these conventions makes it much easier to interpret the plots.

it is apparent that the constant a must be a zero. It is also apparent that the constant b for this case represents the amount the reference plane is displaced from the paraxial image plane. Thus the curve for lateral spherical aberration plotted with respect to the paraxial focus can be expressed by the equation

$$TA' = c \tan^3 U' + d \tan^5 U' + e \tan^7 U' + \cdots \tag{5.8}$$

It is, of course, possible to represent the curve by a power series expansion in terms of the final angle U', or $\sin U'$, or the ray height at the lens (Y), or even the initial slope of the ray at the object (U_0) instead of $\tan U'$. The constants will, of course, be different for each.

For simple uncorrected lenses, the first term of Eq. 5.8 is usually adequate to describe the aberration. For the great majority of corrected lenses the first two terms are dominant; in a few cases three terms (and rarely four) are necessary to satisfactorily represent the aberration. As examples, Figs. 5.6*a* and *b* can be represented by $TA' = c \tan^3 U'$, and this type of aberration is called *third-order spherical*. Figure 5.6*c* however, would require two terms of the expansion to represent it adequately; thus $TA' = c \tan^3 U' + d \tan^5 U'$. The amount of aberration represented by the second term is called the *fifth-order aberration*. Similarly, the aberration represented by the third term of Eq. 5.8 is called the *seventh-order aberration*. The fifth-, seventh-, ninth-, etc., order aberrations are collectively referred to as *higher-order aberrations*.

The ray intercept plot is subject to a number of interesting interpretations. It is immediately apparent that the top-to-bottom extent of the plot gives the size of the image blur. Also, a rotation of the horizontal (abscissa) lines of the graph is equivalent to a refocusing of the image and can be used to determine the effect of refocusing on the size of the blur.

Figure 5.5 shows that the ray intercept plot for a defocused images is a sloping line. If we consider the slope of the curve at any point on an H–$\tan U$ ray intercept plot, the slope is equal to the defocus of a small-diameter bundle of rays centered about the ray represented by that point. In other words, this would represent the focus of the rays passing through a pinhole aperture which was so positioned as to pass the rays at that part of the H–$\tan U$ plot. Similarly, since shifting an aperture stop along the axis is, for an oblique bundle of rays, the equivalent of selecting one part or another of the ray intercept plot, one can understand why shifting the stop can change the field curvature.

The OPD (optical path difference) or wavefront aberration can be derived from an H–$\tan U$ ray intercept plot. The area under the curve between two points is equal to the OPD between the two rays which correspond to the two points. Ordinarily, the reference ray for OPD is either the optical axis or the principal ray (for an oblique bundle). Thus the OPD for a given ray is usually the area under the ray intercept plot between the center point and the ray.

Mathematically speaking, then, the OPD is the integral of the H–$\tan U$ plot and the defocus is the first derivative. The coma is related to the curvature or second derivative of the plot, as a glance at Fig. 5.6*d* will show.

It should be apparent that a ray intercept plot for a given object point can be considered as a power series expansion of the form

$$H' = h + a + bx + cx^2 + dx^3 + ex^4 + fx^5 + \cdots \tag{5.9}$$

where h is the paraxial image height, a is the distortion, and x is the aperture variable (e.g., $\tan U'$). Then the art of interpreting a ray intercept plot becomes analogous to decomposing the plot into its various terms. For example, cx^2 and ex^4 represent third- and fifth-order coma, while dx^3 and fx^5 are the third- and fifth-order spherical. The bx term is due to a defocusing from the paraxial focus and could be due to curvature of field. Note that the constants a, b, c, etc., will be different for points of differing distances from the axis.

Chapter

6

Telescope Objectives*

6.1 The Thin Doublet

The telescope objective can be considered to be a classic example of a thin lens. A thin aplanatic (i.e., free of both coma and spherical aberration) lens cannot be corrected for astigmatism. A thin lens has astigmatism as given by Eq. F.9.4, TAC = $h^2\phi u_k'/2$. This astigmatism cannot be changed by moving the stop. The stop shift equation (Eq. F.10.5) indicates that, if spherical and coma are zero, the astigmatism remains the same regardless of the location of the stop.

Therefore, since the astigmatism must be large and negative per Eq. F.9.4, the image quality away from the optical axis is destined to be poor, and the useful field is limited to a few degrees. However, given sufficient effective degrees of freedom, one *can* control the focal length and can easily correct coma and chromatic and spherical aberration. (Because the lens is thin and has only a small field, distortion and lateral color are usually completely negligible.)

The element powers for a thin achromat of power $\phi = 1/f$ are given by

$$\phi_A = \frac{\phi V_A}{V_A - V_B} \qquad (6.1)$$

$$\phi_B = \frac{\phi V_B}{V_B - V_A} \qquad (6.2)$$

In a thin airspaced doublet, there are four degrees of freedom—the curvatures of the four surfaces. This is just sufficient for the task outlined above. Since the spherical aberration is a quadratic function of

*See W. J. Smith, *Modern Optical Engineering,* McGraw-Hill, New York, 1990, Chap. 12, pp. 372–384, for a discussion of the basics of telescope objective design.

lens shape, for a thin lens there are two solutions to the problem. These are the Gauss form and the Fraunhofer form (or Steinheil, if the flint element faces the distant object).

Several solutions are illustrated here. Figure 6.1 shows a Gauss lens; Fig. 6.2, a Fraunhofer; and Fig. 6.3, a Steinheil, all at a speed of *f*/7. Note that the residual aberrations of the Fraunhofer and the Steinheil forms are quite similar, while the Gauss has somewhat less spherochromatic but much more zonal spherical aberration, and slightly more secondary spectrum. The Gauss solution does not exist if the relative aperture of the lens is too large; the increased element thickness required for the large-diameter elements increases the effective spacing between the elements, and the lowered ray heights on the flint preclude a simultaneous solution for chromatic, coma, and spherical.

As can be seen, these designs have residuals of zonal spherical aberration and spherochromatism, plus secondary spectrum. Which of these constitutes the dominant limitation to performance depends on the aperture, focal length, and spectral bandpass. If the field of view is large enough, the fifth-order coma may also be a problem. There are specific techniques which can be used to attack each of these aberrations; they are outlined in the following sections.

If the objective is to be a cemented doublet, one of the four degrees of freedom cited above must be dedicated to matching the inner (cemented) radii so that they can be cemented together. Thus there are not enough constructional degrees of freedom to simultaneously control the focal length, chromatic, spherical, and coma. However, glass choice can be used as an additional variable parameter. If one chooses glasses with a large difference in V value, there may be no bending at which the spherical aberration is corrected; it is always undercorrected. Conversely, if the glasses have too small a difference in V values, there may be two bendings at which the spherical is corrected, but neither bending will be corrected for coma. Thus the relative glass characteristics can be adjusted to achieve a simultaneous correction. Figures 6.4 and 6.5 illustrate similar lenses at a speed of *f*/3.0, the first cemented and the other airspaced (so as to reduce spherochromatism and zonal spherical as discussed below).

Although seldom used in ordinary telescope objectives, higher-index glasses can reduce the spherical zonal and spherochromatism by a factor of 2 or 3. In general, even for an airspaced doublet there seems to be an optimum mate for a given glass. Figure 6.6 is an airspaced doublet of high-index glasses; compare its aberrations with Fig. 6.7 which uses low-index glass types.

(Text continues on page 72.)

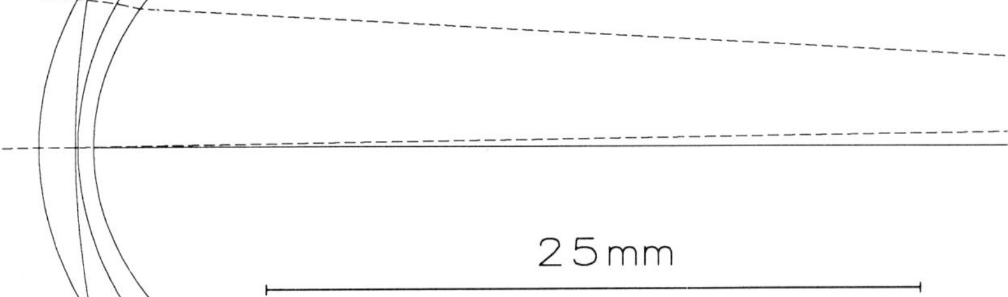

F/7 1degHFOV GAUSS TELESCOPE OBJECTIVE

radius	thickness	mat'l	index	V-no	sa
17.654	1.400	BK7	1.517	64.2	7.2
53.898	0.100	air			7.2
16.530	0.600	SF1	1.717	29.5	7.2
13.075	94.931	air			7.2

EFL = 100
BFL = 94.93
NA = -0.0713 (F/7.0)
GIH = 1.75
PTZ/F = -1.583
VL = 2.10
OD infinite conjugate

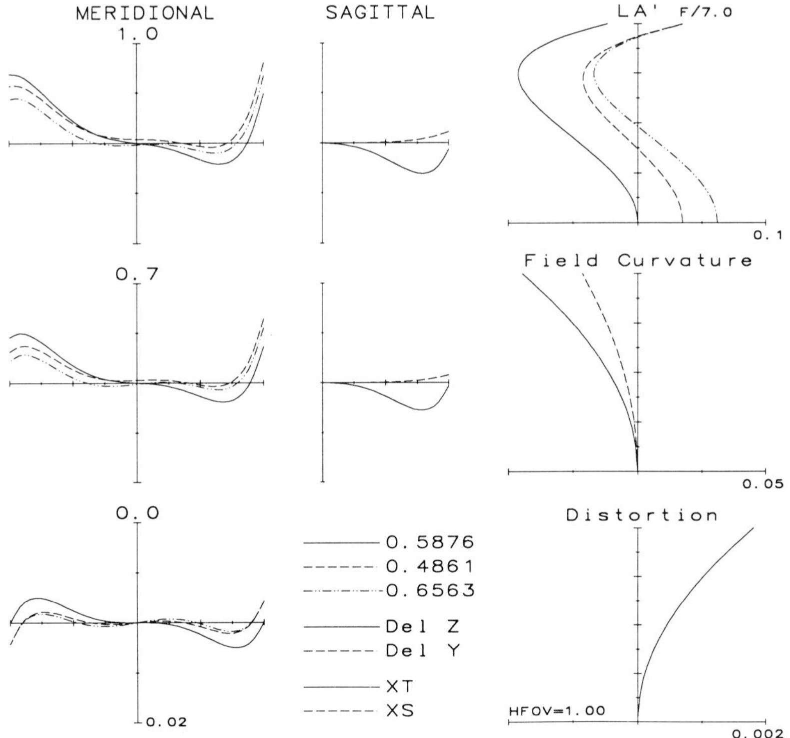

Figure 6.1

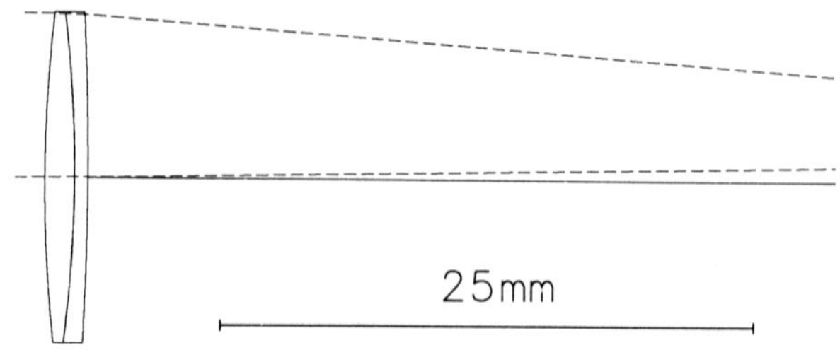

F/7 1degHFOV FRAUNHOFER OBJECTIVE

radius	thickness	mat'l	index	V-no	sa
60.415	1.400	BK7	1.517	64.2	7.2
-52.830	0.012	air			7.2
-51.552	0.600	SF1	1.717	29.5	7.2
-128.126	99.017	air			7.2

EFL = 99.98
BFL = 99.02
NA = -0.0714 (F/7.0)
GIH = 1.75
PTZ/F = -1.38
VL = 2.01
OD infinite conjugate

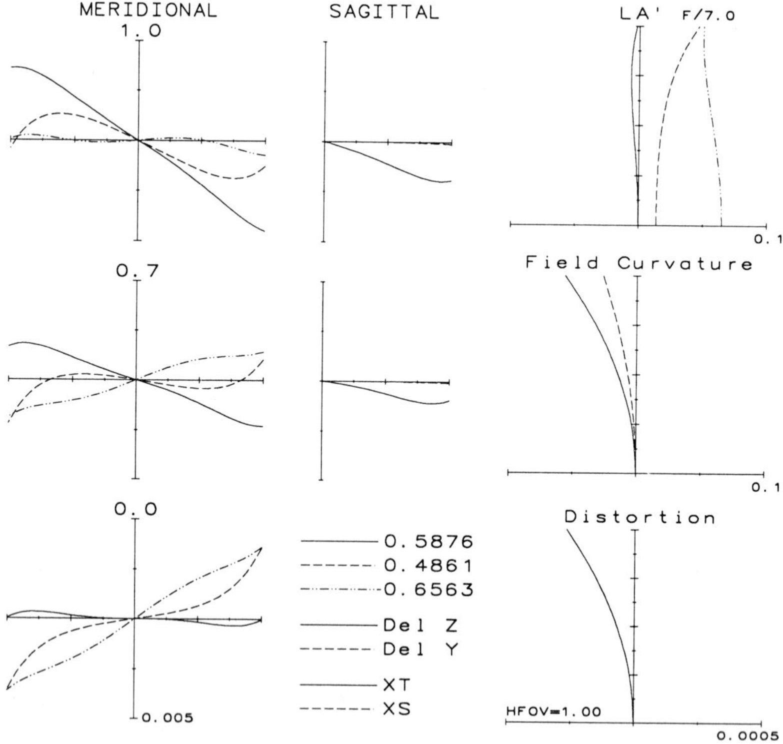

Figure 6.2

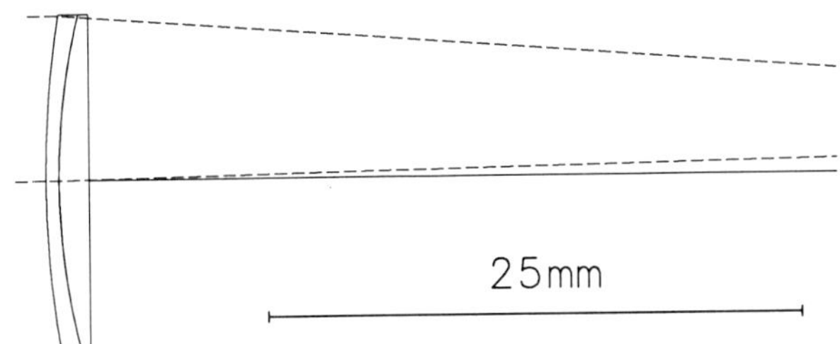

F/7 1degHFOV STEINHEIL OBJECTIVE

radius	thickness	mat'l	index	V-no	sa
41.582	0.600	SF1	1.717	29.5	7.2
27.865	0.019	air			7.2
28.425	1.400	BK7	1.517	64.2	7.2
-1936.302	98.514	air			7.2

EFL = 100
BFL = 98.51
NA = -0.0714 (F/7.0)
GIH = 1.75
PTZ/F = -1.385
VL = 2.02
OD infinite conjugate

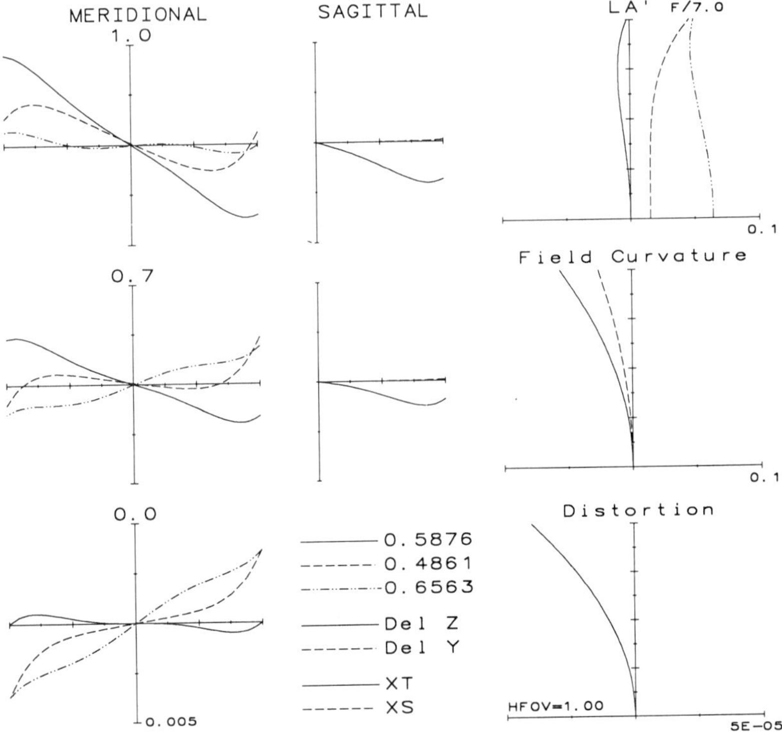

Figure 6.3

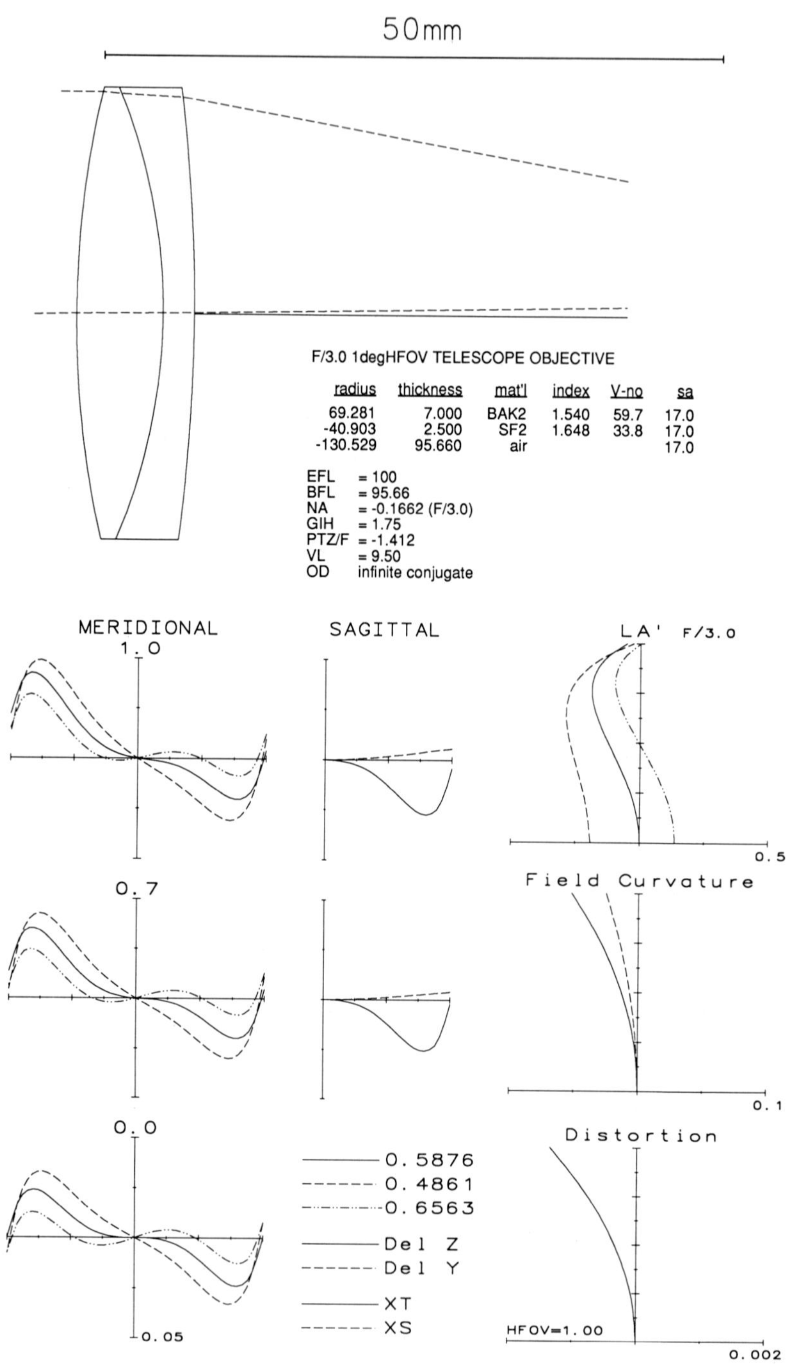

50mm
F/3.0 1degHFOV TELESCOPE OBJECTIVE
radius thickness mat'l index V-no sa
69.281 7.000 BAK2 1.540 59.7 17.0
-40.903 2.500 SF2 1.648 33.8 17.0
-130.529 95.660 air 17.0
EFL = 100
BFL = 95.66
NA = -0.1662 (F/3.0)
GIH = 1.75
PTZ/F = -1.412
VL = 9.50
OD infinite conjugate
MERIDIONAL
1.0
SAGITTAL
LA' F/3.0
0.5
0.7
Field Curvature
0.1
0.0
Distortion
0.5876
0.4861
0.6563
Del Z
Del Y
XT
XS
0.05
HFOV=1.00
0.002

Figure 6.4

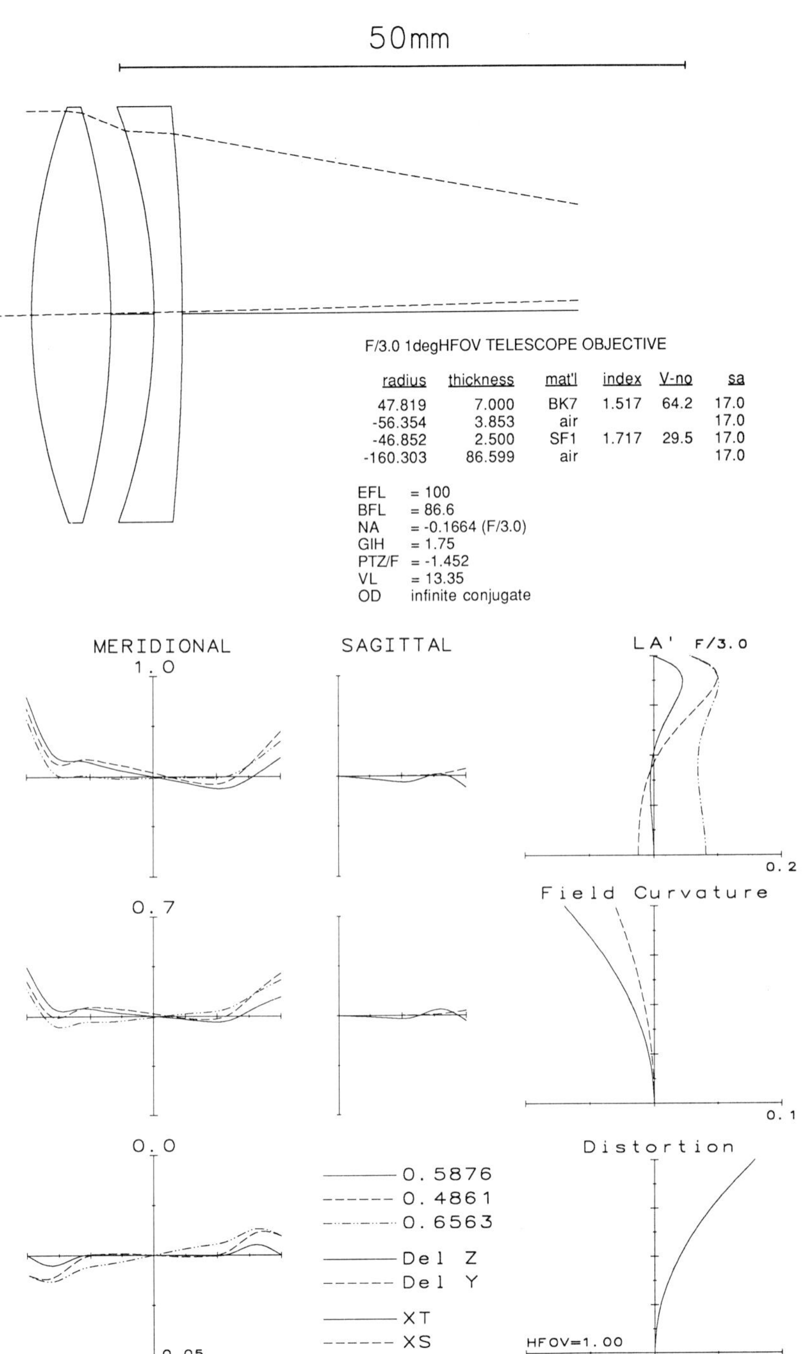

F/3.0 1degHFOV TELESCOPE OBJECTIVE

radius	thickness	mat'l	index	V-no	sa
47.819	7.000	BK7	1.517	64.2	17.0
-56.354	3.853	air			17.0
-46.852	2.500	SF1	1.717	29.5	17.0
-160.303	86.599	air			17.0

EFL = 100
BFL = 86.6
NA = -0.1664 (F/3.0)
GIH = 1.75
PTZ/F = -1.452
VL = 13.35
OD infinite conjugate

Figure 6.5

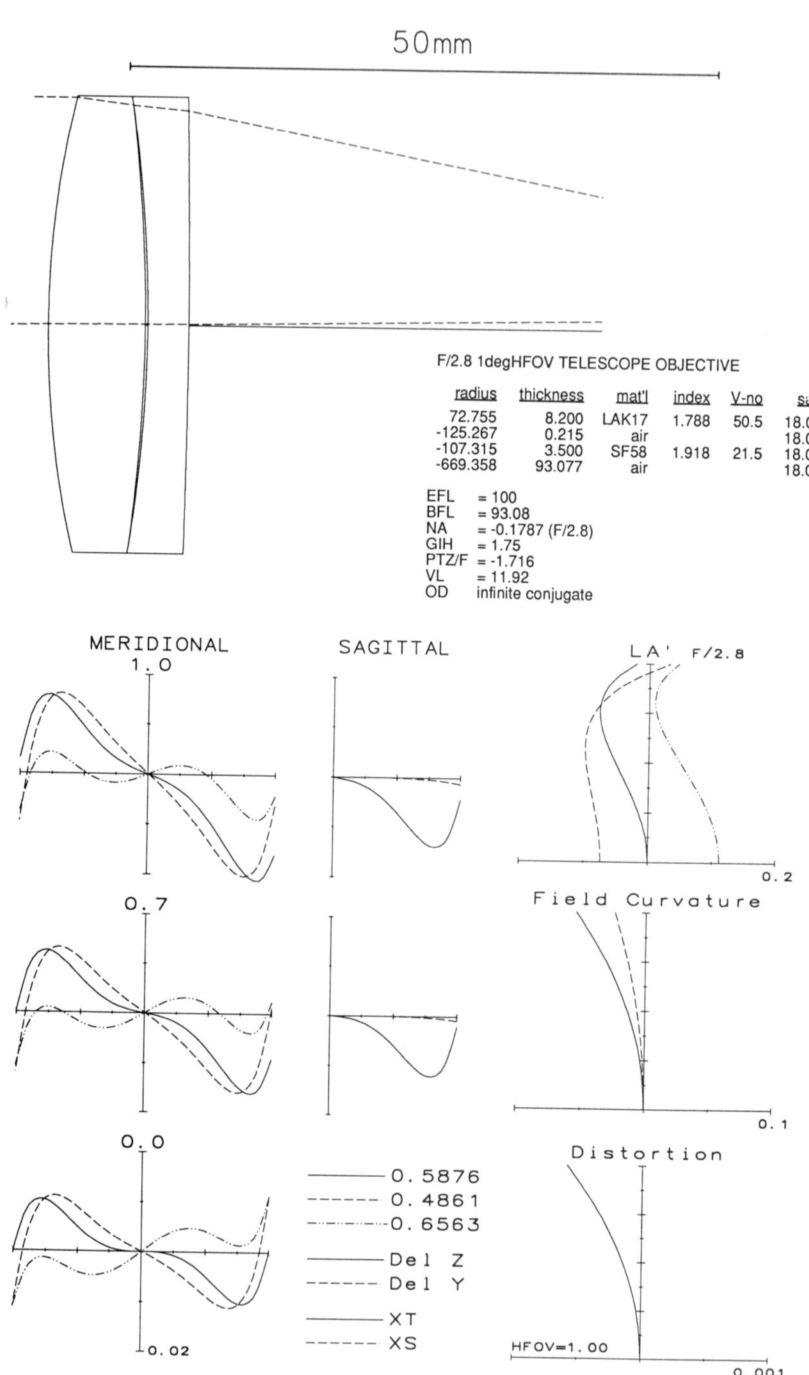

radius	thickness	mat'l	index	V-no	sa
72.755	8.200	LAK17	1.788	50.5	18.0
-125.267	0.215	air			18.0
-107.315	3.500	SF58	1.918	21.5	18.0
-669.358	93.077	air			18.0

Figure 6.6

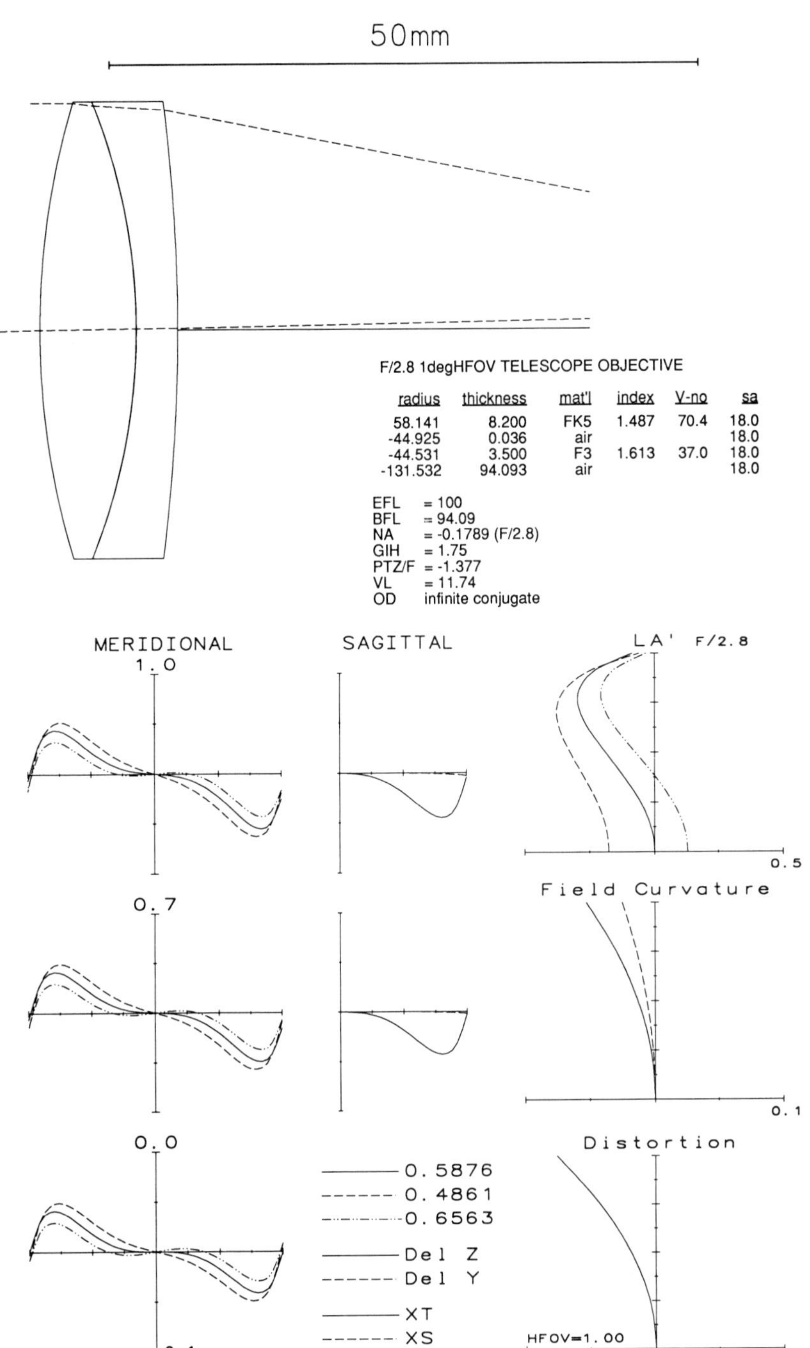
50mm
F/2.8 1degHFOV TELESCOPE OBJECTIVE
radius thickness mat'l index V-no sa
58.141 8.200 FK5 1.487 70.4 18.0
-44.925 0.036 air 18.0
-44.531 3.500 F3 1.613 37.0 18.0
-131.532 94.093 air 18.0
EFL = 100
BFL = 94.09
NA = -0.1789 (F/2.8)
GIH = 1.75
PTZ/F = -1.377
VL = 11.74
OD infinite conjugate
MERIDIONAL
SAGITTAL
1.0
0.7
0.0
0.1
LA' F/2.8
0.5
Field Curvature
0.1
Distortion
HFOV=1.00
0.002
0.5876
0.4861
0.6563
Del Z
Del Y
XT
XS

Figure 6.7

6.2 Secondary Spectrum (Apochromatic Systems)

The secondary spectrum contribution of a thin lens is given by Eq. F.9.9. The secondary spectrum of a doublet (or of any thin achromat using just two glass types) is given by

$$\mathrm{SS} = \frac{f(P_A - P_B)}{V_A - V_B} \tag{6.3}$$

$$= \frac{f(\Delta P)}{\Delta V}$$

where f = focal length
P = partial dispersion
$= (n_F - n_d)/(n_F - n_C)$
V = Abbe V number
$= (n_d - 1)/(n_F - n_C)$

The secondary spectrum can only be reduced by using combinations of glasses which have a lower $\Delta P/\Delta V$ than ordinary glass types. Glasses with unusual partial dispersions can easily be selected from glass catalogs. Most manufacturers include a separate list of their glasses tabulating the amount by which the partial dispersion departs from the normal run of glass. For positive elements, glasses such as FK51, FK52, FK54, PSK53, PK51, LgSK2, the higher-index SF, and the TiF glasses (the last two for use in negative achromatic components) and many crystals such as CaF_2 will tend to reduce the secondary spectrum in any sort of lens. For negative elements, the short flints (KzF and KzSF types) and some lanthanum glasses are useful. Many, or most, of these materials are characterized by poor resistance to atmospheric attack, poor working characteristics in the shop, and frequently by a high price. They tend to be more difficult to manufacture than the ordinary glasses; as a result, their optical quality is sometimes lower than that of normal glasses. In addition, most glass pairs which have well-matched partial dispersions, so that $(P_A - P_B)$ is very small, also tend to have small V-value differences; this means that the element powers required to produce achromatism will be correspondingly large. The result is that, although such glasses can reduce or eliminate secondary spectrum, the large element powers cause greatly increased amounts of spherochromatism and spherical zonal. Such a lens is shown in Fig. 6.8; compare its residuals to the ordinary doublets of the same speed in Figs. 6.2 and 6.3. Obviously, then, such lenses must be used at slower speeds, or designed with many elements to reduce these residuals. Note that the lens of Fig. 6.8 has been

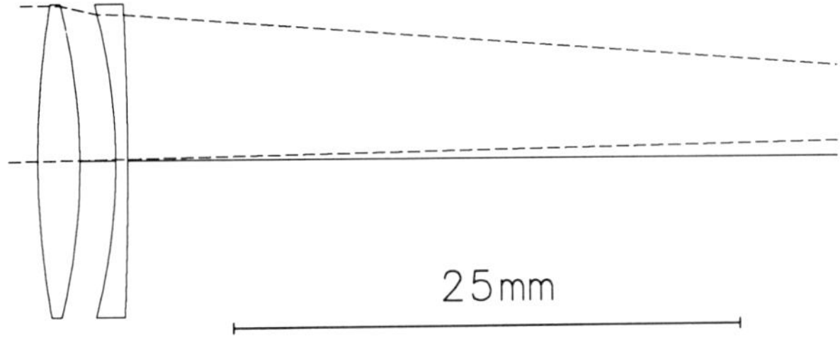

F/7 1degHFOV FRAUNHOFER OBJECTIVE

radius	thickness	mat'l	index	V-no	sa
40.679	2.100	FK54	1.437	90.7	7.2
-27.252	1.789	air			7.2
-25.887	0.600	KZFS2	1.558	54.2	7.2
-141.761	93.440	air			7.2

EFL = 100
BFL = 93.44
NA = -0.0714 (F/7.0)
GIH = 1.75
PTZ/F = -1.364
VL = 4.49
OD infinite conjugate

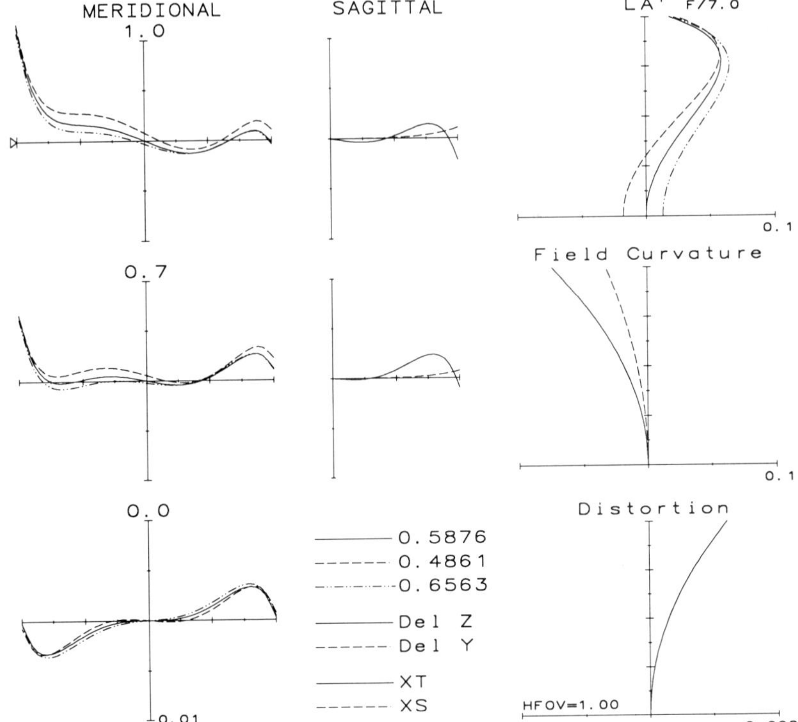

Figure 6.8

spaced to reduce spherochromatic and zonal spherical aberrations, and that the lenses of Figs. 6.2 and 6.3 are not; they are edge-contacted. Even so, the difference in correction is quite apparent.

Figure 6.9 is a plot of partial dispersion versus V value; each dot represents an optical glass. The slope of a line connecting any two glasses is equal to $(P_A - P_B)/(V_A - V_B)$, which, as indicated in Eq. 6.3, determines the amount of secondary spectrum in a thin lens. It is apparent that any pair of glasses selected from along the normal line will have the same secondary spectrum as any other pair. It is also apparent that the glasses with unusual partial dispersions will form doublets with reduced secondary, and that the further they lie away from the normal line, the greater the difference between their V values, the lower the element powers, and the lower the residual aberrations which the lenses will have.

On a P versus V plot such as Fig. 6.9, one can simulate a nonexistent glass anywhere along a straight line connecting the points representing two real glasses. Thus, if a telescope objective is made as a triplet rather than a doublet, it is possible to reduce the $(P_A - P_B)/(V_A - V_B)$ to zero by using two of the glasses to simulate one whose partial dispersion exactly matches that of the third glass. Figures

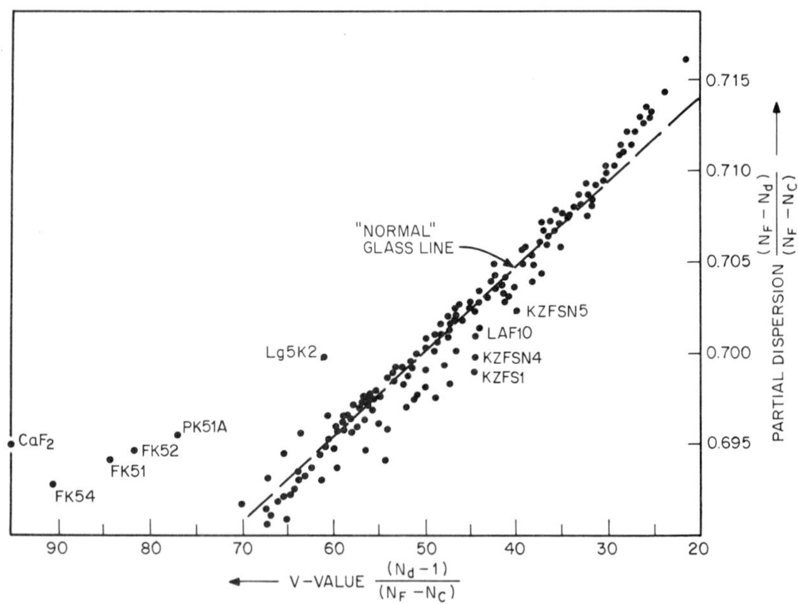

Figure 6.9 Plot of the partial dispersion $P = (n_F - n_d)/(n_F - n_C)$ versus Abbe V number $V = (n_d - 1)/(n_F - n_C)$ for the Schott glass catalog, showing several glasses with unusual partial dispersions which are useful in reducing secondary spectrum.

6.10, 6.11, and 6.12 illustrate such glass combinations in cemented and airspaced (coma-corrected) versions. Note that this technique does not eliminate the need for glasses with unusual partials; it merely facilitates a good match for the partials. By suitable glass choice, it is possible to bring four or even five wavelengths to a common focus; for most applications this is an unnecessary refinement.

Although the secondary spectrum of a lens which is not thin is a somewhat more complex function than the simple relationship given in Eq. 6.3 above, its reduction is effected by the same technique of using appropriate glasses which lie significantly off the normal glass line.

6.3 Spherochromatism

Changing the spacing between two components (of an achromatic system) which are individually uncorrected for axial chromatic aberration will change the spherochromatism of the system. In the case of the doublet telescope objective, the positive element has undercorrected chromatic. Thus the rays between the elements converge more rapidly in blue (or short-wavelength) light than in red light. When the airspace is increased, the ray height at the negative lens is reduced, and it is reduced more for the blue light rays than for the red. This reduction in ray height reduces the overcorrecting spherical aberration contribution from the negative element, and it reduces it more for the blue light than it does for the red. This is, of course, a change in the spherochromatism, and a change in the right direction, since the usual problem is that the blue spherical is overcorrected compared to the red. This effect is illustrated by comparing Fig. 6.4 with Fig. 6.5. Note well that this principle is not limited to doublets or telescope objectives; it can be applied quite generally.

6.4 Zonal Spherical Aberration

Changing the spacing between two components which are individually and oppositely uncorrected for spherical aberration will change the zonal residual aberration. The principle works a lot like a spacing change does with spherochromatism. When the doublet's airspace is increased, the ray heights at the flint are reduced. The undercorrected spherical of the positive element causes the marginal rays between the elements to converge at a disproportionately greater rate than the zonal rays. The overcorrection of the negative element is reduced and it must be reshaped to correct the spherical again. But the overcorrection is reduced more for the marginal ray than for the zonal ray; this reduces the undercorrected zonal residual spherical. This is ordi-

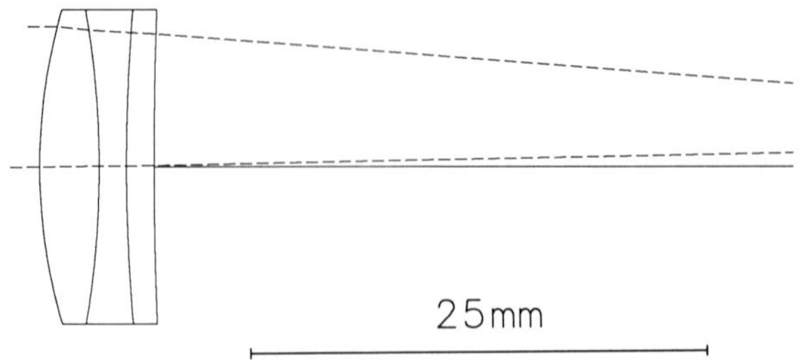

F/7.0 1degHFOV TELESCOPE OBJECTIVE

radius	thickness	mat'l	index	V-no	sa
26.797	3.300	PK51	1.529	77.0	8.0
-42.798	1.500	LAF21	1.788	47.5	8.0
92.002	1.500	SF15	1.699	30.1	8.0
218.030	93.427	air			8.0

EFL = 100
BFL = 93.43
NA = -0.0723 (F/7.0)
GIH = 1.75
PTZ/F = -1.21
VL = 6.30
OD infinite conjugate

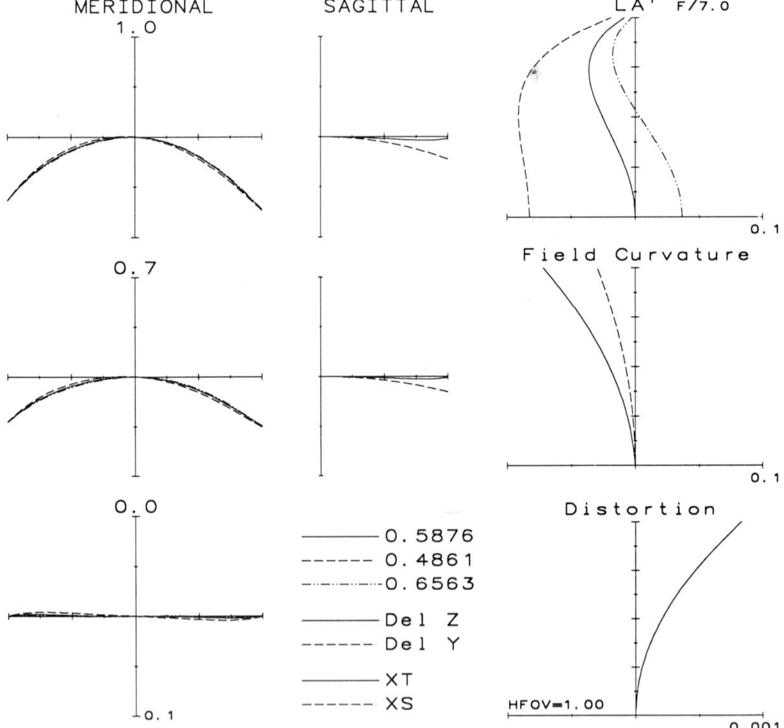

Figure 6.10

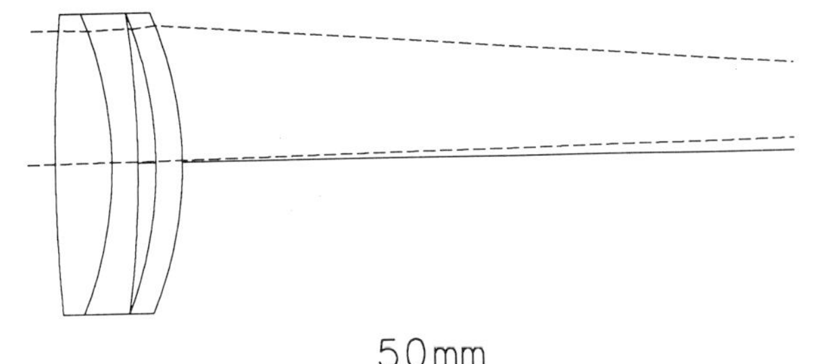

F/7.0 1degHFOV TELESCOPE OBJECTIVE

radius	thickness	mat'l	index	V-no	sa
88.869	3.300	PK51	1.529	77.0	8.0
-19.643	1.500	LAF21	1.788	47.5	8.0
-54.177	1.039	air			8.0
-20.408	1.500	SF15	1.699	30.1	8.0
-19.367	100.880	air			8.0

EFL = 100
BFL = 100.9
NA = -0.0714 (F/7.0)
GIH = 1.75
PTZ/F = -1.21
VL = 7.34
OD infinite conjugate

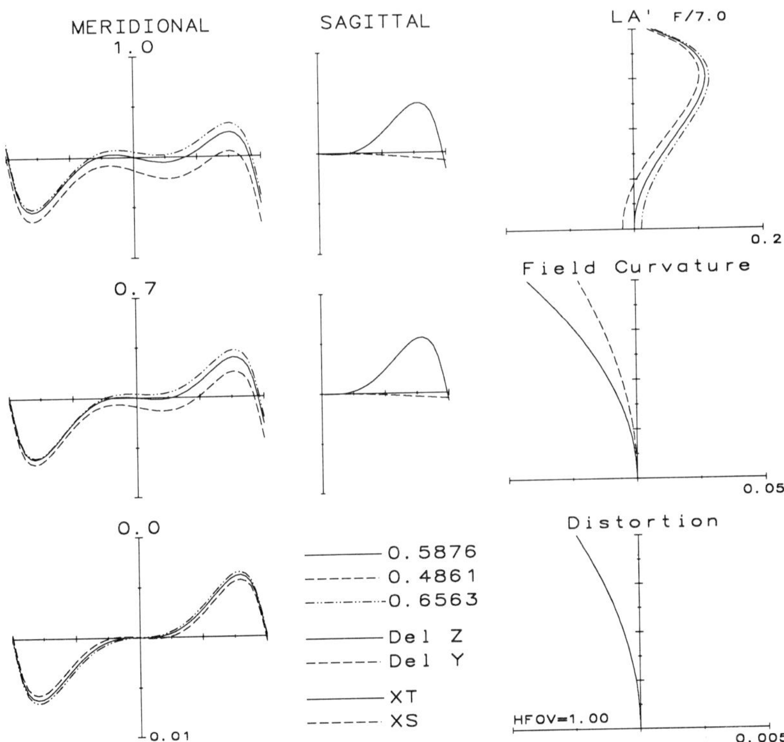

Figure 6.11

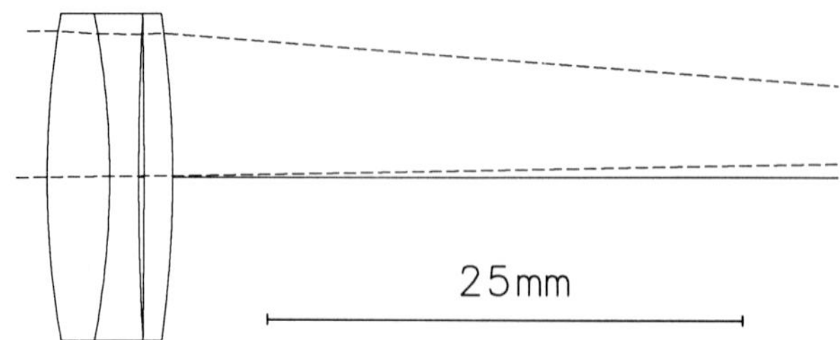

F/7.0 1degHFOV TELESCOPE OBJECTIVE

radius	thickness	mat'l	index	V-no	sa
44.144	3.300	PK51	1.529	77.0	8.0
-39.524	1.500	LSF18	1.913	32.4	8.0
158.460	0.278	air			8.0
-418.801	1.500	SF57	1.847	23.8	8.0
-54.845	97.720	air			0.0

EFL = 100
BFL = 97.72
NA = -0.0714 (F/7.0)
GIH = 1.75
PTZ/F = -1.142
VL = 6.58
OD infinite conjugate

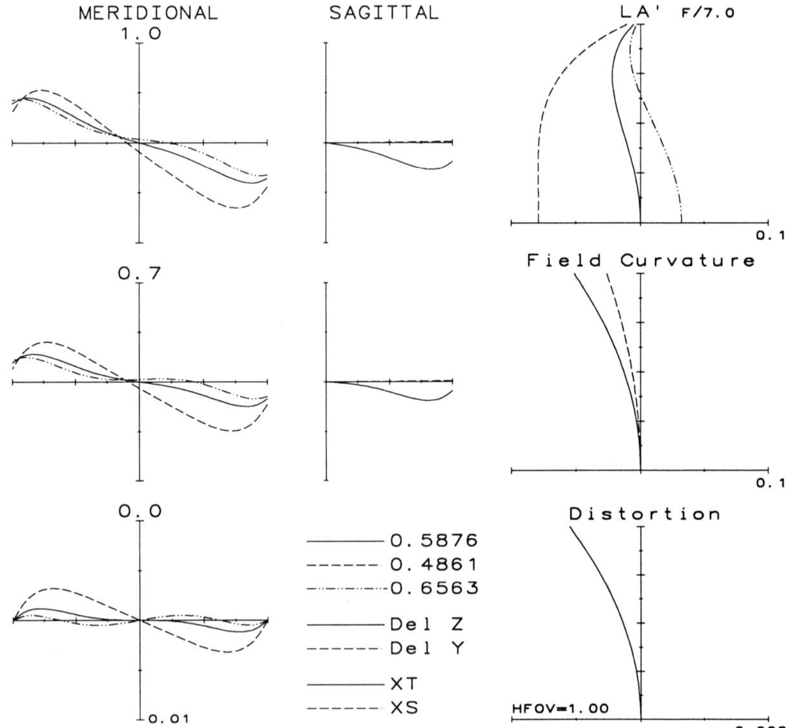

Figure 6.12

narily the change which one would desire. Again, a comparison of Fig. 6.4 with Fig. 6.5 shows this principle in action. This principle can be applied quite generally; it is not limited to telescope objectives.

6.5 Induced Aberrations

The two preceding sections illustrate quite clearly the idea of induced aberrations. The third-order aberration contributions are based solely on the paraxial angles and ray heights which occur at the surface or element contributing the aberration. The higher-order aberrations, however, are affected by the aberrations of the *other* surfaces or elements. In the two cases described above, the change of the doublet airspace affects the spherochromatism and the zonal spherical because the individual elements of the doublet are undercorrected for chromatic and spherical aberrations, respectively. If the elements were fully corrected, the effect of the spacing change would be negligible.

6.6 Three-Element Objectives

The idea of splitting elements to reduce aberrations can be applied to the telescope objective by dividing the crown element into two or more elements. Note that, in general, splitting a positive element will reduce an undercorrected zonal spherical residual and splitting a negative element will reduce a positive or overcorrected zonal. A simple split can reduce the zonal; however, it does nothing for the spherochromatism, which must be controlled as described above by respacing components which are chromatically uncorrected. The triplet telescope objective can be executed as three airspaced elements, or as a singlet crown plus a cemented doublet. The latter form is somewhat easier to fabricate because it is less sensitive to misalignments between the components. Either form has enough effective degrees of freedom to control both spherochromatism and zonal spherical, as well as the primary aberrations: chromatic, coma, and spherical. Figures 6.13 through 6.18 show a series of telescope objectives, all using BK7 (517:642) and SF1 (717:295) glass and all at a speed of *f*/2.8, which is too fast for most applications but which is used here to clearly display the aberrations. Figure 6.13 is the basic edge-contacted doublet, shown here for comparison. In Fig. 6.14 the crown element is split to reduce the zonal and the flint is spaced away to correct the spherochromatism. The result is a lens whose axial aberrations are practically eliminated except for secondary spectrum (which could be reduced by using glasses with unusual partial dispersions). Figure 6.15 shows a different sequence of elements. Here the split has been equally effective in reducing the spherical zonal, but, since the spac-

(Text continues on page 86.)

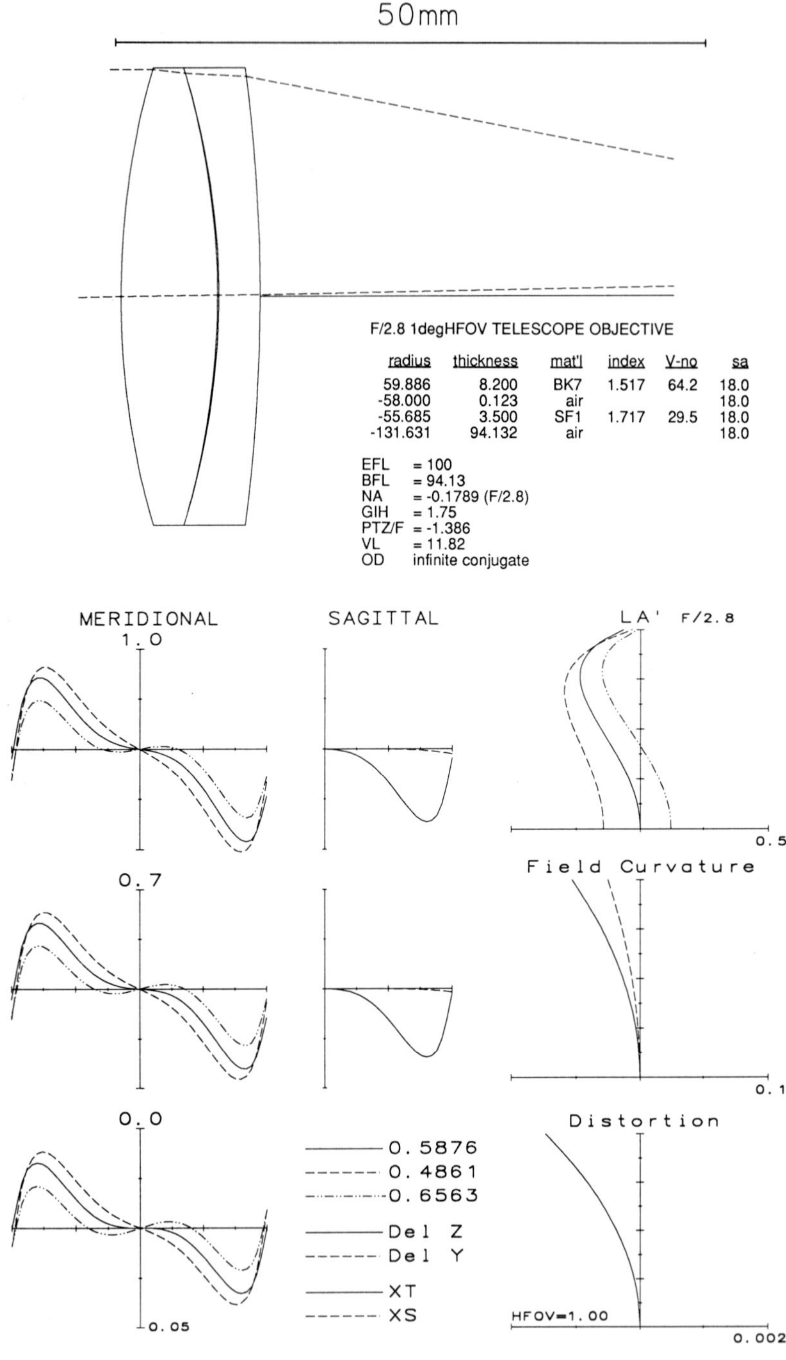

radius	thickness	mat'l	index	V-no	sa
59.886	8.200	BK7	1.517	64.2	18.0
-58.000	0.123	air			18.0
-55.685	3.500	SF1	1.717	29.5	18.0
-131.631	94.132	air			18.0

EFL = 100
BFL = 94.13
NA = -0.1789 (F/2.8)
GIH = 1.75
PTZ/F = -1.386
VL = 11.82
OD infinite conjugate

Figure 6.13

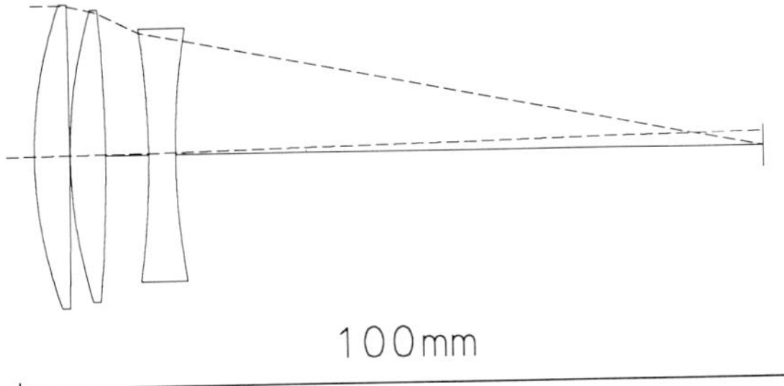

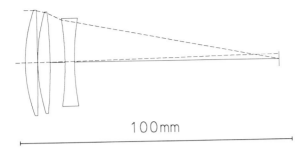

F/2.8 1degHFOV TRIPLET TELESCOPE OBJECTIVE

radius	thickness	mat'l	index	V-no	sa
50.098	4.500	BK7	1.517	64.2	18.0
-983.420	0.100	air			18.0
56.671	4.500	BK7	1.517	64.2	17.3
-171.150	5.571	air			17.3
-97.339	3.500	SF1	1.717	29.5	15.0
81.454	75.132	air			0.0

EFL = 100
BFL = 75.13
NA = -0.1788 (F/2.8)
GIH = 1.75
PTZ/F = -1.749
VL = 18.17
OD infinite conjugate

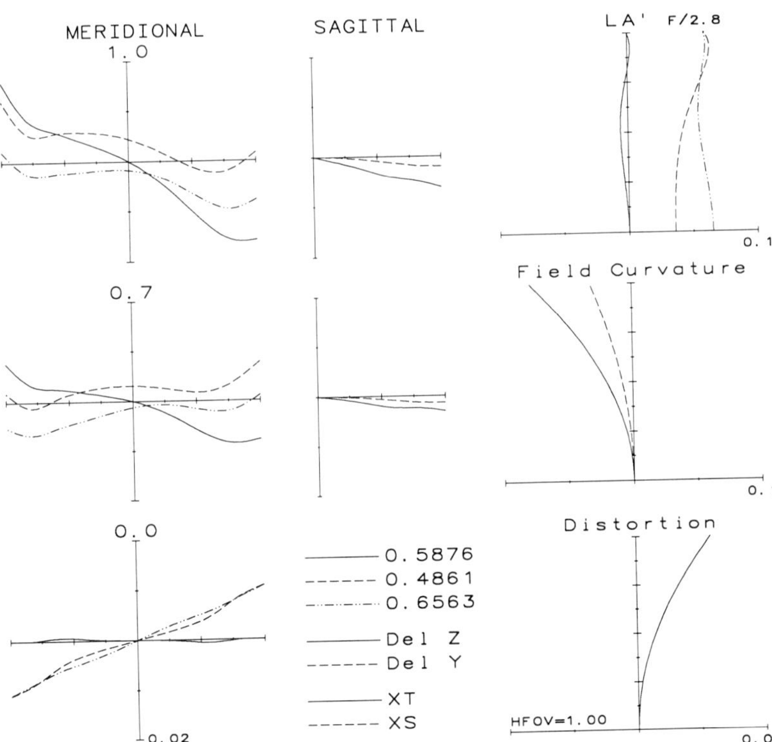

Figure 6.14

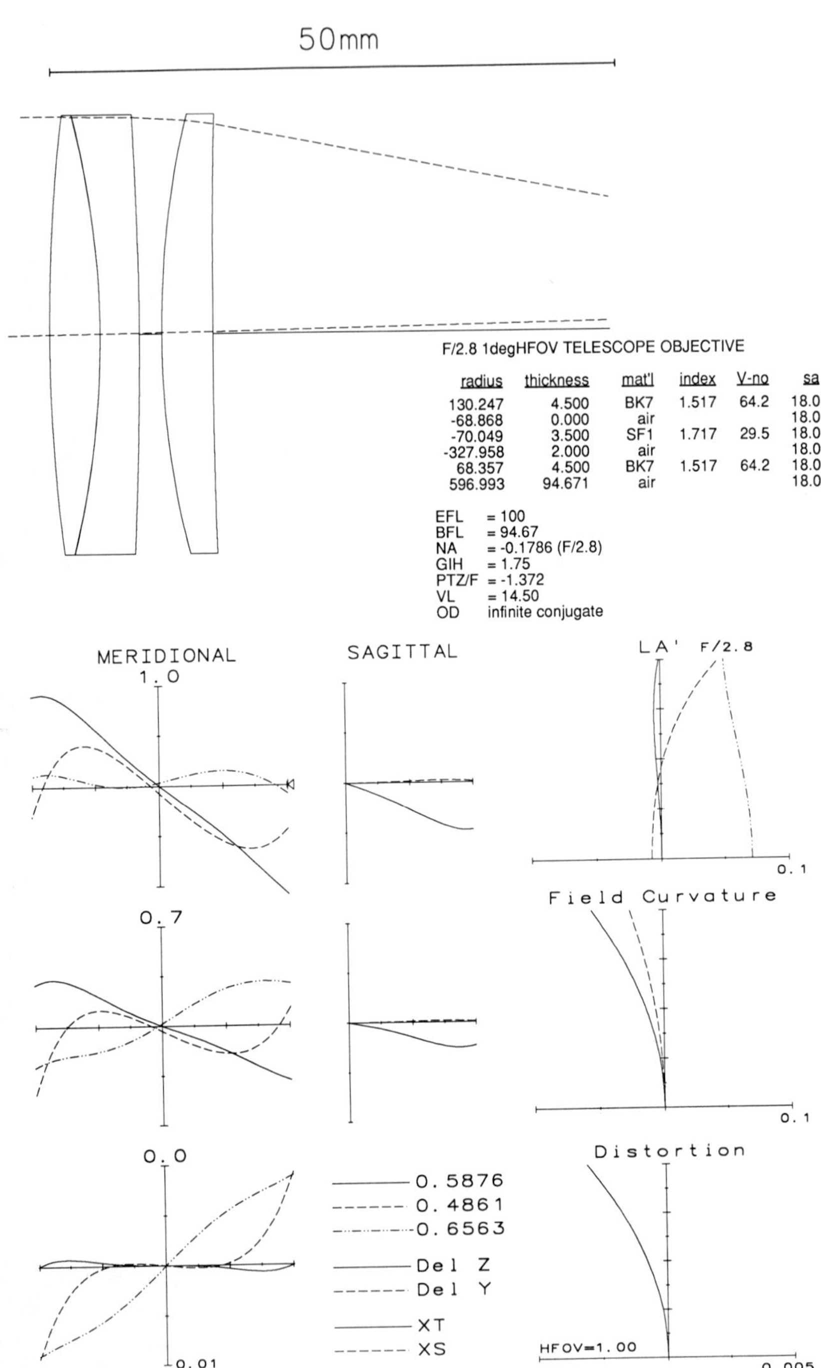

radius	thickness	mat'l	index	V-no	sa
130.247	4.500	BK7	1.517	64.2	18.0
-68.868	0.000	air			18.0
-70.049	3.500	SF1	1.717	29.5	18.0
-327.958	2.000	air			18.0
68.357	4.500	BK7	1.517	64.2	18.0
596.993	94.671	air			18.0

EFL = 100
BFL = 94.67
NA = -0.1786 (F/2.8)
GIH = 1.75
PTZ/F = -1.372
VL = 14.50
OD infinite conjugate

Figure 6.15

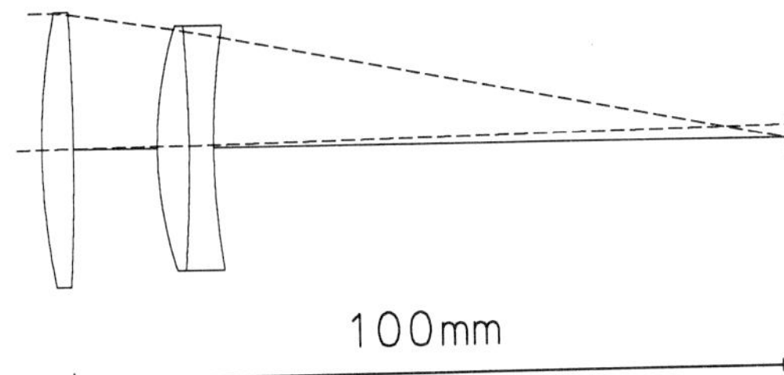

F/2.8 1degHFOV TELESCOPE OBJECTIVE

radius	thickness	mat'l	index	V-no	sa
84.998	4.500	BK7	1.517	64.2	18.0
-317.606	11.928	air			18.0
49.243	4.500	BK7	1.517	64.2	16.0
-195.692	3.500	SF1	1.717	29.5	16.0
93.947	80.694	air			16.0

EFL = 100
BFL = 80.69
NA = -0.1784 (F/2.8)
GIH = 1.75
PTZ/F = -1.394
VL = 24.43
OD infinite conjugate

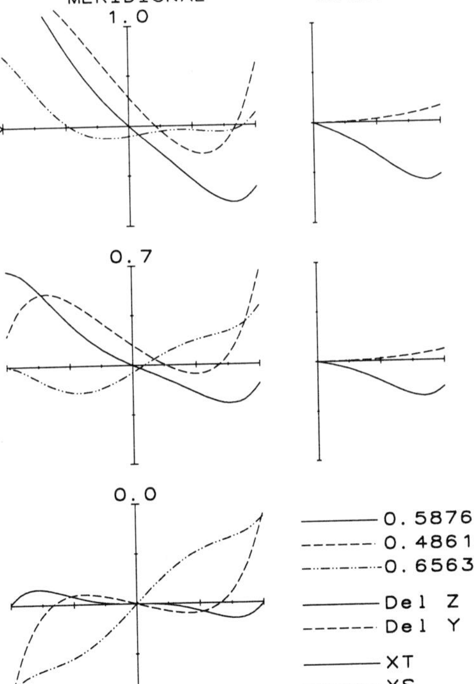

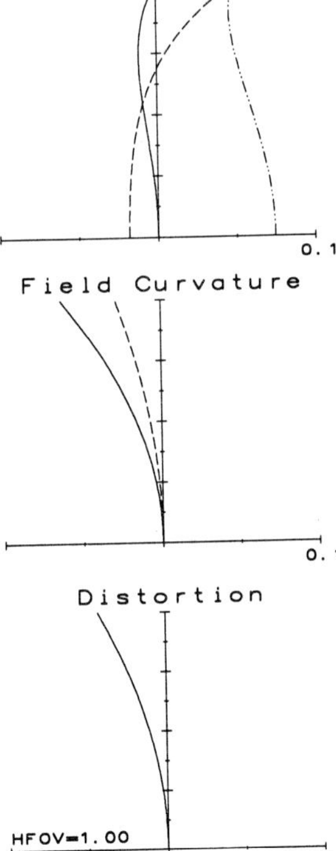

Figure 6.16

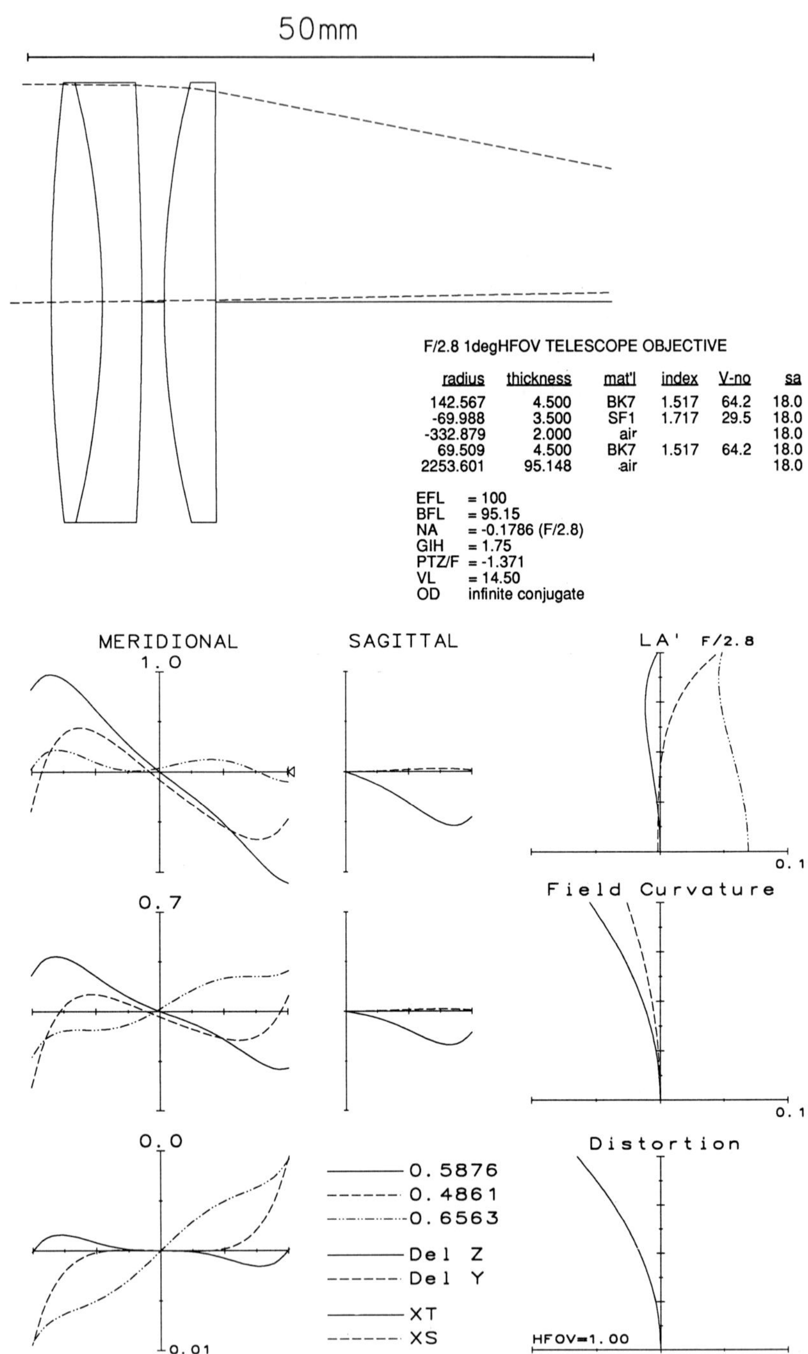

F/2.8 1degHFOV TELESCOPE OBJECTIVE

radius	thickness	mat'l	index	V-no	sa
142.567	4.500	BK7	1.517	64.2	18.0
-69.988	3.500	SF1	1.717	29.5	18.0
-332.879	2.000	air			18.0
69.509	4.500	BK7	1.517	64.2	18.0
2253.601	95.148	air			18.0

EFL = 100
BFL = 95.15
NA = -0.1786 (F/2.8)
GIH = 1.75
PTZ/F = -1.371
VL = 14.50
OD infinite conjugate

Figure 6.17

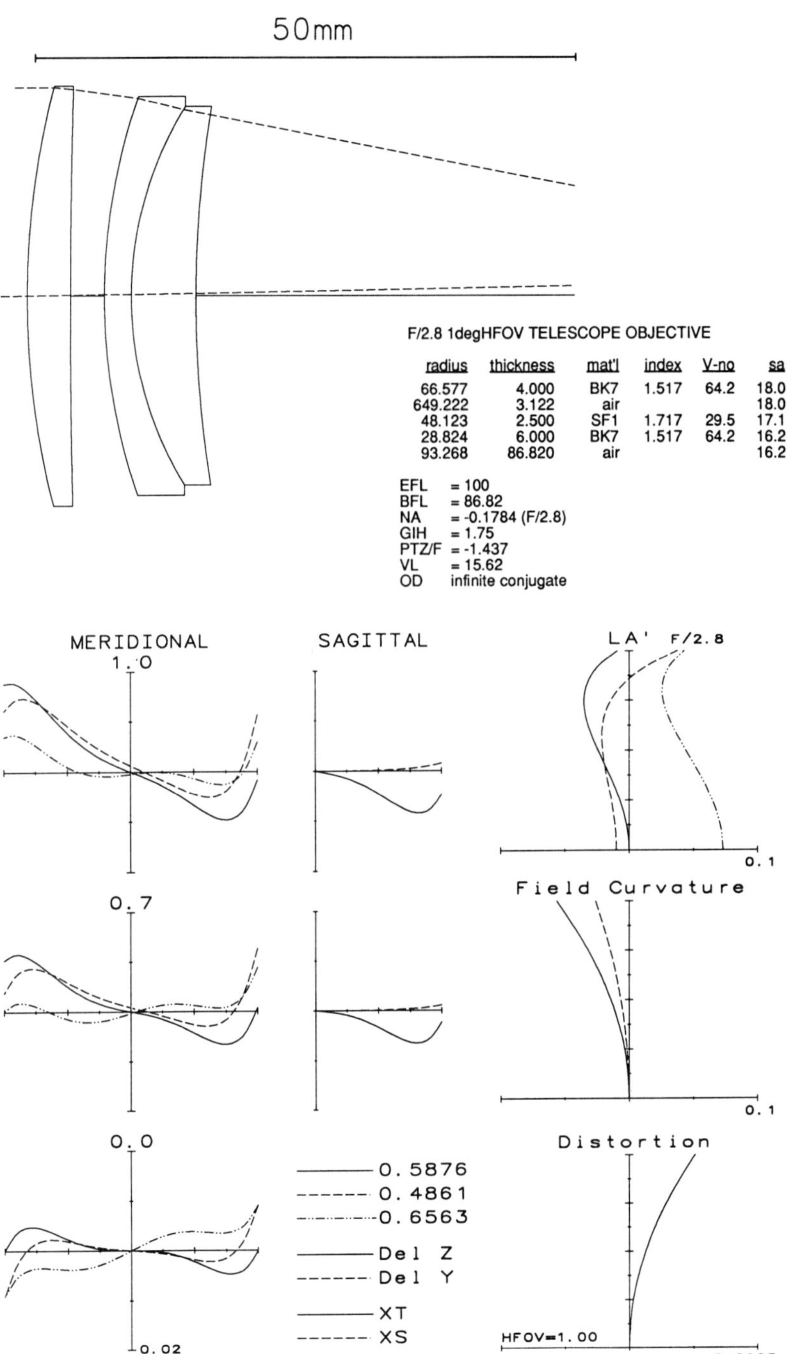

F/2.8 1degHFOV TELESCOPE OBJECTIVE

radius	thickness	mat'l	index	V-no	sa
66.577	4.000	BK7	1.517	64.2	18.0
649.222	3.122	air			18.0
48.123	2.500	SF1	1.717	29.5	17.1
28.824	6.000	BK7	1.517	64.2	16.2
93.268	86.820	air			16.2

EFL = 100
BFL = 86.82
NA = -0.1784 (F/2.8)
GIH = 1.75
PTZ/F = -1.437
VL = 15.62
OD infinite conjugate

Figure 6.18

ing has been kept small, the spherochromatism has not been eliminated. Figures 6.16, 6.17, and 6.18 show various arrangements of triplet objectives where the flint and one of the crown elements have been cemented.

Note that many of the sample telescope objective designs in this chapter are shown at speeds which are quite high for a telescope objective, e.g., *f*/2.8 or *f*/3.0. This has been done so that the aberrations and the techniques used to reduce them are clearly demonstrated. Most telescope objectives are used at speeds which are considerably slower and at which the aberrations are much smaller.

Figures 21.1, 21.2, and 21.3 in Chap. 21 show telescope objectives designed for use in the 8- to 12-μm region of the infrared. Figure 22.9 in Chap. 22 is effectively a telescope objective designed for use in monochromatic (laser) light; both elements are of the same glass.

Chapter

7

Eyepieces and Magnifiers

7.1 Eyepieces

The eyepiece of a telescope or microscope is an unusual optical system in that its pupil must be located completely outside the system. The aperture stop is usually located at the objective lens. The exit pupil is the image of the stop, and it must lie a suitable distance away from the eyepiece so that the eye can be placed at the pupil (in order to see the full field of view). This distance, or *eye relief,* should be at least 9 or 10 mm just to clear the eyelashes; an additional 5 mm eases the situation considerably. An eye relief of 20 mm or so is about the minimum necessary for comfortable use by spectacle wearers. Many systems depart from these guidelines. Any system subject to sudden motion (e.g., a rifle scope, which recoils) needs a much longer eye relief to prevent injury to the eye. For a 22-caliber rifle, the eye relief should be 2 in or more; for a high-powered rifle, the eye relief is typically about 4 or 5 in. It should be obvious that, for a given apparent field, the longer the eye relief, the larger the diameter of the eye lens must be in order to pass the edge-of-the-field rays. At the other extreme, the eyepieces of surveying telescopes, laboratory equipment, and microscopes often have an uncomfortably short eye relief.

Because there is absolutely no symmetry about the stop, distortion, coma, and lateral color tend to be a problem in an eyepiece. Most eyepieces have an amount of distortion which would be considered intolerable in a good camera lens. Several percent is typical, and, in wide-angle eyepieces, a distortion of 8 to 12 percent is not uncommon. The distortion is usually of the f-theta, or f-θ, type, however, and this is more easily tolerated than is that of the opposite sign. Coma and lateral color are ordinarily well corrected, but spherical and axial color are not; they are typically small and are usually balanced out by compensating aberrations in the objective and/or the erector lenses.

Spherical aberration of the pupil and distortion are directly related. For example, if an eyepiece has undercorrected spherical aberration of the chief or principal ray, that ray is usually bent toward the axis at a greater angle than it should be; the greater slope means that the apparent image of this part of the field will be too large, i.e., too far from the axis, and pincushion distortion is the result.

Field curvature and astigmatism are important in eyepieces. For many eyepieces, there is simply no correction of the Petzval curvature; the eyepiece is effectively composed of only positive components, and the result is a strongly inward-curving Petzval surface. Eyepieces which have flatter fields usually achieve them with thick meniscus components (which function just as described in Chap. 11). In some, a concave surface located near the focal plane has a strong field-flattening effect. This arrangement is the equivalent of a negative field lens; in addition to flattening the Petzval field it also lengthens the eye relief (and necessitates a correspondingly larger-diameter eye lens). A separate negative field lens is occasionally encountered when the concomitant increase in the size and weight of the eyepiece is an acceptable tradeoff, as in a tank sight, for example.

For an eyepiece afflicted with an inward-curving Petzval field, the astigmatism should definitely *not* be negative; usually some positive, overcorrecting astigmatism is desirable in order to artificially flatten the field. Often the factor which limits the extent of the angular field of the eyepiece is an overcorrected higher-order astigmatism, which produces both a backward-curving tangential field and too much astigmatism. In many eyepieces the overcorrected astigmatism arises at a diverging cemented surface. An increase in either the surface curvature or the index difference across the surface will increase the overcorrection of the astigmatism; a suitable balance between these two factors is necessary to both flatten the tangential field and restrain the high-order astigmatism.

A high-power microscope objective with an aplanatic front has lateral color, which, given the limitations of the classical design form, cannot be controlled. A *compensating eyepiece* is one which is designed to have a matching amount of lateral color, so that the final image presented to the eye is free of lateral color.

The field curvature of an eyepiece is best evaluated in diopters of defocusing at the eye. In an eyepiece which is designed or analyzed separately from the telescope, the ray trace is conventionally done in reverse, with the object located at infinity on the eye side. Therefore, the field curvature (X_s and X_t) which is calculated this way must be converted from the short conjugate to the long. Using the newtonian imaging equation, we get $x' = -f^2/x$, and the reciprocal of x' (in meters) is the field curvature in diopters. Thus the field curvatures X_s and X_t can be converted to diopters by

$$D = \frac{X}{f^2} \qquad X \text{ and } f \text{ in meters}$$

$$= \frac{1000X}{f^2} \qquad X \text{ and } f \text{ in millimeters}$$

$$= \frac{39.37X}{f^2} \qquad X \text{ and } f \text{ in inches}$$

If D is negative, the eye can accommodate to focus on the image; if it is positive the normal eye cannot focus on the image. In most cases a field curvature of about 1 diopter (of either sign) is acceptable. A positive D of 3 diopters is about the largest tolerable curvature.

The off-axis image quality of most eyepieces is relatively poor. As a result, the outer portions of the field are used primarily to identify and locate objects of interest, which are then brought to the center of the field for a closer examination. Interestingly enough, the human eye shares this characteristic and mode of operation, so that this seems quite natural to most observers.

7.2 Two Magnifier Designs

The first sample lens of this chapter, Fig. 7.1, is a magnifier, suitable for use as a slide viewer, an optical comparitor, or a general-purpose magnifier. Unlike an eyepiece, a good magnifier must be insensitive to the position of the eye, because in a magnifier there is no exit-pupil-image of the objective lens to define the location of the user's eye. This requires a better level of correction, especially for spherical aberration in the magnifier, so that the image doesn't swim or distort as the eye is moved about. The symmetrical eyepiece (Figs. 7.10 through 7.14) and the orthoscopic eyepiece (Fig. 7.8) also work well as magnifiers.

The second sample lens, Fig. 7.2, is also a magnifier, but of a totally different type. It can be considered either as a magnifying glass placed in front of a galilean telescope, or as a reverse telephoto construction. The result is a working distance which is (in this sample) 50 percent longer than the effective focal length. This type of magnifier can also be executed in a much simpler construction.

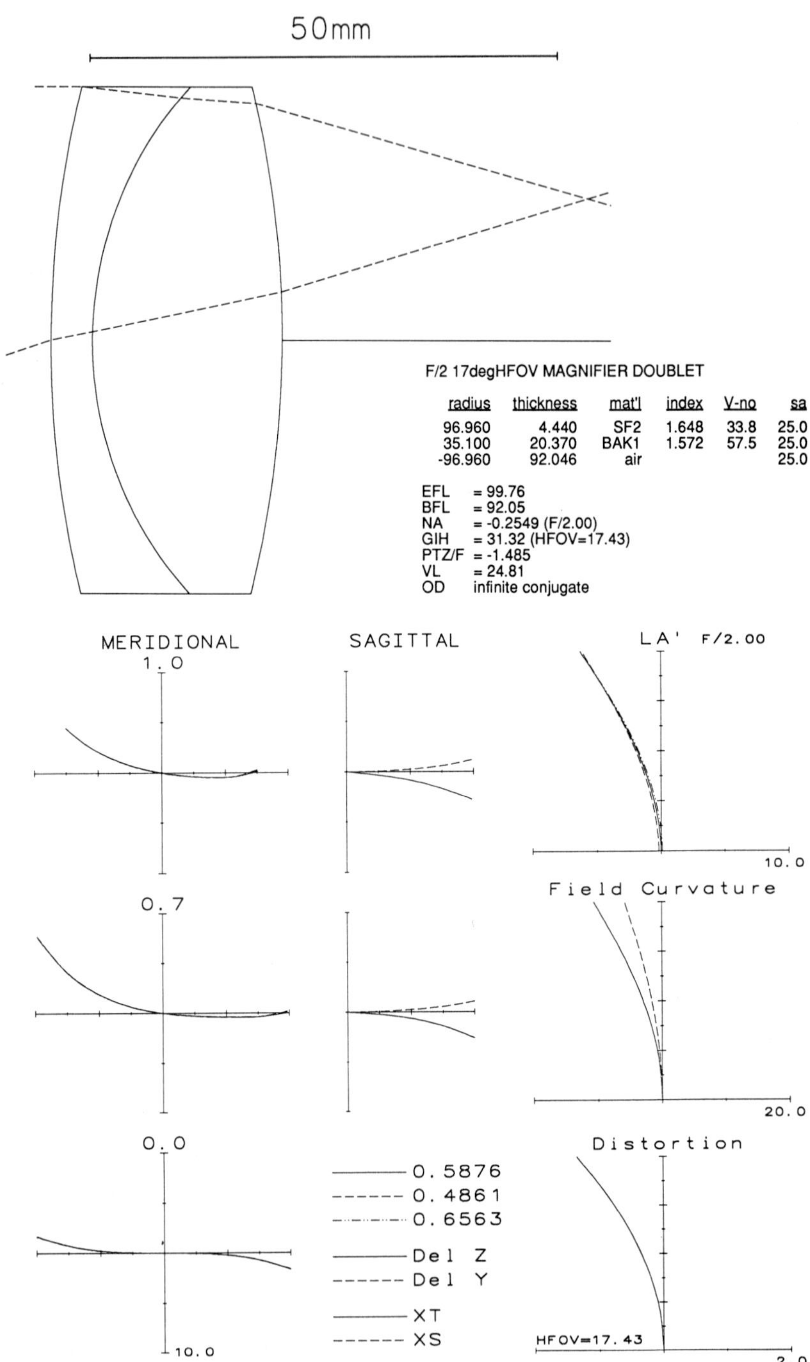
50mm
F/2 17degHFOV MAGNIFIER DOUBLET
radius thickness mat'l index V-no sa
96.960 4.440 SF2 1.648 33.8 25.0
35.100 20.370 BAK1 1.572 57.5 25.0
-96.960 92.046 air 25.0
EFL = 99.76
BFL = 92.05
NA = -0.2549 (F/2.00)
GIH = 31.32 (HFOV=17.43)
PTZ/F = -1.485
VL = 24.81
OD infinite conjugate
MERIDIONAL
1.0
SAGITTAL
LA' F/2.00
10.0
Field Curvature
0.7
20.0
0.0
Distortion
0.5876
0.4861
0.6563
Del Z
Del Y
XT
XS
10.0
HFOV=17.43
2.0

Figure 7.1

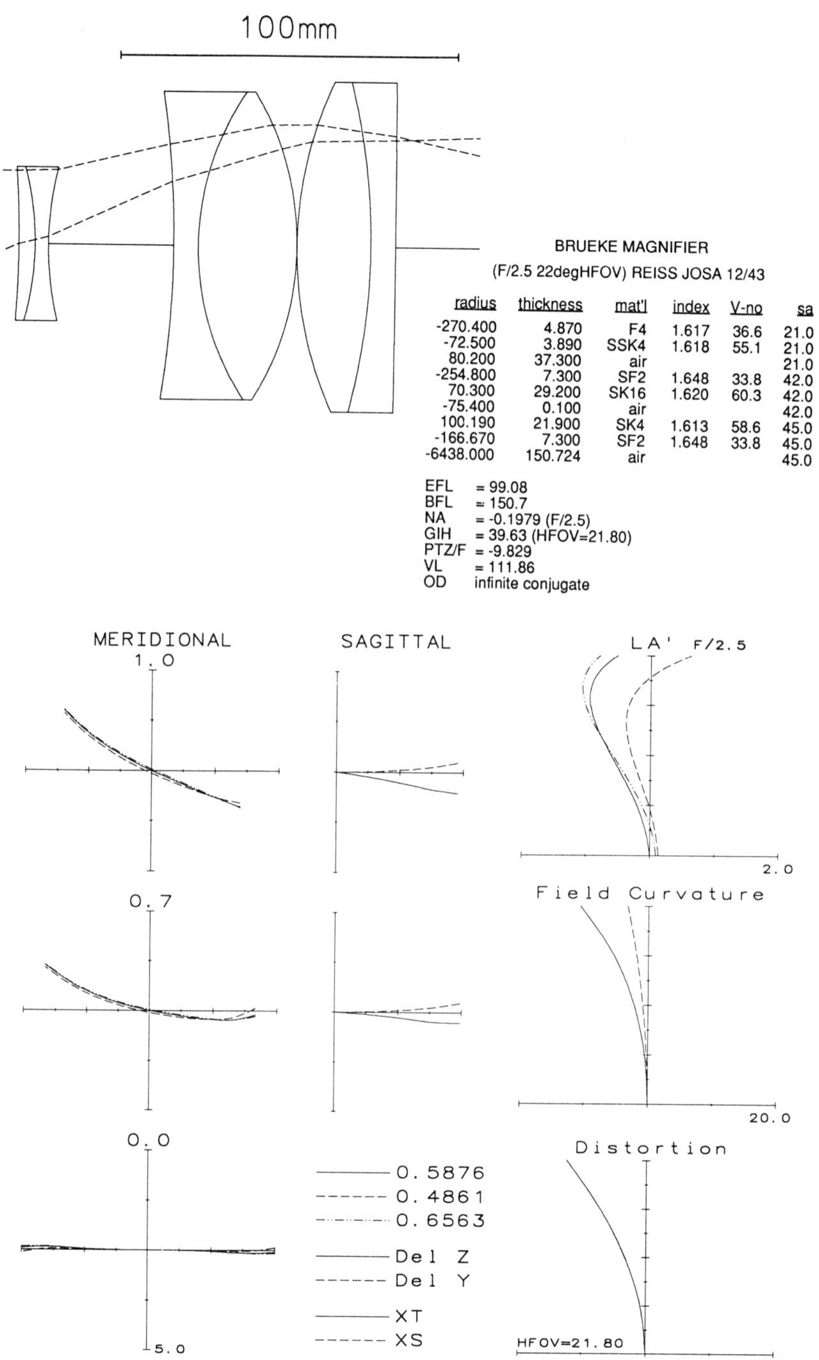

BRUEKE MAGNIFIER

(F/2.5 22degHFOV) REISS JOSA 12/43

radius	thickness	mat'l	index	V-no	sa
-270.400	4.870	F4	1.617	36.6	21.0
-72.500	3.890	SSK4	1.618	55.1	21.0
80.200	37.300	air			21.0
-254.800	7.300	SF2	1.648	33.8	42.0
70.300	29.200	SK16	1.620	60.3	42.0
-75.400	0.100	air			42.0
100.190	21.900	SK4	1.613	58.6	45.0
-166.670	7.300	SF2	1.648	33.8	45.0
-6438.000	150.724	air			45.0

EFL = 99.08
BFL = 150.7
NA = -0.1979 (F/2.5)
GIH = 39.63 (HFOV=21.80)
PTZ/F = -9.829
VL = 111.86
OD infinite conjugate

Figure 7.2

7.3 Simple Two- and Three-Element Eyepieces

Figures 7.3, 7.4, and 7.5 are the classical Huygens, Ramsden, and Kellner eyepieces (respectively). They are simple, inexpensive, and cover modest angular fields. The huygenian has two big disadvantages: its short eye relief and its internal, uncorrected (unsuited for a reticle) image/focal plane. The Ramsden ameliorates these problems, but gives up the correction of lateral chromatic (which the huygenian achieves with the large spacing) in order to do so. The Kellner, with its achromatized eye lens, improves on the Ramsden's color correction, and is widely used in inexpensive binoculars. Figure 7.6 is a close-coupled combination which, in the configuration shown, has a long working distance. It is also used in the reverse orientation (as a binocular eyepiece, for example) as a medium-field, long–eye-relief eyepiece.

7.4 Four-Element Eyepieces

The first of these, Fig. 7.7, is often used as a long– eye-relief microscope eyepiece. It can also be executed with identical meniscus singlets and/or with an equi-convex crown in the cemented doublet. The glasses here are typical, although SF1 (717-293) is often used for the flint.

The classical orthoscopic eyepiece is shown in Fig. 7.8. This construction is quite typical; a plano-convex eye lens of a light barium crown (or a light flint) and a symmetrical cemented triplet. The orthoscopic is noted for its freedom from distortion, although this example is obviously not outstanding in this regard. Like Fig. 7.7, it has a long eye relief.

The eyepiece of Fig. 7.9 can be regarded as a simplified Erfle eyepiece. The thick meniscus doublet helps the Petzval curvature, but is not as effective in correction of the chromatic as the doublet in Fig. 7.7.

The next five lenses are examples of the *symmetrical* or *Ploessl* eyepiece. This is an excellent, versatile, general-purpose, medium-field eyepiece. It has a long eye relief and is relatively insensitive to pupil shift. The angular field coverage is usually limited by higher-order overcorrected astigmatism which causes the tangential field to become very strongly backward-curving toward the edge of the field. Of course, some overcorrected astigmatism is necessary to offset the inward Petzval curvature; the amount is determined by both the shape of the components and the index break at the cemented surfaces.

Figure 7.10 is a classic example, executed in BK7 (517-642) and

(Text continues on page 101.)

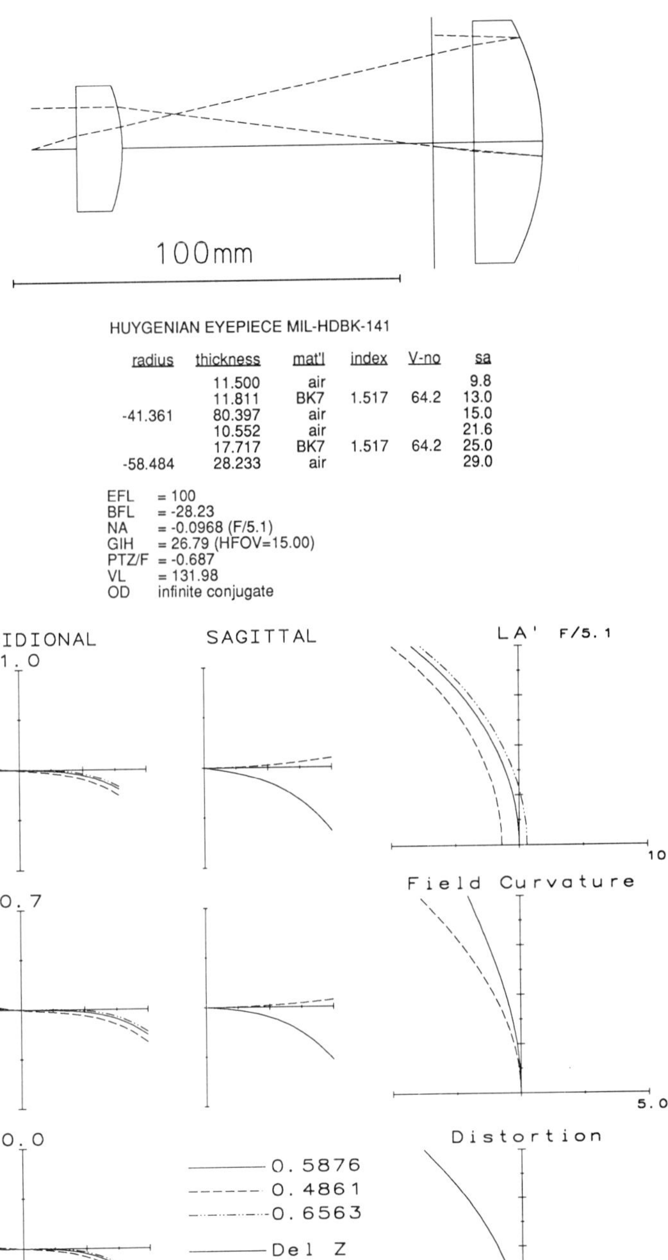

HUYGENIAN EYEPIECE MIL-HDBK-141

radius	thickness	mat'l	index	V-no	sa
	11.500	air			9.8
	11.811	BK7	1.517	64.2	13.0
-41.361	80.397	air			15.0
	10.552	air			21.6
	17.717	BK7	1.517	64.2	25.0
-58.484	28.233	air			29.0

EFL = 100
BFL = -28.23
NA = -0.0968 (F/5.1)
GIH = 26.79 (HFOV=15.00)
PTZ/F = -0.687
VL = 131.98
OD infinite conjugate

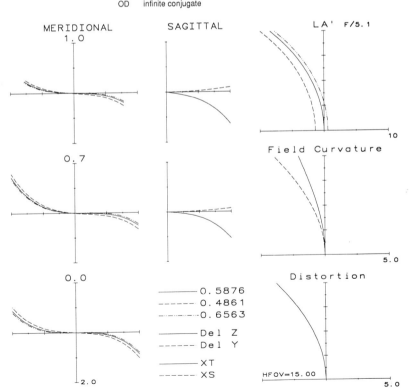

Figure 7.3

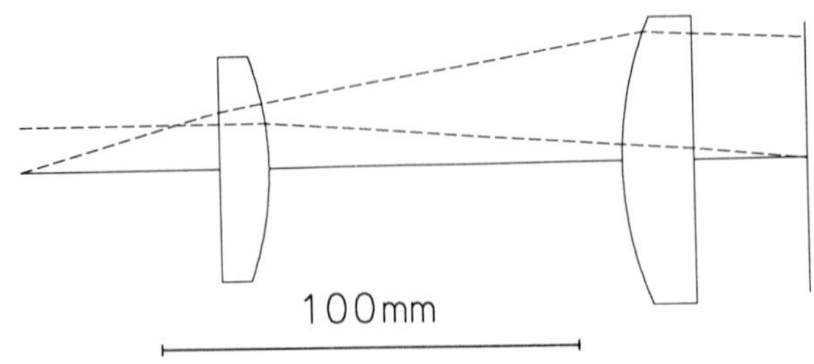

RAMSDEN EYEPIECE MIL-HDBK-141

radius	thickness	mat'l	index	V-no	sa
	47.480	air			10.0
	11.880	BK7	1.517	64.2	23.0
-70.028	84.640	air			25.0
78.787	16.960	BK7	1.517	64.2	32.0
	26.957	air			32.0

EFL = 101.6
BFL = 26.96
NA = -0.0984 (F/5.1)
GIH = 27.22 (HFOV=15.00)
PTZ/F = -1.071
VL = 160.96
OD infinite conjugate

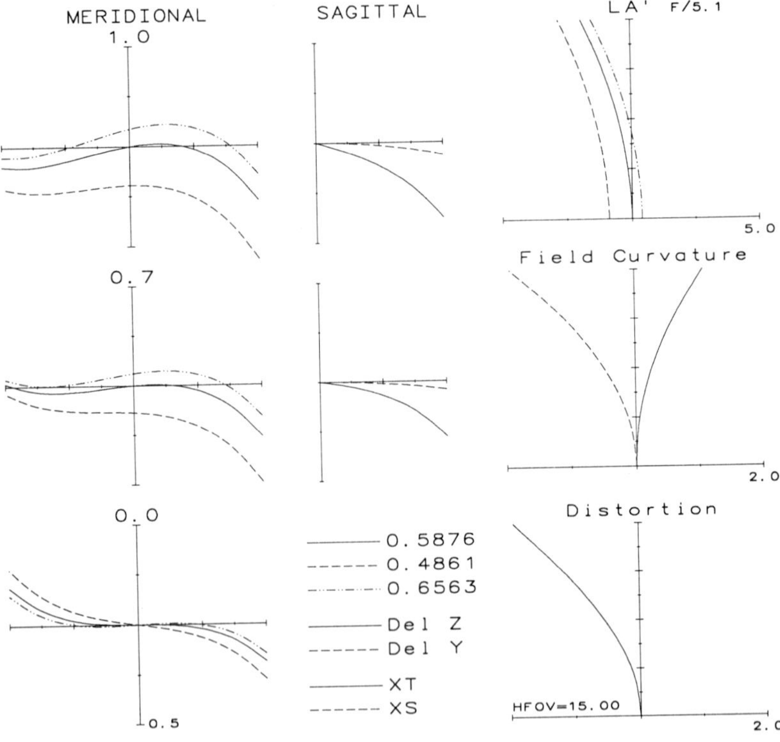

Figure 7.4

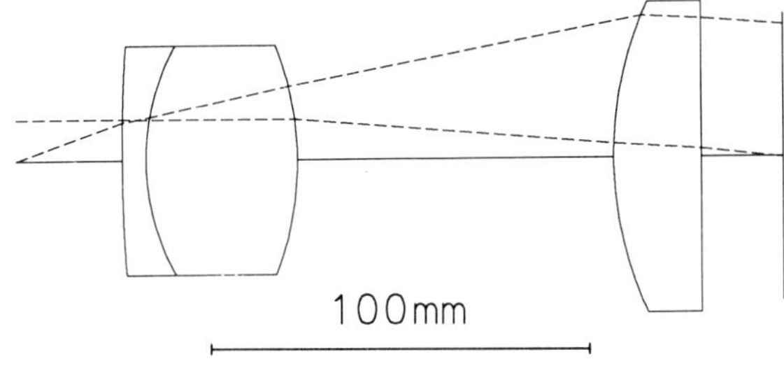

KELLNER EYEPIECE MIL-HDBK-141

radius	thickness	mat'l	index	V-no	sa
	28.200	air			10.0
384.986	6.360	F4	1.617	36.6	28.0
54.105	39.800	BAK2	1.540	59.7	28.0
-72.398	83.560	air			28.0
85.124	23.080	BAK2	1.540	59.7	38.0
	21.266	air			38.0

EFL = 101.7
BFL = 21.27
NA = -0.0982 (F/5.1)
GIH = 33.06 (HFOV=18.00)
PTZ/F = -1.046
VL = 181.00
OD infinite conjugate

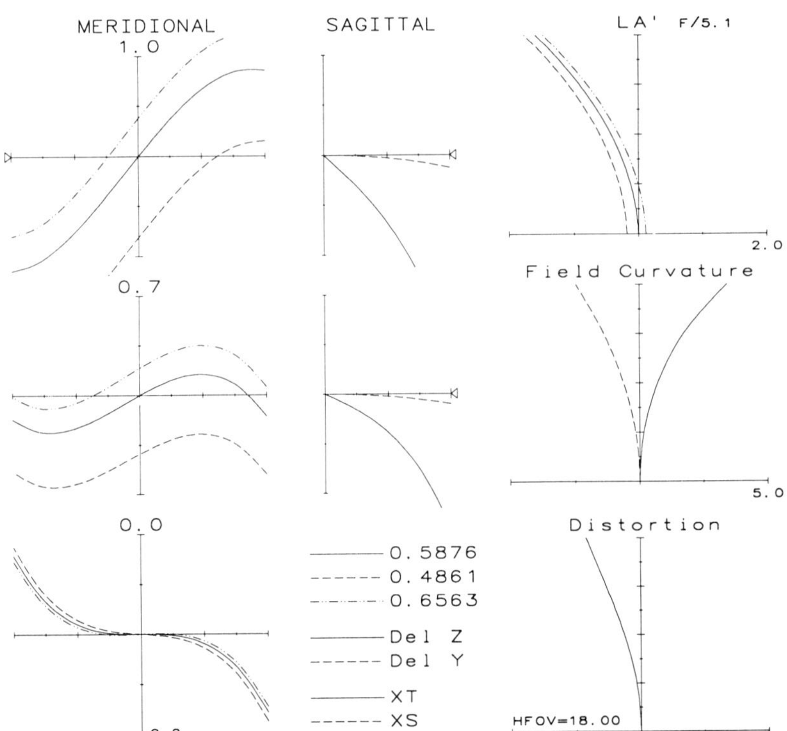

Figure 7.5

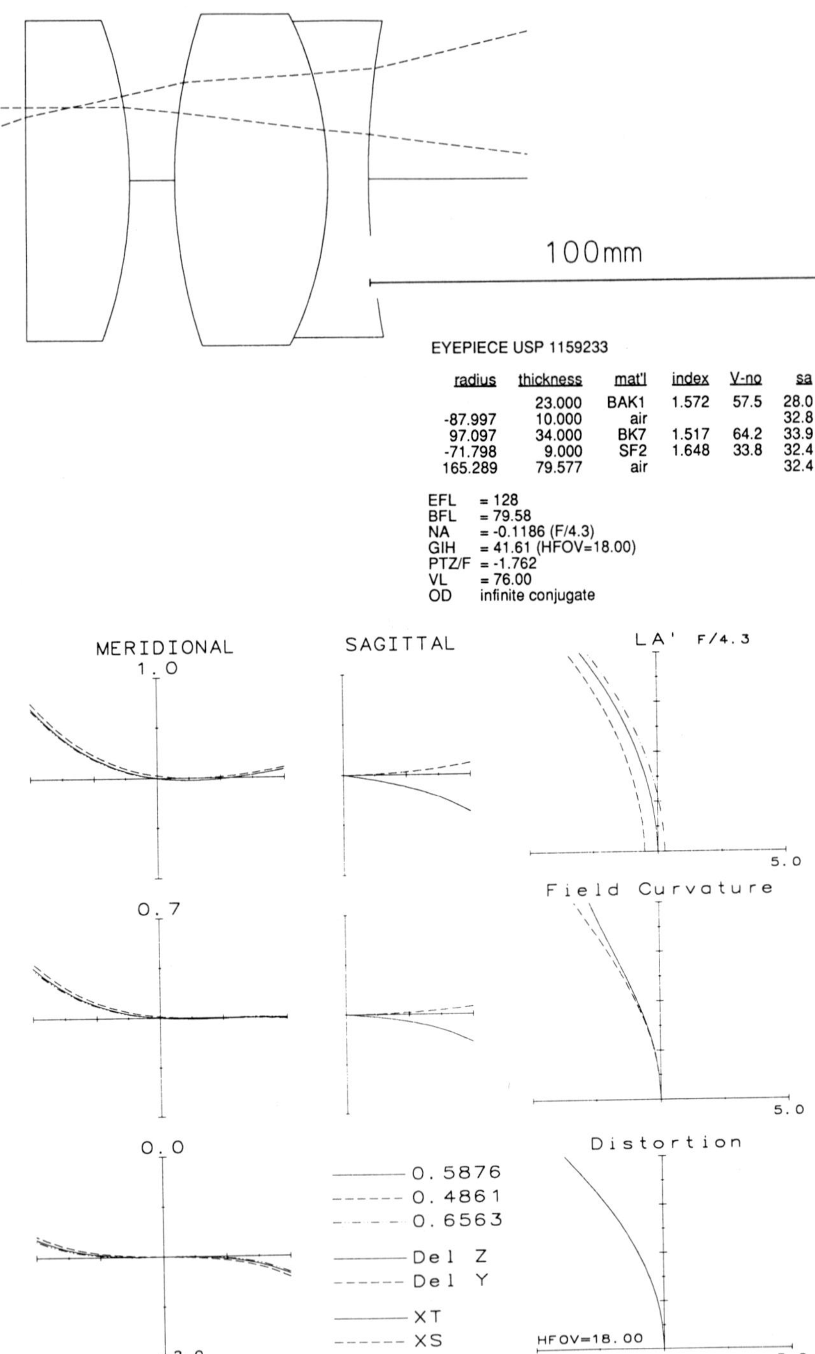

EYEPIECE USP 1159233

radius	thickness	mat'l	index	V-no	sa
	23.000	BAK1	1.572	57.5	28.0
-87.997	10.000	air			32.8
97.097	34.000	BK7	1.517	64.2	33.9
-71.798	9.000	SF2	1.648	33.8	32.4
165.289	79.577	air			32.4

EFL = 128
BFL = 79.58
NA = -0.1186 (F/4.3)
GIH = 41.61 (HFOV=18.00)
PTZ/F = -1.762
VL = 76.00
OD infinite conjugate

Figure 7.6

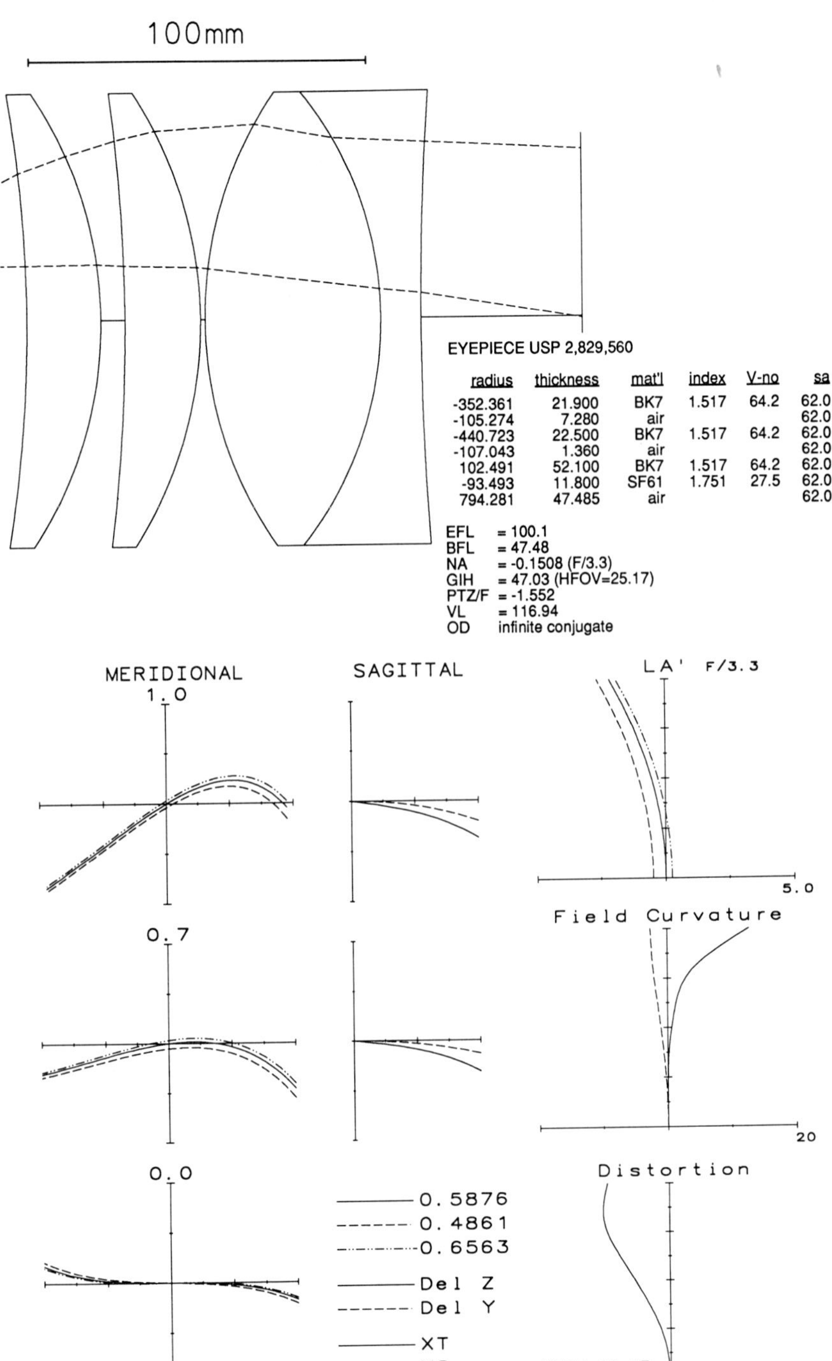

EYEPIECE USP 2,829,560

radius	thickness	mat'l	index	V-no	sa
-352.361	21.900	BK7	1.517	64.2	62.0
-105.274	7.280	air			62.0
-440.723	22.500	BK7	1.517	64.2	62.0
-107.043	1.360	air			62.0
102.491	52.100	BK7	1.517	64.2	62.0
-93.493	11.800	SF61	1.751	27.5	62.0
794.281	47.485	air			62.0

EFL = 100.1
BFL = 47.48
NA = -0.1508 (F/3.3)
GIH = 47.03 (HFOV=25.17)
PTZ/F = -1.552
VL = 116.94
OD infinite conjugate

Figure 7.7

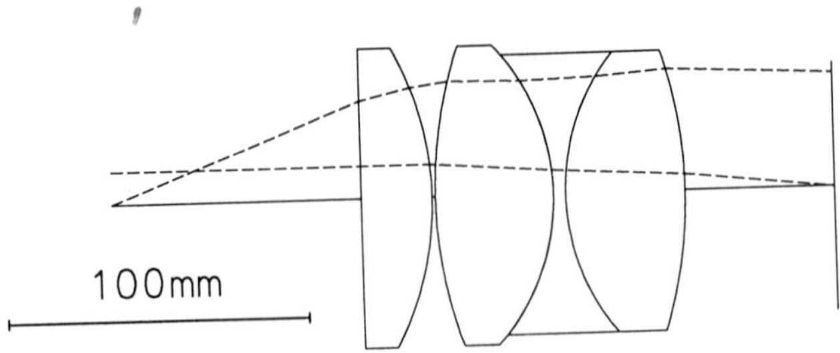

ORTHOSCOPIC MIL-HDBK-141

radius	thickness	mat'l	index	V-no	sa
	82.560	air			10.0
	23.280	BAK1	1.572	57.5	40.0
-90.950	1.100	air			46.0
129.490	39.680	KF3	1.515	54.7	46.0
-63.690	4.050	F3	1.613	37.0	43.0
63.690	39.680	KF3	1.515	54.7	43.0
-129.490	49.185	air			43.0

EFL = 100.1
BFL = 49.19
NA = -0.1001 (F/5.0)
GIH = 36.44 (HFOV=20.00)
PTZ/F = -1.257
VL = 190.35
OD infinite conjugate

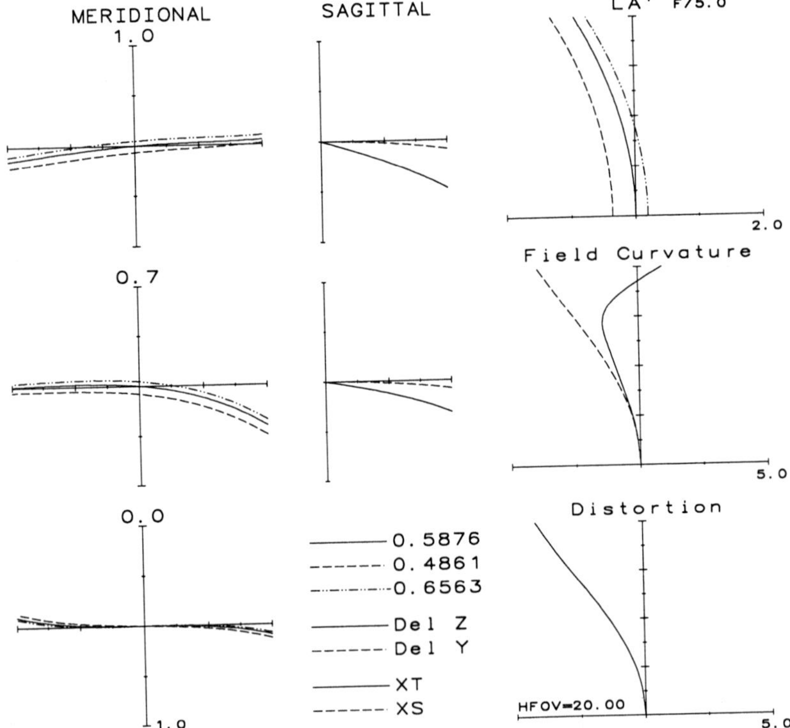

Figure 7.8

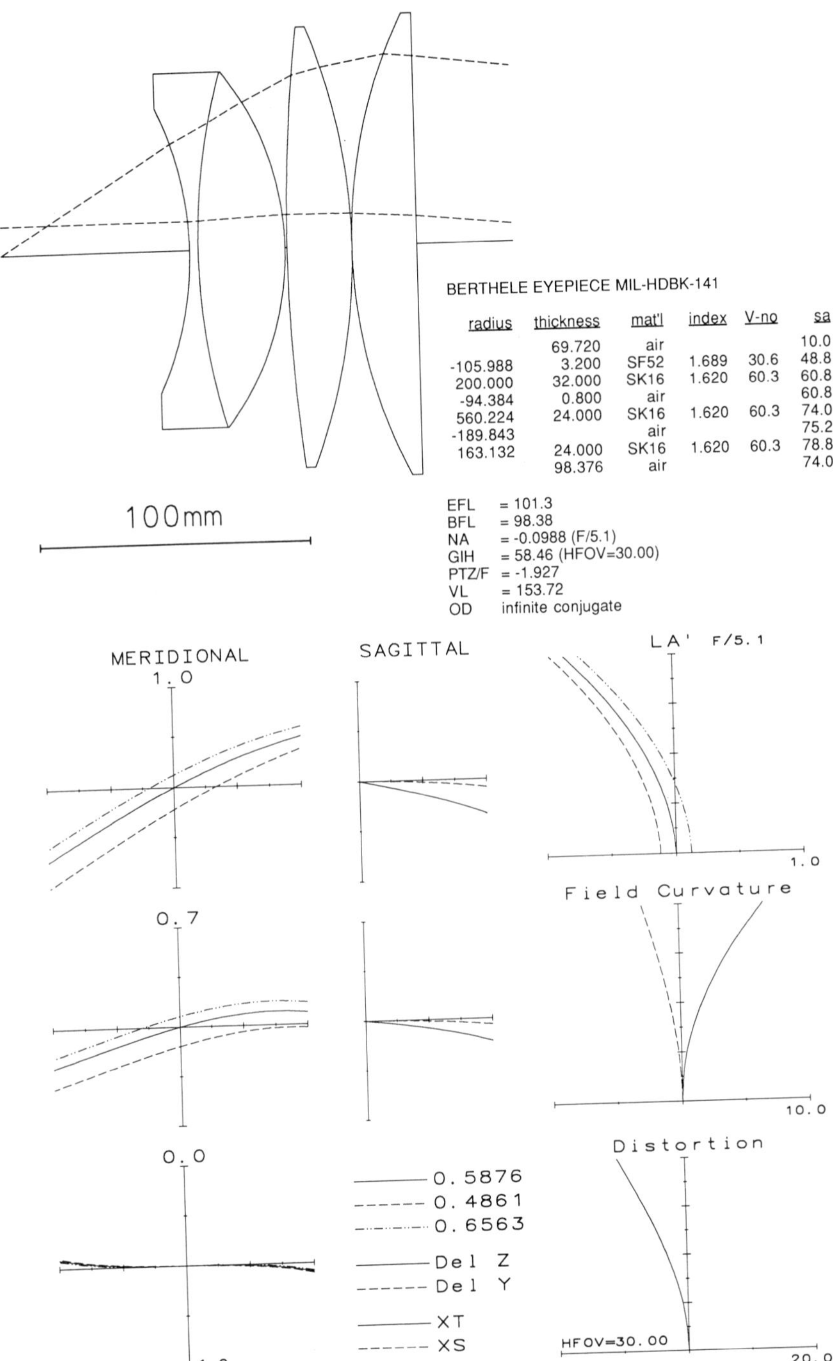

BERTHELE EYEPIECE MIL-HDBK-141

radius	thickness	mat'l	index	V-no	sa
	69.720	air			10.0
-105.988	3.200	SF52	1.689	30.6	48.8
200.000	32.000	SK16	1.620	60.3	60.8
-94.384	0.800	air			60.8
560.224	24.000	SK16	1.620	60.3	74.0
-189.843		air			75.2
163.132	24.000	SK16	1.620	60.3	78.8
	98.376	air			74.0

EFL = 101.3
BFL = 98.38
NA = -0.0988 (F/5.1)
GIH = 58.46 (HFOV=30.00)
PTZ/F = -1.927
VL = 153.72
OD infinite conjugate

Figure 7.9

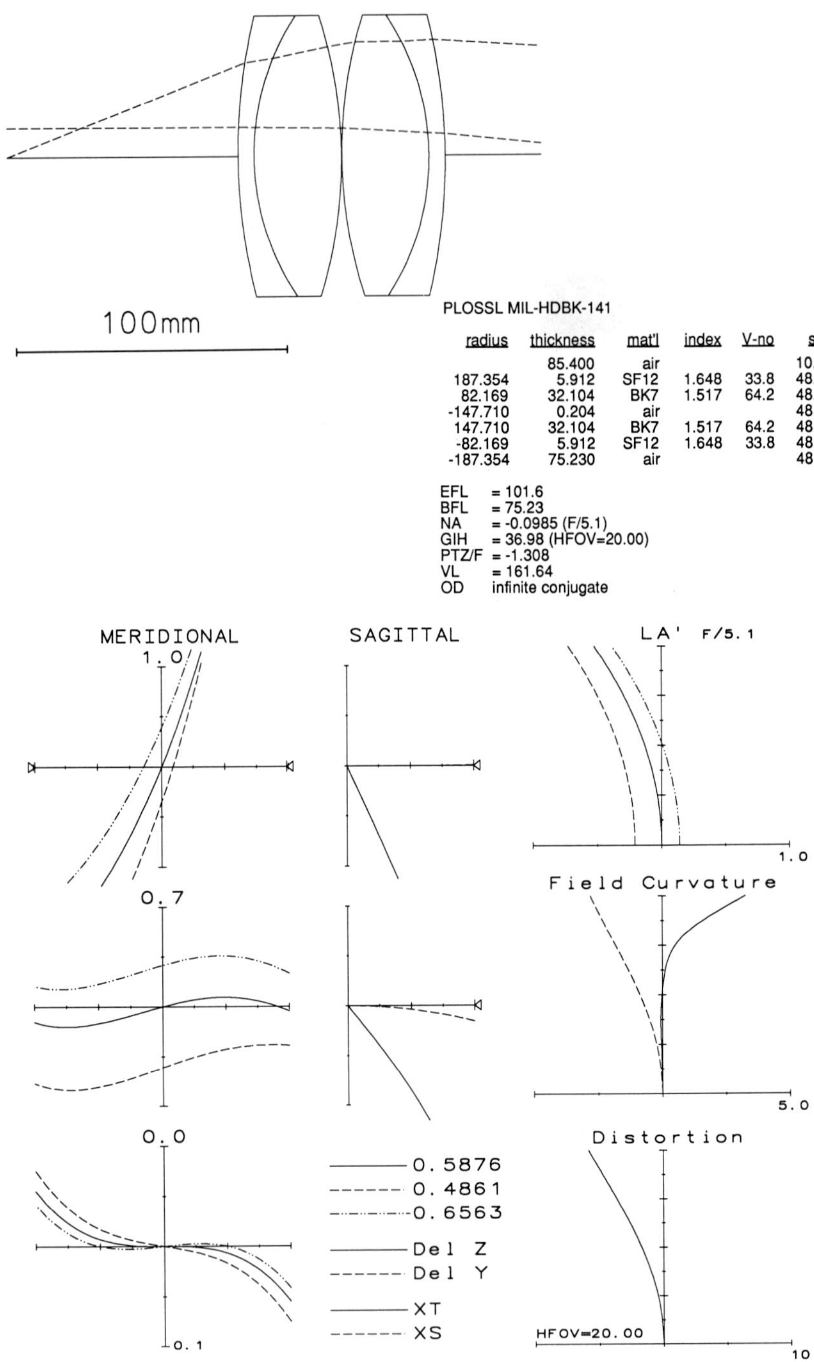

PLOSSL MIL-HDBK-141

radius	thickness	mat'l	index	V-no	sa
	85.400	air			10.0
187.354	5.912	SF12	1.648	33.8	48.0
82.169	32.104	BK7	1.517	64.2	48.0
-147.710	0.204	air			48.0
147.710	32.104	BK7	1.517	64.2	48.0
-82.169	5.912	SF12	1.648	33.8	48.0
-187.354	75.230	air			48.0

EFL = 101.6
BFL = 75.23
NA = -0.0985 (F/5.1)
GIH = 36.98 (HFOV=20.00)
PTZ/F = -1.308
VL = 161.64
OD infinite conjugate

Figure 7.10

SF12 (648-338) glasses. Figure 7.11 utilizes higher-index glasses (SK1, 610-567, and SF61, 751-275) to achieve a modestly improved performance. Figure 7.12 uses the classical glasses, but features economical equi-convex crown elements. Figures 7.13 and 7.14 are departures from strict symmetry: the first keeps a symmetrical arrangement of the glasses but has plano-concave flints and one equi-convex crown; the second uses four different glasses, but manages both equi-convex crowns and plano-concave flints. This design would call for very careful production control in the shop to avoid mixing the elements, although the tooling required is obviously minimal.

The last eyepiece of this section (Fig. 7.15) is composed of two identical, nearly plano-convex doublets. The design data for the pupil position is obviously incorrect, as can be seen by the path of the principal ray in the drawing, but this configuration, as well as a version with both elements reversed, is often a useful design form for rifle scopes.

7.5 Five-Element Eyepieces

The eye lens of Fig. 7.16 has a parabolic surface; the glasses of this lens are all of relatively high index. As is typical of the eyepieces in this section, the angular field is quite large (70° in this case; the others range from 60° to 75°). Note also that the distortion and field curvature in wide-field eyepieces both tend to be very large. The von Hofe (Fig. 7.17) shows a 75° field, but uses a rather expensive cemented triplet to achieve it. Figures 7.18, 7.19, and 7.20 are three examples of the five-element version of the Erfle construction. This is probably the most widely used wide-angle eyepiece. It utilizes what is, in effect, a negative field lens to flatten the field and lengthen the eye relief; it does so by placing the strong concave surface of the meniscus doublet very close to the focal plane. Note the high-index glasses of Fig. 7.20.

7.6 Six- and Seven-Element Eyepieces

The internal focusing eyepiece of Fig. 7.21 is focused, not by moving the whole eyepiece, but by moving just the doublet and the triplet, while maintaining the eye-lens-to-focal plane spacing constant. This allows a more easily sealed construction for the mechanism of the assembly. At the scale of the data, the two components can be shifted about 1 unit toward the eye and about 23 units away; the amount of focusing that this produces is obviously a function of the focal length to which the design is scaled. At a focal length of about 1.5 in, this shift will focus through about 4 diopters at the eye.

Figures 7.22 and 7.23 are quite similar to each other, except for the orientation of the field lens doublet, although the Fig. 7.23 design cov-

(Text continues on page 115.)

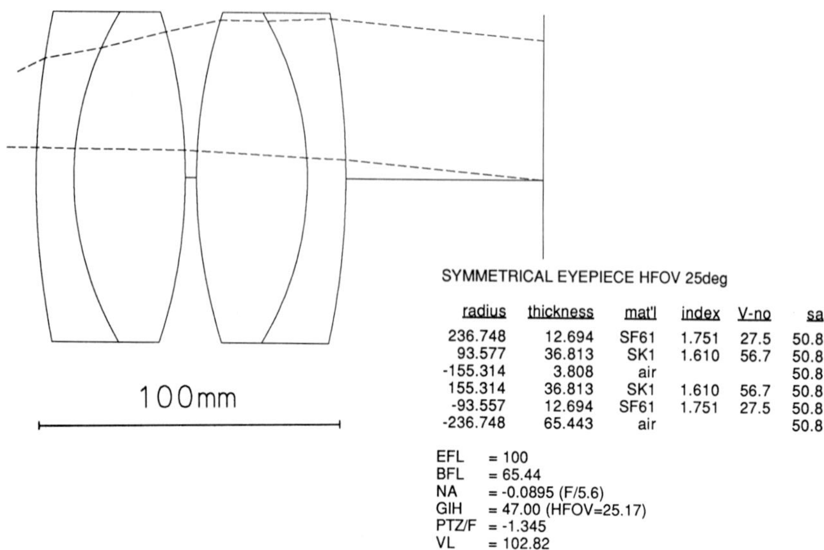

SYMMETRICAL EYEPIECE HFOV 25deg

radius	thickness	mat'l	index	V-no	sa
236.748	12.694	SF61	1.751	27.5	50.8
93.577	36.813	SK1	1.610	56.7	50.8
-155.314	3.808	air			50.8
155.314	36.813	SK1	1.610	56.7	50.8
-93.557	12.694	SF61	1.751	27.5	50.8
-236.748	65.443	air			50.8

EFL = 100
BFL = 65.44
NA = -0.0895 (F/5.6)
GIH = 47.00 (HFOV=25.17)
PTZ/F = -1.345
VL = 102.82
OD infinite conjugate

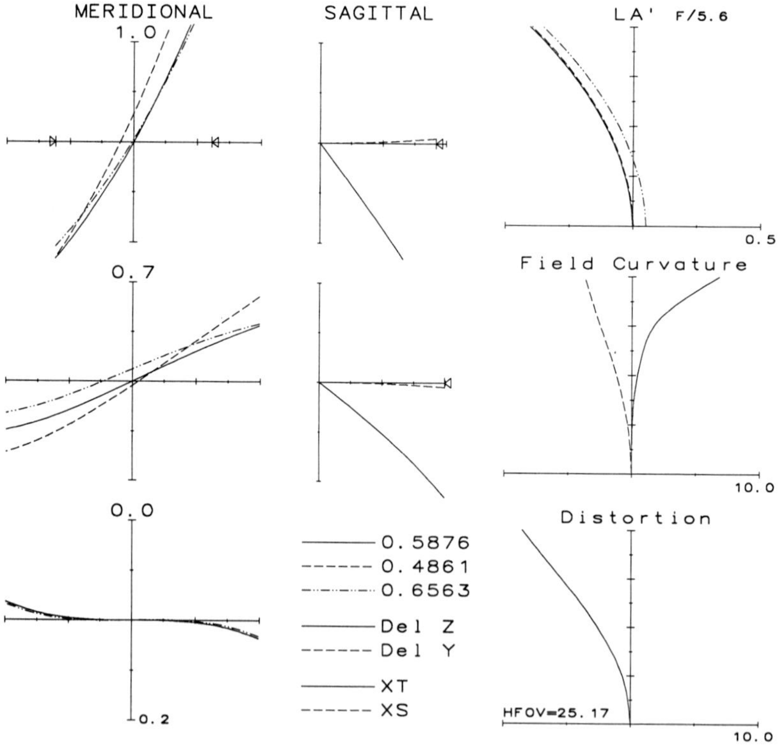

Figure 7.11

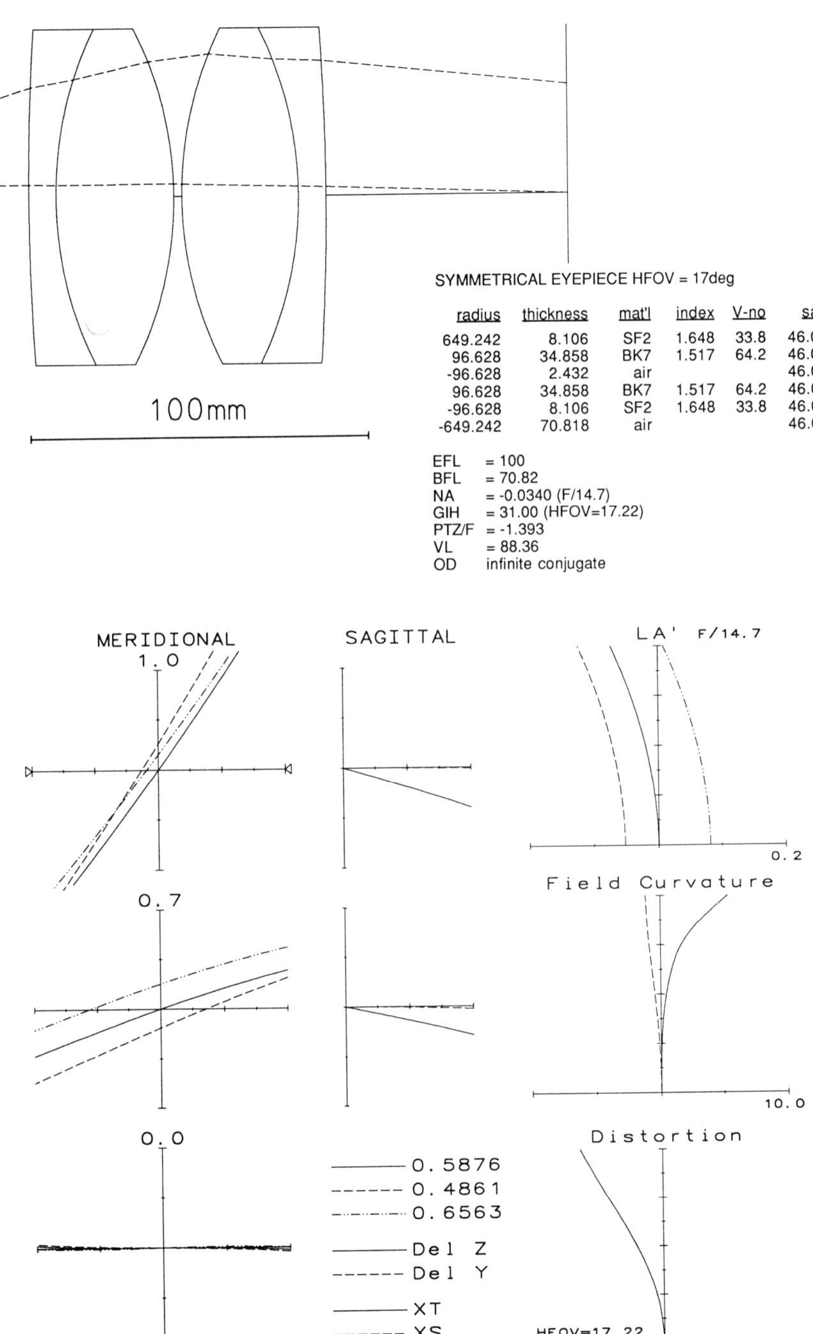

SYMMETRICAL EYEPIECE HFOV = 17deg

radius	thickness	mat'l	index	V-no	sa
649.242	8.106	SF2	1.648	33.8	46.0
96.628	34.858	BK7	1.517	64.2	46.0
-96.628	2.432	air			46.0
96.628	34.858	BK7	1.517	64.2	46.0
-96.628	8.106	SF2	1.648	33.8	46.0
-649.242	70.818	air			46.0

EFL = 100
BFL = 70.82
NA = -0.0340 (F/14.7)
GIH = 31.00 (HFOV=17.22)
PTZ/F = -1.393
VL = 88.36
OD infinite conjugate

Figure 7.12

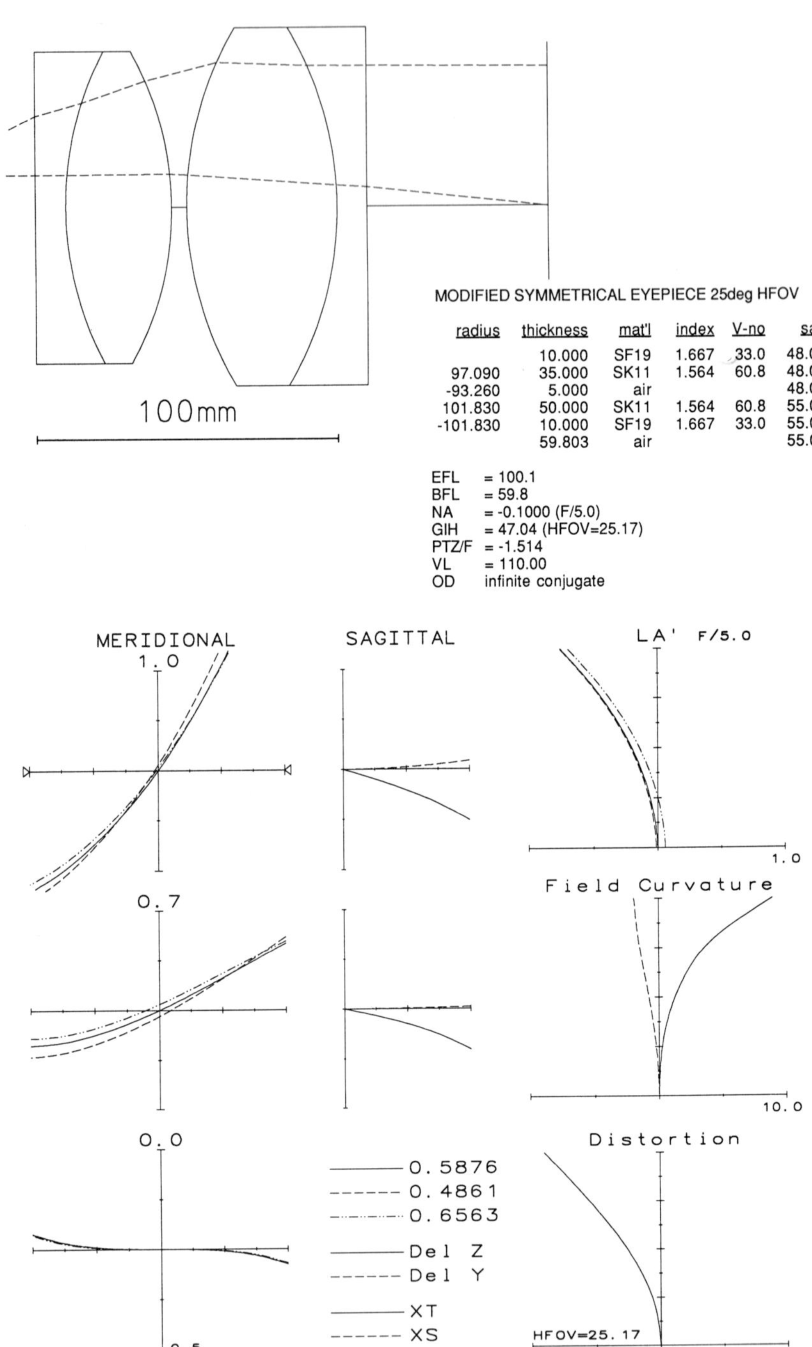

MODIFIED SYMMETRICAL EYEPIECE 25deg HFOV

radius	thickness	mat'l	index	V-no	sa
	10.000	SF19	1.667	33.0	48.0
97.090	35.000	SK11	1.564	60.8	48.0
-93.260	5.000	air			48.0
101.830	50.000	SK11	1.564	60.8	55.0
-101.830	10.000	SF19	1.667	33.0	55.0
	59.803	air			55.0

EFL = 100.1
BFL = 59.8
NA = -0.1000 (F/5.0)
GIH = 47.04 (HFOV=25.17)
PTZ/F = -1.514
VL = 110.00
OD infinite conjugate

Figure 7.13

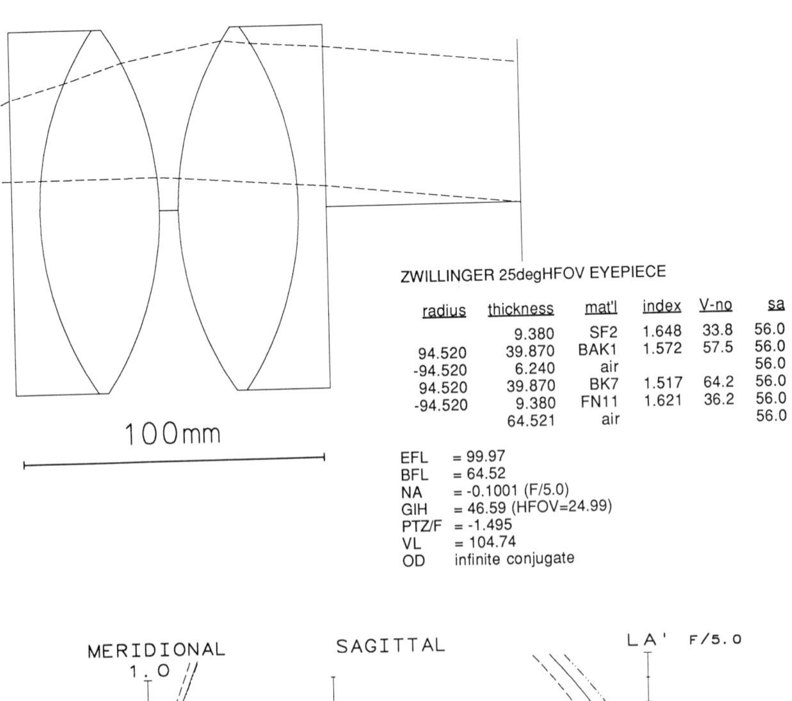

ZWILLINGER 25degHFOV EYEPIECE

radius	thickness	mat'l	index	V-no	sa
	9.380	SF2	1.648	33.8	56.0
94.520	39.870	BAK1	1.572	57.5	56.0
-94.520	6.240	air			56.0
94.520	39.870	BK7	1.517	64.2	56.0
-94.520	9.380	FN11	1.621	36.2	56.0
	64.521	air			56.0

EFL = 99.97
BFL = 64.52
NA = -0.1001 (F/5.0)
GIH = 46.59 (HFOV=24.99)
PTZ/F = -1.495
VL = 104.74
OD infinite conjugate

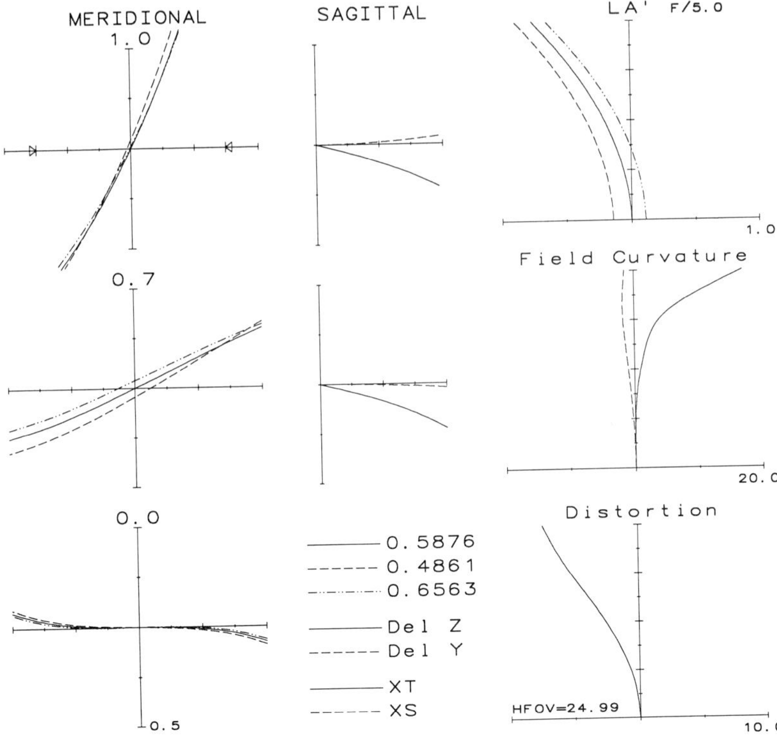

Figure 7.14

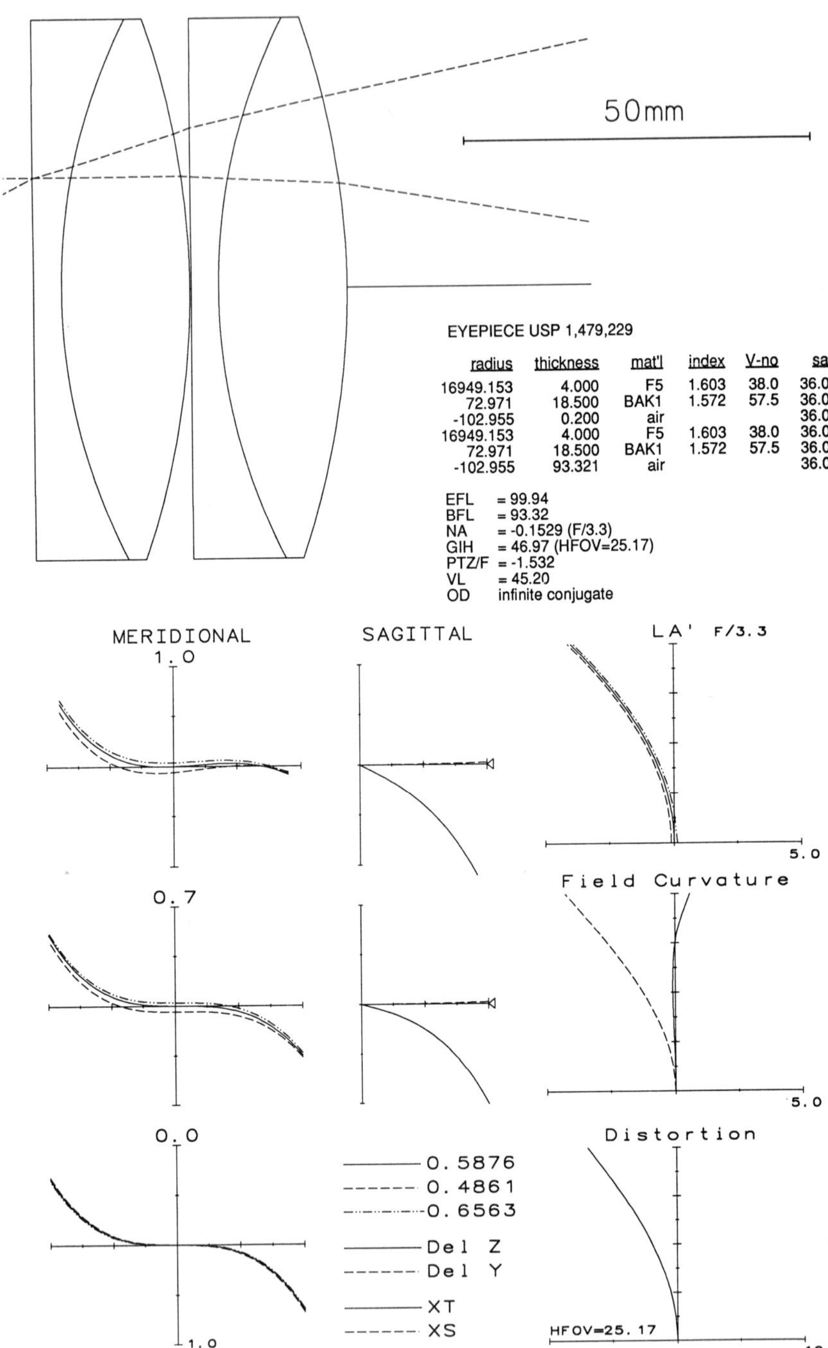

radius	thickness	mat'l	index	V-no	sa
16949.153	4.000	F5	1.603	38.0	36.0
72.971	18.500	BAK1	1.572	57.5	36.0
-102.955	0.200	air			36.0
16949.153	4.000	F5	1.603	38.0	36.0
72.971	18.500	BAK1	1.572	57.5	36.0
-102.955	93.321	air			36.0

EFL = 99.94
BFL = 93.32
NA = -0.1529 (F/3.3)
GIH = 46.97 (HFOV=25.17)
PTZ/F = -1.532
VL = 45.20
OD infinite conjugate

Figure 7.15

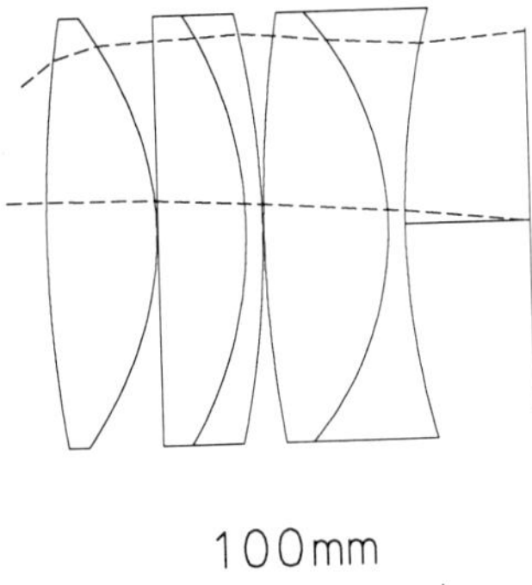

PARABOLOID EYEPIECE 35degHFOV

radius	thickness	mat'l	index	V-no	sa
395.500	37.940	LAK14	1.697	55.4	68.4
-94.240	0.500	air			68.4
kappa		-1.000			
	30.570	LAK14	1.697	55.4	68.4
-123.700	5.820	SF11	1.785	25.8	68.4
-282.600	0.500	air			68.4
395.500	42.630	LAK14	1.697	55.4	68.4
-98.940	5.820	SF11	1.785	25.8	68.4
249.700	43.070	air			59.0

EFL = 99.9
BFL = 43.07
NA = -0.1007 (F/5.0)
GIH = 69.93 (HFOV=34.99)
PTZ/F = -1.776
VL = 123.78
OD infinite conjugate

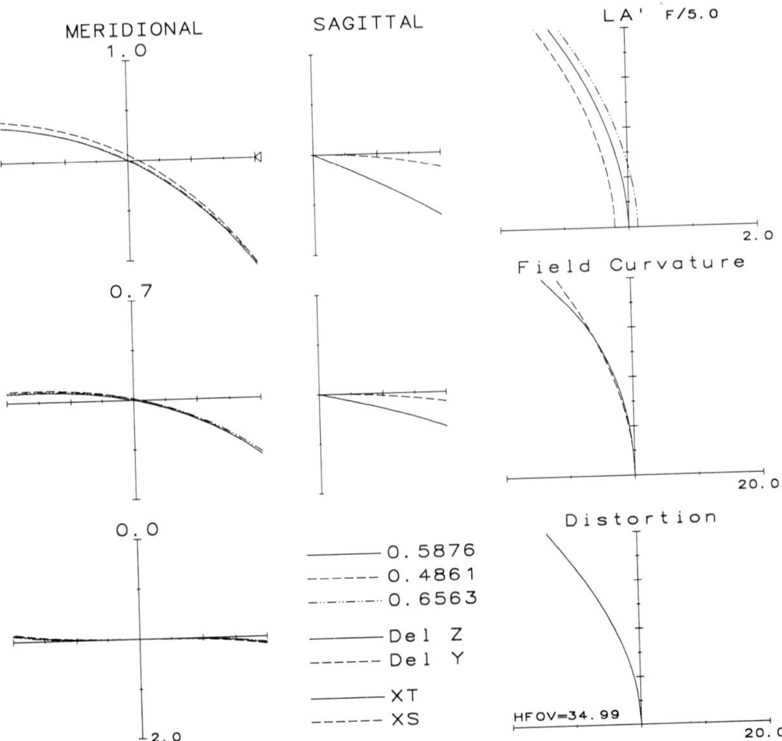

Figure 7.16

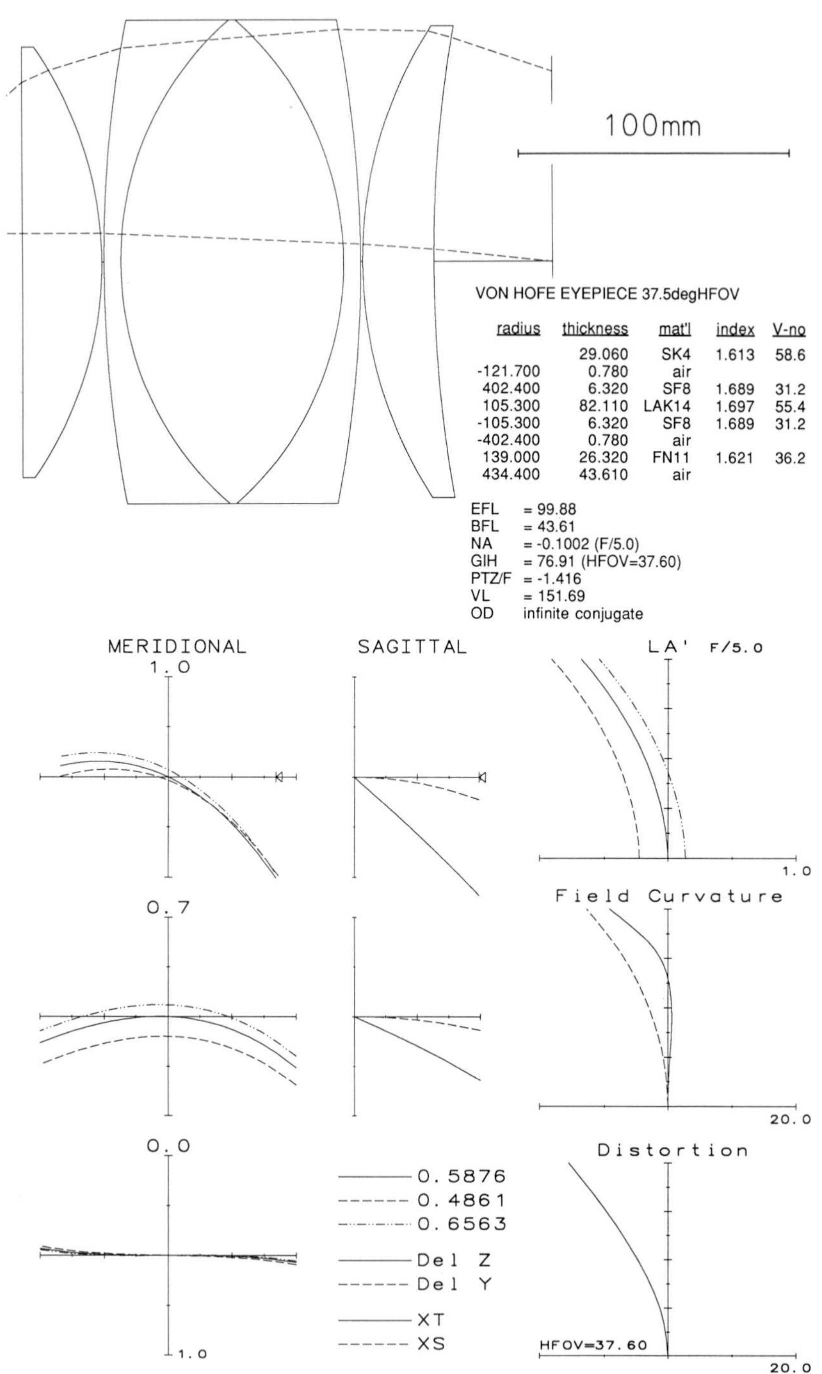

radius	thickness	mat'l	index	V-no
	29.060	SK4	1.613	58.6
-121.700	0.780	air		
402.400	6.320	SF8	1.689	31.2
105.300	82.110	LAK14	1.697	55.4
-105.300	6.320	SF8	1.689	31.2
-402.400	0.780	air		
139.000	26.320	FN11	1.621	36.2
434.400	43.610	air		

EFL = 99.88
BFL = 43.61
NA = -0.1002 (F/5.0)
GIH = 76.91 (HFOV=37.60)
PTZ/F = -1.416
VL = 151.69
OD infinite conjugate

Figure 7.17

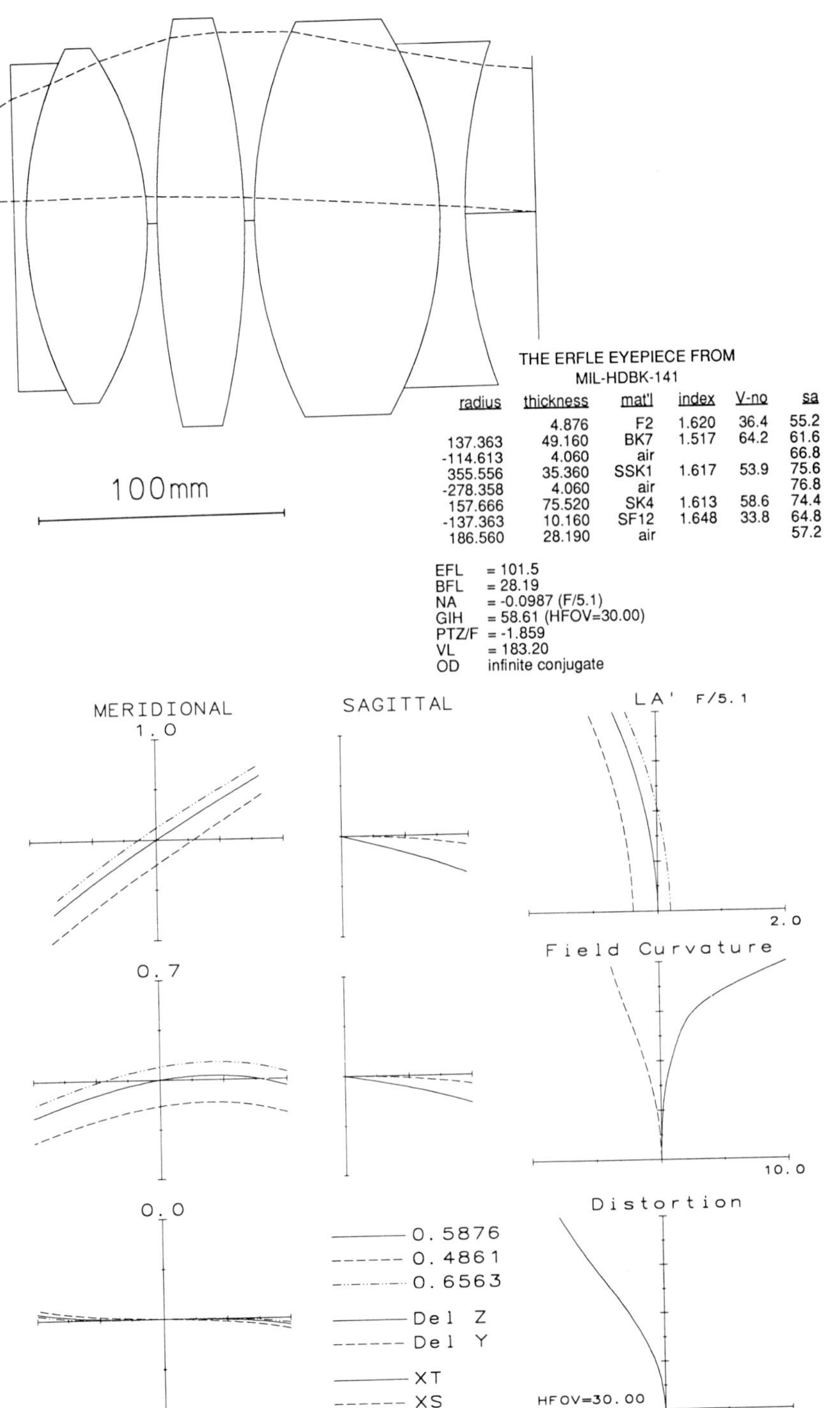

THE ERFLE EYEPIECE FROM
MIL-HDBK-141

radius	thickness	mat'l	index	V-no	sa
	4.876	F2	1.620	36.4	55.2
137.363	49.160	BK7	1.517	64.2	61.6
-114.613	4.060	air			66.8
355.556	35.360	SSK1	1.617	53.9	75.6
-278.358	4.060	air			76.8
157.666	75.520	SK4	1.613	58.6	74.4
-137.363	10.160	SF12	1.648	33.8	64.8
186.560	28.190	air			57.2

EFL = 101.5
BFL = 28.19
NA = -0.0987 (F/5.1)
GIH = 58.61 (HFOV=30.00)
PTZ/F = -1.859
VL = 183.20
OD infinite conjugate

Figure 7.18

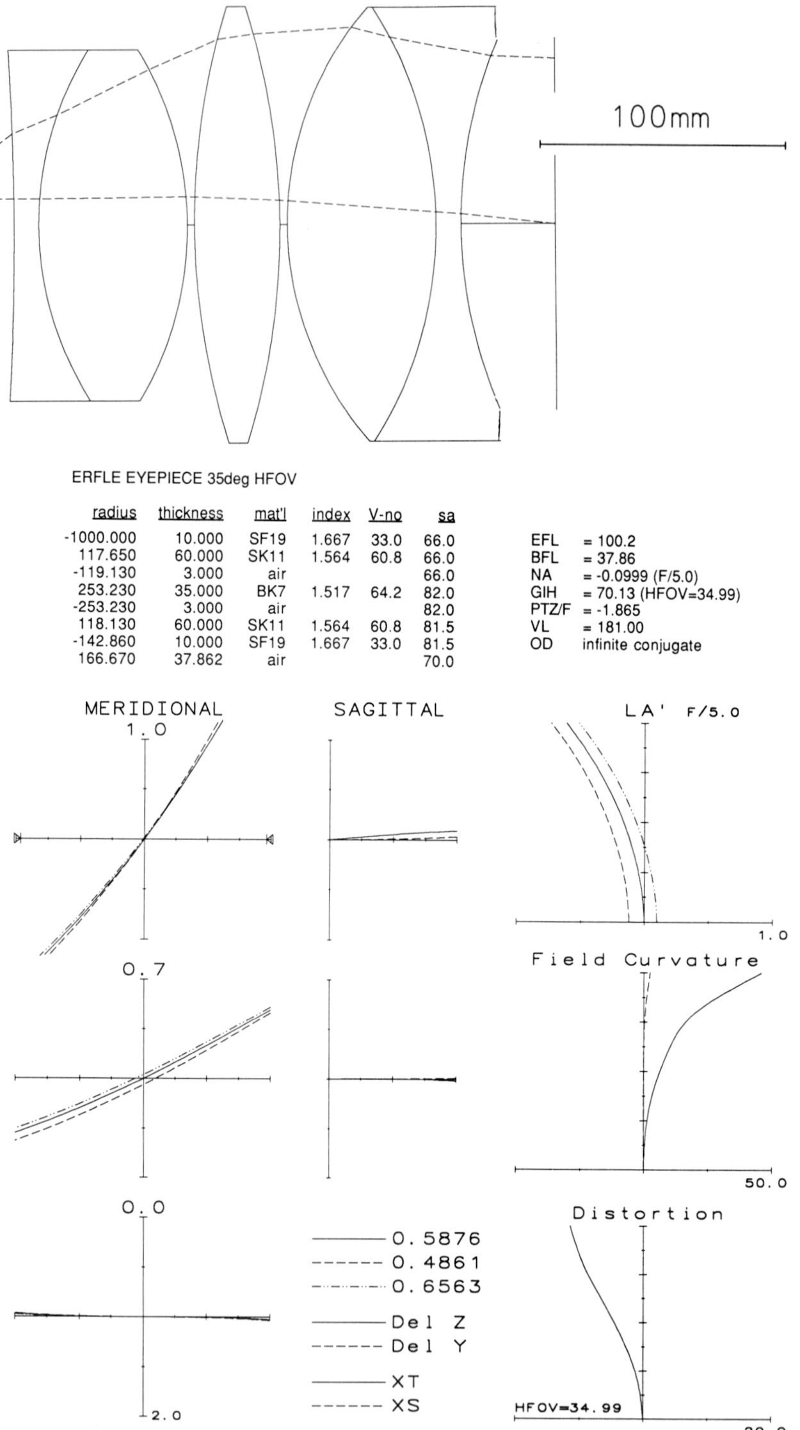

ERFLE EYEPIECE 35deg HFOV

radius	thickness	mat'l	index	V-no	sa
-1000.000	10.000	SF19	1.667	33.0	66.0
117.650	60.000	SK11	1.564	60.8	66.0
-119.130	3.000	air			66.0
253.230	35.000	BK7	1.517	64.2	82.0
-253.230	3.000	air			82.0
118.130	60.000	SK11	1.564	60.8	81.5
-142.860	10.000	SF19	1.667	33.0	81.5
166.670	37.862	air			70.0

EFL = 100.2
BFL = 37.86
NA = -0.0999 (F/5.0)
GIH = 70.13 (HFOV=34.99)
PTZ/F = -1.865
VL = 181.00
OD infinite conjugate

Figure 7.19

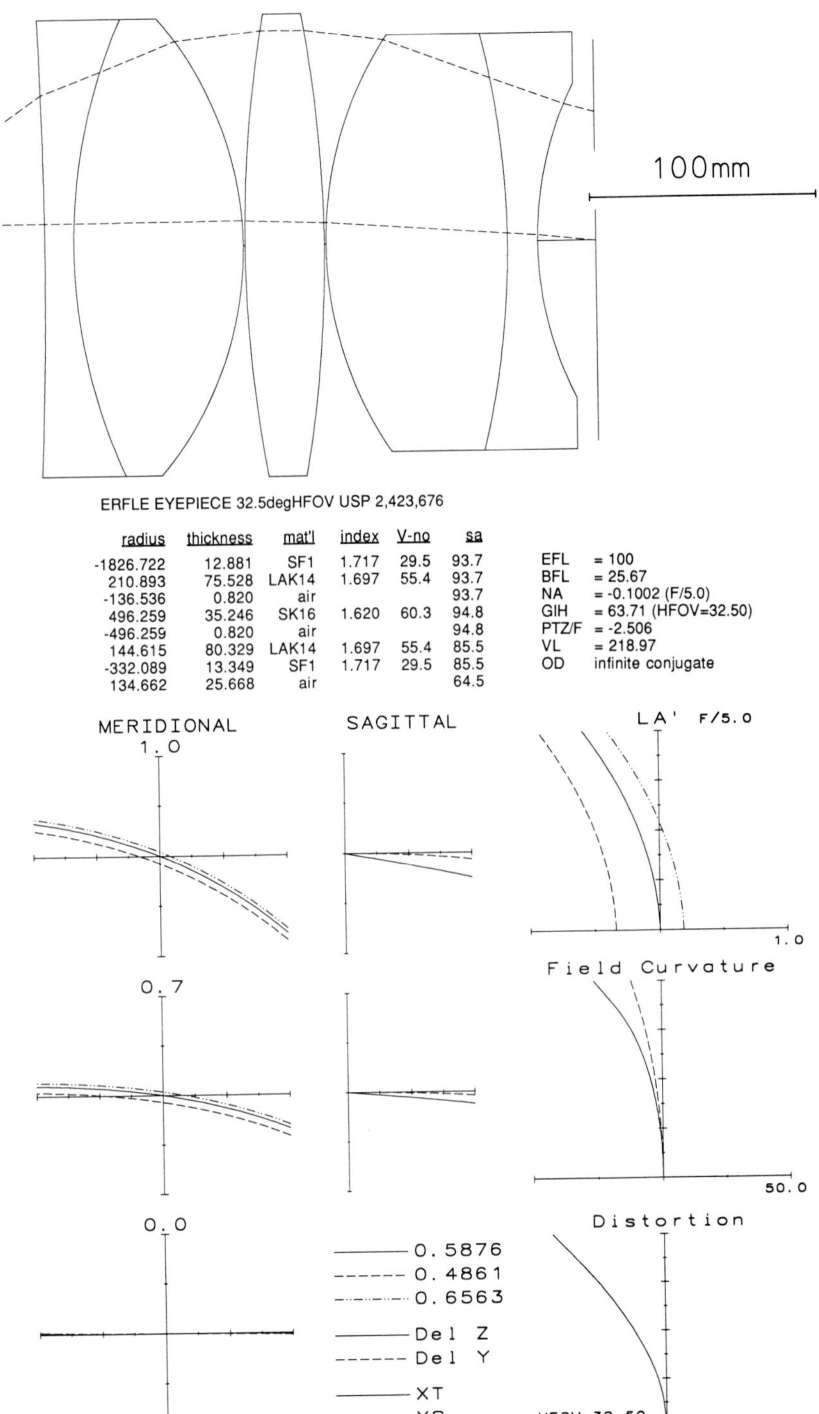

ERFLE EYEPIECE 32.5degHFOV USP 2,423,676

radius	thickness	mat'l	index	V-no	sa
-1826.722	12.881	SF1	1.717	29.5	93.7
210.893	75.528	LAK14	1.697	55.4	93.7
-136.536	0.820	air			93.7
496.259	35.246	SK16	1.620	60.3	94.8
-496.259	0.820	air			94.8
144.615	80.329	LAK14	1.697	55.4	85.5
-332.089	13.349	SF1	1.717	29.5	85.5
134.662	25.668	air			64.5

EFL	= 100
BFL	= 25.67
NA	= -0.1002 (F/5.0)
GIH	= 63.71 (HFOV=32.50)
PTZ/F	= -2.506
VL	= 218.97
OD	infinite conjugate

Figure 7.20

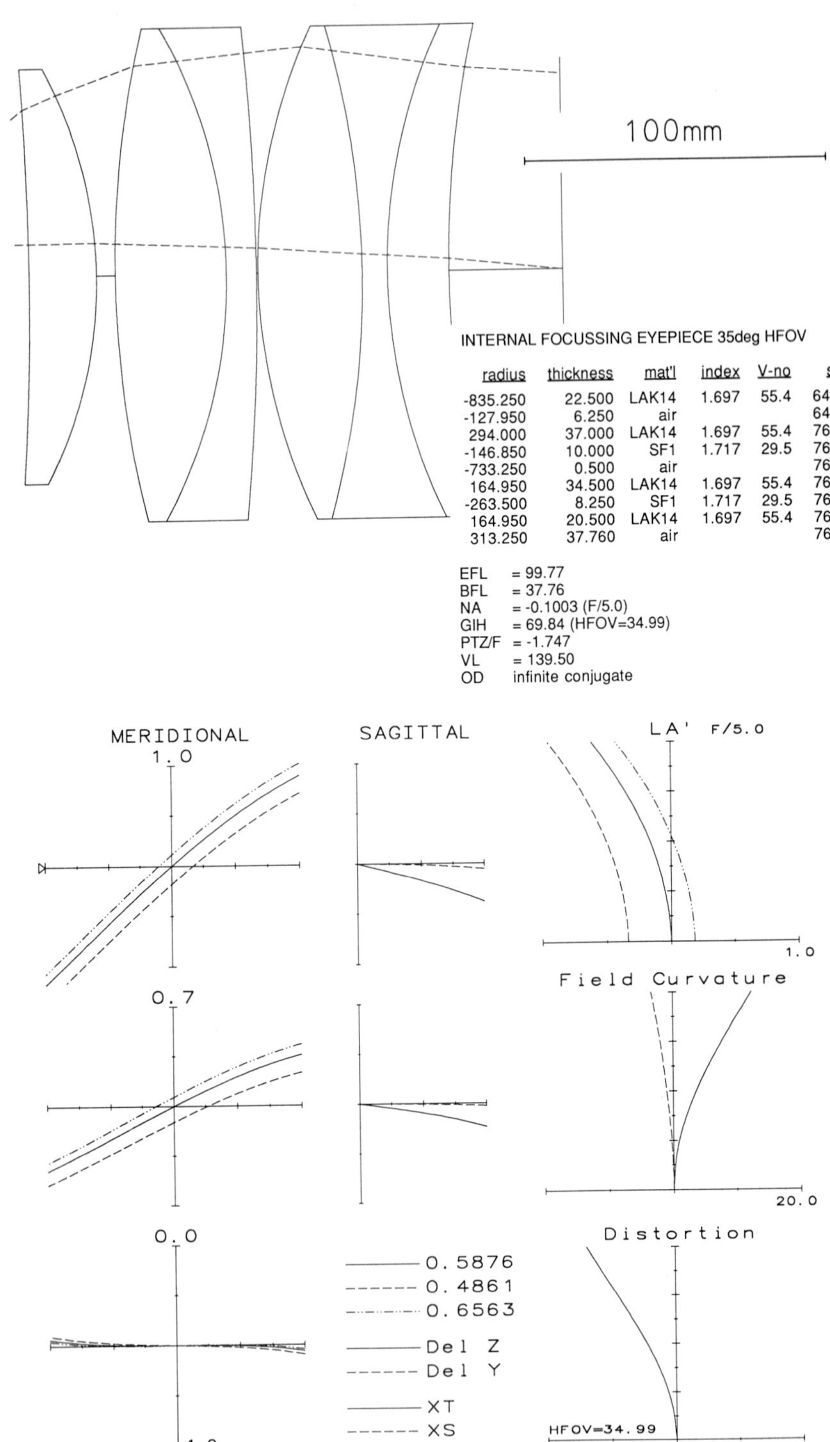

INTERNAL FOCUSSING EYEPIECE 35deg HFOV

radius	thickness	mat'l	index	V-no	sa
-835.250	22.500	LAK14	1.697	55.4	64.0
-127.950	6.250	air			64.0
294.000	37.000	LAK14	1.697	55.4	76.0
-146.850	10.000	SF1	1.717	29.5	76.0
-733.250	0.500	air			76.0
164.950	34.500	LAK14	1.697	55.4	76.0
-263.500	8.250	SF1	1.717	29.5	76.0
164.950	20.500	LAK14	1.697	55.4	76.0
313.250	37.760	air			76.0

EFL = 99.77
BFL = 37.76
NA = -0.1003 (F/5.0)
GIH = 69.84 (HFOV=34.99)
PTZ/F = -1.747
VL = 139.50
OD infinite conjugate

Figure 7.21

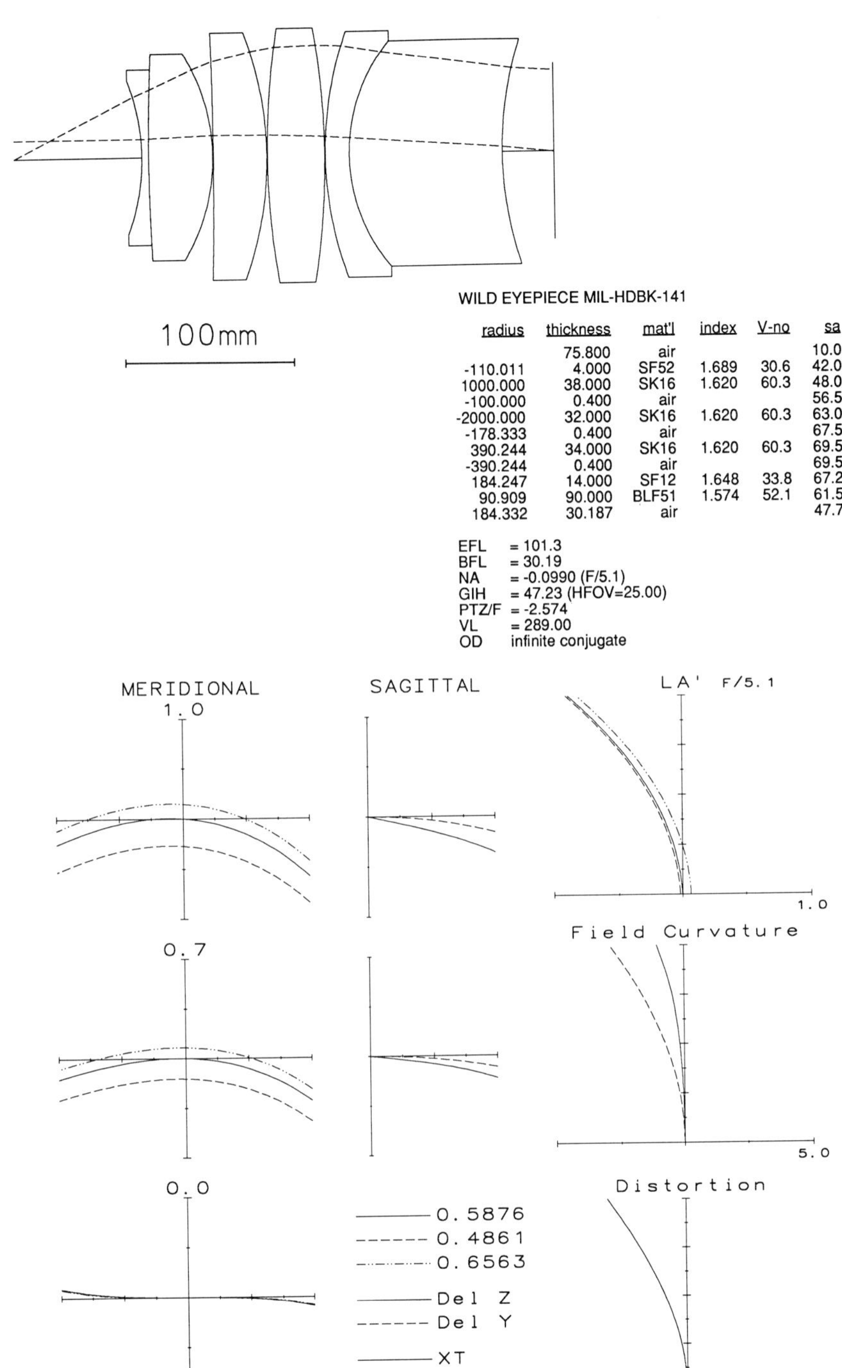

WILD EYEPIECE MIL-HDBK-141

radius	thickness	mat'l	index	V-no	sa
	75.800	air			10.0
-110.011	4.000	SF52	1.689	30.6	42.0
1000.000	38.000	SK16	1.620	60.3	48.0
-100.000	0.400	air			56.5
-2000.000	32.000	SK16	1.620	60.3	63.0
-178.333	0.400	air			67.5
390.244	34.000	SK16	1.620	60.3	69.5
-390.244	0.400	air			69.5
184.247	14.000	SF12	1.648	33.8	67.2
90.909	90.000	BLF51	1.574	52.1	61.5
184.332	30.187	air			47.7

EFL = 101.3
BFL = 30.19
NA = -0.0990 (F/5.1)
GIH = 47.23 (HFOV=25.00)
PTZ/F = -2.574
VL = 289.00
OD infinite conjugate

Figure 7.22

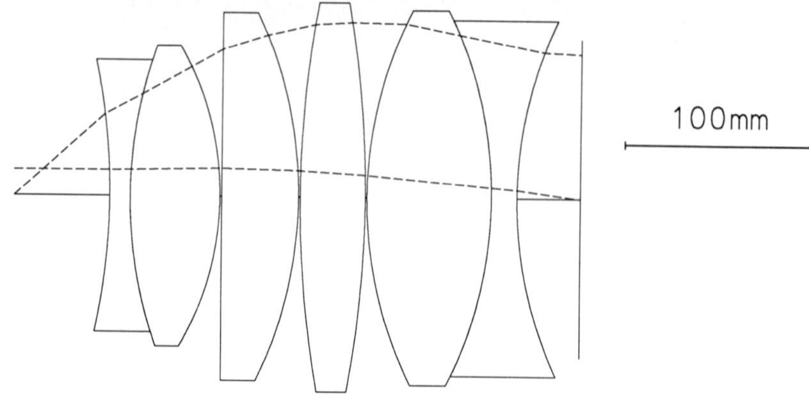

WRIGHT H. SCIDMORE; USP 3390935; 80 DEG. EYEPIECE LENS

radius	thickness	mat'l	index	V-no	sa
	49.250	air			0.0
-280.800	10.700	SF12	1.648	33.8	50.0
201.800	46.100	SK16	1.620	60.3	65.0
-134.600	0.900	air			71.8
	39.900	SK16	1.620	60.3	87.9
-187.500	0.900	air			87.9
476.600	33.700	SK16	1.620	60.3	92.9
-476.600	0.900	air			92.9
187.500	63.900	SK16	1.620	60.3	89.6
-187.500	13.300	SF12	1.648	33.8	85.0
187.500	32.389	air			72.3

EFL = 99.83
BFL = 32.39
NA = -0.1256 (F/4.0)
GIH = 83.76 (HFOV=40.00)
PTZ/F = -2.044
VL = 259.55
OD infinite conjugate

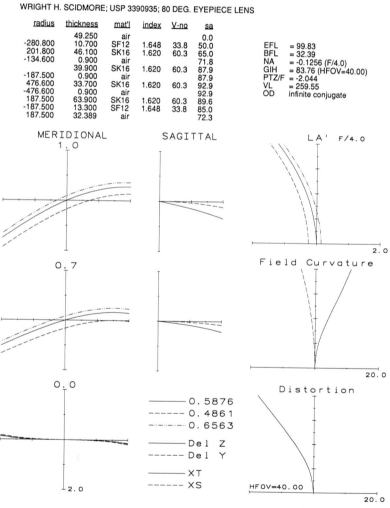

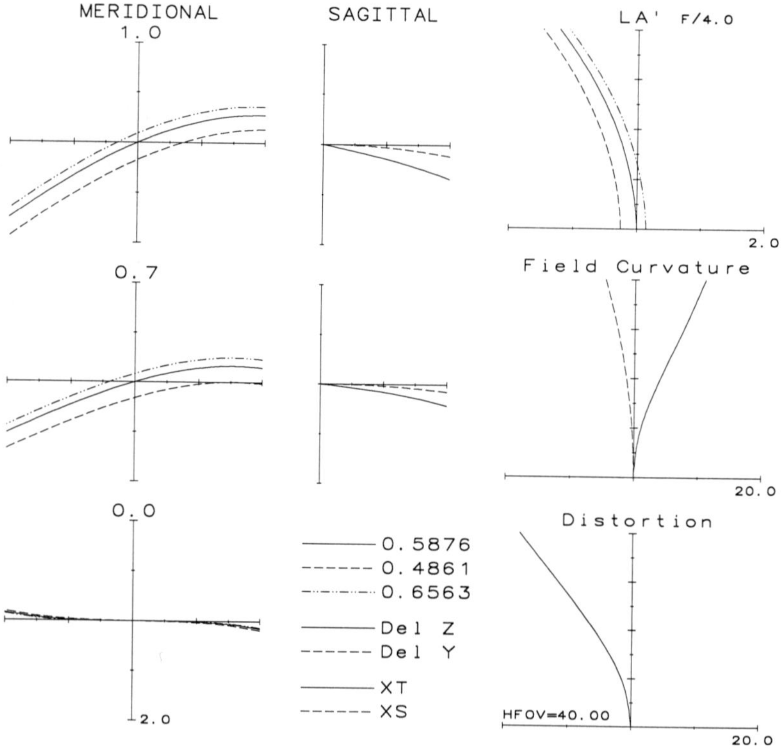

MERIDIONAL 1.0 — SAGITTAL — LA' F/4.0 (2.0)

0.7 — Field Curvature (20.0)

0.0 (2.0) — Distortion (HFOV=40.00, 20.0)

0.5876 (solid)
0.4861 (dashed)
0.6563 (dash-dot)
Del Z (solid)
Del Y (dashed)
XT (solid)
XS (dashed)

Figure 7.23

ers a rather astonishing 80° total field of view. Both systems can be regarded as elaborations of the five-element Erfle, with a split or doubled inner singlet. Note that in these designs both the field lens and the eye lens are meniscus. Figure 7.24 has an explicit negative field lens to flatten the field and lengthen the eye relief. Its plano side also serves as a reticle surface, eliminating the need for a separate element to carry the reticle pattern.

Figures 7.25 through 7.28 are six- and seven-element variations on the basic Erfle eyepiece, the first three with relatively moderate fields of 50 or 60°; the last has a rather immoderate total field of 90°. In practice, a field this large tends to be rather difficult to use, since, as the eye rotates in its socket, it is not easy to keep the pupil of the eye and the exit pupil of the instrument in proper alignment. Note also that spherical aberration of the pupil becomes both more important and more of a design problem as wider fields are covered.

The last eyepiece of this chapter, the Nagler eyepiece of Fig. 7.29, achieves a 90° field and a similar level of correction, but uses a unique construction, incorporating a negative achromatic field lens beyond the focal plane. Pupil aberration can be a problem in this eyepiece also.

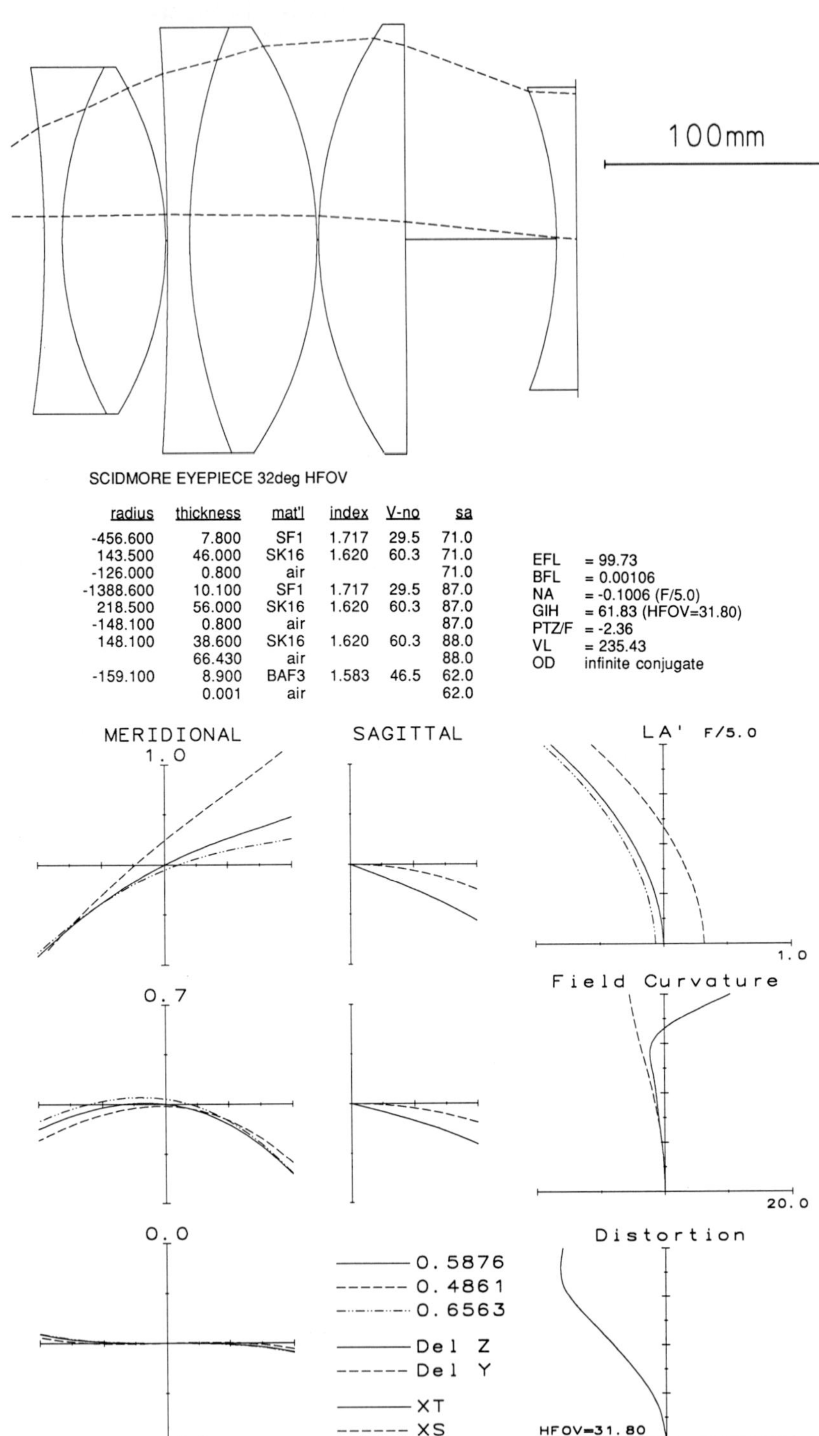

SCIDMORE EYEPIECE 32deg HFOV

radius	thickness	mat'l	index	V-no	sa
-456.600	7.800	SF1	1.717	29.5	71.0
143.500	46.000	SK16	1.620	60.3	71.0
-126.000	0.800	air			71.0
-1388.600	10.100	SF1	1.717	29.5	87.0
218.500	56.000	SK16	1.620	60.3	87.0
-148.100	0.800	air			87.0
148.100	38.600	SK16	1.620	60.3	88.0
	66.430	air			88.0
-159.100	8.900	BAF3	1.583	46.5	62.0
	0.001	air			62.0

EFL = 99.73
BFL = 0.00106
NA = -0.1006 (F/5.0)
GIH = 61.83 (HFOV=31.80)
PTZ/F = -2.36
VL = 235.43
OD infinite conjugate

Figure 7.24

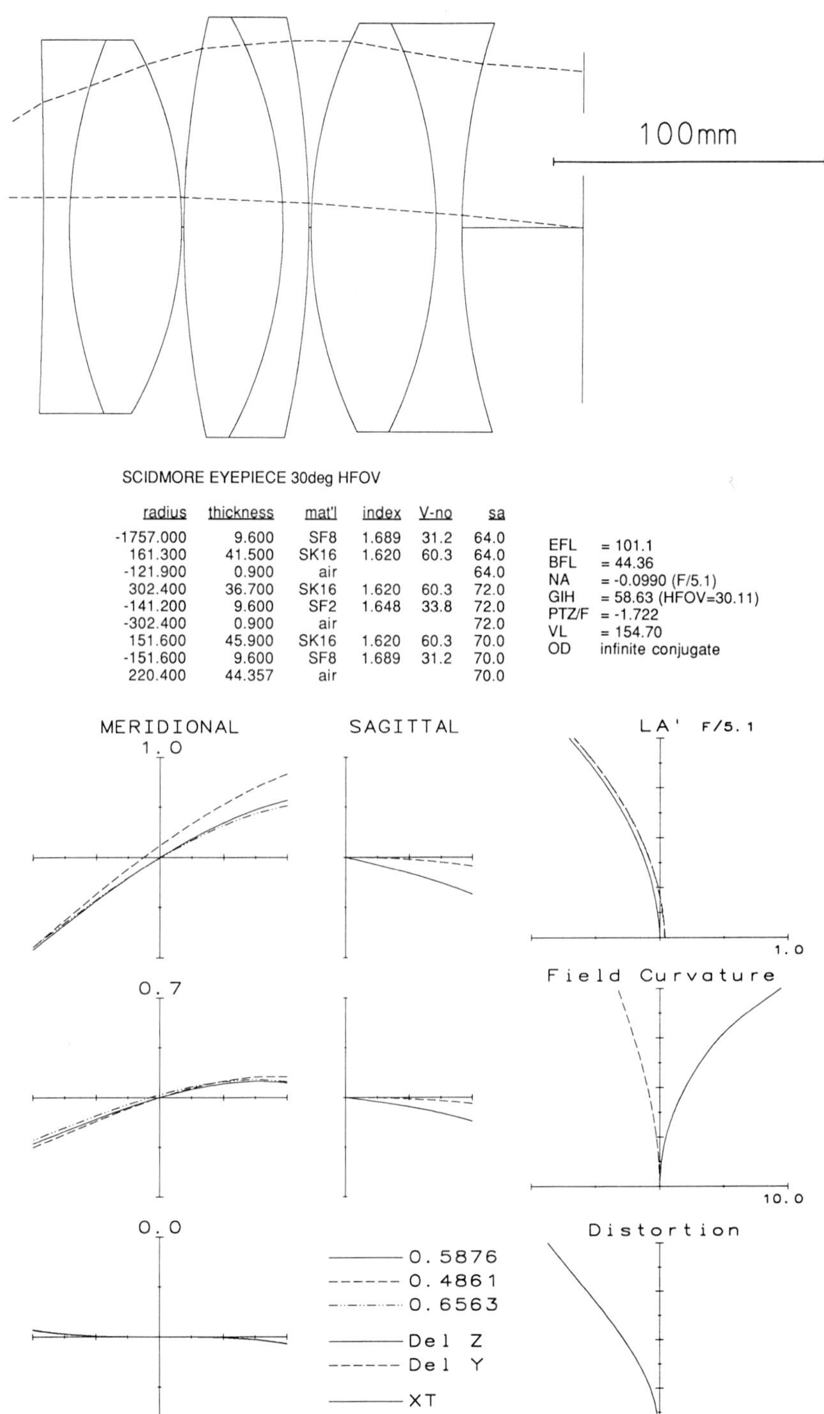

SCIDMORE EYEPIECE 30deg HFOV

radius	thickness	mat'l	index	V-no	sa
-1757.000	9.600	SF8	1.689	31.2	64.0
161.300	41.500	SK16	1.620	60.3	64.0
-121.900	0.900	air			64.0
302.400	36.700	SK16	1.620	60.3	72.0
-141.200	9.600	SF2	1.648	33.8	72.0
-302.400	0.900	air			72.0
151.600	45.900	SK16	1.620	60.3	70.0
-151.600	9.600	SF8	1.689	31.2	70.0
220.400	44.357	air			70.0

EFL = 101.1
BFL = 44.36
NA = -0.0990 (F/5.1)
GIH = 58.63 (HFOV=30.11)
PTZ/F = -1.722
VL = 154.70
OD infinite conjugate

Figure 7.25

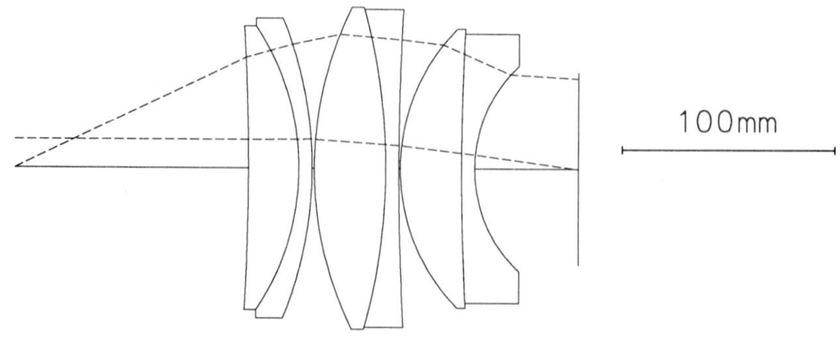

HARUO ABE; USP 3586418; 47.7 DEG. EYEPIECE LENS #2

radius	thickness	mat'l	index	V-no	sa
	109.700	air			0.0
-923.100	23.600	TAF1	1.773	49.6	60.5
-105.000	6.200	FD6	1.805	25.5	61.9
-161.200	1.100	air			65.4
151.000	33.400	TAF1	1.773	49.6	70.1
-241.100	6.200	FD4	1.755	27.5	69.3
1232.200	0.700	air			66.6
82.800	28.800	TAF1	1.773	49.6	60.8
1056.700	6.200	FD4	1.755	27.5	58.8
58.900	48.504	air			44.8

EFL = 100.6
BFL = 48.5
NA = -0.1243 (F/4.0)
GIH = 44.79 (HFOV=24.00)
PTZ/F = -3.667
VL = 215.90
OD infinite conjugate

MERIDIONAL 1.0 SAGITTAL LA' F/4.0

1.0

0.7 Field Curvature

2.0

0.0 Distortion

0.2

0.5876
0.4861
0.6563
Del Z
Del Y
XT
XS

HFOV=24.00

20

Figure 7.26

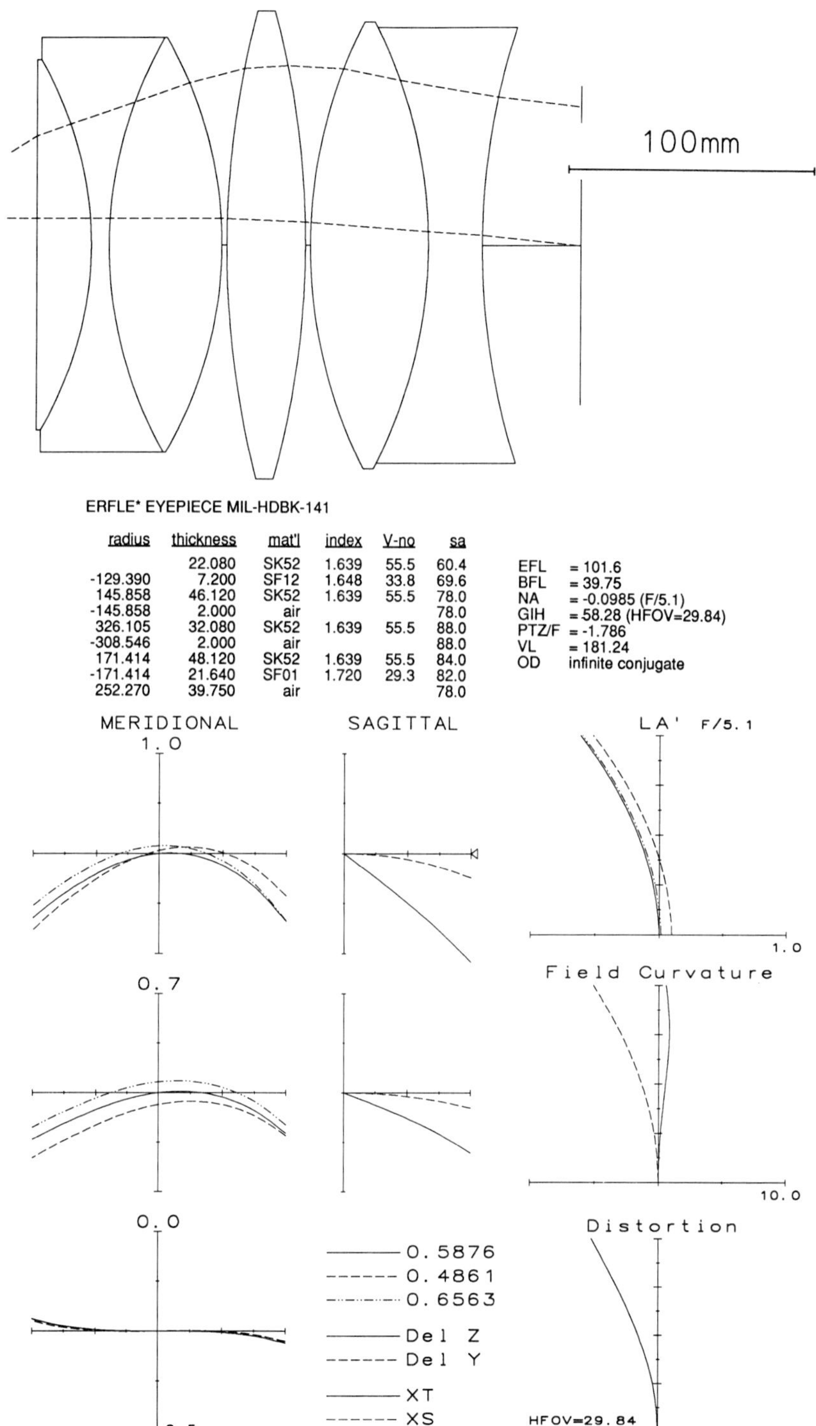

ERFLE* EYEPIECE MIL-HDBK-141

radius	thickness	mat'l	index	V-no	sa
	22.080	SK52	1.639	55.5	60.4
-129.390	7.200	SF12	1.648	33.8	69.6
145.858	46.120	SK52	1.639	55.5	78.0
-145.858	2.000	air			78.0
326.105	32.080	SK52	1.639	55.5	88.0
-308.546	2.000	air			88.0
171.414	48.120	SK52	1.639	55.5	84.0
-171.414	21.640	SF01	1.720	29.3	82.0
252.270	39.750	air			78.0

EFL = 101.6
BFL = 39.75
NA = -0.0985 (F/5.1)
GIH = 58.28 (HFOV=29.84)
PTZ/F = -1.786
VL = 181.24
OD infinite conjugate

Figure 7.27

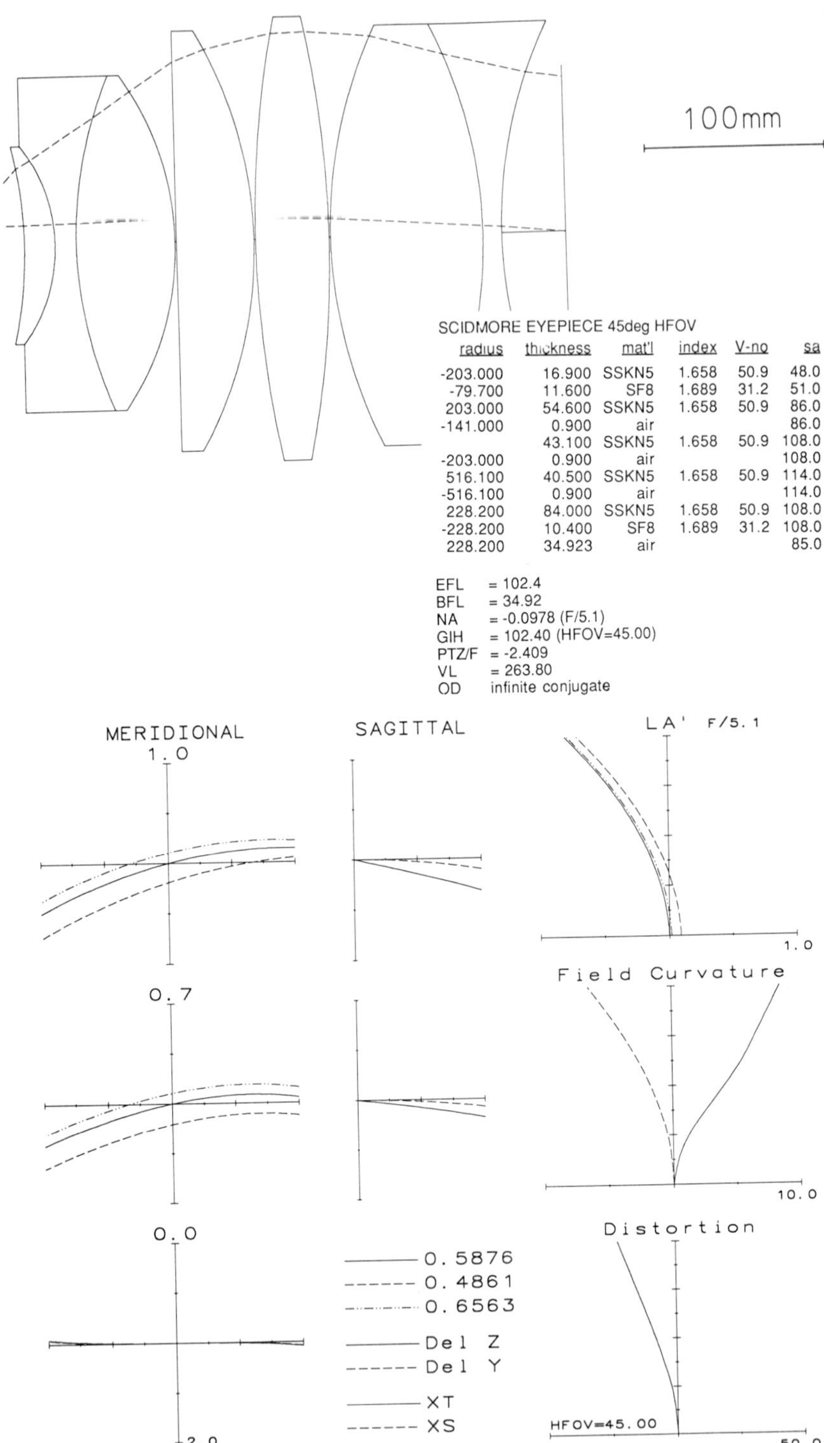

SCIDMORE EYEPIECE 45deg HFOV

radius	thickness	mat'l	index	V-no	sa
-203.000	16.900	SSKN5	1.658	50.9	48.0
-79.700	11.600	SF8	1.689	31.2	51.0
203.000	54.600	SSKN5	1.658	50.9	86.0
-141.000	0.900	air			86.0
	43.100	SSKN5	1.658	50.9	108.0
-203.000	0.900	air			108.0
516.100	40.500	SSKN5	1.658	50.9	114.0
-516.100	0.900	air			114.0
228.200	84.000	SSKN5	1.658	50.9	108.0
-228.200	10.400	SF8	1.689	31.2	108.0
228.200	34.923	air			85.0

EFL = 102.4
BFL = 34.92
NA = -0.0978 (F/5.1)
GIH = 102.40 (HFOV=45.00)
PTZ/F = -2.409
VL = 263.80
OD infinite conjugate

Figure 7.28

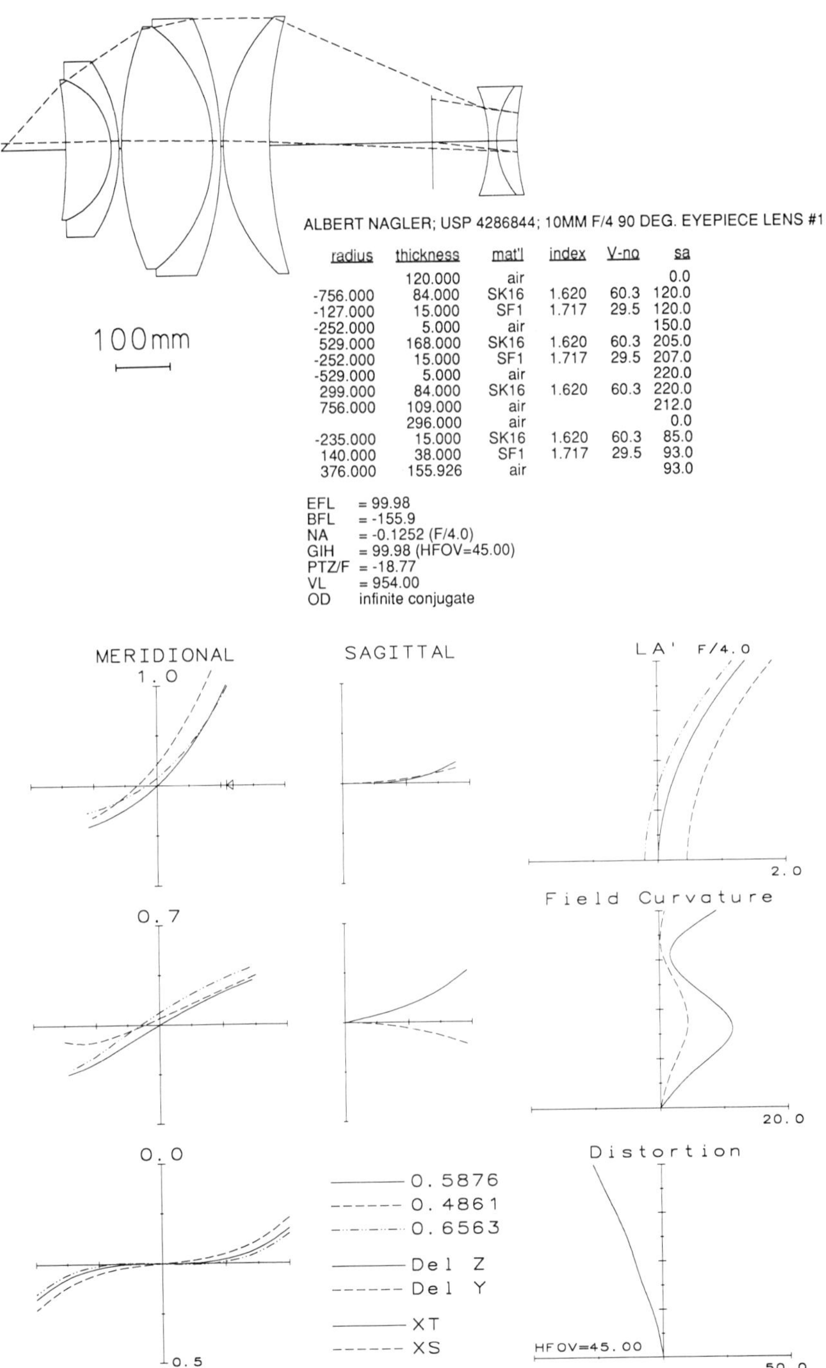

ALBERT NAGLER; USP 4286844; 10MM F/4 90 DEG. EYEPIECE LENS #1

radius	thickness	mat'l	index	V-no	sa
	120.000	air			0.0
-756.000	84.000	SK16	1.620	60.3	120.0
-127.000	15.000	SF1	1.717	29.5	120.0
-252.000	5.000	air			150.0
529.000	168.000	SK16	1.620	60.3	205.0
-252.000	15.000	SF1	1.717	29.5	207.0
-529.000	5.000	air			220.0
299.000	84.000	SK16	1.620	60.3	220.0
756.000	109.000	air			212.0
	296.000	air			0.0
-235.000	15.000	SK16	1.620	60.3	85.0
140.000	38.000	SF1	1.717	29.5	93.0
376.000	155.926	air			93.0

EFL = 99.98
BFL = -155.9
NA = -0.1252 (F/4.0)
GIH = 99.98 (HFOV=45.00)
PTZ/F = -18.77
VL = 954.00
OD infinite conjugate

Figure 7.29

Chapter

8

Cooke Triplet Anastigmats*

8.1 Airspaced Triplet Anastigmats

The Cooke triplet (Figs. 8.2 to 8.18) is an especially interesting design form for several reasons. It possesses just enough effective degrees of freedom to control or correct all the primary aberrations. Because of the nonlinearity of the relationships between the aberrations and the design variables, there are (theoretically) at least eight potential solutions for the primary aberrations, two or three of which may be useful.

The basic method used to flatten the Petzval field curvature in *all* anastigmats is the longitudinal separation of positive power from negative power. This separation may be between surfaces, between elements, or between components. Since the contribution to the Petzval curvature is $\text{TPC} = (n' - n)ch^2 n_k' u_k'/2nn'$ for a surface (per Eq. F.8.6), or $\text{TPC} = h^2 \phi u_k'/2n$ for a thin element (per Eq. F.9.5), it is obviously independent of the height at which the rays strike. The contribution to the system power, however, is $y(n' - n)c$ for a surface, or $y\phi$ for a component. Thus, increasing the spacing of positive power away from negative power, which lowers the relative ray height y on the negative, will reduce the (negative) power contribution of the surfaces without changing their Petzval contribution. The result is an effective net positive power without the undesirable excess of inward Petzval curvature which would otherwise accompany positive power.

All anastigmats make use of this principle. Some incorporate thick meniscus components to separate positive convex surfaces from negative concave surfaces. Examples are the older anastigmats such as the Protar and Dagor (Chap. 11), the Ernostar and Sonnar types (Chap.

*See W. J. Smith, *Modern Optical Engineering*, McGraw-Hill, New York, 1990, Chap. 12, pp. 384–392, for a complete discussion of the elements of Cooke triplet design.

14), and the widely used and powerful Biotar or double-Gauss form (Chap. 17). Others, such as the Cooke triplet and its modifications, or the retrofocus forms (Chap. 9), utilize spaced-apart components to achieve the same ends. The Cooke triplet can be viewed as the trunk of the family tree of the airspaced anastigmats.

For the Cooke triplet the powers and spaces must satisfy the following relationships:

For lens power (efl):

$$\phi = \frac{1}{y_A} \Sigma y_i \phi_i$$

For axial chromatic:

$$\mathrm{TA_{ch}A} = \frac{1}{u'_k} \Sigma \frac{y_i^2 \phi_i}{V_i}$$

For lateral chromatic:

$$\mathrm{T_{ch}A} = \frac{1}{u'_k} \Sigma \frac{y_i y_{Pi} \phi_i}{V_i}$$

For Petzval curvature:

$$\mathrm{TPC} = \frac{h^2}{2u'_k} \Sigma \frac{\phi_i}{n_i}$$

Since the triplet has three element powers and two airspaces which are effective constructional variables (in regard to the primary aberrations, the element thicknesses are only weakly effective and duplicate the airspaces), these requirements are easily satisfied. Two element powers and the two spaces are dedicated to this end. The remaining element power and the three shapes (bendings) of the elements are then available to control the primary spherical, coma, astigmatism, and distortion.

Despite the neat congruence of eight effective variables and eight aberrations (if we include efl), the satisfactory solution to the Cooke triplet problem, i.e., the simultaneous correction or control of the eight characteristics, requires a delicate balance between the aberrations. For example, the Petzval curvature must be slightly inward-curving with the Petzval radius $\rho = -k(\text{efl})$, where K is ordinarily in the range of about 2 to 6. In general, K should be small for high-speed, narrow-angle triplets and large for low-speed, wide-angle lenses. In addition, the third-order astigmatism must be slightly overcorrected to offset the undercorrected fifth-order astigmatism.

8.2 Glass Choice

The choice of glass types is an important degree of freedom and has a significant effect on the characteristics of the triplet design. The glass for the positive elements should be a dense barium or lanthanum crown type; an index of 1.6 or more is almost essential. A triplet using an ordinary low-index crown glass (or even acrylic plastic) for the positive elements is, of course, possible, but the result is poor unless the aperture or field (or both) is small. The other important factor in glass choice is the use of the relative difference in V values between the crown and flint elements as a means to adjust the vertex length of the triplet to its optimum value.

8.3 Vertex Length and Residual Aberrations

In general, the longer an anastigmat is (i.e., the greater its vertex length), the smaller the field angle it will cover and the less zonal spherical aberration it will have. High-speed lenses of small angular coverage tend to have a large vertex length, whereas slow-speed, wide-angle lenses tend to be relatively short. Figure 8.1 is a plot of the zonal spherical and the angular coverage of typical anastigmats as a function of vertex length for a large number of published lens designs. Each dot in the figure represents an ordinary anastigmat lens. Note that the figure is not limited to Cooke triplets. Lenses with three, four, five, six, and seven elements, with no restrictions on glass type, designed by about 50 different designers, are included. The correlation, while not exact, is obvious. The lines in the figure show the results of a study* in which the design characteristics were carefully controlled, so that the effects of vertex length on the lens characteristics were clearly demonstrated.

In Cooke triplets (and lenses derived from them) the vertex length can be controlled by the choice of glass types used in the design. The axial chromatic aberration contribution of a thin element varies with $y^2\phi/V$, as in Eq. F.9.7. If we require a specific amount of chromatic aberration from an element (in order to correct the chromatic for the entire lens), it is apparent that y, the marginal-axial ray height, must be smaller if we use a

*W. Smith, "Control of Residual Aberrations in the Design of Anastigmat Objectives" *J. Opt. Soc. Am.*, vol. 48, 1958, pp. 98–105.

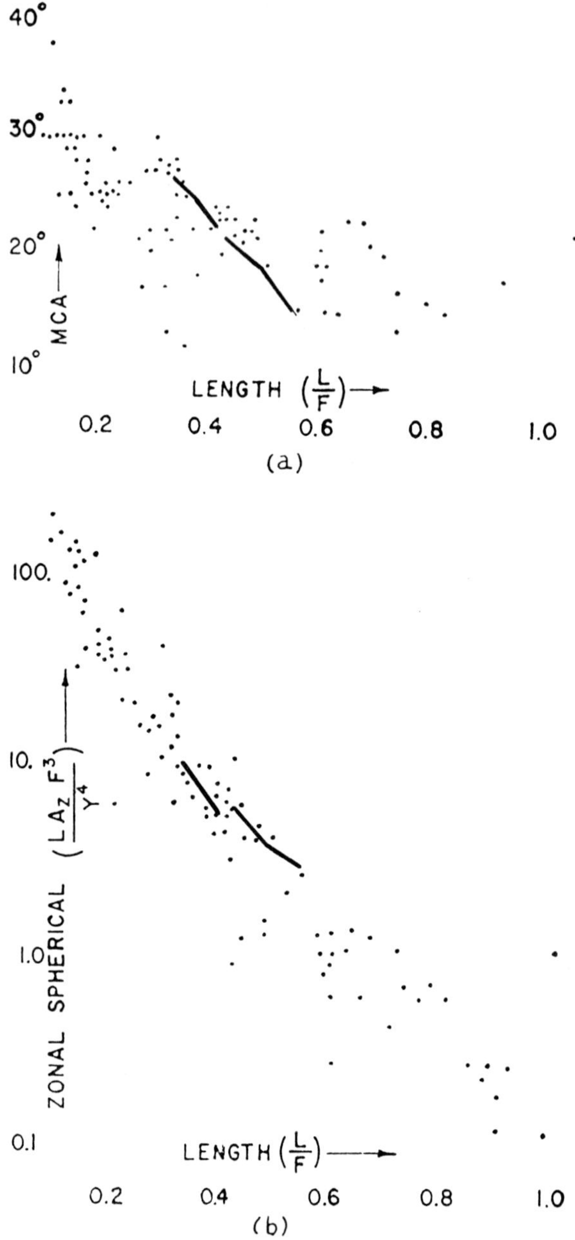

Figure 8.1 (*a*) Plot of angular field coverage or maximum coverage angle (MCA) as a function of vertex length (*L/f*) for about 85 different anastigmat designs. Lenses with three to seven elements, without restriction as to glass types, age, application or design configuration are included. (*b*) Plot of fifth-order spherical (zonal) aberration as a function of the vertex length for the same lenses as in *a*. The solid lines are for a four-element Dogmar/Celor-type design in which the conditions (Petzval, chromatic, etc. correction) were carefully controlled so that the vertex length was the primary variable.

glass with a low V value than if we use a higher V-value glass. In the triplet, the ray height at the center flint will be reduced if the spacing between the elements is increased. Thus a greater vertex length produces a lower ray height at the flint and requires a lower V value for the flint element (or a higher V value for the positive crown elements) in order to maintain chromatic correction.

Therefore, to produce a long, high-speed, narrow-field (triplet) anastigmat, glasses with a large V-value difference between (positive) crown and (negative) flint elements are appropriate. For short, slow, wide-angle systems, a small V value difference is suitable. By allowing the glass types to vary during the optimization process, an automatic design program can select the appropriate V values to produce a vertex length which is optimally suited to the aperture and field desired. Obviously, the starting glasses for the design should be chosen with this relationship in mind. It is usually preferable to allow the flint glass to vary and to select the crown glass from those in the upper left region of the glass map (Fig. 2.2) on the basis of cost and stability. The optimum flint glass is almost always on the glass line, because a low index in a negative element is beneficial to the Petzval correction. These glass-line glasses are mostly inexpensive, durable, available, and workable types. It is usually quite safe to allow the design program to select from among these glasses.

Figure 2.3 gives a rough idea of the limit of the design capabilities of the Cooke triplet, as well as other design forms. In general, as with most lenses, when the speed of the triplet is increased, the angular coverage must be reduced (and vice versa) in order to maintain a given level of image quality.

8.4 Other Design Considerations

A number of other generalizations also apply to the Cooke triplet:

1. The higher the index of the positive elements, the better. Actually, it is primarily the *difference* between the index of the crown and flint which affects the lens performance, but since the index of the flint cannot be reduced significantly below the glass line, and since the flint V value should be chosen to produce the optimum vertex length, the flint index is almost predestined once the crown glass type is chosen; thus the index difference is effectively determined by the index of the crown elements.
2. Allowing a more inward-curving Petzval field will result in lower powers for all the elements when the other primary aberrations are corrected. The effect of the lower element powers is that the residual aberrations are smaller—there is less zonal spherical aberration, less high-order coma, astigmatism, etc.

3. Allowing the axial chromatic to be somewhat undercorrected has the same beneficial effect as described above for the Petzval undercorrection.

Figures 8.2 through 8.5 show a series of four Cooke triplets with various combinations of field and aperture (*f*/6.3, 27°; *f*/4.0, 23°; *f*/3.0, 19°; *f*/2.5, 16°) which were chosen to lie along a line well above and approximately parallel to the triplet area of Fig. 2.2. All four have SK4 crown elements and all were designed with the same optimization program and merit function. (Because the combinations of field and speed of these lenses are well beyond the normal capabilities of the triplet form, these are not very good lenses.) Notice the relationships between:

a. The vertex length

b. The flint glass *V* value

c. The field and aperture

which clearly illustrate the correlations shown in Fig. 8.1 and the discussion in Sec. 8.3.

Figures 8.6 through 8.18 present a selection of Cooke triplets of various aperture and field combinations. Figures 8.6, 8.7, and 8.8 show three lenses with similar, but not identical, combinations of aperture and field. Note that the speed of *f*/2.5 in Fig. 8.6 has produced a very long lens (55 percent of the focal length) and the lens has achieved a good balance of correction. Note the unusually thick elements; it would be an interesting exercise to attempt to reduce the thickness of the two front elements and simultaneously maintain the same level of correction. A comparison of Figs. 8.7 and 8.8 indicates the improvement in field curvature (while maintaining the axial correction) which results from the use of a higher-index crown glass in Fig. 8.8.

The lenses of Fig. 8.9 and 8.10 cover a somewhat larger field angle than the preceding lenses, approaching what used to be referred to as a *normal* lens coverage. The aberration corrections are similar, but Fig. 8.9 has a speed of *f*/2.8 versus *f*/3.0 for Fig. 8.10; the added speed capability of Fig. 8.9 comes from the higher-index crown glass used.

Figures 8.11 and 8.12 are triplets of a speed and angular coverage which are typical of 35-mm slide-projection lenses. While the glasses used are different, the glasses are probably about the same in their aberration effects. Figure 8.12, which is significantly slower at *f*/3.5, is about twice as well-corrected (except for the lateral color) as the faster (*f*/2.9) Fig. 8.11.

Figure 8.13, at a speed of *f*/4.4 and a normal coverage angle is typical of a well-corrected triplet camera lens. Figure 8.14 is an ex-

(*Text continues on page 146.*)

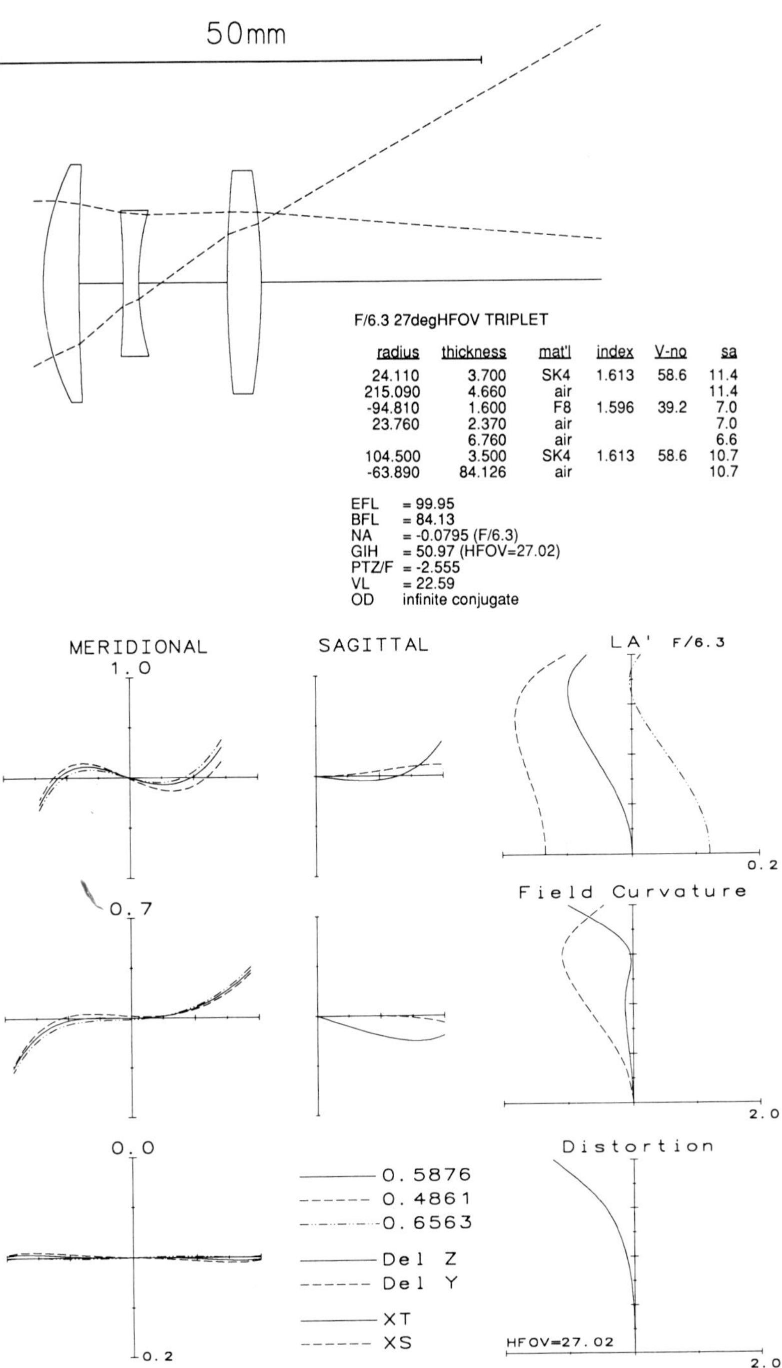
F/6.3 27degHFOV TRIPLET

radius	thickness	mat'l	index	V-no	sa
24.110	3.700	SK4	1.613	58.6	11.4
215.090	4.660	air			11.4
-94.810	1.600	F8	1.596	39.2	7.0
23.760	2.370	air			7.0
	6.760	air			6.6
104.500	3.500	SK4	1.613	58.6	10.7
-63.890	84.126	air			10.7

EFL = 99.95
BFL = 84.13
NA = -0.0795 (F/6.3)
GIH = 50.97 (HFOV=27.02)
PTZ/F = -2.555
VL = 22.59
OD infinite conjugate

Figure 8.2

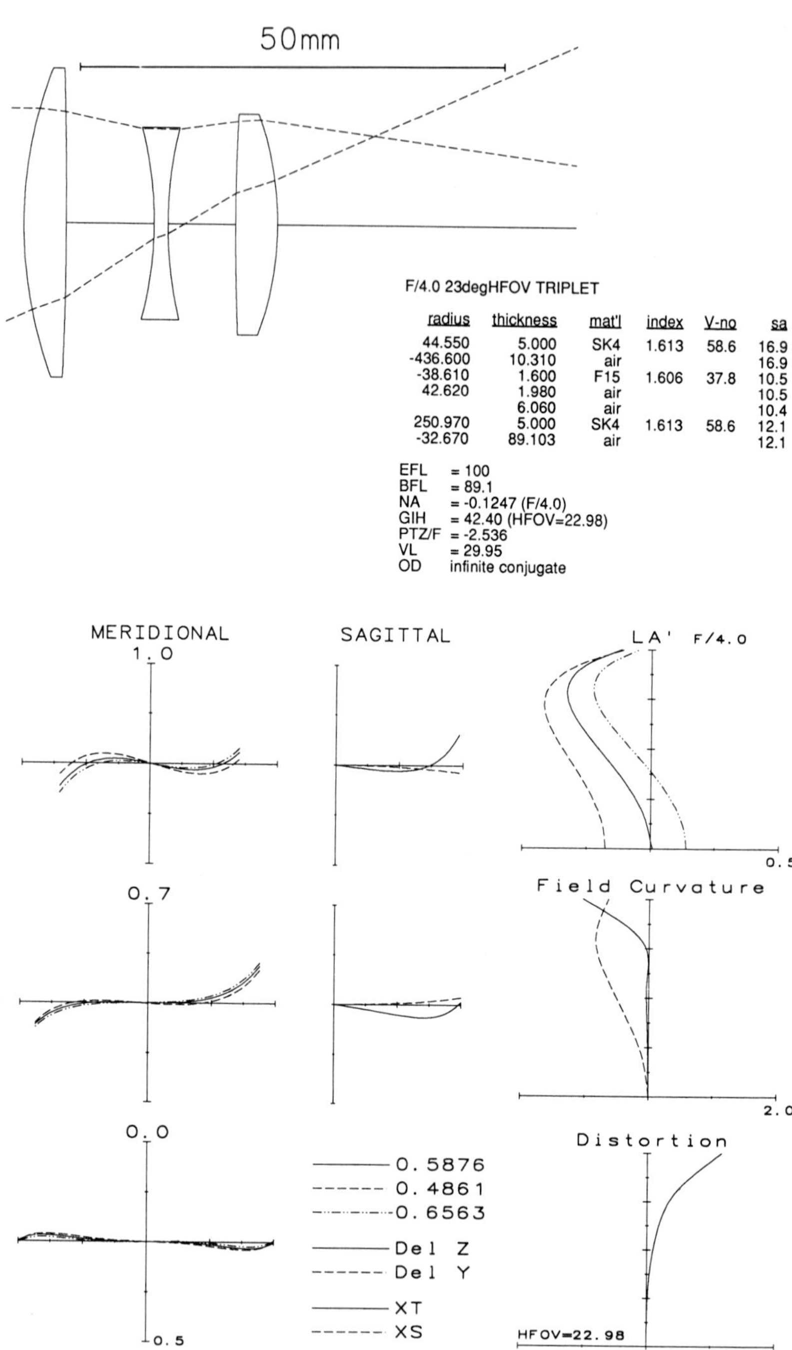

F/4.0 23degHFOV TRIPLET

radius	thickness	mat'l	index	V-no	sa
44.550	5.000	SK4	1.613	58.6	16.9
-436.600	10.310	air			16.9
-38.610	1.600	F15	1.606	37.8	10.5
42.620	1.980	air			10.5
	6.060	air			10.4
250.970	5.000	SK4	1.613	58.6	12.1
-32.670	89.103	air			12.1

EFL = 100
BFL = 89.1
NA = -0.1247 (F/4.0)
GIH = 42.40 (HFOV=22.98)
PTZ/F = -2.536
VL = 29.95
OD infinite conjugate

Figure 8.3

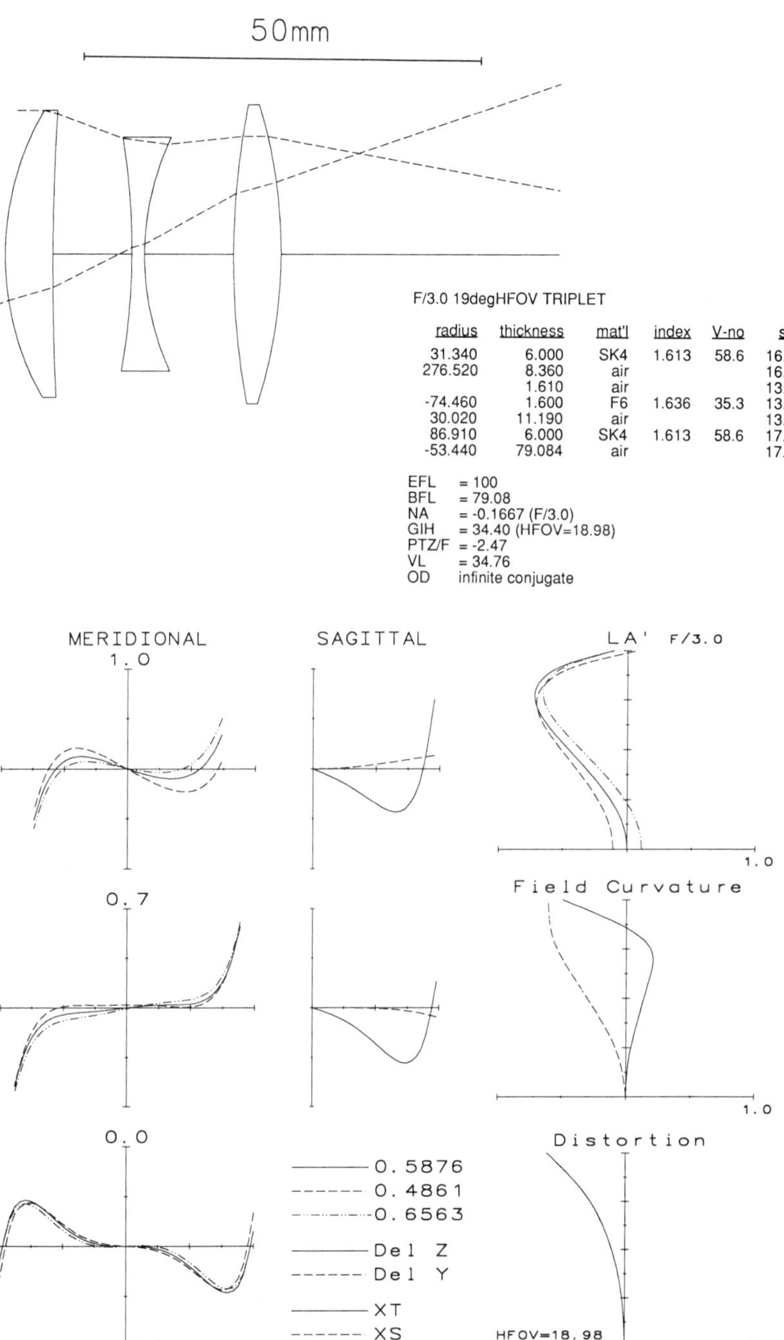

F/3.0 19degHFOV TRIPLET

radius	thickness	mat'l	index	V-no	sa
31.340	6.000	SK4	1.613	58.6	16.7
276.520	8.360	air			16.7
	1.610	air			13.6
-74.460	1.600	F6	1.636	35.3	13.6
30.020	11.190	air			13.6
86.910	6.000	SK4	1.613	58.6	17.4
-53.440	79.084	air			17.4

EFL = 100
BFL = 79.08
NA = -0.1667 (F/3.0)
GIH = 34.40 (HFOV=18.98)
PTZ/F = -2.47
VL = 34.76
OD infinite conjugate

Figure 8.4

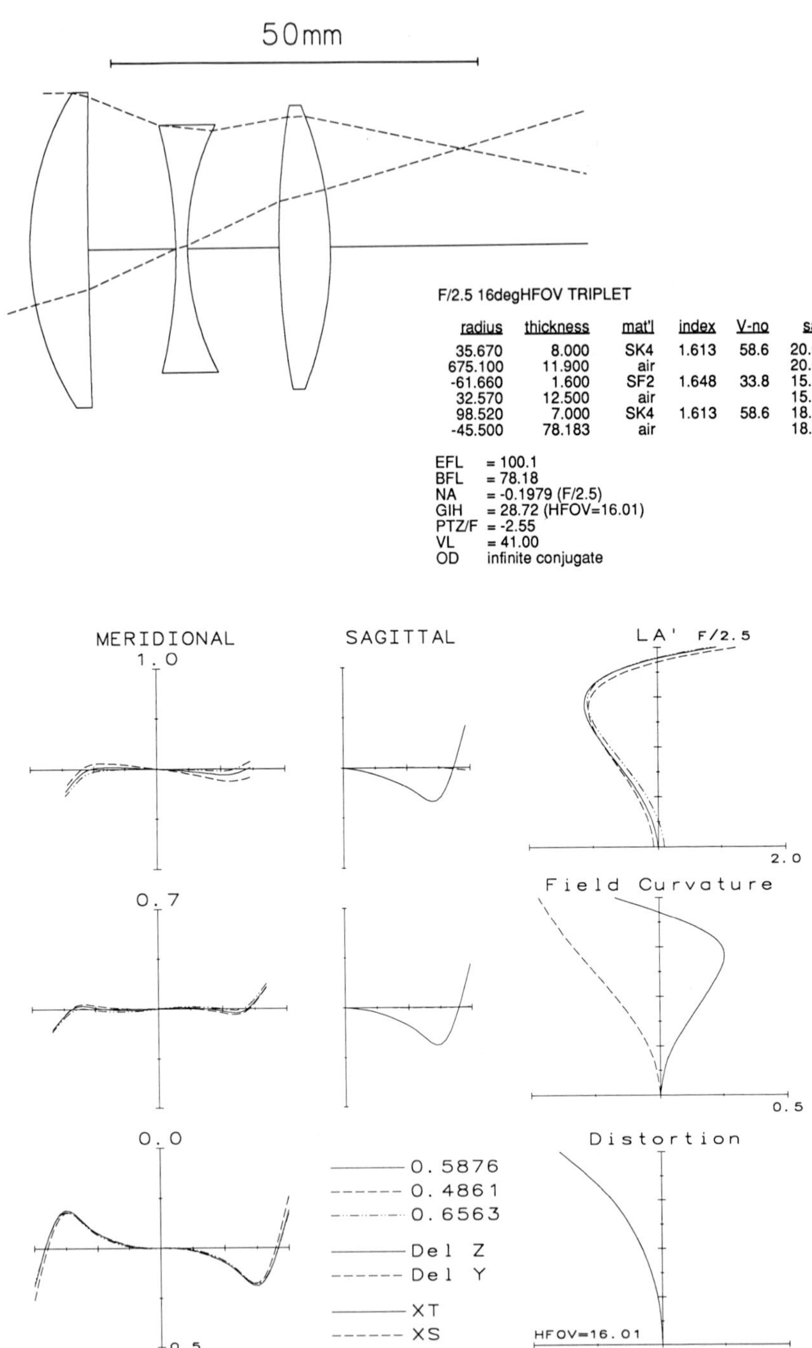

F/2.5 16degHFOV TRIPLET

radius	thickness	mat'l	index	V-no	sa
35.670	8.000	SK4	1.613	58.6	20.0
675.100	11.900	air			20.0
-61.660	1.600	SF2	1.648	33.8	15.7
32.570	12.500	air			15.7
98.520	7.000	SK4	1.613	58.6	18.0
-45.500	78.183	air			18.0

EFL = 100.1
BFL = 78.18
NA = -0.1979 (F/2.5)
GIH = 28.72 (HFOV=16.01)
PTZ/F = -2.55
VL = 41.00
OD infinite conjugate

Figure 8.5

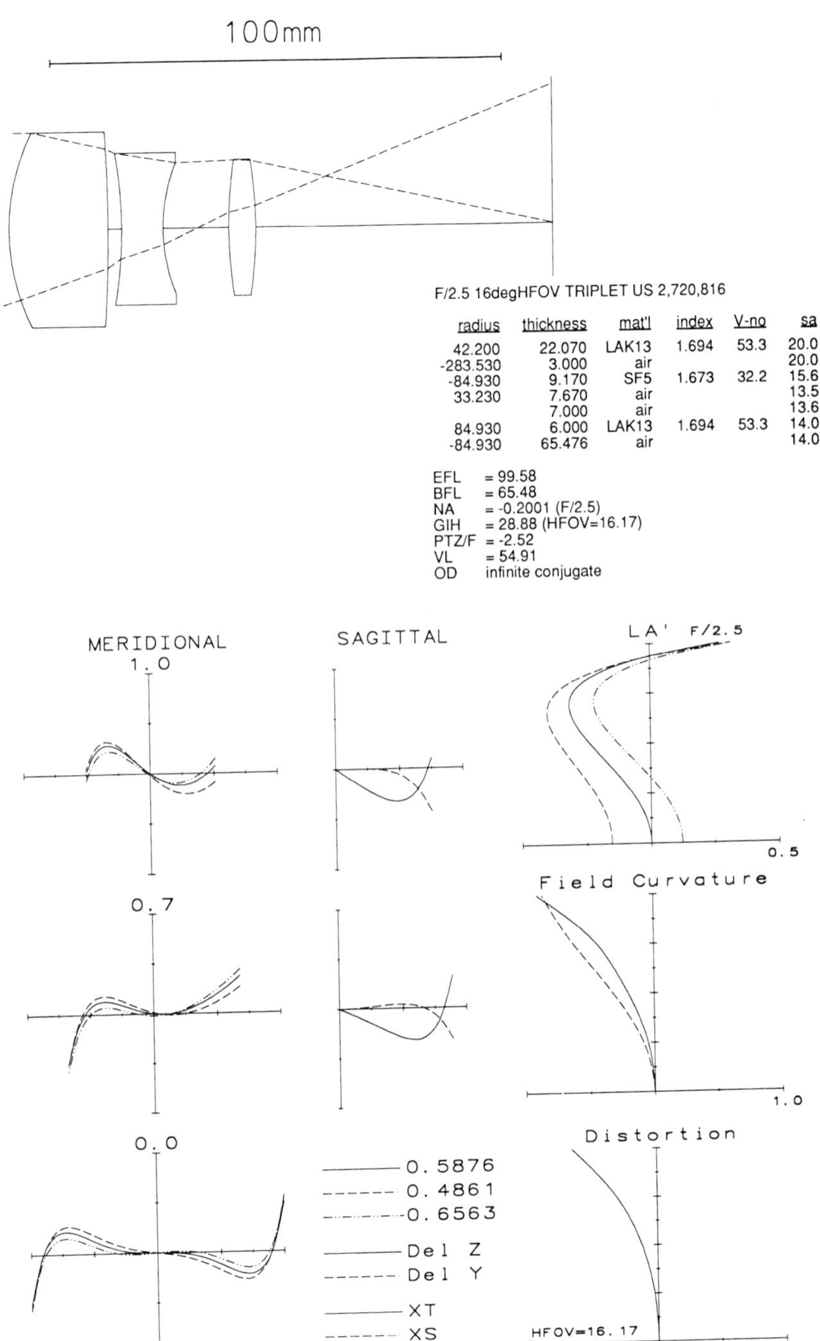

F/2.5 16degHFOV TRIPLET US 2,720,816

radius	thickness	mat'l	index	V-no	sa
42.200	22.070	LAK13	1.694	53.3	20.0
-283.530	3.000	air			20.0
-84.930	9.170	SF5	1.673	32.2	15.6
33.230	7.670	air			13.5
	7.000	air			13.6
84.930	6.000	LAK13	1.694	53.3	14.0
-84.930	65.476	air			14.0

EFL = 99.58
BFL = 65.48
NA = -0.2001 (F/2.5)
GIH = 28.88 (HFOV=16.17)
PTZ/F = -2.52
VL = 54.91
OD infinite conjugate

Figure 8.6

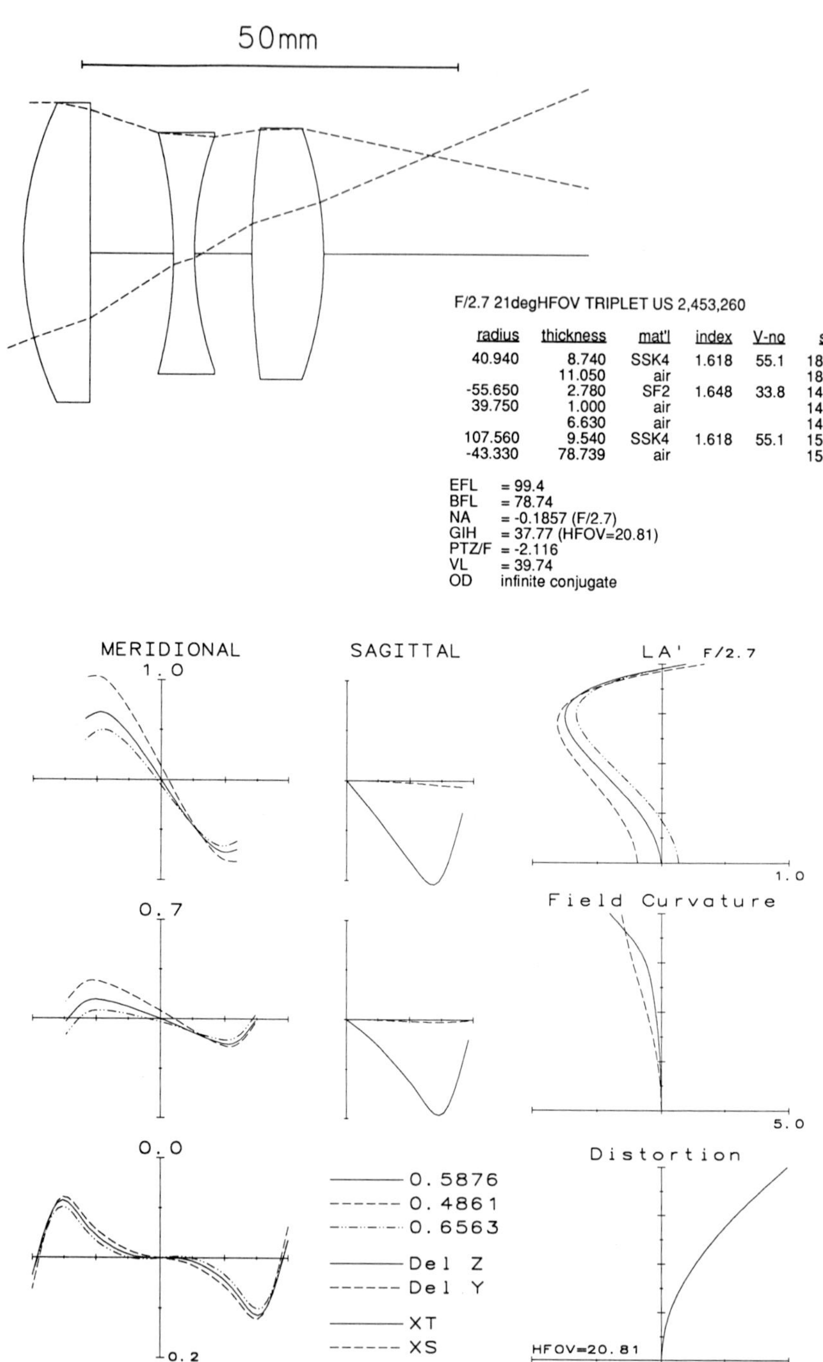

F/2.7 21degHFOV TRIPLET US 2,453,260

radius	thickness	mat'l	index	V-no	sa
40.940	8.740	SSK4	1.618	55.1	18.5
	11.050	air			18.5
-55.650	2.780	SF2	1.648	33.8	14.9
39.750	1.000	air			14.4
	6.630	air			14.1
107.560	9.540	SSK4	1.618	55.1	15.5
-43.330	78.739	air			15.5

EFL = 99.4
BFL = 78.74
NA = -0.1857 (F/2.7)
GIH = 37.77 (HFOV=20.81)
PTZ/F = -2.116
VL = 39.74
OD infinite conjugate

Figure 8.7

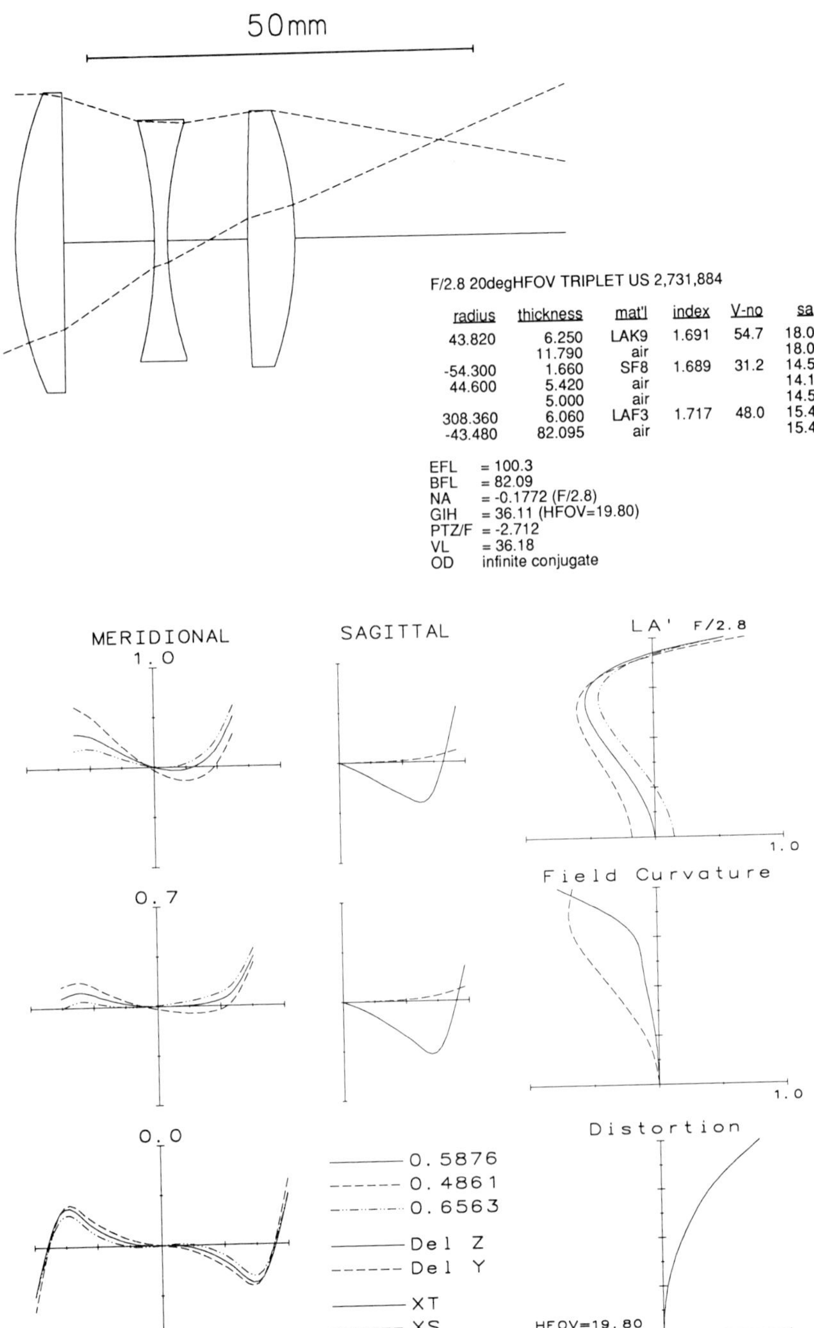

F/2.8 20degHFOV TRIPLET US 2,731,884

radius	thickness	mat'l	index	V-no	sa
43.820	6.250	LAK9	1.691	54.7	18.0
	11.790	air			18.0
-54.300	1.660	SF8	1.689	31.2	14.5
44.600	5.420	air			14.1
	5.000	air			14.5
308.360	6.060	LAF3	1.717	48.0	15.4
-43.480	82.095	air			15.4

EFL = 100.3
BFL = 82.09
NA = -0.1772 (F/2.8)
GIH = 36.11 (HFOV=19.80)
PTZ/F = -2.712
VL = 36.18
OD infinite conjugate

Figure 8.8

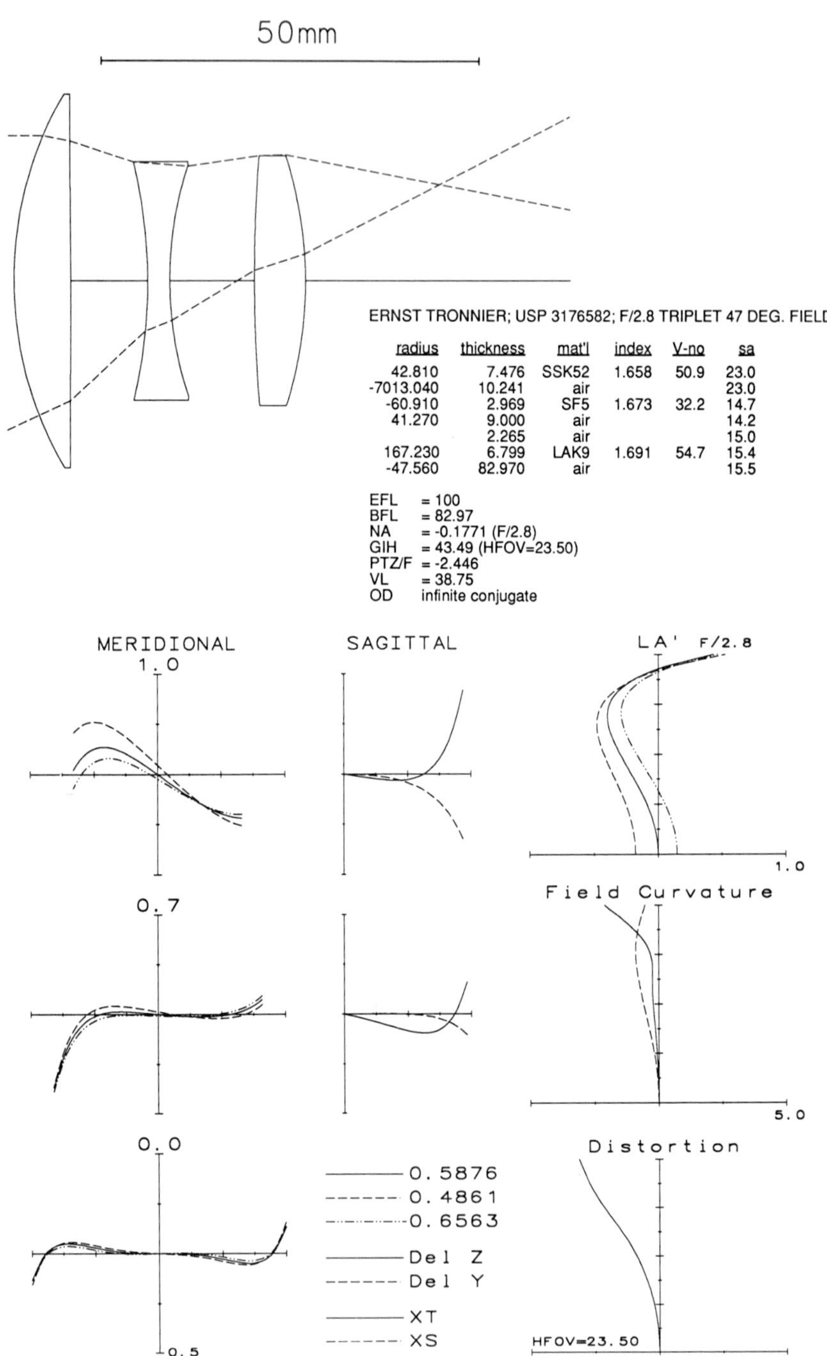

ERNST TRONNIER; USP 3176582; F/2.8 TRIPLET 47 DEG. FIELD

radius	thickness	mat'l	index	V-no	sa
42.810	7.476	SSK52	1.658	50.9	23.0
-7013.040	10.241	air			23.0
-60.910	2.969	SF5	1.673	32.2	14.7
41.270	9.000	air			14.2
	2.265	air			15.0
167.230	6.799	LAK9	1.691	54.7	15.4
-47.560	82.970	air			15.5

EFL = 100
BFL = 82.97
NA = -0.1771 (F/2.8)
GIH = 43.49 (HFOV=23.50)
PTZ/F = -2.446
VL = 38.75
OD infinite conjugate

Figure 8.9

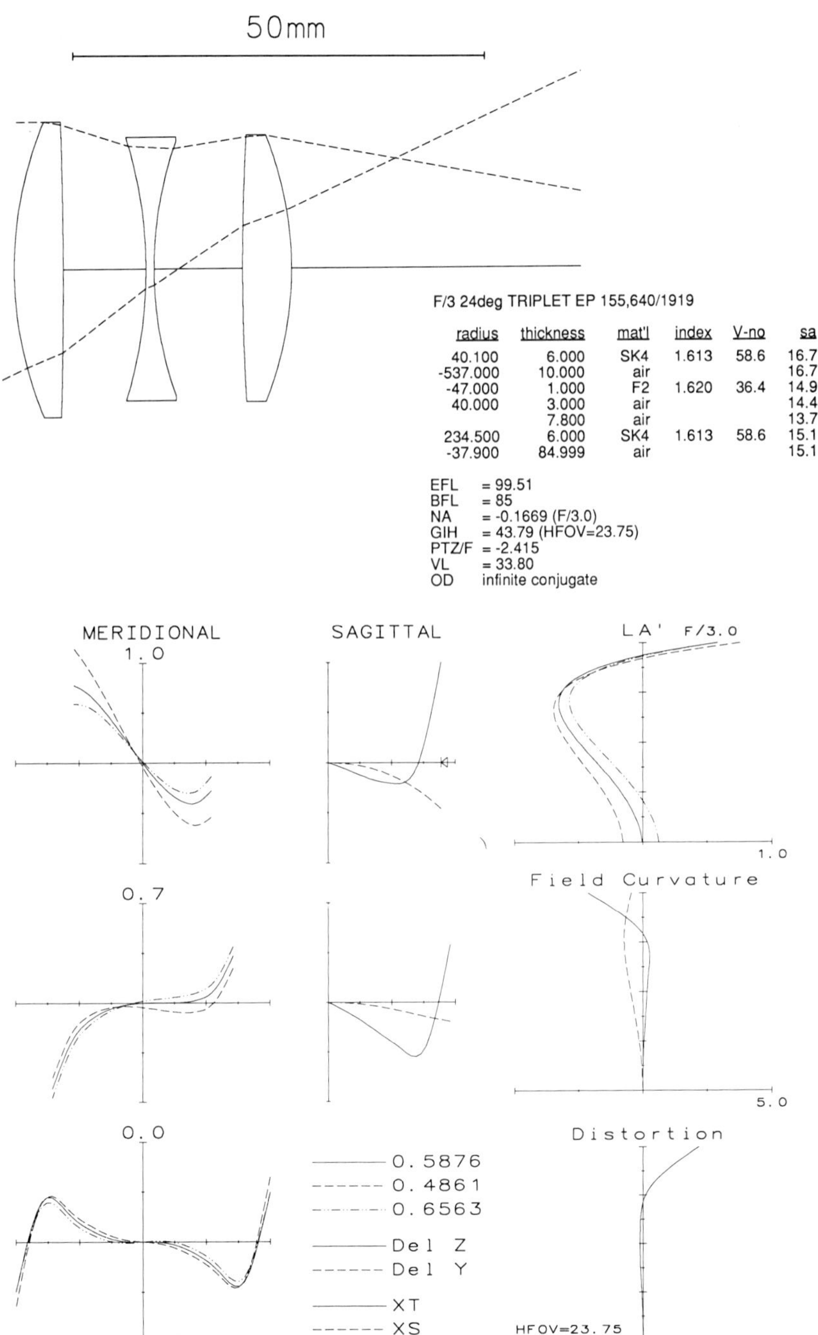

F/3 24deg TRIPLET EP 155,640/1919

radius	thickness	mat'l	index	V-no	sa
40.100	6.000	SK4	1.613	58.6	16.7
-537.000	10.000	air			16.7
-47.000	1.000	F2	1.620	36.4	14.9
40.000	3.000	air			14.4
	7.800	air			13.7
234.500	6.000	SK4	1.613	58.6	15.1
-37.900	84.999	air			15.1

EFL = 99.51
BFL = 85
NA = -0.1669 (F/3.0)
GIH = 43.79 (HFOV=23.75)
PTZ/F = -2.415
VL = 33.80
OD infinite conjugate

Figure 8.10

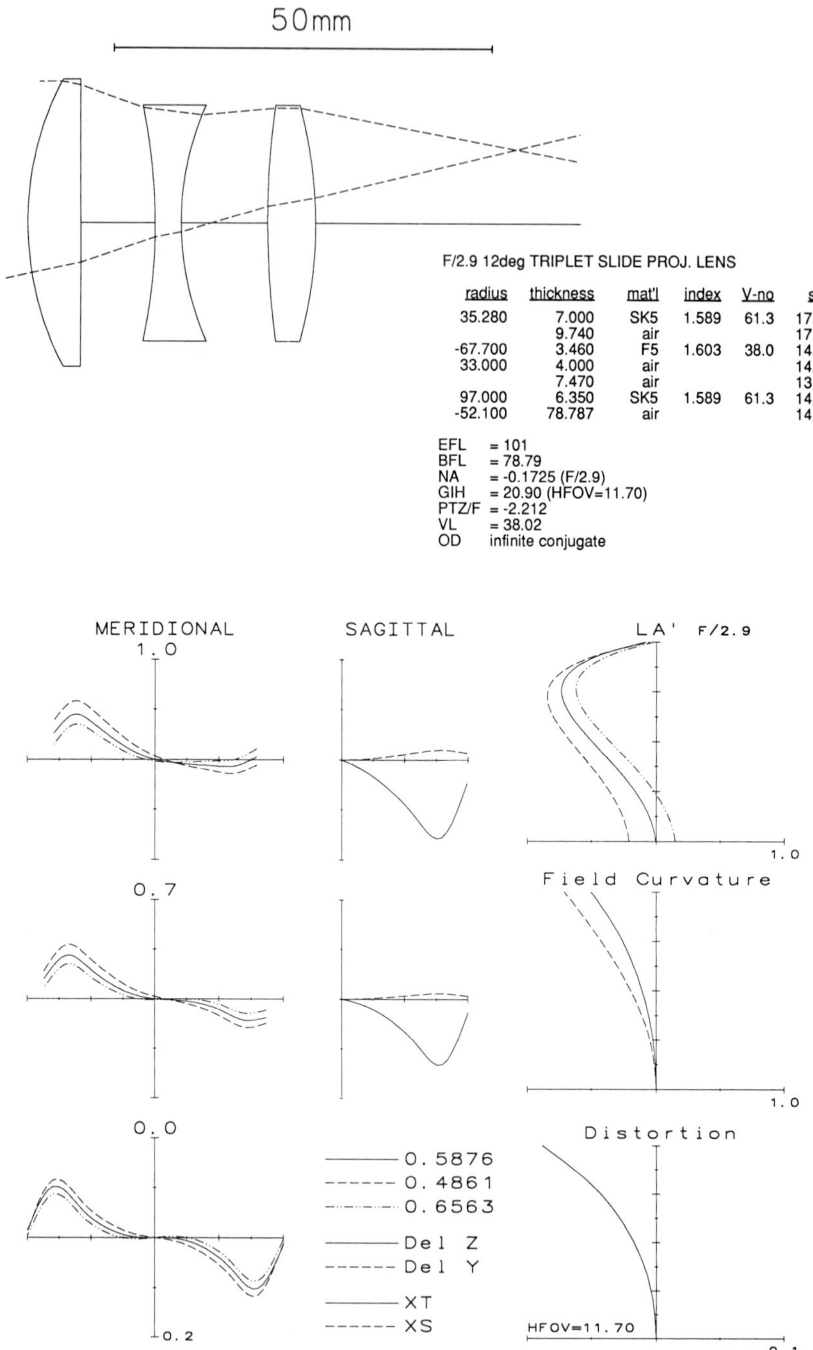

F/2.9 12deg TRIPLET SLIDE PROJ. LENS

radius	thickness	mat'l	index	V-no	sa
35.280	7.000	SK5	1.589	61.3	17.7
	9.740	air			17.7
-67.700	3.460	F5	1.603	38.0	14.5
33.000	4.000	air			14.5
	7.470	air			13.4
97.000	6.350	SK5	1.589	61.3	14.5
-52.100	78.787	air			14.5

EFL = 101
BFL = 78.79
NA = -0.1725 (F/2.9)
GIH = 20.90 (HFOV=11.70)
PTZ/F = -2.212
VL = 38.02
OD infinite conjugate

Figure 8.11

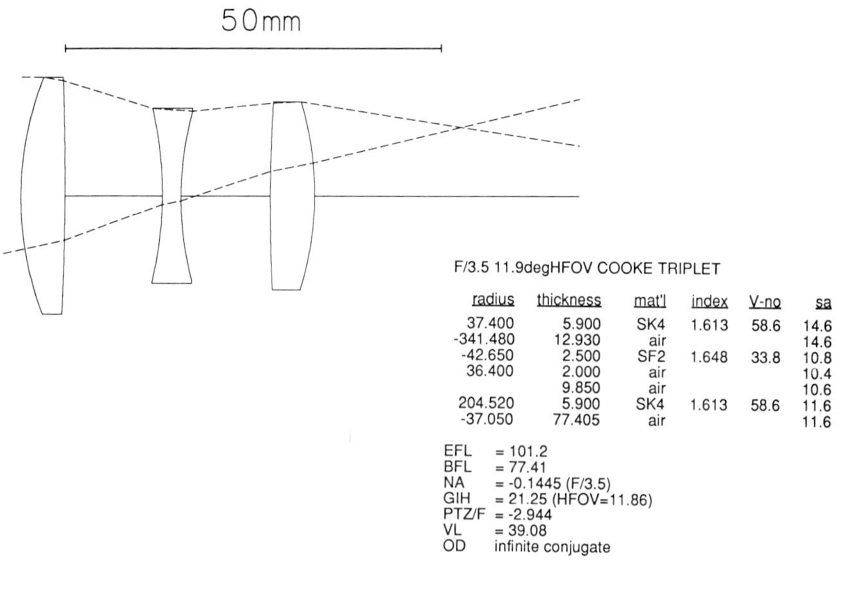

F/3.5 11.9degHFOV COOKE TRIPLET

radius	thickness	mat'l	index	V-no	sa
37.400	5.900	SK4	1.613	58.6	14.6
-341.480	12.930	air			14.6
-42.650	2.500	SF2	1.648	33.8	10.8
36.400	2.000	air			10.4
	9.850	air			10.6
204.520	5.900	SK4	1.613	58.6	11.6
-37.050	77.405	air			11.6

EFL = 101.2
BFL = 77.41
NA = -0.1445 (F/3.5)
GIH = 21.25 (HFOV=11.86)
PTZ/F = -2.944
VL = 39.08
OD infinite conjugate

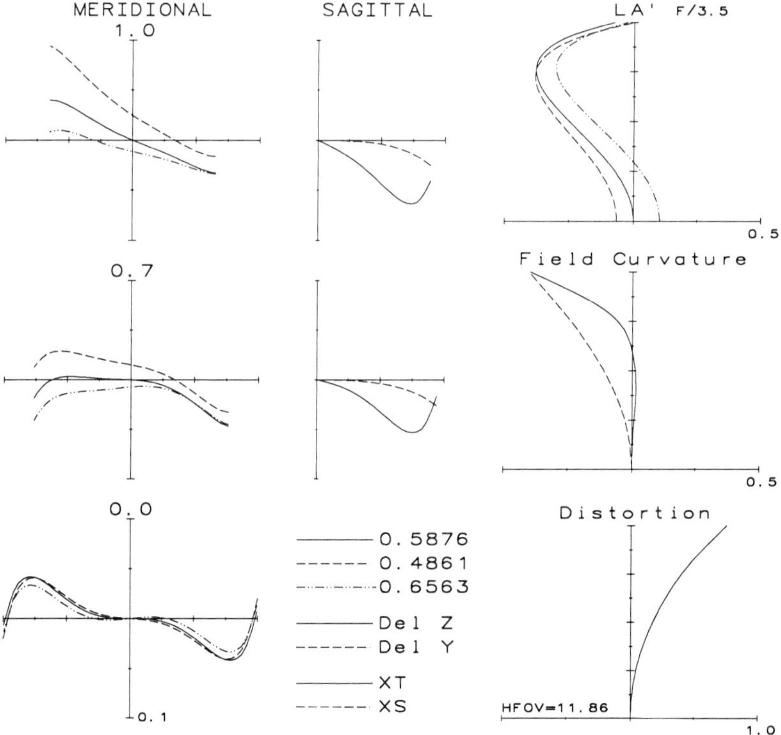

Figure 8.12

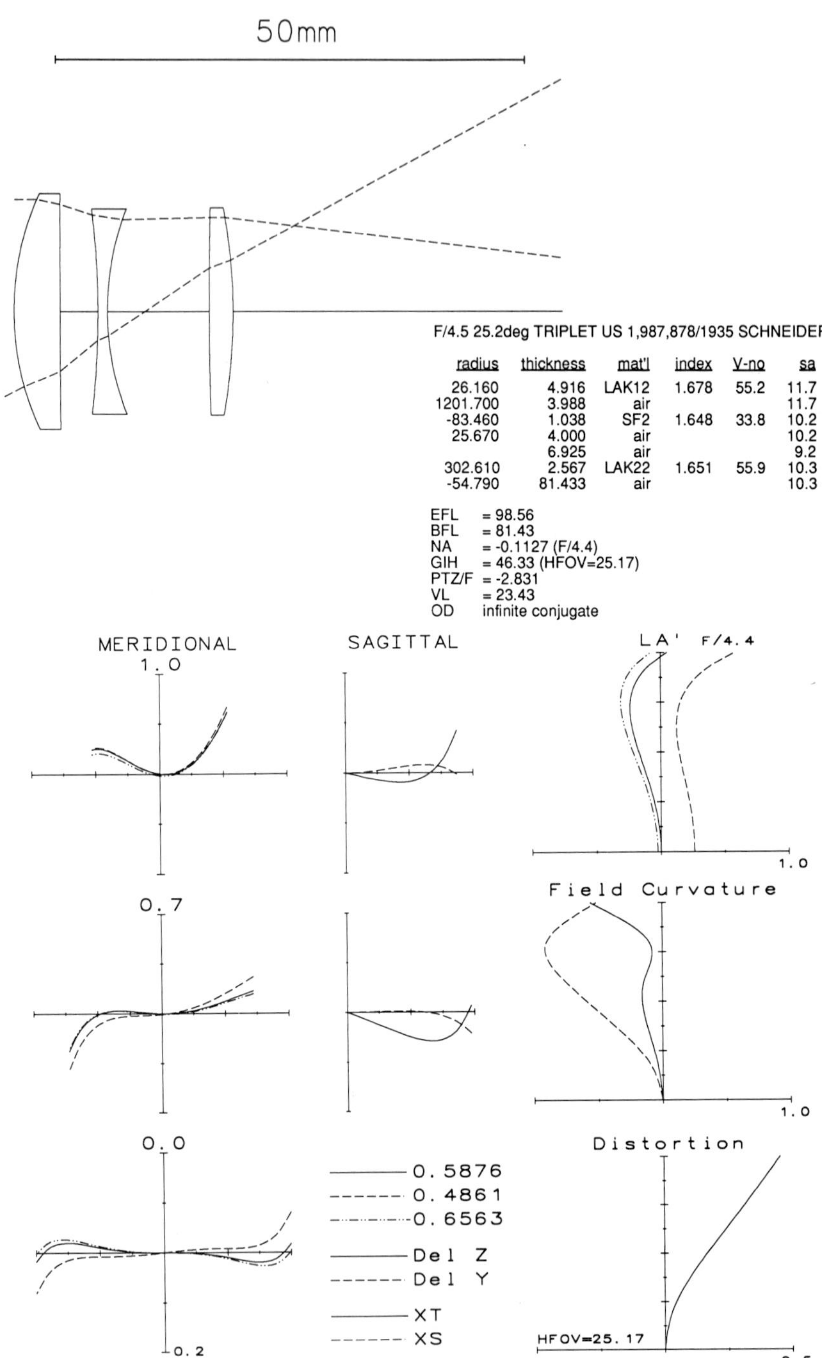

F/4.5 25.2deg TRIPLET US 1,987,878/1935 SCHNEIDER

radius	thickness	mat'l	index	V-no	sa
26.160	4.916	LAK12	1.678	55.2	11.7
1201.700	3.988	air			11.7
-83.460	1.038	SF2	1.648	33.8	10.2
25.670	4.000	air			10.2
	6.925	air			9.2
302.610	2.567	LAK22	1.651	55.9	10.3
-54.790	81.433	air			10.3

EFL = 98.56
BFL = 81.43
NA = -0.1127 (F/4.4)
GIH = 46.33 (HFOV=25.17)
PTZ/F = -2.831
VL = 23.43
OD infinite conjugate

Figure 8.13

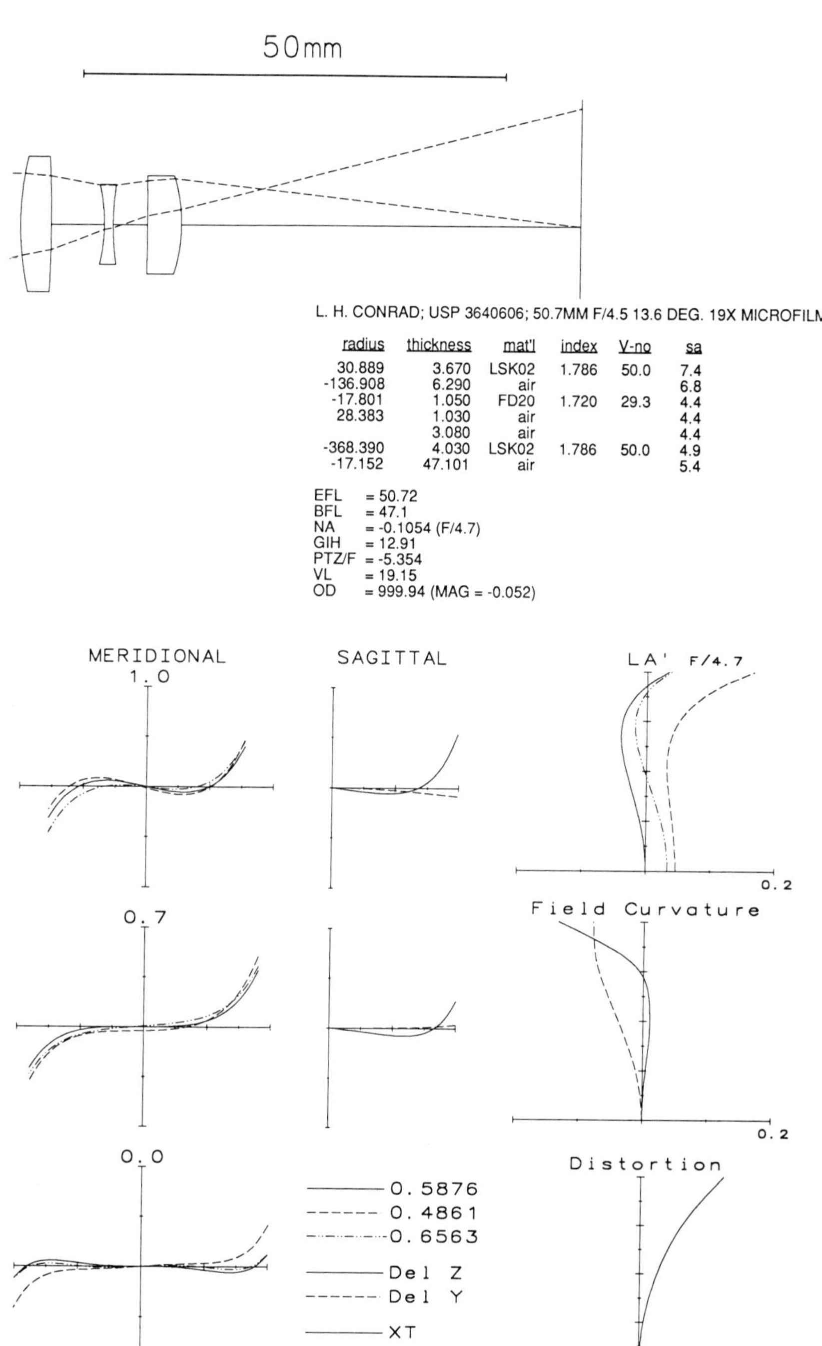

L. H. CONRAD; USP 3640606; 50.7MM F/4.5 13.6 DEG. 19X MICROFILM

radius	thickness	mat'l	index	V-no	sa
30.889	3.670	LSK02	1.786	50.0	7.4
-136.908	6.290	air			6.8
-17.801	1.050	FD20	1.720	29.3	4.4
28.383	1.030	air			4.4
	3.080	air			4.4
-368.390	4.030	LSK02	1.786	50.0	4.9
-17.152	47.101	air			5.4

EFL = 50.72
BFL = 47.1
NA = -0.1054 (F/4.7)
GIH = 12.91
PTZ/F = -5.354
VL = 19.15
OD = 999.94 (MAG = -0.052)

Figure 8.14

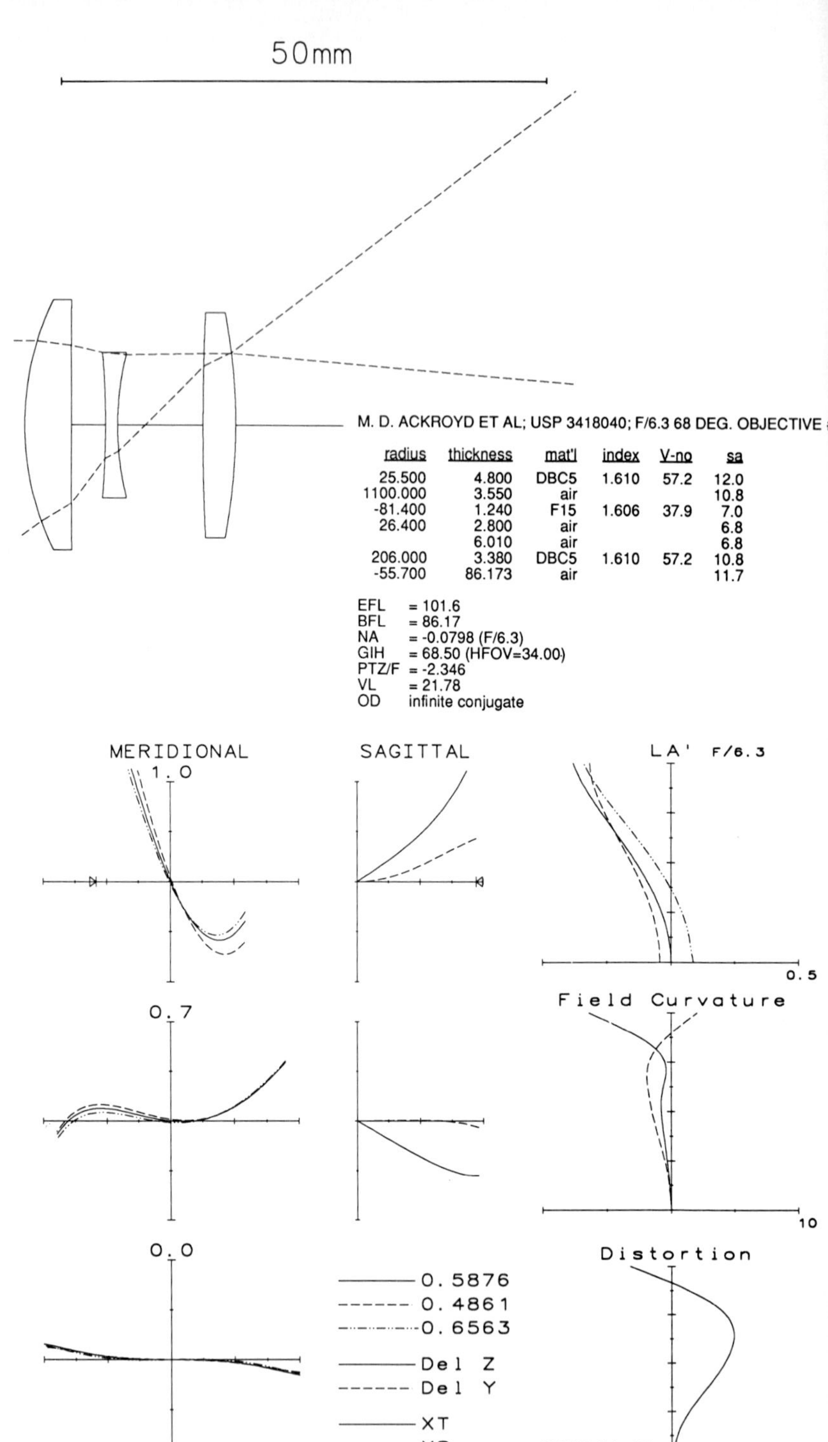

M. D. ACKROYD ET AL; USP 3418040; F/6.3 68 DEG. OBJECTIVE #

radius	thickness	mat'l	index	V-no	sa
25.500	4.800	DBC5	1.610	57.2	12.0
1100.000	3.550	air			10.8
-81.400	1.240	F15	1.606	37.9	7.0
26.400	2.800	air			6.8
	6.010	air			6.8
206.000	3.380	DBC5	1.610	57.2	10.8
-55.700	86.173	air			11.7

EFL = 101.6
BFL = 86.17
NA = -0.0798 (F/6.3)
GIH = 68.50 (HFOV=34.00)
PTZ/F = -2.346
VL = 21.78
OD infinite conjugate

Figure 8.15

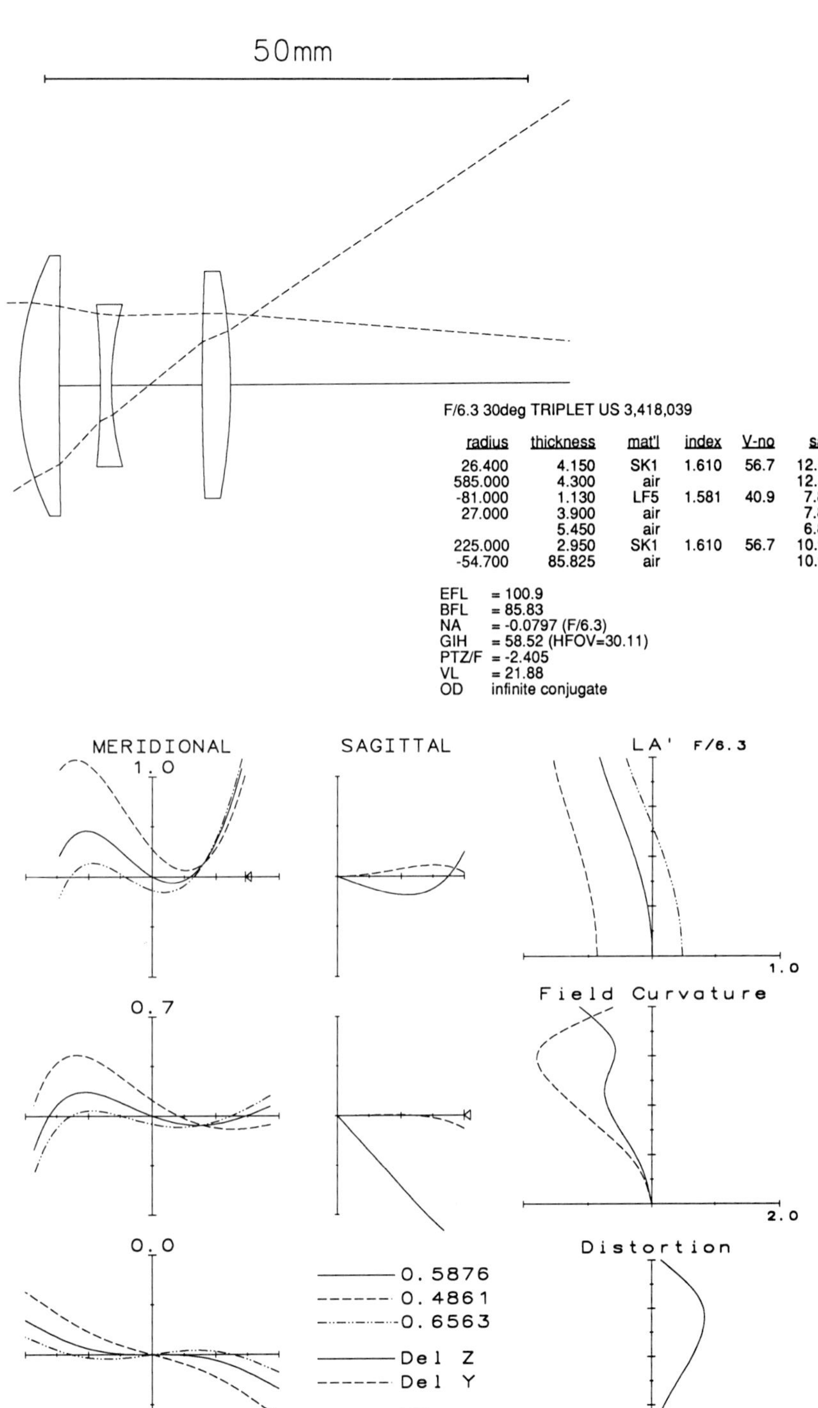

F/6.3 30deg TRIPLET US 3,418,039

radius	thickness	mat'l	index	V-no	sa
26.400	4.150	SK1	1.610	56.7	12.5
585.000	4.300	air			12.5
-81.000	1.130	LF5	1.581	40.9	7.8
27.000	3.900	air			7.8
	5.450	air			6.8
225.000	2.950	SK1	1.610	56.7	10.9
-54.700	85.825	air			10.9

EFL = 100.9
BFL = 85.83
NA = -0.0797 (F/6.3)
GIH = 58.52 (HFOV=30.11)
PTZ/F = -2.405
VL = 21.88
OD infinite conjugate

Figure 8.16

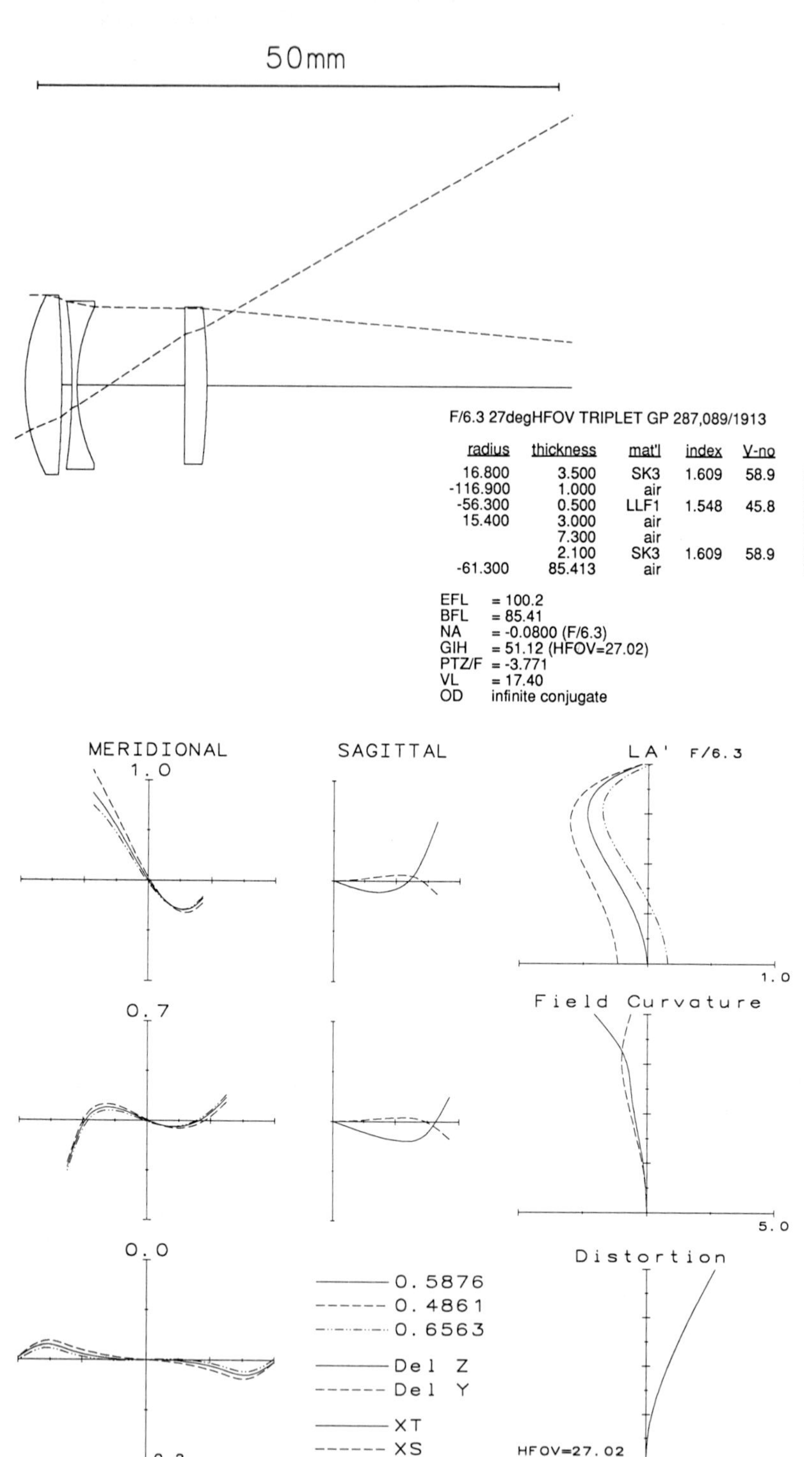

F/6.3 27degHFOV TRIPLET GP 287,089/1913

radius	thickness	mat'l	index	V-no	sa
16.800	3.500	SK3	1.609	58.9	8.0
-116.900	1.000	air			8.0
-56.300	0.500	LLF1	1.548	45.8	7.5
15.400	3.000	air			7.0
	7.300	air			7.0
	2.100	SK3	1.609	58.9	7.0
-61.300	85.413	air			7.0

EFL = 100.2
BFL = 85.41
NA = -0.0800 (F/6.3)
GIH = 51.12 (HFOV=27.02)
PTZ/F = -3.771
VL = 17.40
OD infinite conjugate

Figure 8.17

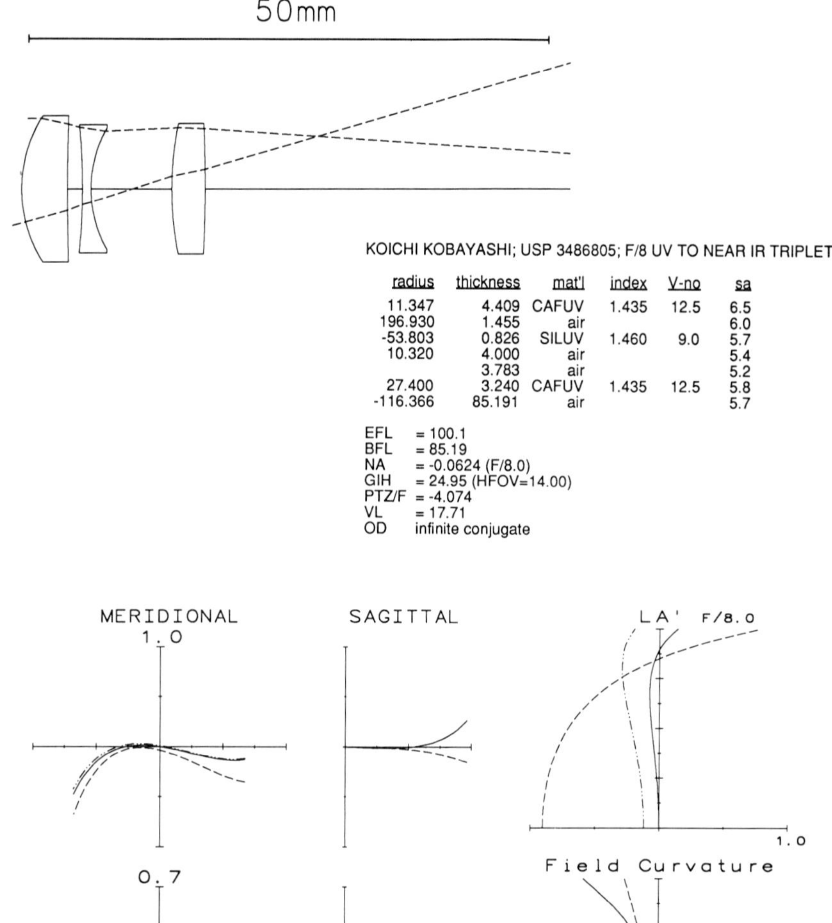

KOICHI KOBAYASHI; USP 3486805; F/8 UV TO NEAR IR TRIPLET

radius	thickness	mat'l	index	V-no	sa
11.347	4.409	CAFUV	1.435	12.5	6.5
196.930	1.455	air			6.0
-53.803	0.826	SILUV	1.460	9.0	5.7
10.320	4.000	air			5.4
	3.783	air			5.2
27.400	3.240	CAFUV	1.435	12.5	5.8
-116.366	85.191	air			5.7

EFL = 100.1
BFL = 85.19
NA = -0.0624 (F/8.0)
GIH = 24.95 (HFOV=14.00)
PTZ/F = -4.074
VL = 17.71
OD infinite conjugate

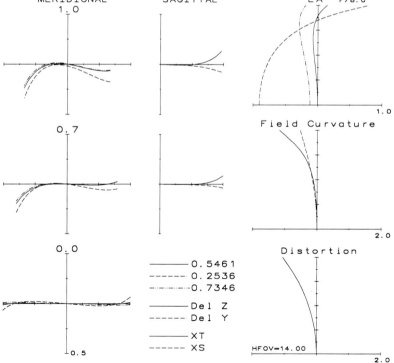

Figure 8.18

tremely well-corrected lens designed for microfilm work; note the very high index glasses and the fact that the field angle is about half that of Fig. 8.13; these factors combine to produce a level of correction which is well above that of any of the other lenses in this chapter. Note also that the front airspace is longer than the rear and that R3 is shorter than R4; this is the reverse of most Cooke triplets.

Figures 8.15, 8.16, and 8.17 are all at a speed of *f*/6.3, and all use similar crown glass types; all have similar levels of correction. Note that Figs. 8.15 and 8.16 have almost identical vertex lengths, but use quite different flint glasses; one result of this is that the higher *V* value of the Fig. 8.16 flint causes the lens to have undercorrected axial chromatic. The higher *V* value of the flint in Fig. 8.17 produces a shorter vertex length when the chromatic is corrected; the lower flint index and the reduced angular coverage produce a somewhat better level of correction than in the other two.

Figure 8.18 is an unusual design with calcium fluoride (CaF_2) crowns and a fused quartz (SiO_2) flint. Presumably these materials were chosen to produce a system which would transmit well into the ultraviolet wavelengths. The speed (*f*/8) and the half-field angle (14°) are quite modest and allow a very acceptable level of correction in spite of the extremely low index of all three elements. The very wide spectral bandwidth (0.25 to 0.73 μm) produces an unpleasantly large amount of spherochromatism. Note also the unusual feature of a reversed secondary spectrum, with the combined long- and short-wavelength focus falling closer to the lens than that of the middle wavelength.

Chapter

9

Reverse Telephoto (Retrofocus and Fish-Eye) Lenses

9.1 The Reverse Telephoto Principle

The arrangement of a positive component following a negative component, as shown in Fig. 9.1, has a (thin-lens) back focal length which is longer than the effective focal length. The necessary power and spacing arrangements can be determined from

$$F_A = \frac{(D - F)}{F - B} \tag{9.1}$$

$$F_B = \frac{-DB}{F - B - D} \tag{9.2}$$

An extremely long back focal length B is possible; however, the Petzval field then has a tendency to become strongly overcorrected (backward-curving). Obviously, the optimum arrangement (from a purely Petzval standpoint) is approximated by the one in which the focal lengths of the two components are nearly equal (and of opposite sign). This will occur when D is chosen so that $D = (F - B)^2/F$; then $F_A = F - B$ and $F_B = B - F$.

However, the retrofocus construction typically departs somewhat from this arrangement, and, as indicated above, the usual problem is a backward-curving Petzval surface. Thus ordinary glasses (low-index crowns and high-index flints) are appropriate for the positive rear component, because this combination increases its negative, inward-curving contribution to the Petzval sum. For the negative front component, the crown glasses (which are used for the negative elements) should be high-index and the flint glasses (in the positive elements)

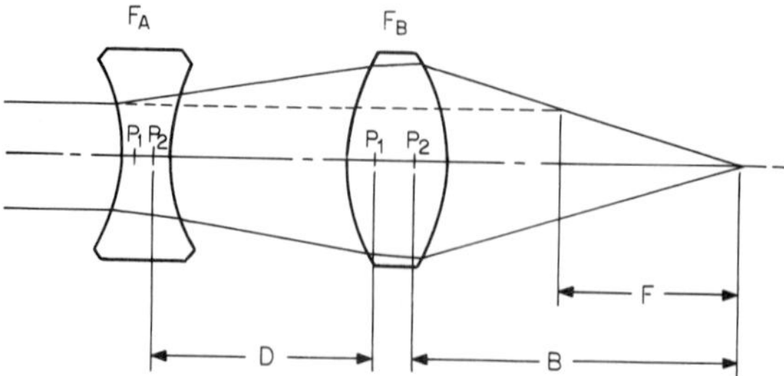

Figure 9.1 The basic power arrangement for a reversed telephoto lens yields a back focal length which is longer than the effective focal length.

should be low-index; this reduces the overcorrected Petzval contribution from this component.

It is apparent that there is absolutely no hope of any semblance of symmetry in this type of lens. Thus we can expect the correction of coma, distortion, and lateral color to be difficult. In many other design types, these aberrations are reduced or corrected by an approximately symmetrical arrangement of the elements about the aperture stop. In order to correct both the axial and lateral chromatic of the retrofocus, both components must be individually achromatized. (Although in a system which has only a modestly extended back focus, the front negative component may be weak enough that a high-V-value crown singlet will be acceptable.) The simplest *fully* corrected form is thus a pair of achromatic components. Usually the aperture stop is at the rear component; the natural shape for the front negative achromat is then that of a meniscus, concave toward the stop.

In an unsymmetrical system, as noted above, coma, distortion, and lateral color are difficult to control. Of these, only distortion does not affect the image definition. If it can be tolerated, allowing a few percent distortion will sometimes permit a better level of correction for all the other aberrations. Thus it is often worthwhile in the course of the optimization process to greatly reduce the weight on distortion in the merit function to see if an overall improvement will result.

9.2 The Basic Retrofocus Lens

Figure 9.2 is a relatively simple retrofocus design with an airspaced doublet achromatic front and a rear component which is a split-rear-crown triplet form. Designed as a single-frame 35-mm projection lens, it covers a modest 37° field at a speed of *f*/3. Note that the glass types are all rather ordinary. As is true of most retrofocus designs, this lens

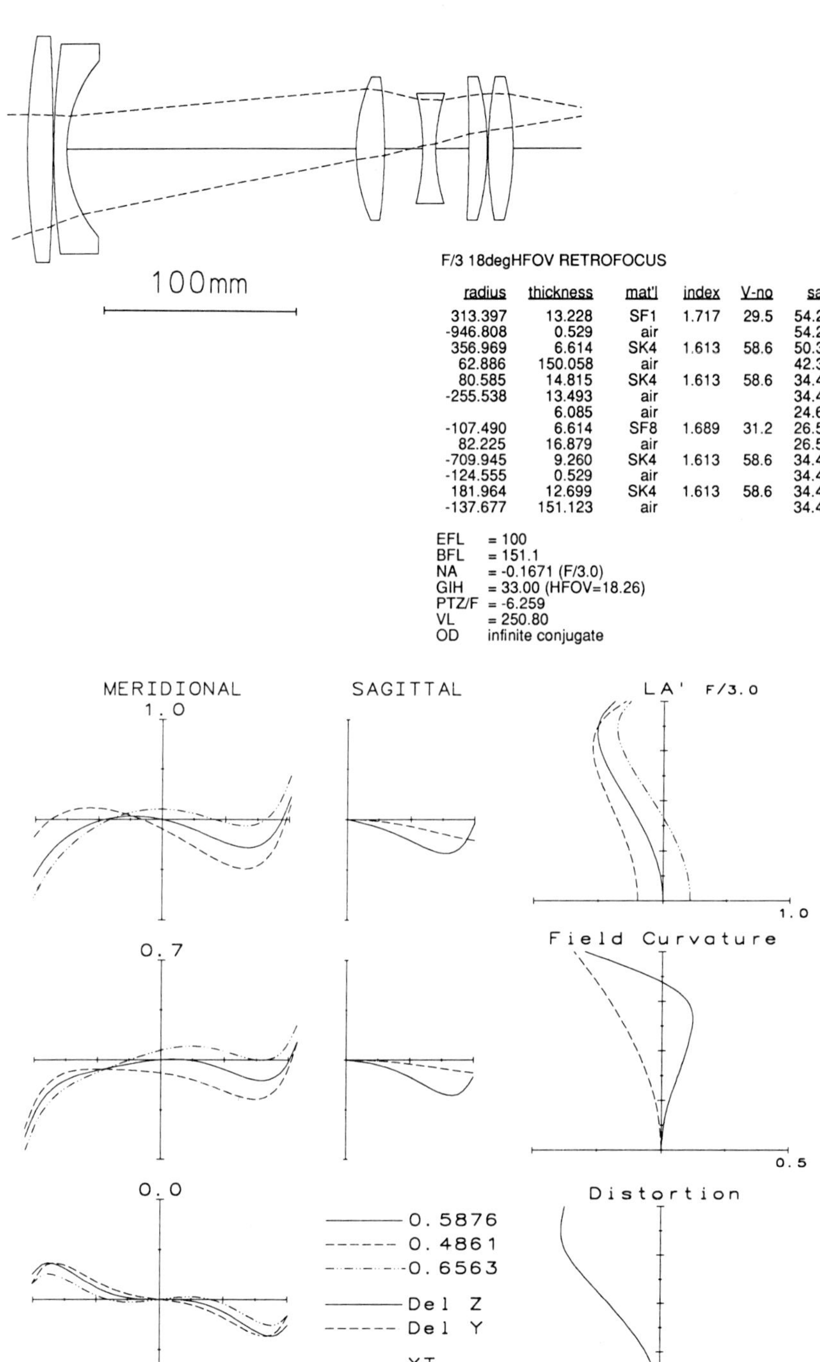

F/3 18degHFOV RETROFOCUS

radius	thickness	mat'l	index	V-no	sa
313.397	13.228	SF1	1.717	29.5	54.2
-946.808	0.529	air			54.2
356.969	6.614	SK4	1.613	58.6	50.3
62.886	150.058	air			42.3
80.585	14.815	SK4	1.613	58.6	34.4
-255.538	13.493	air			34.4
	6.085	air			24.6
-107.490	6.614	SF8	1.689	31.2	26.5
82.225	16.879	air			26.5
-709.945	9.260	SK4	1.613	58.6	34.4
-124.555	0.529	air			34.4
181.964	12.699	SK4	1.613	58.6	34.4
-137.677	151.123	air			34.4

EFL = 100
BFL = 151.1
NA = -0.1671 (F/3.0)
GIH = 33.00 (HFOV=18.26)
PTZ/F = -6.259
VL = 250.80
OD infinite conjugate

Figure 9.2

is quite sensitive to changes in object distance, and the design should be modified if it is to be used at a short conjugate distance. In this particular case, shortening both the second and third airspaces can rebalance the lens reasonably well for use at close object distances.

Figures 9.3 and 9.4 are two philosophically similar lenses which are quite well corrected for high speeds (*f*/1.2) and relatively modest fields. Note the use of a significant airspace in the negative front component. The rear components are almost a simple stack of positive power, although both designs make use of very similar constructions for the three elements nearest the short conjugate to increase the speed. The negligible back focus distance which accompanies this is very unusual for a retrofocus lens. The red-blue lateral color is corrected, but there is a quite noticeable amount of secondary spectrum of lateral color apparent in both lenses.

Figures 9.5 and 9.6 each have seven elements, but are totally different designs. The front doublet of Fig. 9.5 has reversed glass types; the usual arrangement of glass in a negative achromatic doublet is for the negative element to be made of a low-dispersion, high-*V*-value (crown) glass and the positive element to be made of a high-dispersion, low-*V*-value (flint) glass. Figure 9.6 covers a wider field (64° versus 42°) at a higher speed (*f*/2.0 versus *f*/2.8); for this reason, its aberration correction is not as good and its back focus is significantly shorter. Note that both designs have fronts which are almost afocal (the first three elements in Fig. 9.5 and the first four in Fig. 9.6).

The rear component of Fig. 9.7 is quite similar to that of Fig. 9.6, but the front negative component of Fig. 9.7 (and also that of Fig. 9.8) is a singlet of crown glass. These simple constructions achieve angular coverages of 64° and 74° at a speed of *f*/2.8 with back focal lengths which exceed their focal lengths by 10 and 20 percent, respectively.

Fields of 80° to 90° are covered by Figs. 9.9 to 9.12. Note the similarity of the construction of the first four elements of Figs. 9.9, 9.10, and 9.12, although the higher speeds of Figs. 9.10 and 9.12 require a more complex configuration to correct the aberrations over the larger (*f*/2.0 versus *f*/2.8) aperture. Figure 9.11 manages to cover a surprising 90° total field using only a singlet negative front.

9.3 Fish-Eye, or Extreme Wide-Angle Reverse Telephoto, Lenses

When the reverse telephoto form is carried to extremes, a total field of 180° or more can be covered. In such a system, the front elements are very strongly bent negative meniscus elements, concave toward the image and the aperture stop. These elements have very heavily overcorrected spherical aberration of the pupil; this serves to deviate

(*Text continues on page 161.*)

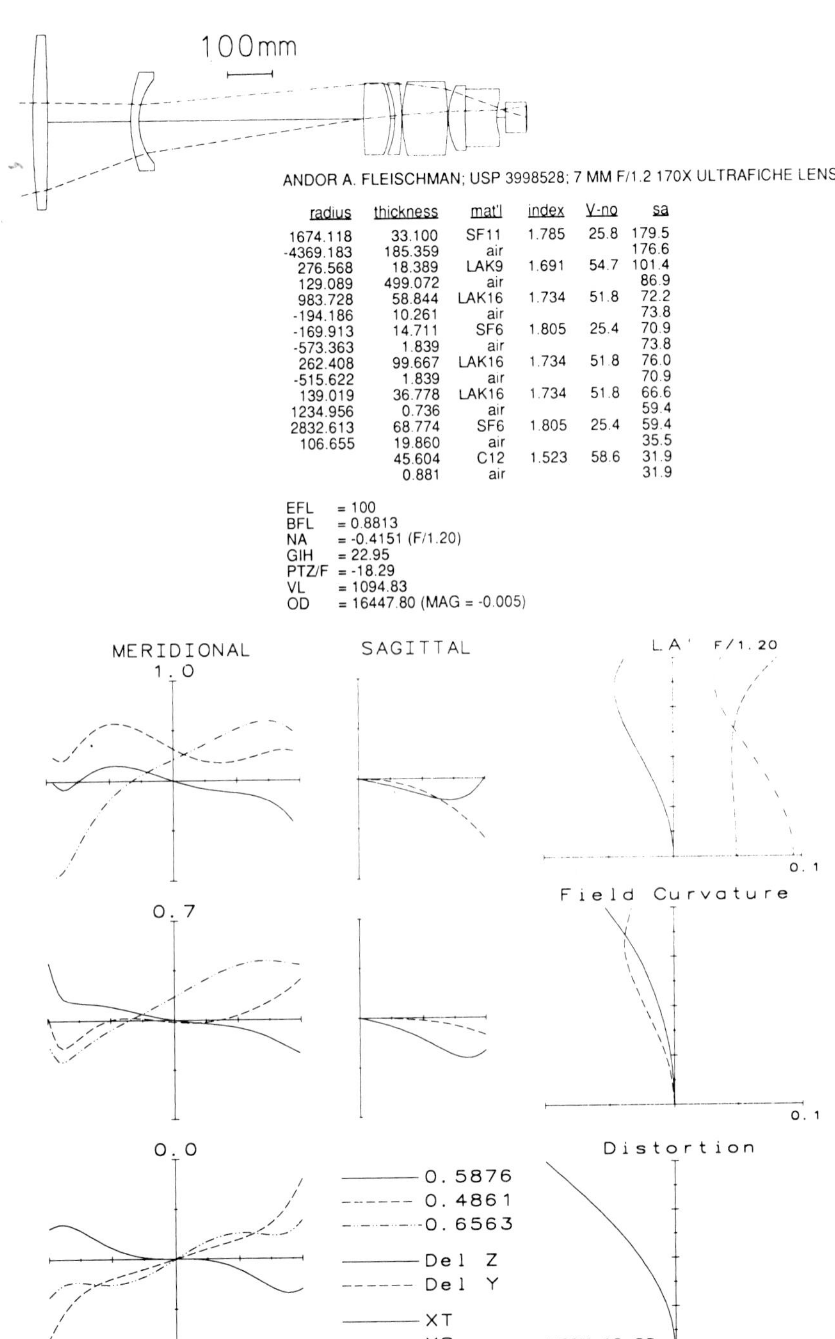

ANDOR A. FLEISCHMAN; USP 3998528; 7 MM F/1.2 170X ULTRAFICHE LENS

radius	thickness	mat'l	index	V-no	sa
1674.118	33.100	SF11	1.785	25.8	179.5
-4369.183	185.359	air			176.6
276.568	18.389	LAK9	1.691	54.7	101.4
129.089	499.072	air			86.9
983.728	58.844	LAK16	1.734	51.8	72.2
-194.186	10.261	air			73.8
-169.913	14.711	SF6	1.805	25.4	70.9
-573.363	1.839	air			73.8
262.408	99.667	LAK16	1.734	51.8	76.0
-515.622	1.839	air			70.9
139.019	36.778	LAK16	1.734	51.8	66.6
1234.956	0.736	air			59.4
2832.613	68.774	SF6	1.805	25.4	59.4
106.655	19.860	air			35.5
	45.604	C12	1.523	58.6	31.9
	0.881	air			31.9

EFL = 100
BFL = 0.8813
NA = -0.4151 (F/1.20)
GIH = 22.95
PTZ/F = -18.29
VL = 1094.83
OD = 16447.80 (MAG = -0.005)

Figure 9.3

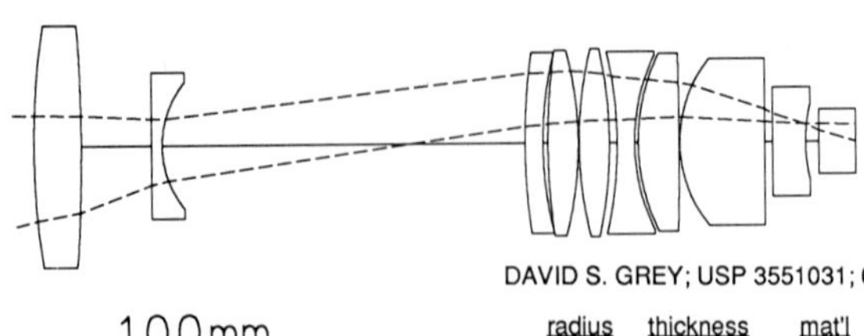

DAVID S. GREY; USP 3551031; 6.17 MM F/1.2 150X PROJECTION LENS

radius	thickness	mat'l	index	V-no	sa
965.187	67.529	SF01	1.720	29.3	158.0
-2016.769	99.425	air			147.0
	14.873	UBK7	1.517	64.3	95.4
124.807	347.488	air			82.5
	161.667	air			72.7
808.499	25.058	SF01	1.720	29.3	114.8
460.979	7.162	air			118.0
821.238	42.163	LAK9	1.691	54.8	118.0
-445.265	1.487	air			121.0
484.065	42.034	LAK9	1.691	54.8	122.5
-537.027	10.961	air			122.5
-494.525	24.961	SF01	1.720	29.3	119.0
242.566	5.254	air			114.0
242.566	58.265	LAK9	1.691	54.8	116.0
-4390.080	1.083	air			116.0
163.316	119.327	LAK9	1.691	54.8	109.0
	10.848	air			79.0
	46.043	SF01	1.720	29.3	71.0
199.514	18.700	air			49.0
	50.408	C12	1.523	58.6	42.0
	0.044	air			42.0

EFL = 100
BFL = 0.04442
NA = -0.4164 (F/1.20)
GIH = 23.08
PTZ/F = -66.51
VL = 1154.74
OD = 14296.25 (MAG = -0.006)

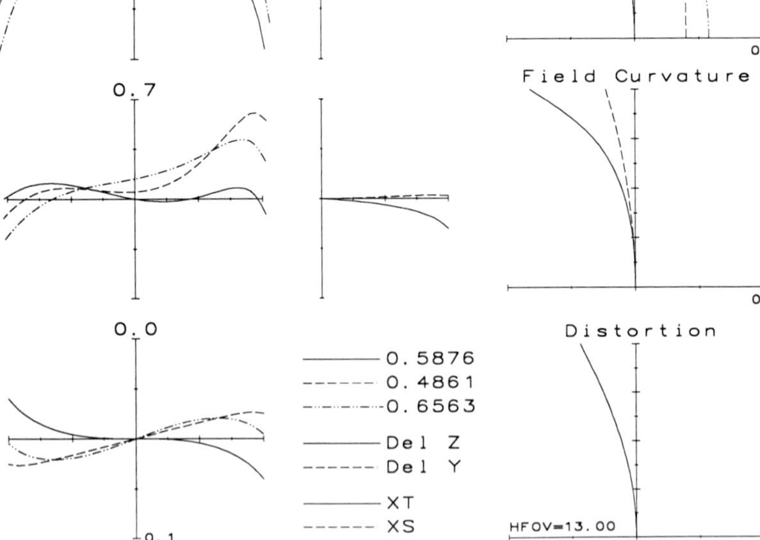

Figure 9.4

100mm

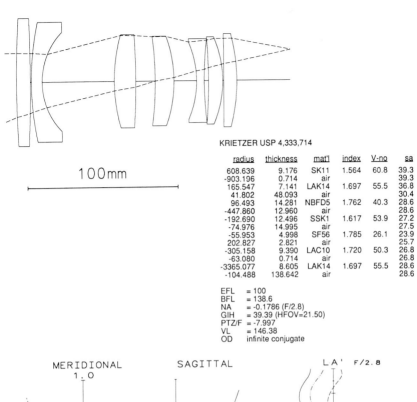

KRIETZER USP 4,333,714

radius	thickness	mat'l	index	V-no	sa
608.639	9.176	SK11	1.564	60.8	39.3
-903.196	0.714	air			39.3
165.547	7.141	LAK14	1.697	55.5	36.8
41.802	48.093	air			30.4
96.493	14.281	NBFD5	1.762	40.3	28.6
-447.860	12.960	air			28.6
-192.690	12.496	SSK1	1.617	53.9	27.2
-74.976	14.995	air			27.5
-55.953	4.998	SF56	1.785	26.1	23.9
202.827	2.821	air			25.7
-305.158	9.390	LAC10	1.720	50.3	26.8
-63.080	0.714	air			26.8
-3365.077	8.605	LAK14	1.697	55.5	28.6
-104.488	138.642	air			28.6

EFL = 100
BFL = 138.6
NA = -0.1786 (F/2.8)
GIH = 39.39 (HFOV=21.50)
PTZ/F = -7.997
VL = 146.38
OD infinite conjugate

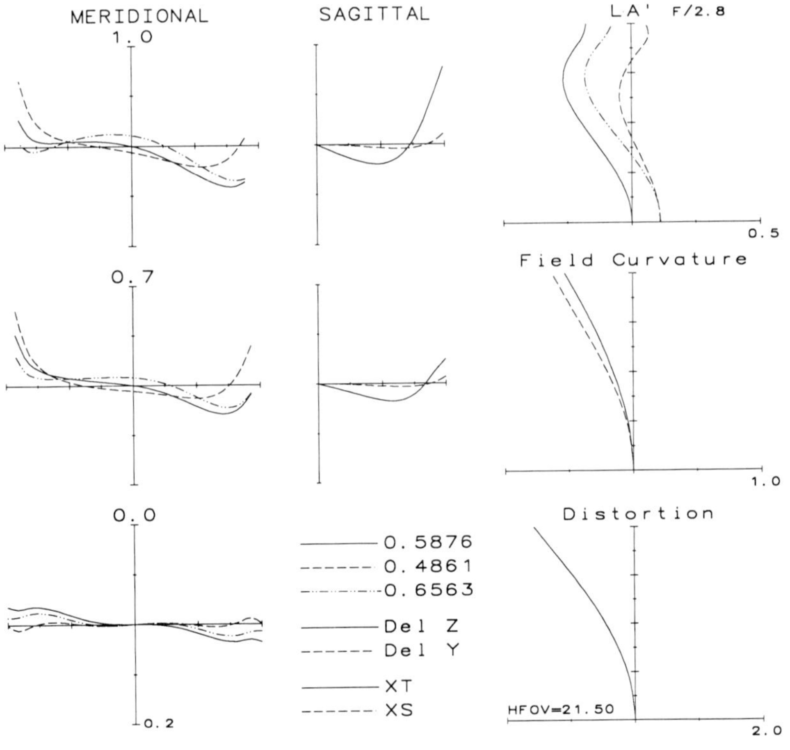

MERIDIONAL
SAGITTAL
1.0
0.7
0.0
0.2
LA' F/2.8
0.5
Field Curvature
1.0
Distortion
HFOV=21.50
2.0
0.5876
0.4861
0.6563
Del Z
Del Y
XT
XS

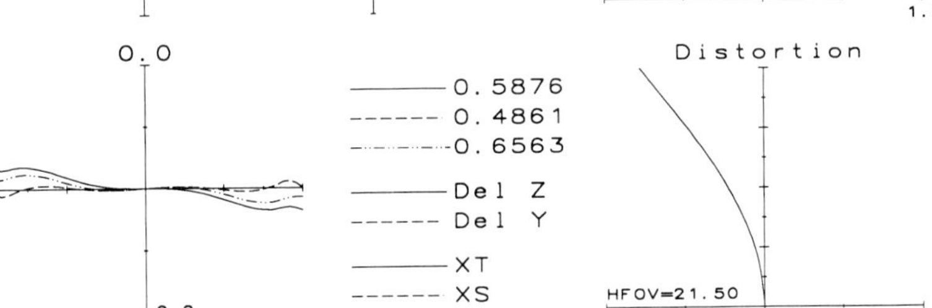

Figure 9.5

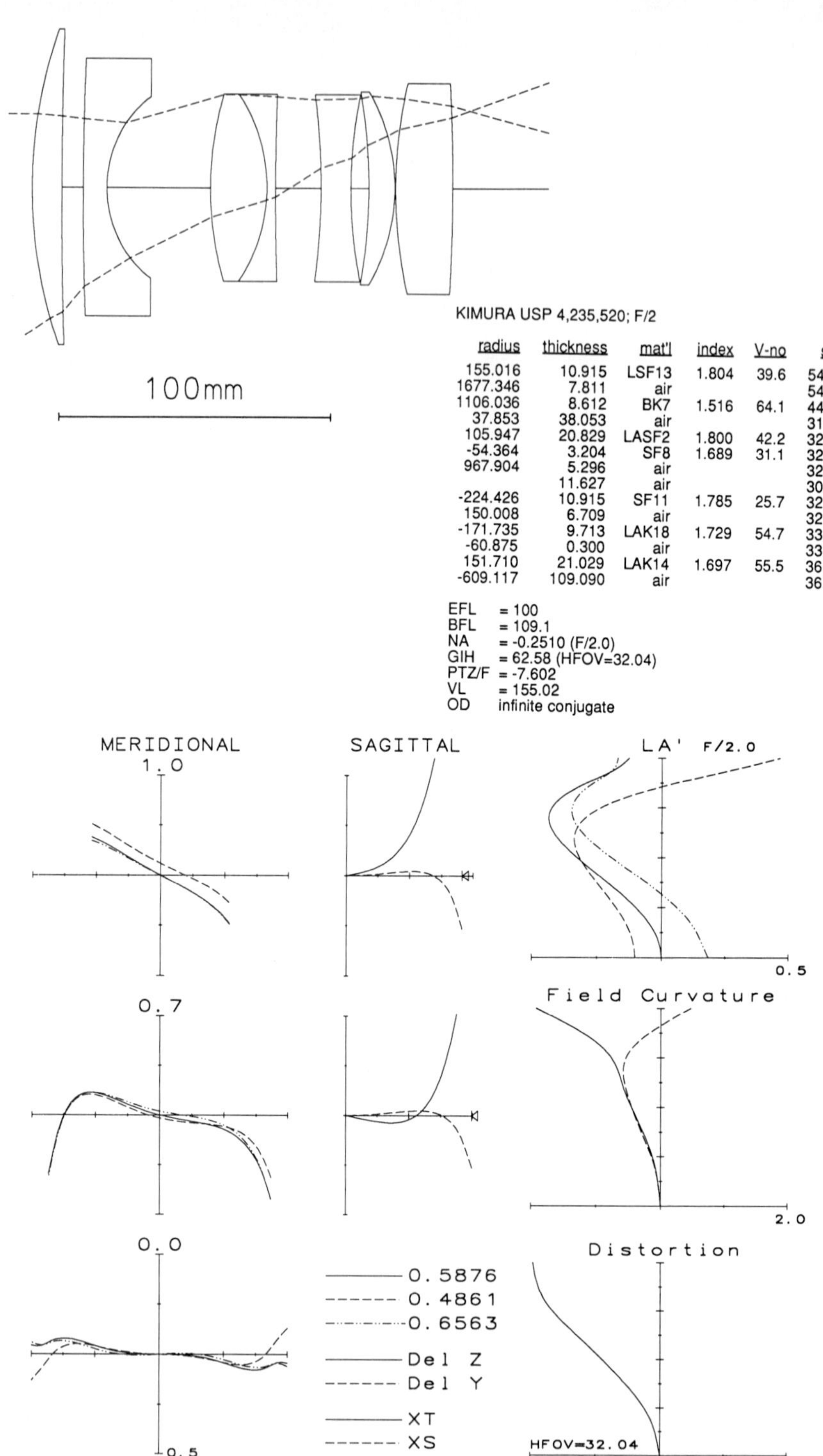

KIMURA USP 4,235,520; F/2

radius	thickness	mat'l	index	V-no	sa
155.016	10.915	LSF13	1.804	39.6	54.0
1677.346	7.811	air			54.0
1106.036	8.612	BK7	1.516	64.1	44.0
37.853	38.053	air			31.0
105.947	20.829	LASF2	1.800	42.2	32.0
-54.364	3.204	SF8	1.689	31.1	32.0
967.904	5.296	air			32.0
	11.627	air			30.8
-224.426	10.915	SF11	1.785	25.7	32.0
150.008	6.709	air			32.0
-171.735	9.713	LAK18	1.729	54.7	33.0
-60.875	0.300	air			33.0
151.710	21.029	LAK14	1.697	55.5	36.0
-609.117	109.090	air			36.0

EFL = 100
BFL = 109.1
NA = -0.2510 (F/2.0)
GIH = 62.58 (HFOV=32.04)
PTZ/F = -7.602
VL = 155.02
OD infinite conjugate

Figure 9.6

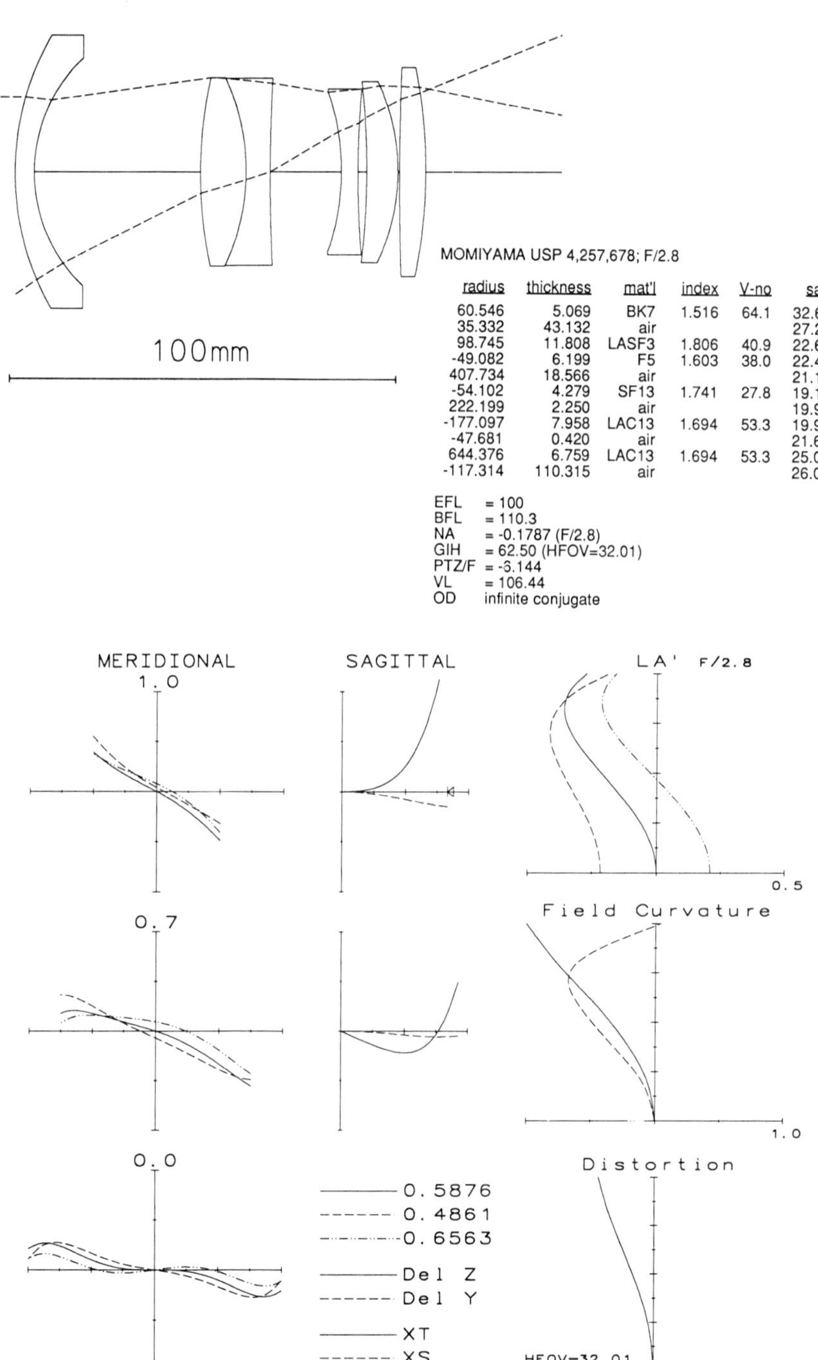

MOMIYAMA USP 4,257,678; F/2.8

radius	thickness	mat'l	index	V-no	sa
60.546	5.069	BK7	1.516	64.1	32.6
35.332	43.132	air			27.2
98.745	11.808	LASF3	1.806	40.9	22.6
-49.082	6.199	F5	1.603	38.0	22.4
407.734	18.566	air			21.1
-54.102	4.279	SF13	1.741	27.8	19.1
222.199	2.250	air			19.9
-177.097	7.958	LAC13	1.694	53.3	19.9
-47.681	0.420	air			21.6
644.376	6.759	LAC13	1.694	53.3	25.0
-117.314	110.315	air			26.0

EFL = 100
BFL = 110.3
NA = -0.1787 (F/2.8)
GIH = 62.50 (HFOV=32.01)
PTZ/F = -3.144
VL = 106.44
OD infinite conjugate

Figure 9.7

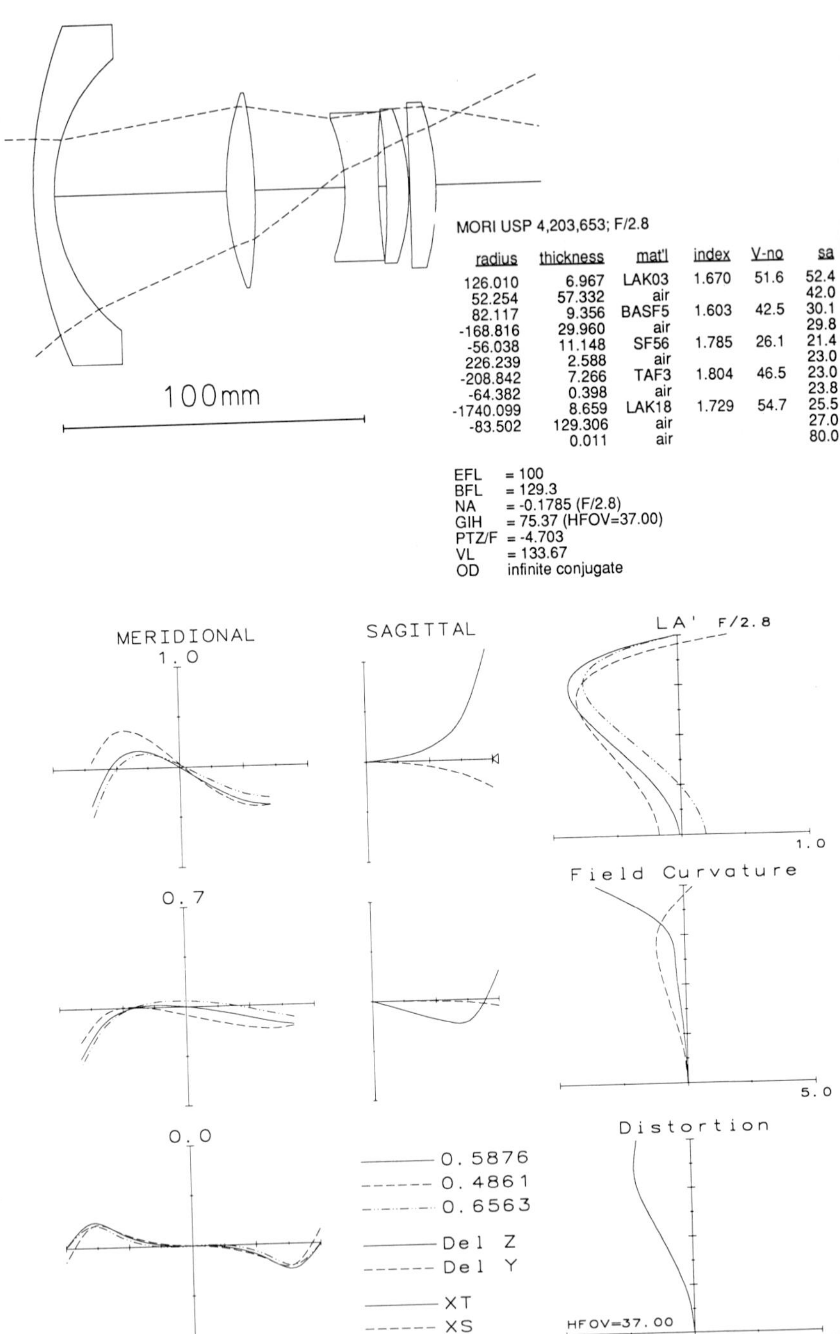

MORI USP 4,203,653; F/2.8

radius	thickness	mat'l	index	V-no	sa
126.010	6.967	LAK03	1.670	51.6	52.4
52.254	57.332	air			42.0
82.117	9.356	BASF5	1.603	42.5	30.1
-168.816	29.960	air			29.8
-56.038	11.148	SF56	1.785	26.1	21.4
226.239	2.588	air			23.0
-208.842	7.266	TAF3	1.804	46.5	23.0
-64.382	0.398	air			23.8
-1740.099	8.659	LAK18	1.729	54.7	25.5
-83.502	129.306	air			27.0
	0.011	air			80.0

EFL = 100
BFL = 129.3
NA = -0.1785 (F/2.8)
GIH = 75.37 (HFOV=37.00)
PTZ/F = -4.703
VL = 133.67
OD infinite conjugate

Figure 9.8

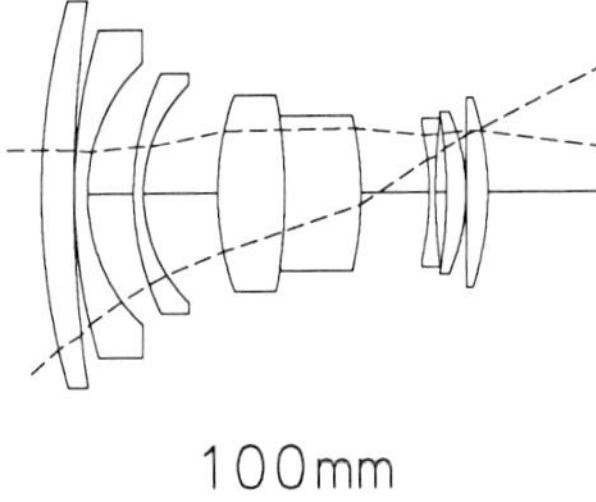

IKUO MORI; USP 3635546; F/4 80 DEG. WIDE ANGLE OBJECTIVE #1

radius	thickness	mat'l	index	V-no	sa
180.000	9.750	LAKN7	1.652	58.5	53.0
339.300	0.180	air			49.0
140.970	3.970	LAKN7	1.652	58.5	45.0
47.830	13.720	air			36.0
71.300	2.890	LAKN7	1.652	58.5	33.0
40.250	21.840	air			30.0
76.610	19.860	BAF10	1.670	47.2	27.0
-144.400	22.380	LACL4	1.670	51.7	21.3
-92.600	5.400	air			17.9
	15.000	air			17.2
-75.270	1.800	SF4	1.755	27.6	18.5
161.190	3.250	air			20.5
-124.000	5.780	SK2	1.607	56.7	20.7
-52.710	0.180	air			22.3
6317.000	6.500	LAKN6	1.642	58.0	26.1
-76.790	132.775	air			26.7

EFL = 97.35
BFL = 132.8
NA = -0.1255 (F/4.0)
GIH = 81.68 (HFOV=40.00)
PTZ/F = -5.842
VL = 132.50
OD infinite conjugate

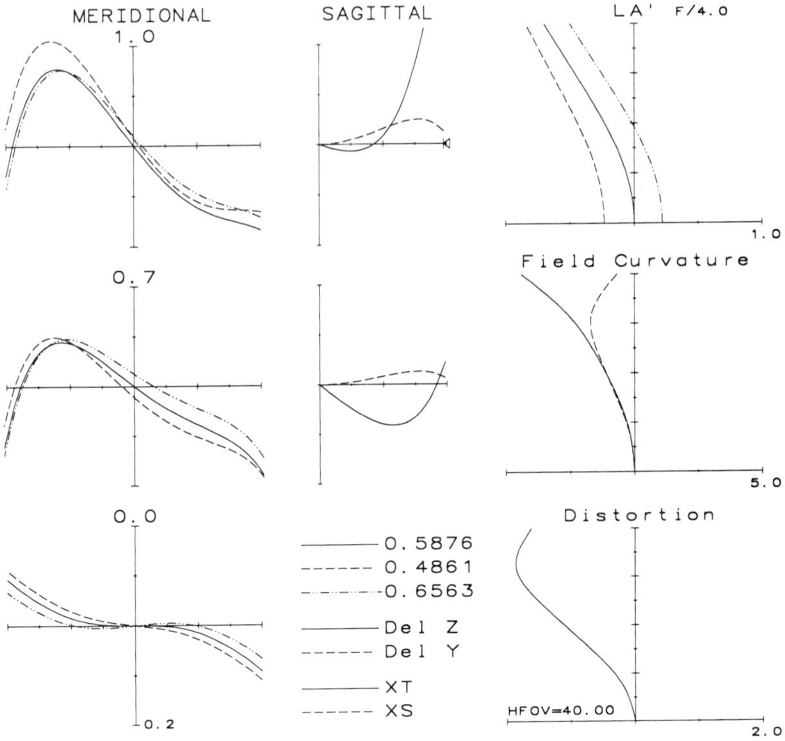

MERIDIONAL 1.0 SAGITTAL LA' F/4.0 1.0
0.7 Field Curvature 5.0
0.0 0.2 Distortion HFOV=40.00 2.0

0.5876
0.4861
0.6563
Del Z
Del Y
XT
XS

Figure 9.9

100mm

EFL = 100.2
BFL = 157.8
NA = -0.2494 (F/2.0)
GIH = 90.25 (HFOV=42.00)
PTZ/F = -7.945
VL = 228.37
OD infinite conjugate

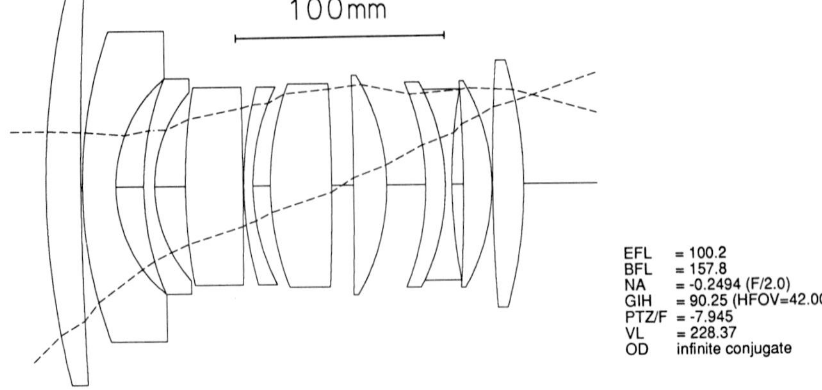

T. TSUNASHIMA; USP 4163603; F/2 84 DEG. WIDE ANGLE OBJECTIVE #1

radius	thickness	mat'l	index	V-no	sa
346.939	16.610	LACL7	1.670	57.3	88.0
1632.653	0.780	air			83.0
193.878	15.840	TAF1	1.773	49.6	69.0
60.943	13.270	air			48.0
120.408	5.140	LAK14	1.697	55.5	48.0
60.310	14.650	air			42.0
239.559	27.710	BAFD6	1.668	41.9	40.0
-2487.845	0.410	air			44.0
152.959	4.370	LAK14	1.697	55.5	44.0
91.073	8.490	air			44.0
121.633	29.390	LAF7	1.749	35.0	44.2
-906.759	5.000	air			45.3
	5.490	air			43.6
-1791.461	15.840	BCD11	1.564	60.8	48.9
-90.808	18.610	air			47.0
-112.057	9.510	LAF10	1.743	49.3	45.7
-91.073	3.180	FD6	1.805	25.5	42.5
220.816	5.550	air			42.5
-676.045	13.470	LAF10	1.743	49.3	46.0
-87.347	0.410	air			46.0
808.163	14.650	LAK14	1.697	55.5	55.0
-190.857	157.794	air			55.0

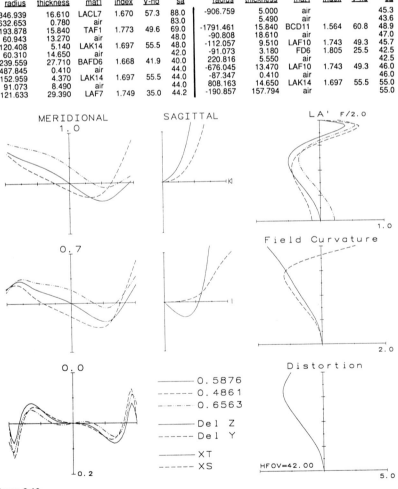

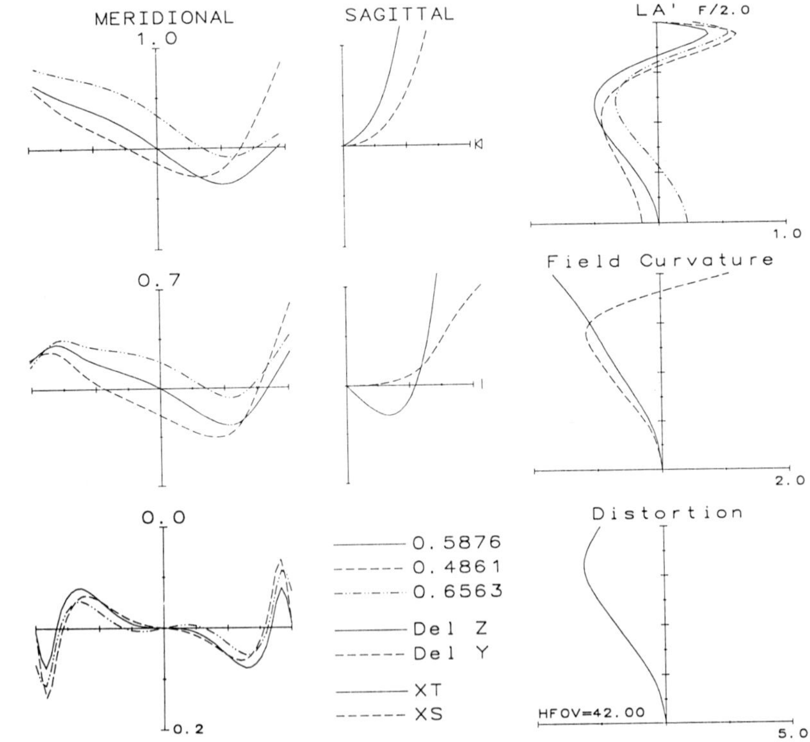

MERIDIONAL
1.0
SAGITTAL
LA' F/2.0
1.0
0.7
Field Curvature
2.0
0.0
0.2
0.5876
0.4861
0.6563
Del Z
Del Y
XT
XS
Distortion
HFOV=42.00
5.0

Figure 9.10

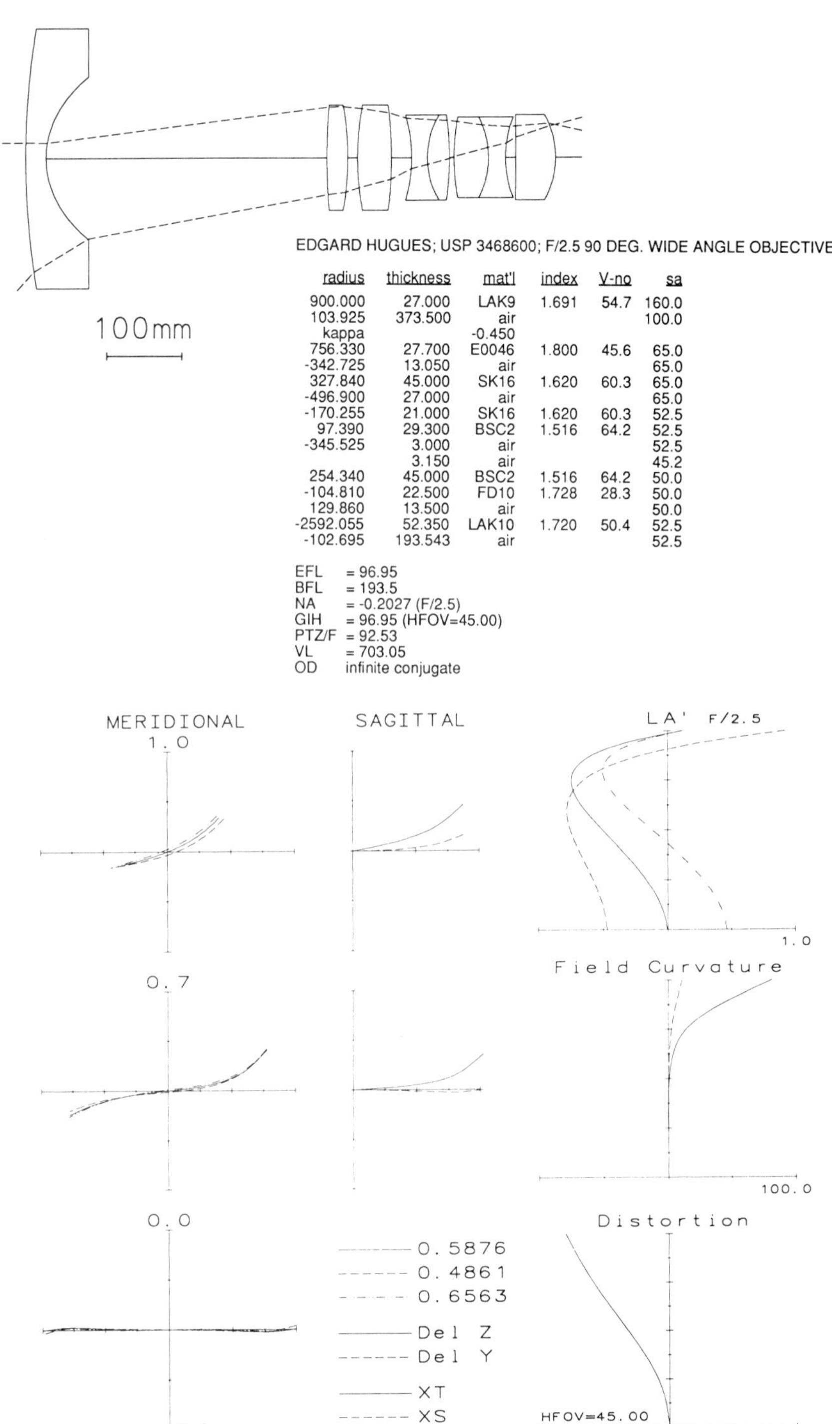

EDGARD HUGUES; USP 3468600; F/2.5 90 DEG. WIDE ANGLE OBJECTIVE

radius	thickness	mat'l	index	V-no	sa
900.000	27.000	LAK9	1.691	54.7	160.0
103.925	373.500	air			100.0
kappa		-0.450			
756.330	27.700	E0046	1.800	45.6	65.0
-342.725	13.050	air			65.0
327.840	45.000	SK16	1.620	60.3	65.0
-496.900	27.000	air			65.0
-170.255	21.000	SK16	1.620	60.3	52.5
97.390	29.300	BSC2	1.516	64.2	52.5
-345.525	3.000	air			52.5
	3.150	air			45.2
254.340	45.000	BSC2	1.516	64.2	50.0
-104.810	22.500	FD10	1.728	28.3	50.0
129.860	13.500	air			50.0
-2592.055	52.350	LAK10	1.720	50.4	52.5
-102.695	193.543	air			52.5

EFL = 96.95
BFL = 193.5
NA = -0.2027 (F/2.5)
GIH = 96.95 (HFOV=45.00)
PTZ/F = 92.53
VL = 703.05
OD infinite conjugate

Figure 9.11

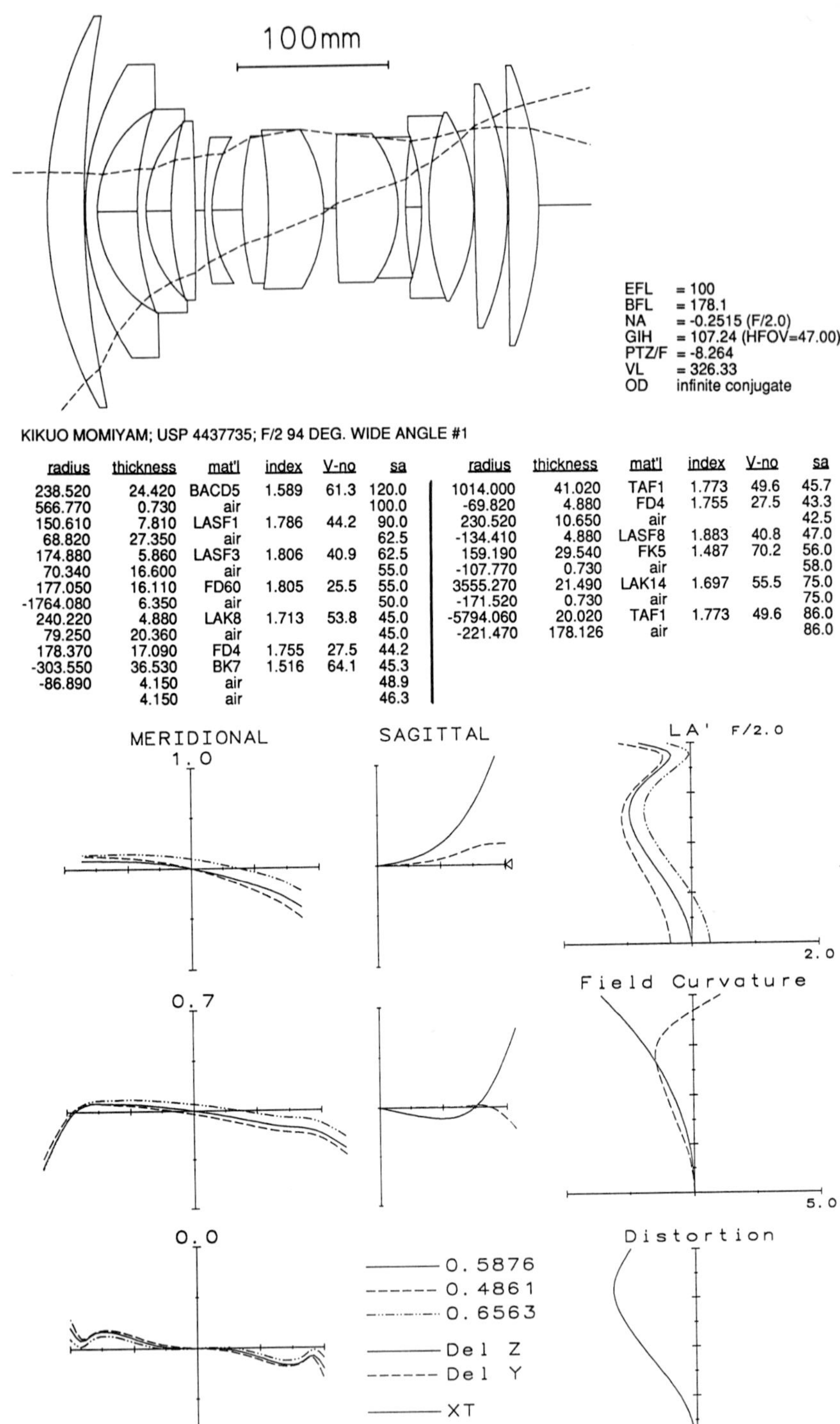

KIKUO MOMIYAM; USP 4437735; F/2 94 DEG. WIDE ANGLE #1

radius	thickness	mat'l	index	V-no	sa
238.520	24.420	BACD5	1.589	61.3	120.0
566.770	0.730	air			100.0
150.610	7.810	LASF1	1.786	44.2	90.0
68.820	27.350	air			62.5
174.880	5.860	LASF3	1.806	40.9	62.5
70.340	16.600	air			55.0
177.050	16.110	FD60	1.805	25.5	55.0
-1764.080	6.350	air			50.0
240.220	4.880	LAK8	1.713	53.8	45.0
79.250	20.360	air			45.0
178.370	17.090	FD4	1.755	27.5	44.2
-303.550	36.530	BK7	1.516	64.1	45.3
-86.890	4.150	air			48.9
	4.150	air			46.3
1014.000	41.020	TAF1	1.773	49.6	45.7
-69.820	4.880	FD4	1.755	27.5	43.3
230.520	10.650	air			42.5
-134.410	4.880	LASF8	1.883	40.8	47.0
159.190	29.540	FK5	1.487	70.2	56.0
-107.770	0.730	air			58.0
3555.270	21.490	LAK14	1.697	55.5	75.0
-171.520	0.730	air			75.0
-5794.060	20.020	TAF1	1.773	49.6	86.0
-221.470	178.126	air			86.0

Figure 9.12

the high-obliquity principal rays through a large angle, thereby directing them into the stop. In the process, a large amount of barrel distortion is introduced. It is, of course, apparent that if a field of 180° or more is to be imaged on a finite-sized flat surface, distortion is inevitable. In an extreme wide-angle lens such as this, the focal length ceases to have its usual meaning of $H' = f \tan \theta$, because the large distortion negates this relationship except for very small angles. Some lenses are made to follow an $H' = f\,\theta$ rule; occasionally a lens is designed to $H' = f \sin \theta$, which can result in a more uniform illumination over the field. In any case, the barrel distortion helps to offset the illumination falloff which usually results from the cosine-fourth-power rule.

The origin of the fish-eye lens was probably the Hill sky lens, which was conceived for the purpose of photographing the entire sky from horizon to horizon; this obviously required a field of view of 180° or more. The original sky lens was a strong meniscus negative front lens with one or two elements forming a positive rear component which was located behind the aperture stop. The name *fish-eye* comes from the fact that the view from under water into air is 180°, compared to a field of only 97° in the water, much like that of a fish's eye.

In short-focal-length, wide-angle lenses, the longitudinal chromatic and spherochromatic aberration and secondary spectrum are usually not a severe problem, simply because of the short focal length. However, the secondary spectrum of *lateral* color may become serious and chromatic variation of distortion (or higher-order lateral color) can also be troublesome because of the extremely wide angles involved.

In a lens which covers a field of 180° or more, it is obvious that the entrance pupil (the image of the aperture stop) *must* rotate and move if a light beam at 90° to the axis is to enter the lens. Figure 9.13 shows the situation in one such design (Fig. 9.17). It is also fairly common for the entrance pupil to increase in size as the field angle increases; this, of course, helps to offset the cosine-fourth effect and thereby produces a more uniform illumination across the field. The large shifts of the position of the entrance pupil can cause problems in the course of the design process if the oblique rays are aimed at the paraxial image of the aperture stop, as is the case in many automatic design programs. In Fig. 9.13 it is apparent that, for the large-obliquity ray bundles, the actual entrance pupil is well forward of the paraxial pupil, and, if these rays are aimed at the paraxial image of the stop, they will fail spectacularly. One way of handling this problem that will work with many automatic design programs is as follows: First, use a very long radius parabola or similar surface centered on the lens as the object surface (rather than a plane). This is necessary in order to get a field of 180° or more. Second, set up a separate configuration (just as you

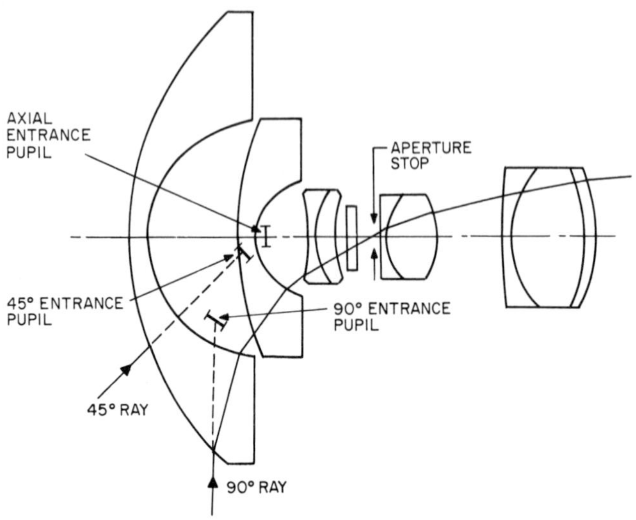

Figure 9.13 In a fish-eye lens, the overcorrected spherical aberration of the principal ray at the front components produces a shift and a rotation of the entrance pupil so that the lens can image a field of 180° or more. (*After R. Kingslake, Optical System Design, Academic Press, N.Y., 1983.*)

would for a zoom lens) for each field point, with each configuration having its own entrance pupil position. Use the height of the principal ray at the aperture stop as an item in the merit function with a target of zero, and allow the entrance pupil position to vary independently in each configuration in order to satisfy this target. In the early stages of the design process, it may be necessary to adjust the size of the entrance pupil as well, so that the upper and lower rim rays pass through the aperture stop at the correct heights.

In many of the early fish-eye designs, the front component consisted of simple negative elements. If this form is used, the lateral color will not be corrected. As can be seen from the examples, a negative achromat or hyperchromat can be used to control the lateral color.

Figures 9.14 through 9.18 show an assortment of extreme retrofocus lenses which cover angular fields that are large enough to put them into the fish-eye category.

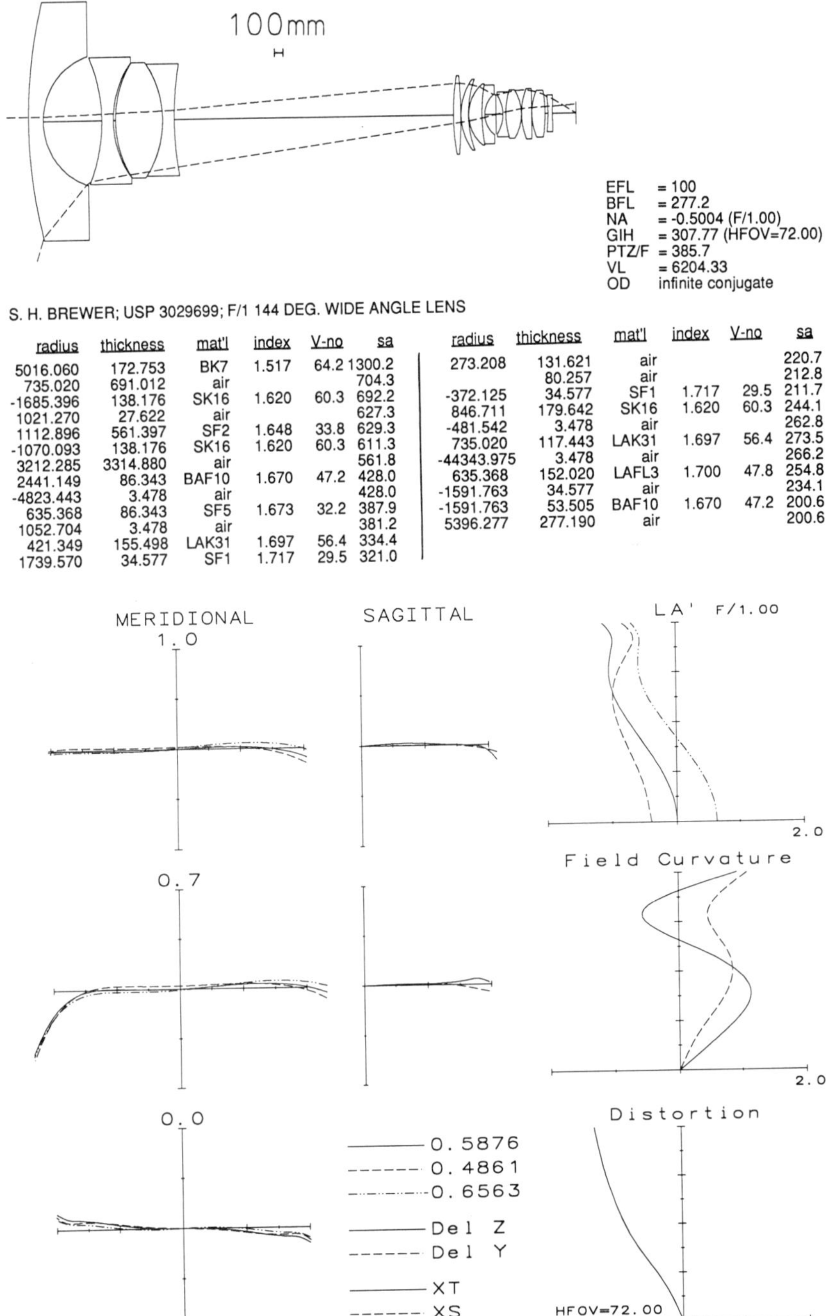

EFL = 100
BFL = 277.2
NA = -0.5004 (F/1.00)
GIH = 307.77 (HFOV=72.00)
PTZ/F = 385.7
VL = 6204.33
OD infinite conjugate

S. H. BREWER; USP 3029699; F/1 144 DEG. WIDE ANGLE LENS

radius	thickness	mat'l	index	V-no	sa
5016.060	172.753	BK7	1.517	64.2	1300.2
735.020	691.012	air			704.3
-1685.396	138.176	SK16	1.620	60.3	692.2
1021.270	27.622	air			627.3
1112.896	561.397	SF2	1.648	33.8	629.3
-1070.093	138.176	SK16	1.620	60.3	611.3
3212.285	3314.880	air			561.8
2441.149	86.343	BAF10	1.670	47.2	428.0
-4823.443	3.478	air			428.0
635.368	86.343	SF5	1.673	32.2	387.9
1052.704	3.478	air			381.2
421.349	155.498	LAK31	1.697	56.4	334.4
1739.570	34.577	SF1	1.717	29.5	321.0
273.208	131.621	air			220.7
	80.257	air			212.8
-372.125	34.577	SF1	1.717	29.5	211.7
846.711	179.642	SK16	1.620	60.3	244.1
-481.542	3.478	air			262.8
735.020	117.443	LAK31	1.697	56.4	273.5
-44343.975	3.478	air			266.2
635.368	152.020	LAFL3	1.700	47.8	254.8
-1591.763	34.577	air			234.1
-1591.763	53.505	BAF10	1.670	47.2	200.6
5396.277	277.190	air			200.6

Figure 9.14

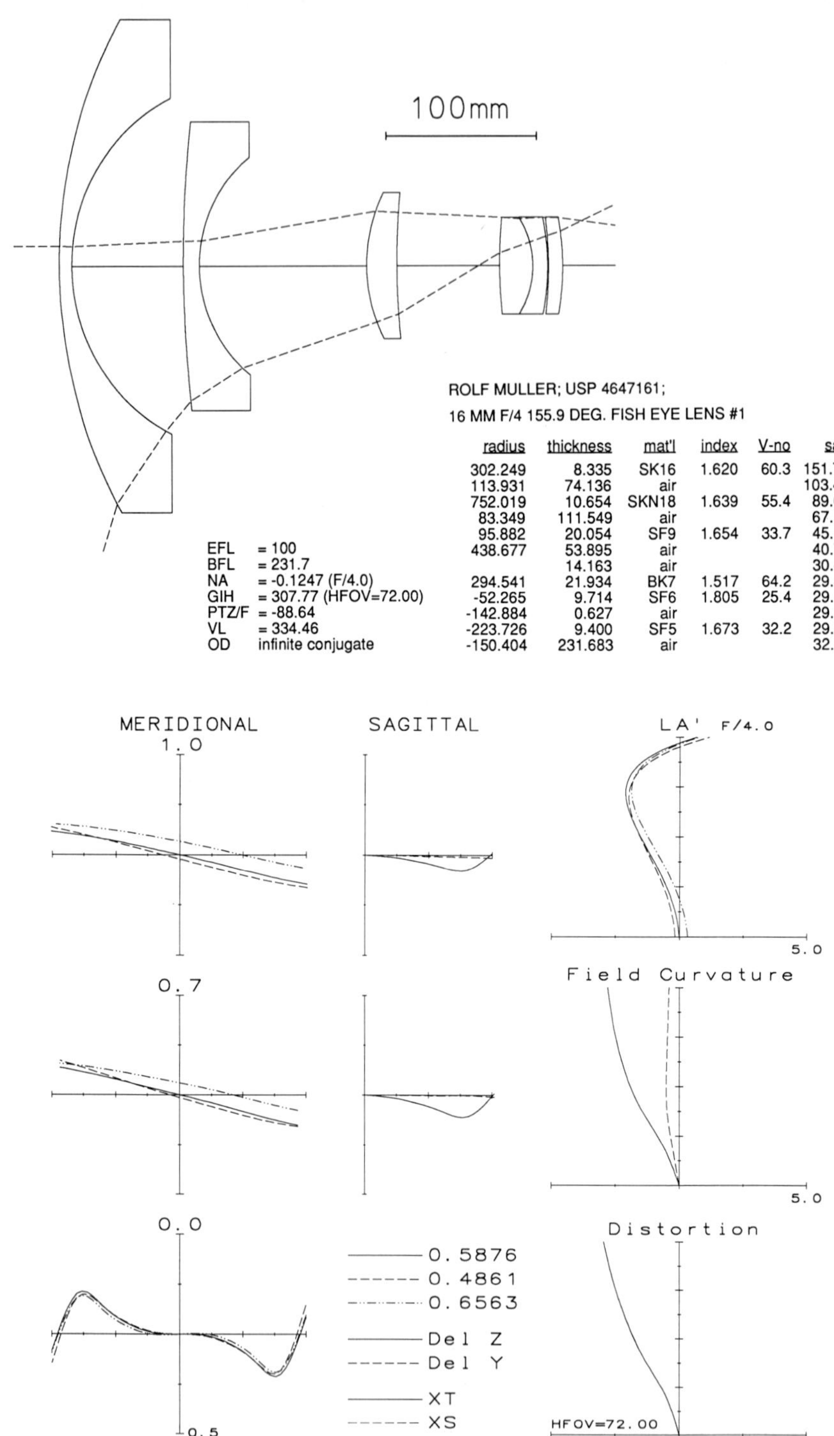

ROLF MULLER; USP 4647161;

16 MM F/4 155.9 DEG. FISH EYE LENS #1

radius	thickness	mat'l	index	V-no	sa
302.249	8.335	SK16	1.620	60.3	151.7
113.931	74.136	air			103.4
752.019	10.654	SKN18	1.639	55.4	89.0
83.349	111.549	air			67.1
95.882	20.054	SF9	1.654	33.7	45.1
438.677	53.895	air			40.7
	14.163	air			30.4
294.541	21.934	BK7	1.517	64.2	29.8
-52.265	9.714	SF6	1.805	25.4	29.2
-142.884	0.627	air			29.8
-223.726	9.400	SF5	1.673	32.2	29.8
-150.404	231.683	air			32.6

EFL = 100
BFL = 231.7
NA = -0.1247 (F/4.0)
GIH = 307.77 (HFOV=72.00)
PTZ/F = -88.64
VL = 334.46
OD infinite conjugate

Figure 9.15

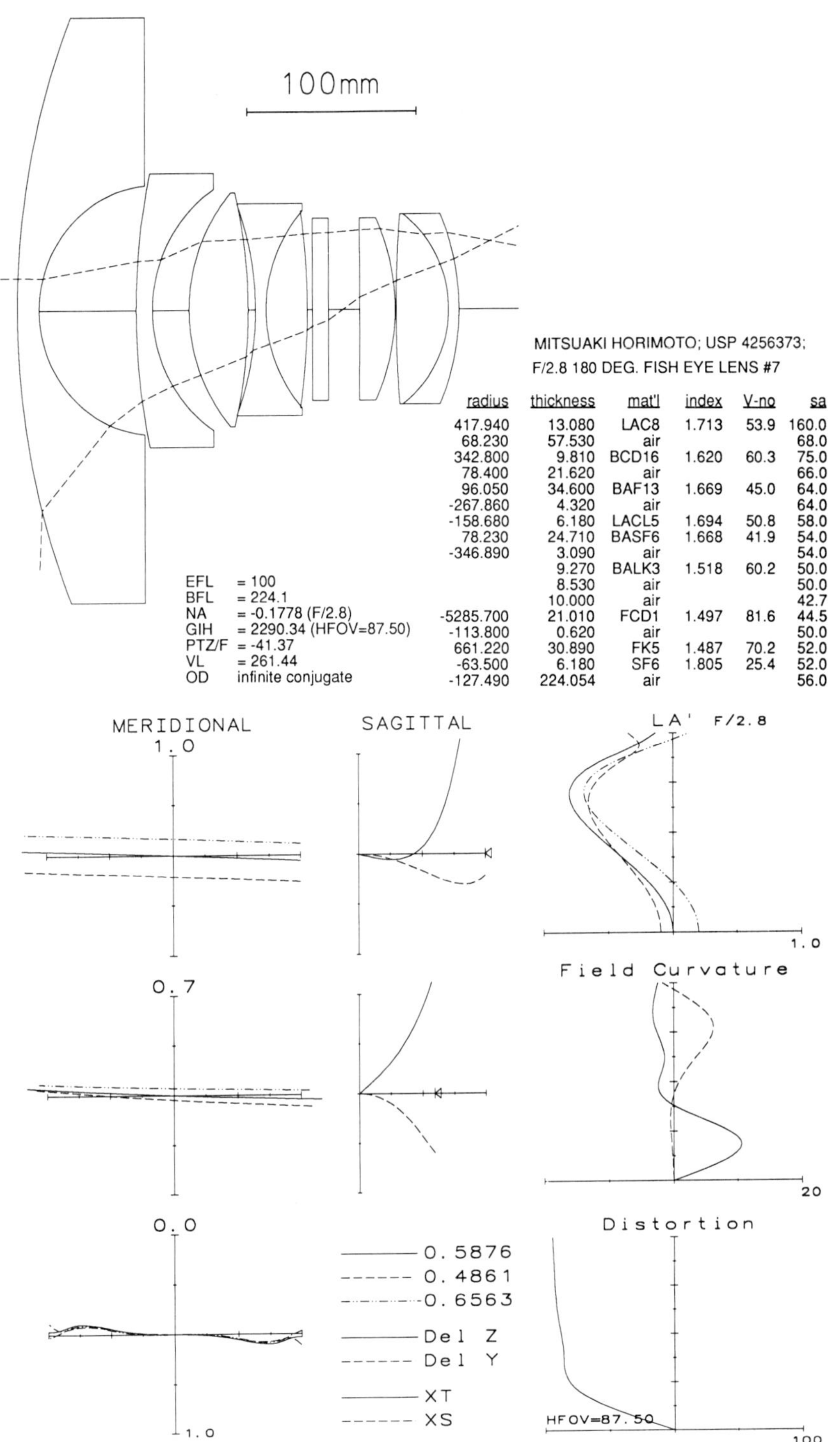

MITSUAKI HORIMOTO; USP 4256373;
F/2.8 180 DEG. FISH EYE LENS #7

radius	thickness	mat'l	index	V-no	sa
417.940	13.080	LAC8	1.713	53.9	160.0
68.230	57.530	air			68.0
342.800	9.810	BCD16	1.620	60.3	75.0
78.400	21.620	air			66.0
96.050	34.600	BAF13	1.669	45.0	64.0
-267.860	4.320	air			64.0
-158.680	6.180	LACL5	1.694	50.8	58.0
78.230	24.710	BASF6	1.668	41.9	54.0
-346.890	3.090	air			54.0
	9.270	BALK3	1.518	60.2	50.0
	8.530	air			50.0
	10.000	air			42.7
-5285.700	21.010	FCD1	1.497	81.6	44.5
-113.800	0.620	air			50.0
661.220	30.890	FK5	1.487	70.2	52.0
-63.500	6.180	SF6	1.805	25.4	52.0
-127.490	224.054	air			56.0

EFL = 100
BFL = 224.1
NA = -0.1778 (F/2.8)
GIH = 2290.34 (HFOV=87.50)
PTZ/F = -41.37
VL = 261.44
OD infinite conjugate

Figure 9.16

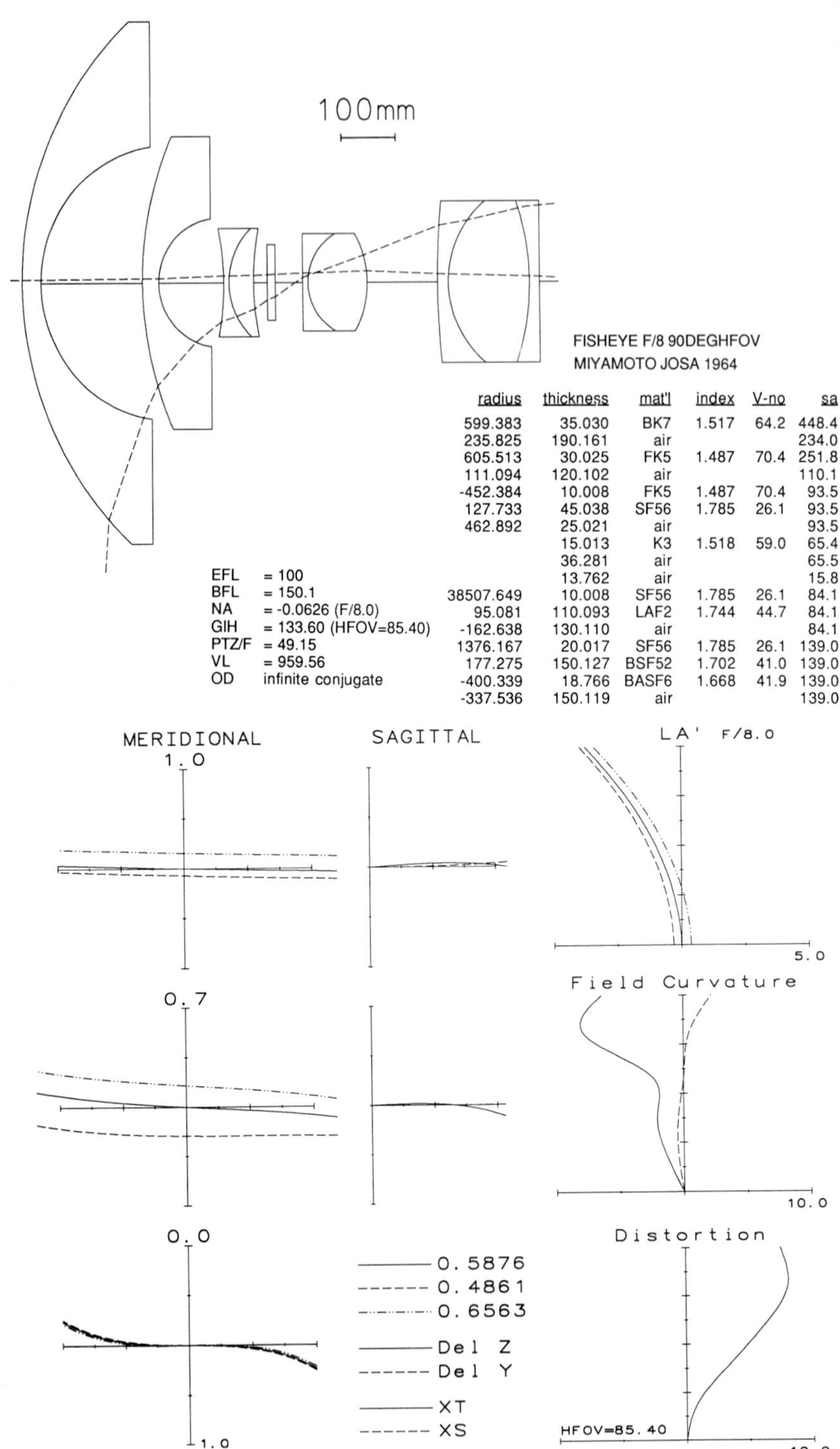

radius	thickness	mat'l	index	V-no	sa
599.383	35.030	BK7	1.517	64.2	448.4
235.825	190.161	air			234.0
605.513	30.025	FK5	1.487	70.4	251.8
111.094	120.102	air			110.1
-452.384	10.008	FK5	1.487	70.4	93.5
127.733	45.038	SF56	1.785	26.1	93.5
462.892	25.021	air			93.5
	15.013	K3	1.518	59.0	65.4
	36.281	air			65.5
	13.762	air			15.8
38507.649	10.008	SF56	1.785	26.1	84.1
95.081	110.093	LAF2	1.744	44.7	84.1
-162.638	130.110	air			84.1
1376.167	20.017	SF56	1.785	26.1	139.0
177.275	150.127	BSF52	1.702	41.0	139.0
-400.339	18.766	BASF6	1.668	41.9	139.0
-337.536	150.119	air			139.0

EFL	= 100
BFL	= 150.1
NA	= -0.0626 (F/8.0)
GIH	= 133.60 (HFOV=85.40)
PTZ/F	= 49.15
VL	= 959.56
OD	infinite conjugate

Figure 9.17

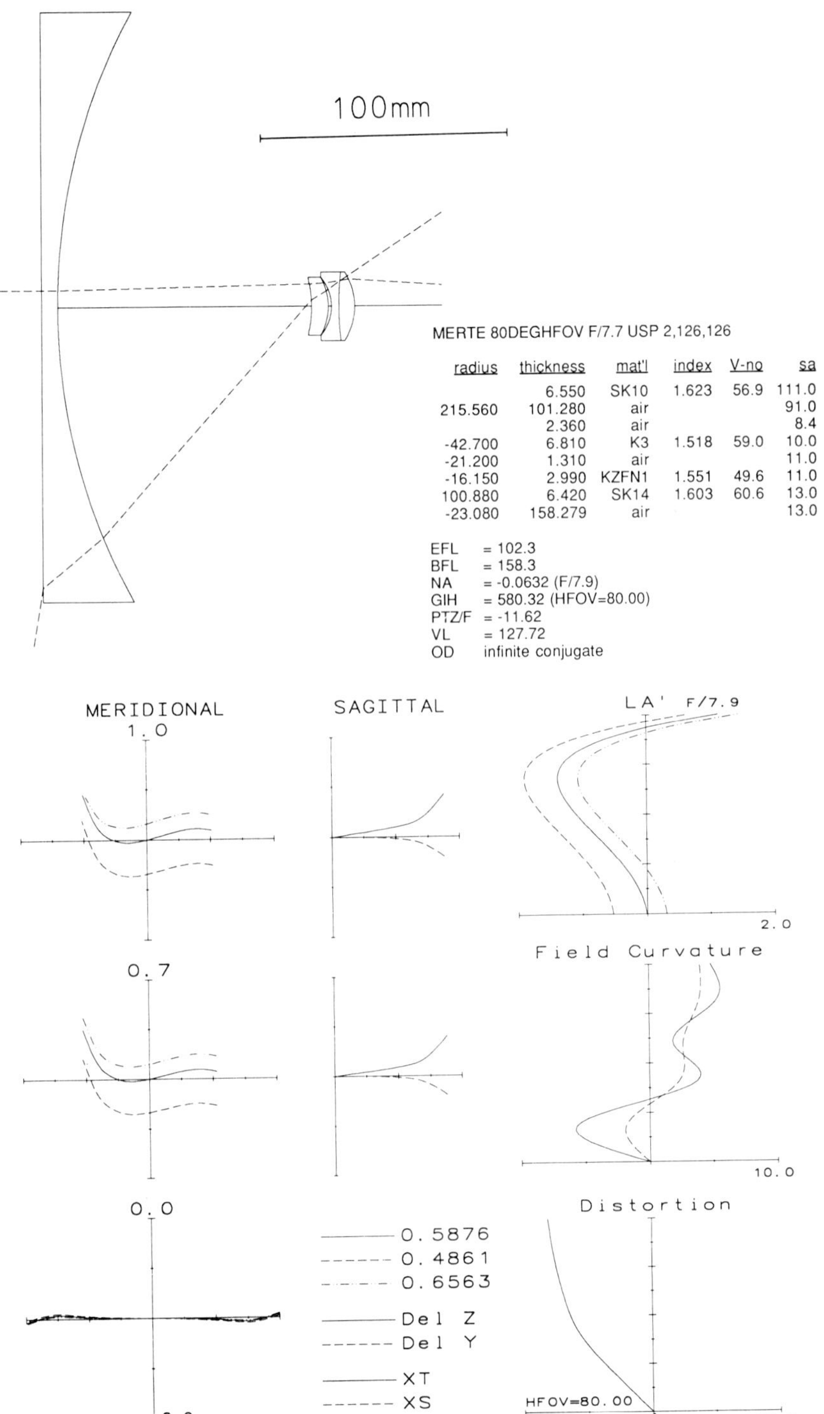

MERTE 80DEGHFOV F/7.7 USP 2,126,126

radius	thickness	mat'l	index	V-no	sa
	6.550	SK10	1.623	56.9	111.0
215.560	101.280	air			91.0
	2.360	air			8.4
-42.700	6.810	K3	1.518	59.0	10.0
-21.200	1.310	air			11.0
-16.150	2.990	KZFN1	1.551	49.6	11.0
100.880	6.420	SK14	1.603	60.6	13.0
-23.080	158.279	air			13.0

EFL = 102.3
BFL = 158.3
NA = -0.0632 (F/7.9)
GIH = 580.32 (HFOV=80.00)
PTZ/F = -11.62
VL = 127.72
OD infinite conjugate

Figure 9.18

Chapter

10

Telephoto Lenses

10.1 The Basic Telephoto

The arrangement shown in Fig. 10.1, with a positive component followed by a negative component, can produce a compact system with an effective focal length F which is longer than the overall length L of the lens. The ratio of L/F is called the *telephoto ratio,* and a lens for which this ratio is less than unity is classified as a telephoto lens. The smaller the ratio, the more difficult the lens is to design. Note that many camera lenses which are sold as telephoto lenses are simply long-focal-length lenses and are not true telephotos.

Many of the comments in Chap. 9 regarding retrofocus or reverse telephoto lenses are equally applicable to the telephoto lens. Equations 9.1 and 9.2 may also be applied to the telephoto. The usual Petzval problem is with a backward-curving field, just as with the retrofocus, and the same glass choices are appropriate for the telephoto. Since the system is unsymmetrical, each component must be individually achromatized if both axial and lateral color are to be cor-

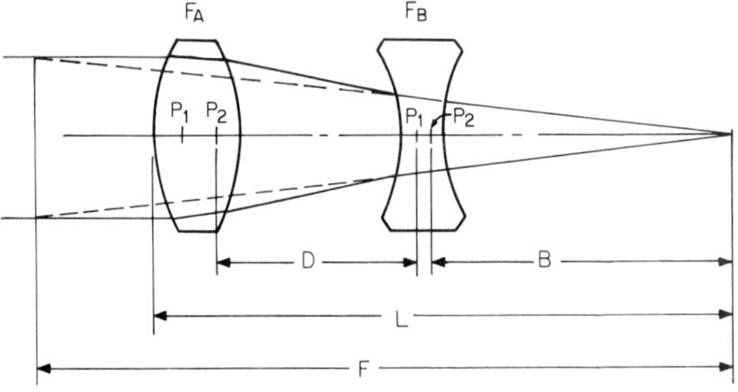

Figure 10.1 The basic power arrangement for a telephoto lens yields a compact lens with an overall length which is less than its effective focal length.

rected. The aperture stop is usually at the front member or part way toward the rear. Since a telephoto lens usually covers only a relatively small angular field, coma, distortion, and lateral color (which in many lenses are reduced by an approximate symmetry about the stop) are not as troublesome as they would be with a wider field.

10.2 Close-up or Macro lenses

The correction of a long-focal-length unsymmetrical lens is usually quite sensitive to a change in object distance, and, for most telephoto lenses, the image quality deteriorates severely when they are focused on nearby objects. Note that this effect varies inversely with the object distance expressed in focal length units; i.e., for a given design type, the image quality may remain acceptable as long as the object distance exceeds some number of focal lengths. Thus, for a given object distance, this effect is more of a problem for a long-focal-length lens than for a short. Since retrofocus lenses tend to have short focal lengths, this problem is somewhat less frequently encountered, in spite of their asymmetry.

Many newer telephoto lenses and the specialized close-focusing lenses (called *macro* lenses) utilize a floating component or separately moving elements to maintain the aberration correction when the lens is focused at a close distance. For many lenses, the spherical aberration and the astigmatism become undercorrected at close conjugates. Thus a relative motion of the elements to increase the marginal ray height on a negative (or overcorrecting) element/component can be used to stabilize the spherical. The astigmatism can be controlled by a motion which increases the height of the chief ray on a component which contributes overcorrected astigmatism, or which reduces it on an undercorrecting one.

The design of such a system is carried out just like the design of a zoom lens. Two (or more) configurations are set up, one with a long (perhaps infinite) object conjugate distance and the other with a short one. The computer then uses the same lens elements with different spacings for each configuration and optimizes the merit function for both configurations simultaneously.

Figures 19.3 to 19.6, and 20.5 show nontelephoto designs with macro features.

10.3 Sample Telephoto Designs

Figures 10.2 and 10.3 show two very basic telephoto lenses; each consists of just two cemented or closely airspaced achromatic doublets, about as simple a construction as possible. Figure 10.2 covers less

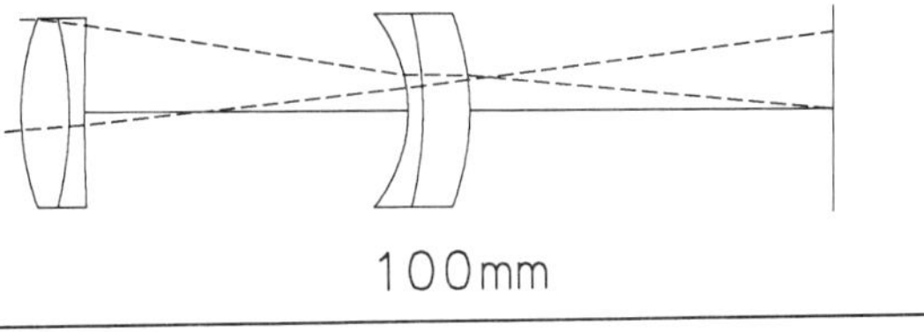

KINGSLAKE TELEPHOTO MODIFIED BY HOPKINS

radius	thickness	mat'l	index	V-no	sa
24.607	5.080	BK7	1.517	64.2	9.2
-36.347	1.600	F2	1.620	36.4	9.2
212.138	12.300	air			9.0
	21.699	air			6.7
-14.123	1.520	BK7	1.517	64.2	9.4
-38.904	4.800	F2	1.620	36.4	9.4
-25.814	37.934	air			9.4

EFL = 101.6
BFL = 37.93
NA = -0.0893 (F/5.6)
GIH = 7.44
PTZ/F = -19.38
VL = 47.00
OD infinite conjugate

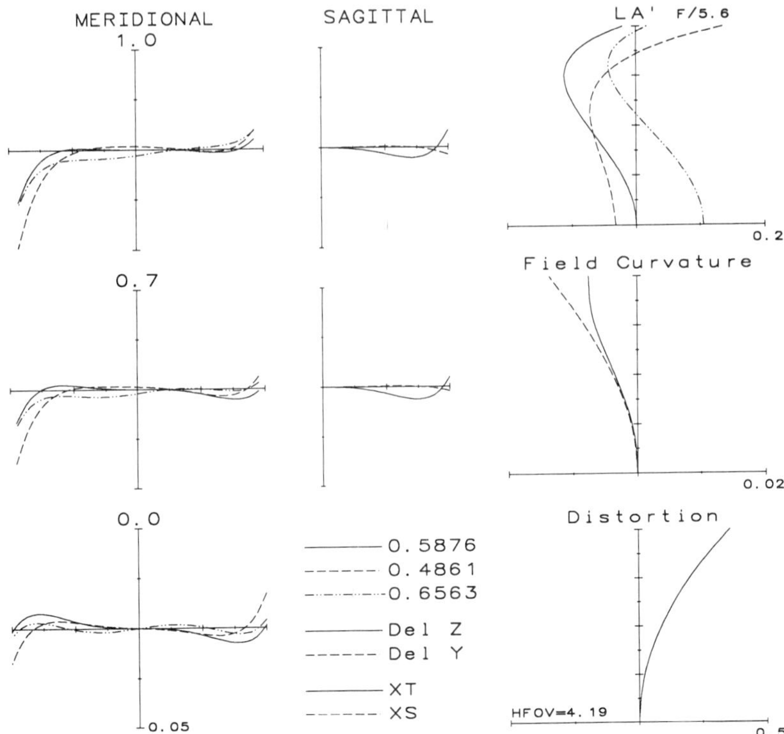

Figure 10.2

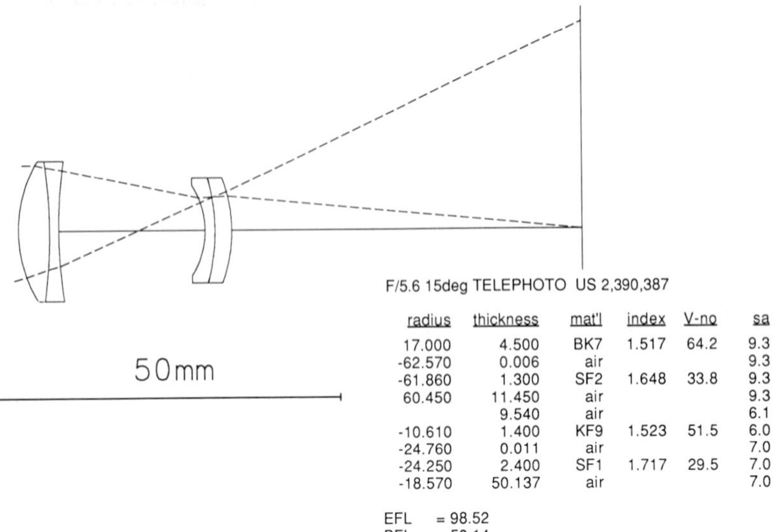

F/5.6 15deg TELEPHOTO US 2,390,387

radius	thickness	mat'l	index	V-no	sa
17.000	4.500	BK7	1.517	64.2	9.3
-62.570	0.006	air			9.3
-61.860	1.300	SF2	1.648	33.8	9.3
60.450	11.450	air			9.3
	9.540	air			6.1
-10.610	1.400	KF9	1.523	51.5	6.0
-24.760	0.011	air			7.0
-24.250	2.400	SF1	1.717	29.5	7.0
-18.570	50.137	air			7.0

EFL = 98.52
BFL = 50.14
NA = -0.0888 (F/5.6)
GIH = 26.60 (HFOV=15.11)
PTZ/F = 16.55
VL = 30.61
OD infinite conjugate

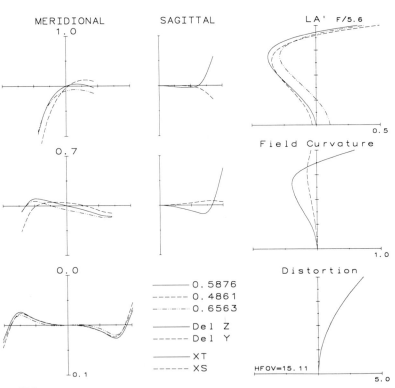

MERIDIONAL
SAGITTAL
LA' F/5.6
1.0
0.5
0.7
Field Curvature
1.0
0.0
Distortion
0.1
0.5876
0.4861
0.6563
Del Z
Del Y
XT
XS
HFOV=15.11
5.0

Figure 10.3

than 9° at $f/5.6$ and has a telephoto ratio of 0.85; it uses BK7 and F2 glasses and was done as an illustrative design exercise. Figure 10.3, at the same speed, covers 3 times as large a field with a telephoto ratio of 0.81. It utilizes heavier flint glasses (higher index and lower V value) to achieve a modestly improved performance.

Figure 10.4 covers a 30° field at $f/4.5$ with excellent distortion correction, illustrating the benefits derived from the added degrees of freedom gained by splitting the cemented doublets into widely airspaced components. The large telephoto ratio of 0.91 and the modestly high-index glasses are also helpful.

Figure 10.5 illustrates the use of unusual partial dispersion glasses (as described in Chap. 6) to reduce the secondary spectrum. The term *superachromat* implies that at least four wavelengths are brought to a common focus, whereas the term *apochromat* indicates that three wavelengths are corrected. Notice, however, that the spherochromatism and zonal spherical aberration in this lens are much larger than the axial chromatic aberration; these are the aberrations which will determine the limiting performance of this lens.

Figures 10.6, 10.7, and 10.8 each have five elements and illustrate some of the different ways that the inherent capabilities of this configuration can be utilized. In Figs. 10.6 and 10.8, the crown element of the front doublet is split into two elements to reduce the zonal spherical aberration (among others). Figure 10.6 is the result of a classroom exercise which specified a 200-mm $f/5.0$ lens with a telephoto ratio of 0.80 for a 35-mm camera. It uses quite ordinary glasses and achieves an excellent level of performance. Figure 10.7 uses high-index glass and a different arrangement to get to a speed of $f/4.0$, but falls a bit short in performance and telephoto ratio (at 0.91). Figure 10.8 uses unusual partial dispersion glasses and breaks the contact in the front doublet, to achieve what is (potentially) a high level of correction, although the telephoto ratio is only a modest 0.95.

Figure 10.9 uses seven elements to produce a well-corrected $f/5.6$, 6° field lens with an extremely short telephoto ratio of 0.66. Notice the overcorrected Petzval field, with $\rho/f = +2.1$; this is one reason that small telephoto ratios are troublesome.

With a telephoto ratio of 1.06, Fig. 10.10 doesn't really qualify as a true telephoto, but at a speed of $f/1.8$ and a field of 18°, it is an interesting lens, even if it is difficult to classify.

Figures 10.11 and 10.12 show an internal-focusing telephoto with a modest ratio of 0.92. The front component is fixed and the lens is focused for close-ups by moving the rear component toward the image plane. This could be considered as a sort of macro-style lens.

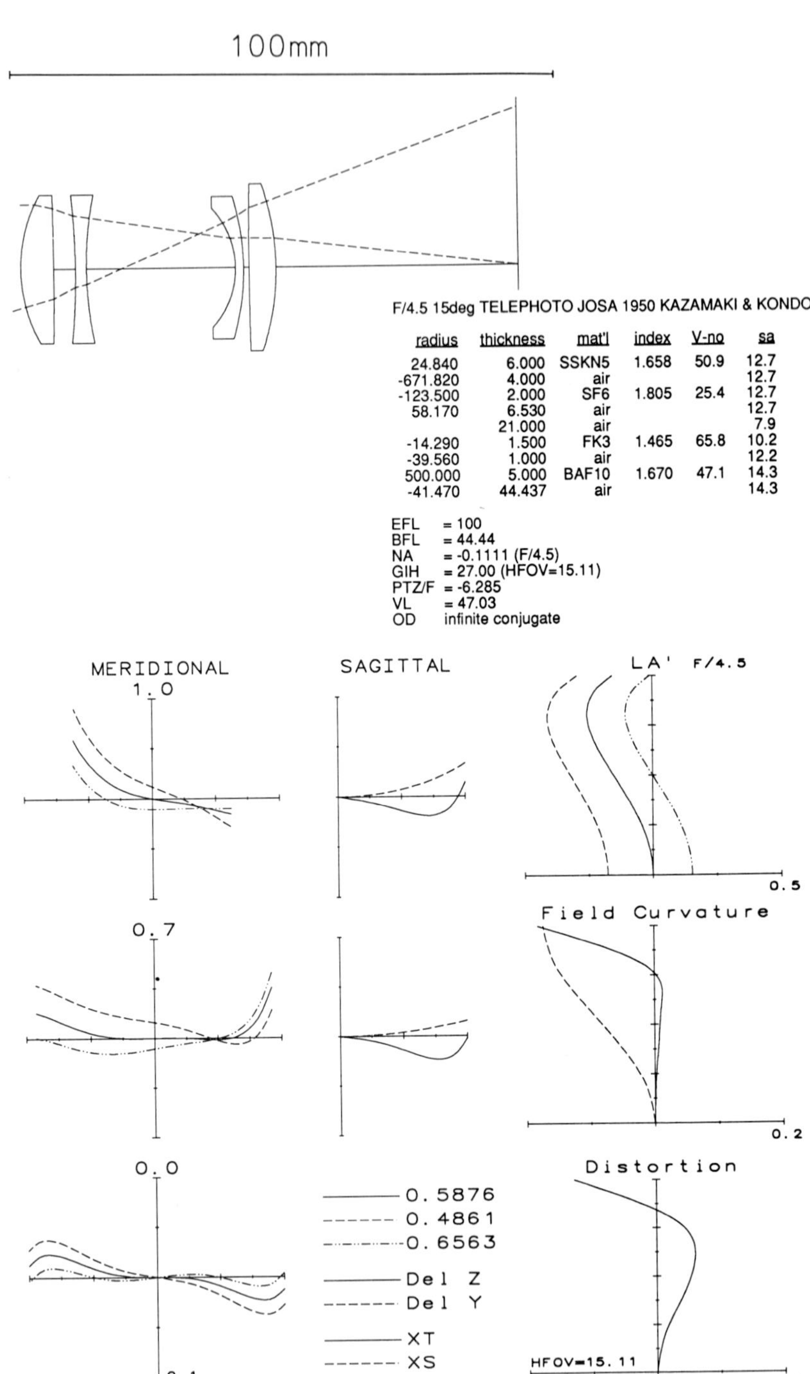

F/4.5 15deg TELEPHOTO JOSA 1950 KAZAMAKI & KONDO

radius	thickness	mat'l	index	V-no	sa
24.840	6.000	SSKN5	1.658	50.9	12.7
-671.820	4.000	air			12.7
-123.500	2.000	SF6	1.805	25.4	12.7
58.170	6.530	air			12.7
	21.000	air			7.9
-14.290	1.500	FK3	1.465	65.8	10.2
-39.560	1.000	air			12.2
500.000	5.000	BAF10	1.670	47.1	14.3
-41.470	44.437	air			14.3

EFL = 100
BFL = 44.44
NA = -0.1111 (F/4.5)
GIH = 27.00 (HFOV=15.11)
PTZ/F = -6.285
VL = 47.03
OD infinite conjugate

Figure 10.4

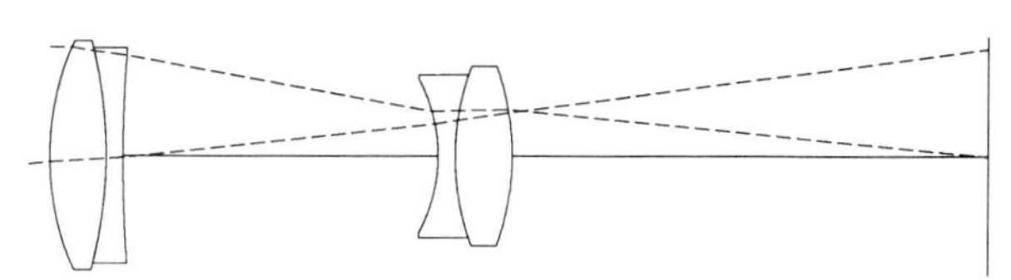

SIGLER; SUPER ACHROMAT; TELEPHOTO EFL=254

radius	thickness	mat'l	index	V-no	sa
21.851	5.008	PK51	1.529	77.0	9.5
-34.546	1.502	KZFS9	1.599	46.9	8.9
108.705	1.127	air			8.3
	26.965	air			8.1
-12.852	1.502	KZFS1	1.613	44.3	6.3
19.813	5.008	BASF5	1.603	42.5	6.7
-20.378	42.174	air			7.4

EFL = 100
BFL = 42.17
NA = -0.0898 (F/5.6)
GIH = 8.75
PTZ/F = -40.38
VL = 41.11
OD infinite conjugate

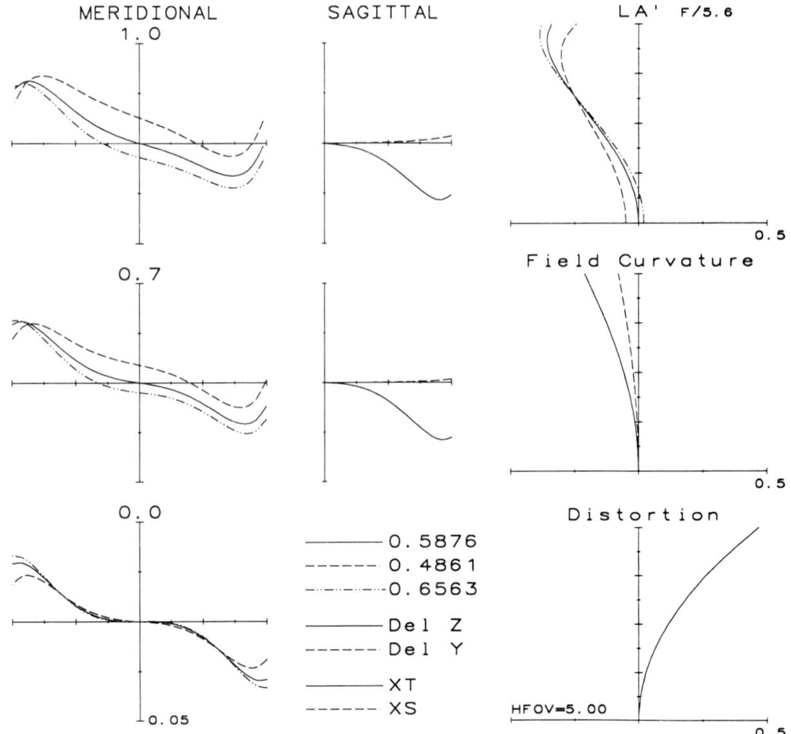

Figure 10.5

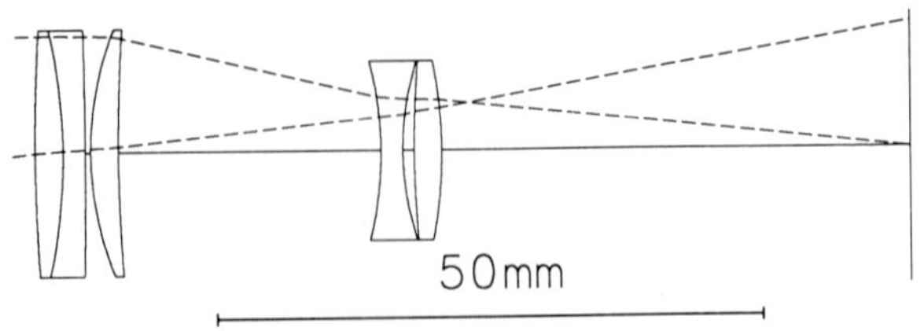

F/5 6deg TELEPHOTO

radius	thickness	mat'l	index	V-no	sa
149.035	2.500	SK4	1.613	58.6	10.5
-46.003	2.000	SF14	1.762	26.5	10.5
-477.921	0.500	air			10.5
26.522	2.500	SK4	1.613	58.6	10.5
132.322	24.060	air			10.5
-28.605	2.000	SK4	1.613	58.6	7.6
22.989	1.050	air			7.6
82.834	2.500	F5	1.603	38.0	7.6
-36.911	42.897	air			7.6

EFL = 100
BFL = 42.9
NA = -0.1000 (F/5.0)
GIH = 10.50 (HFOV=5.99)
PTZ/F = 7.68
VL = 37.11
OD infinite conjugate

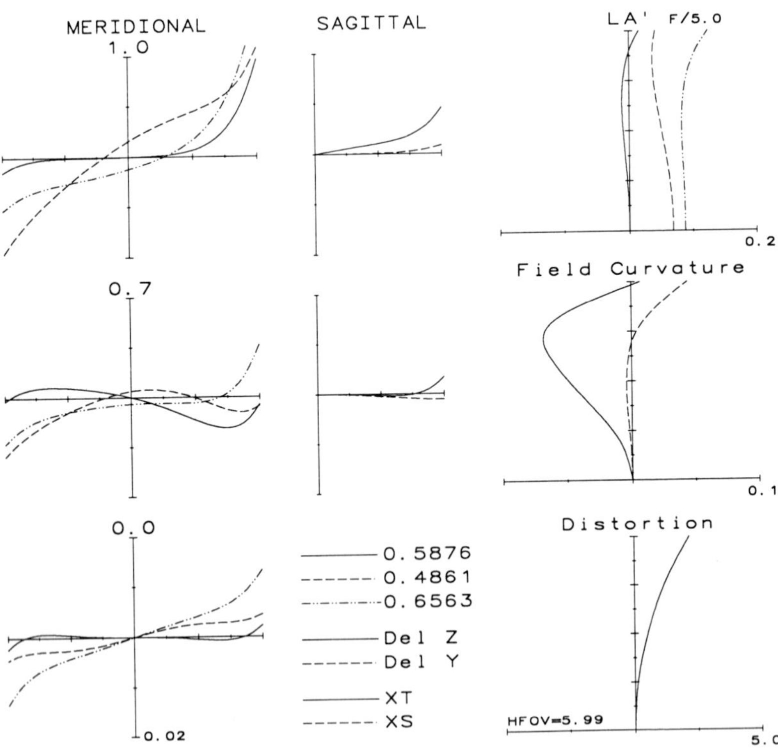

Figure 10.6

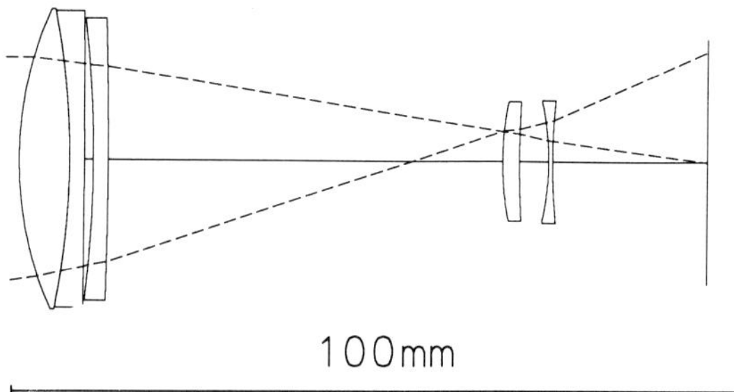

J. EGGERT ET AL; USP 3388956; F/4 15 DEG. TELEPHOTO #1

radius	thickness	mat'l	index	V-no	sa
42.156	6.676	LAKN6	1.642	58.0	18.4
-88.214	1.945	SF10	1.728	28.4	18.2
-1800.073	1.183	air			17.4
-137.288	1.945	SF9	1.654	33.7	17.3
-530.649	40.290	air			16.8
	11.913	air			5.4
34.202	2.234	LAF11	1.757	31.7	7.3
117.496	3.859	air			7.3
-31.069	0.526	BSF10	1.650	39.1	7.3
69.782	20.320	air			7.5

EFL = 99.99
BFL = 20.32
NA = -0.1256 (F/4.0)
GIH = 13.16 (HFOV=7.50)
PTZ/F = 4.272
VL = 70.57
OD infinite conjugate

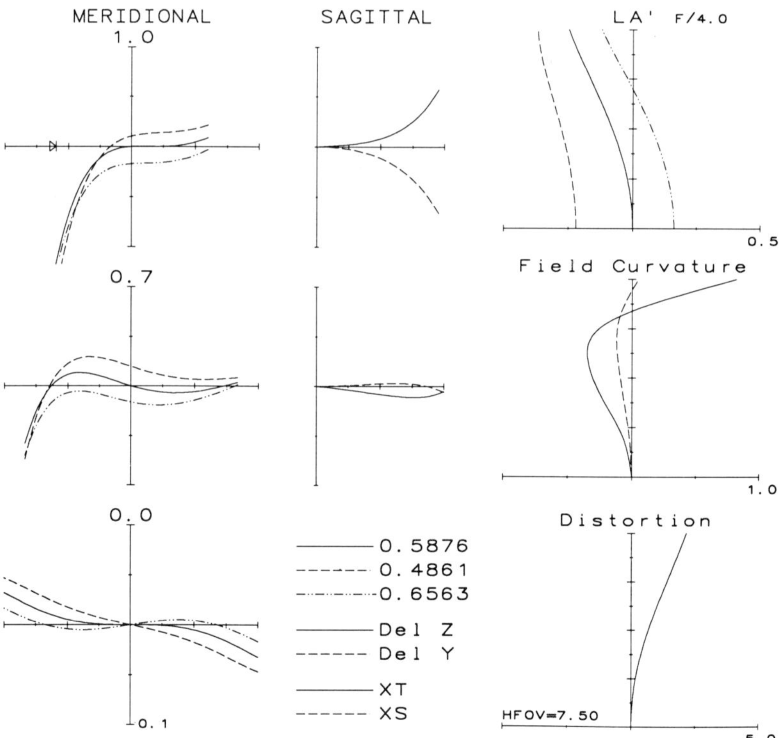

Figure 10.7

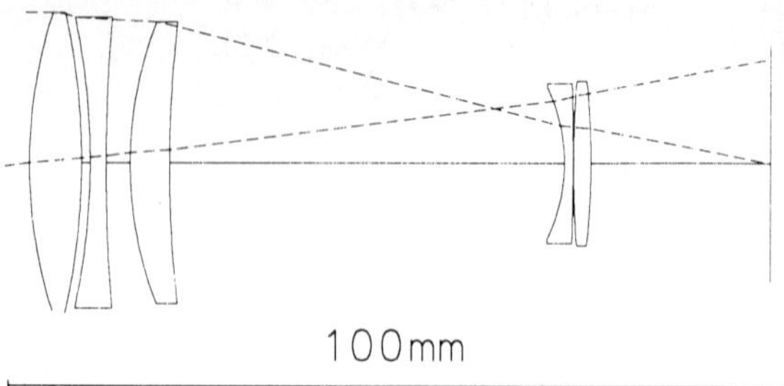

SEI MATSUI; USP 4338001; F/2.8 14 DEG. TELEPHOTO LENS #1

radius	thickness	mat'l	index	V-no	sa
54.585	6.667	FCD1	1.497	81.6	17.9
-77.813	1.111	air			17.7
-76.698	2.056	LAFN7	1.750	34.9	17.3
207.222	3.056	air			17.0
43.208	5.111	BED5	1.658	50.9	16.8
134.444	50.667	air			16.2
-19.462	1.111	K3	1.518	59.0	9.0
-305.556	0.056	air			9.5
121.887	2.222	TAF2	1.794	45.4	9.7
-89.277	22.862	air			9.8

EFL = 100
BFL = 22.86
NA = -0.1790 (F/2.8)
GIH = 12.28 (HFOV=7.00)
PTZ/F = -9.12
VL = 72.06
OD infinite conjugate

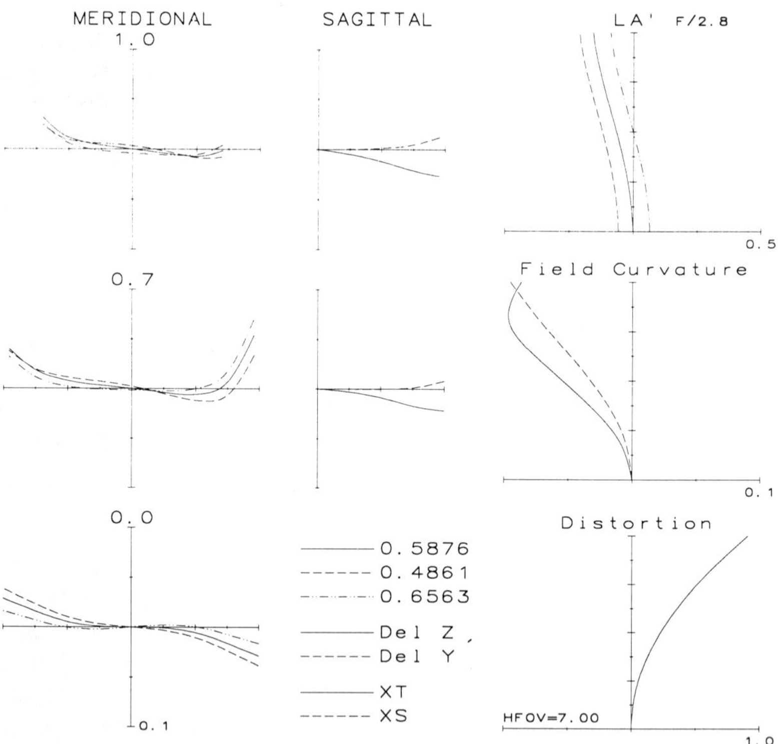

Figure 10.8

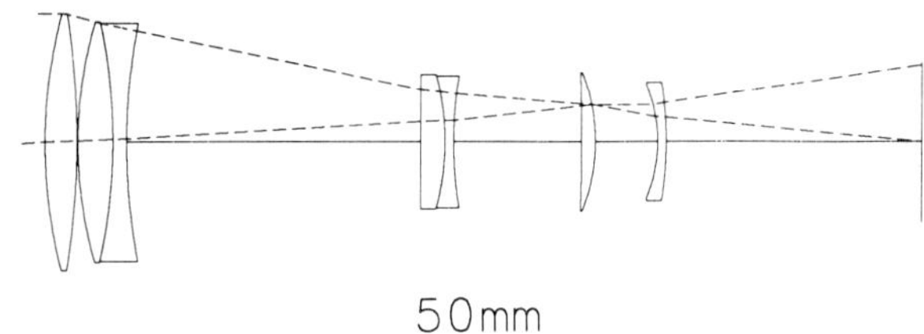

MELVYN H. KREITZER; USP 4359272; 390 MM F/5.6 6 DEG. TEL. #1

radius	thickness	mat'l	index	V-no	sa
33.072	2.386	C3	1.518	59.0	8.9
-53.387	0.077	air			8.9
27.825	2.657	C3	1.518	59.0	8.4
-35.934	1.025	LAF7	1.749	35.0	8.3
40.900	22.084	air			7.8
	1.794	FD110	1.785	25.7	4.7
-16.775	0.641	TAFD5	1.835	43.0	4.6
27.153	9.607	air			4.5
-120.757	1.035	CF6	1.517	52.2	4.8
-12.105	4.705	air			4.8
-9.386	0.641	TAF1	1.773	49.6	4.0
-24.331	18.960	air			4.1

EFL = 100
BFL = 18.96
NA = -0.0892 (F/5.6)
GIH = 5.24
PTZ/F = 2.097
VL = 46.65
OD infinite conjugate

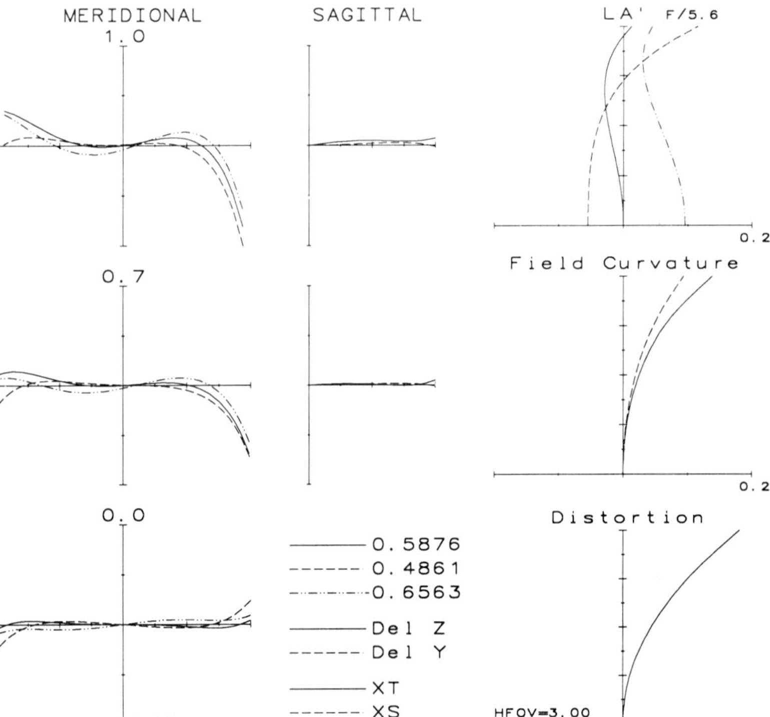

Figure 10.9

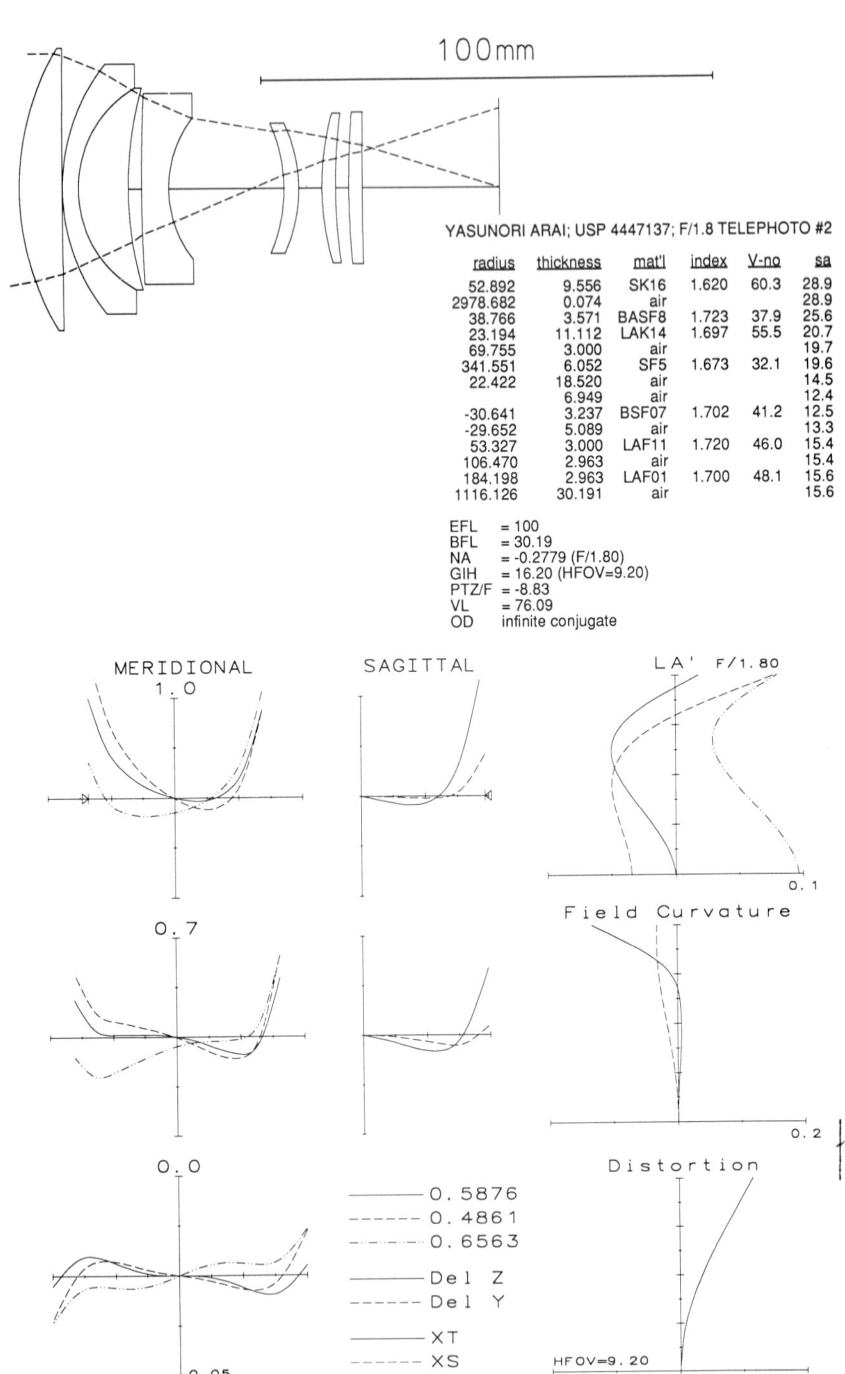

YASUNORI ARAI; USP 4447137; F/1.8 TELEPHOTO #2

radius	thickness	mat'l	index	V-no	sa
52.892	9.556	SK16	1.620	60.3	28.9
2978.682	0.074	air			28.9
38.766	3.571	BASF8	1.723	37.9	25.6
23.194	11.112	LAK14	1.697	55.5	20.7
69.755	3.000	air			19.7
341.551	6.052	SF5	1.673	32.1	19.6
22.422	18.520	air			14.5
	6.949	air			12.4
-30.641	3.237	BSF07	1.702	41.2	12.5
-29.652	5.089	air			13.3
53.327	3.000	LAF11	1.720	46.0	15.4
106.470	2.963	air			15.4
184.198	2.963	LAF01	1.700	48.1	15.6
1116.126	30.191	air			15.6

EFL = 100
BFL = 30.19
NA = -0.2779 (F/1.80)
GIH = 16.20 (HFOV=9.20)
PTZ/F = -8.83
VL = 76.09
OD infinite conjugate

Figure 10.10

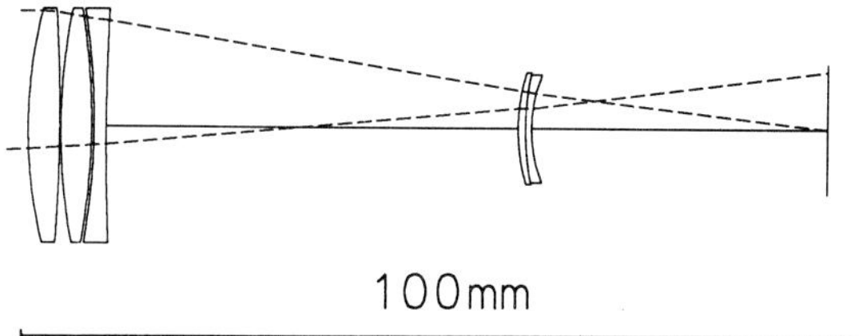

MOMIYAMA USP 4,037,935; F/4.5 (LONG EFL POSITION)

radius	thickness	mat'l	index	V-no	sa
48.796	3.645	CAF	1.434	94.9	12.6
-168.096	0.140	air			12.6
62.820	3.505	FK5	1.487	70.2	12.6
-70.733	0.303	air			12.6
-73.188	1.402	LASF3	1.806	40.9	12.6
250.187	22.431	air			12.6
	25.150	air			7.9
19.535	0.841	SF4	1.755	27.5	6.0
27.456	0.701	LSF16	1.772	49.6	5.8
14.524	34.173	air			5.7

EFL = 112.2
BFL = 34.17
NA = -0.1099 (F/4.5)
GIH = 6.06
PTZ/F = -4.07
VL = 58.12
OD infinite conjugate

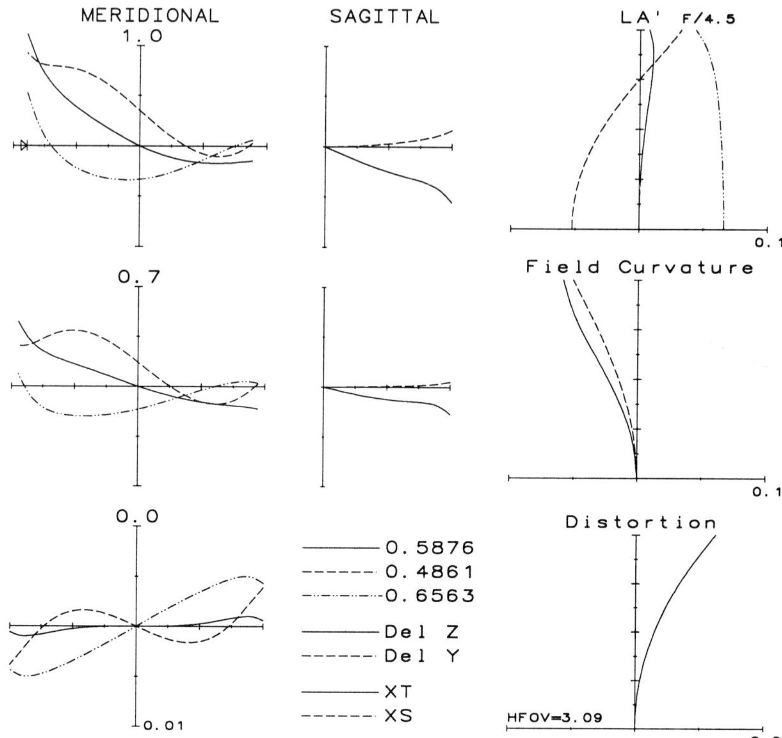

Figure 10.11

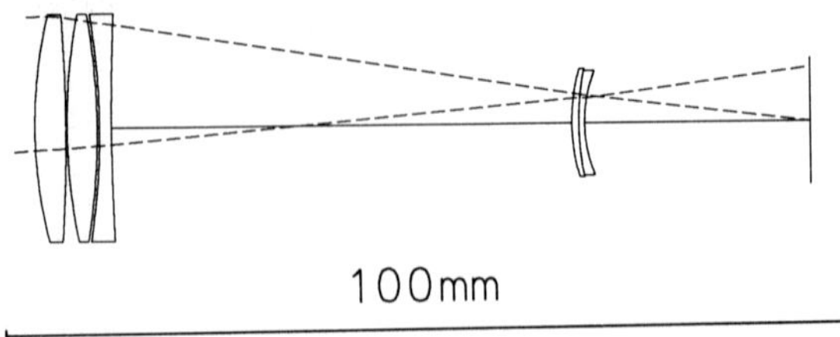

MOMIYAMA USP 4,037,935; F/4.5 (SHORT EFL POSITION)

radius	thickness	mat'l	index	V-no	sa
48.796	3.645	CAF	1.434	94.9	12.6
-168.096	0.140	air			12.6
62.820	3.505	FK5	1.487	70.2	12.6
-70.733	0.303	air			12.6
-73.188	1.402	LASF3	1.806	40.9	12.6
250.187	22.431	air			12.6
	32.440	air			7.9
19.535	0.841	SF4	1.755	27.5	6.0
27.456	0.701	LSF16	1.772	49.6	5.8
14.524	26.657	air			5.7

EFL = 100
BFL = 26.66
NA = -0.1139 (F/4.4)
GIH = 5.93
PTZ/F = -4.605
VL = 65.41
OD = 2243.07 (MAG = -0.048)

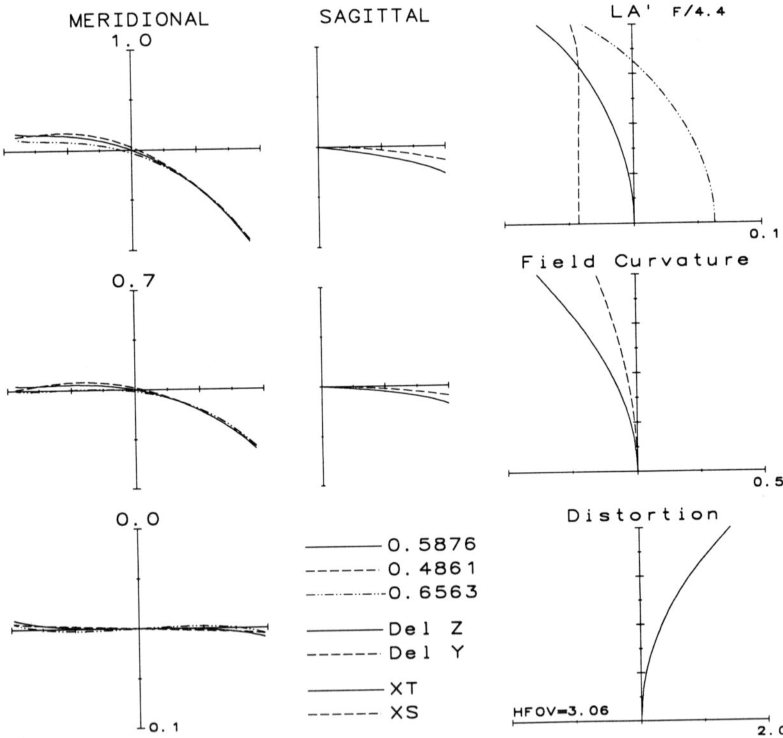

Figure 10.12

Chapter

11

Double-Meniscus Anastigmats

11.1 Meniscus Components

All anastigmats achieve a flat field by the separation of positive and negative power (surface, element, or component power). The earliest anastigmats flattened the field by using a thick meniscus, which separated the positive and negative outer surfaces. At the time there were no antireflection coatings, so designers tried to minimize the number of air-glass surfaces so that the surface reflections would not cause unacceptable ghost images or reduced contrast. For this reason, most of the early anastigmats were limited to two cemented meniscus components.

11.2 The Hypergon, Topogon, and Metrogon

The *Hypergon* lens of Fig. 11.1 can be considered the progenitor of this class. It consists of two identical meniscus elements, symmetrical about a central stop. The concave and convex radii differ by less than 0.7 percent (0.5 percent in some versions) so that the Petzval contributions of the convex surfaces are almost completely offset by the Petzval of the concave surfaces. The astigmatism is controlled by the distance of the lens from the stop, and the symmetrical construction almost completely eliminates the coma, distortion, and lateral color. The lens covers an astonishing field of 135°. Of course, the spherical aberration of these strongly bent meniscus elements is tremendous, and for this reason the lens must be used at a very low speed. The other major defect of this lens is the falloff in illumination with field angle. The cosine-fourth effect at 67° reduces the edge-of-field illumination to less than 2.5 percent of that at the center. (See Chaps. 9 and 18 for the improvement in illumination uniformity which results from

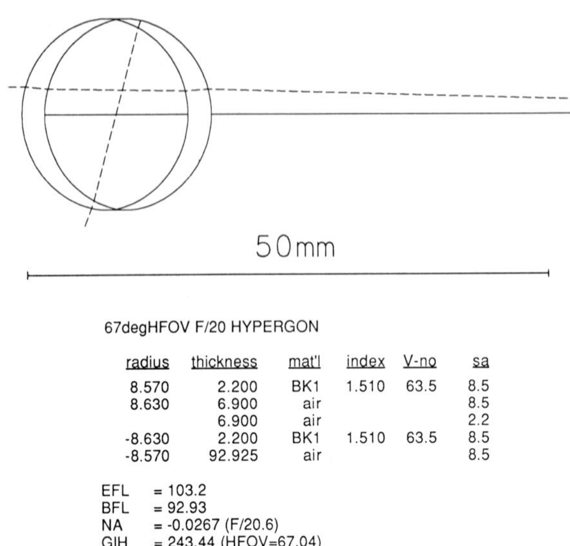

67degHFOV F/20 HYPERGON

radius	thickness	mat'l	index	V-no	sa
8.570	2.200	BK1	1.510	63.5	8.5
8.630	6.900	air			8.5
	6.900	air			2.2
-8.630	2.200	BK1	1.510	63.5	8.5
-8.570	92.925	air			8.5

EFL = 103.2
BFL = 92.93
NA = -0.0267 (F/20.6)
GIH = 243.44 (HFOV=67.04)
PTZ/F = -21.46
VL = 18.20
OD infinite conjugate

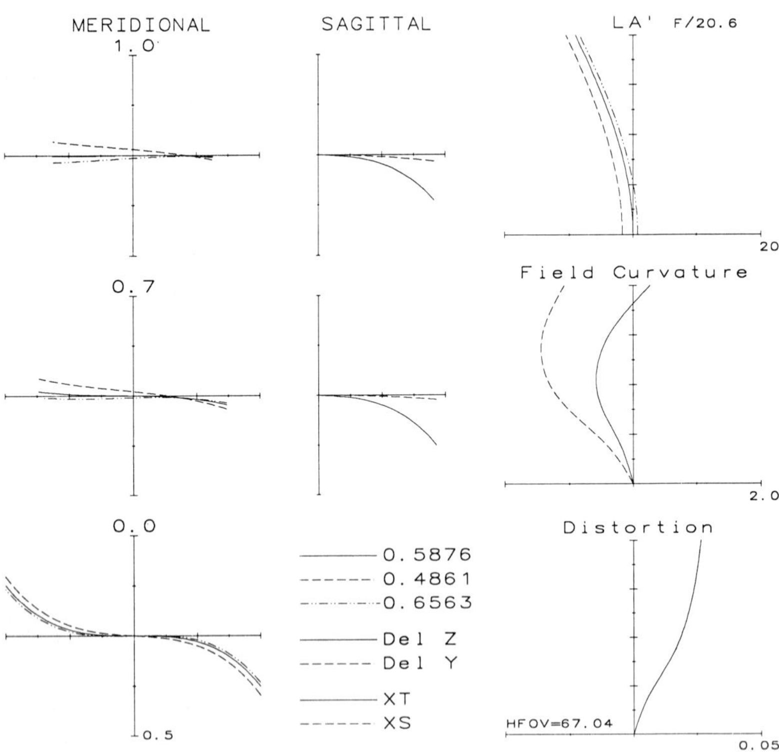

Figure 11.1

the use of negative—rather than positive—outer elements in a wide-angle lens.) The illumination can be evened out either by a gradient filter or by a rotating star-pinwheel device which is introduced in front of the lens part way through the long (in the early days of photography) exposure.

The obvious way to improve the Hypergon is to add negative flint elements to correct the spherical aberration and axial chromatic aberration. The *Topogon* of Fig. 11.2 covers a field of about 100° at a speed of *f*/6.3, using dense barium crown and extra-dense flint glasses, retaining the symmetrical construction and the strong meniscus configuration for all of the elements. In the *Metrogon,* Fig. 11.3, the front crown has been split into two elements, and the sample design achieves an excellent correction for distortion over a 100° field.

11.3 Protar, Dagor, and Convertible Lenses

The *Protar* (Fig. 11.4) is based on the combination of a weak, meniscus, old achromat front member and a strong, meniscus, new achromat rear member. See Sec. 3.5 for a discussion of old versus new achromats. The diverging cemented surface of the front old achromat is used to correct the spherical aberration; in the rear doublet, the low-index flint, high-index crown combination of the new achromat, along with the thick meniscus construction, is used to flatten the Petzval surface. The cemented interfaces and the stop position are used to control the astigmatism. Note that the divergent cemented surface in the front doublet is concave to the stop, and that the convergent cemented surface in the rear is convex to the stop. This is also true in the Dagor lens of Fig. 11.5.

The *Dagor* (Fig. 11.5) combines both the old achromat and the new achromat into a cemented triplet construction. If we visualize the central negative element of the triplet split into two parts, then the outer high-index element and the outer part of the (medium-index) middle element can be seen to make up a new achromat. The inner low-index crown element and the other part of the middle element make up the old achromat. The symmetrical construction about the stop minimizes the coma and distortion, while the spacing from the stop and the cemented surfaces control the astigmatism.

Since either the front or the rear half of this sort of lens can be designed to perform reasonably well by itself, a *convertible lens* can be constructed to yield two different focal lengths when one uses either both components or one alone. If a hemisymmetrical construction (with the halves similar, but of different focal lengths) is used, then three different focal lengths are available.

Many elaborations on this theme were designed, with as many as

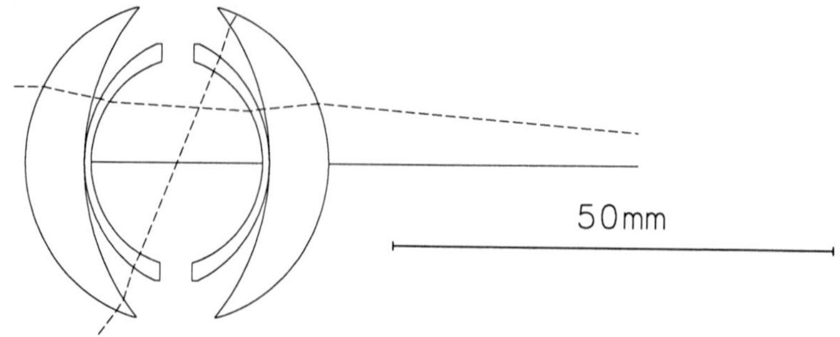

ROBERT RICHTER; USP 2031792; 66 MM F/6.3 100 DEG. FIELD EX. #1

radius	thickness	mat'l	index	V-no	sa
16.875	6.660	SK16	1.620	60.3	16.3
24.825	0.030	air			16.3
13.641	0.750	SF1	1.717	29.5	12.4
11.025	9.735	air			10.6
	9.735	air			5.9
-11.025	0.750	SF1	1.717	29.5	10.6
-13.641	0.030	air			12.4
-24.825	6.660	SK16	1.620	60.3	16.3
-16.875	77.809	air			16.3

EFL = 98.96
BFL = 77.81
NA = -0.0787 (F/6.3)
GIH = 117.93 (HFOV=50.00)
PTZ/F = 7.73e+05
VL = 34.35
OD infinite conjugate

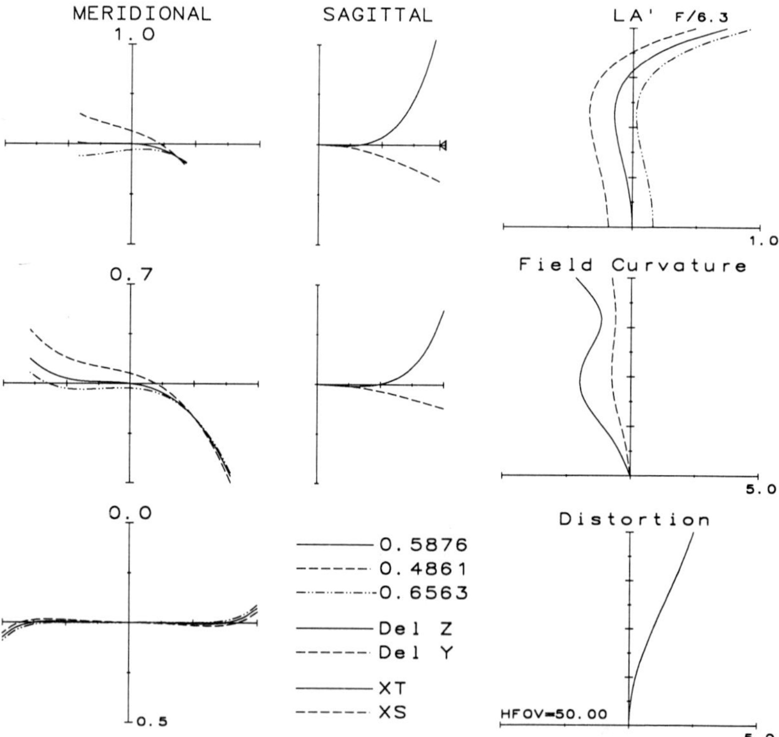

Figure 11.2

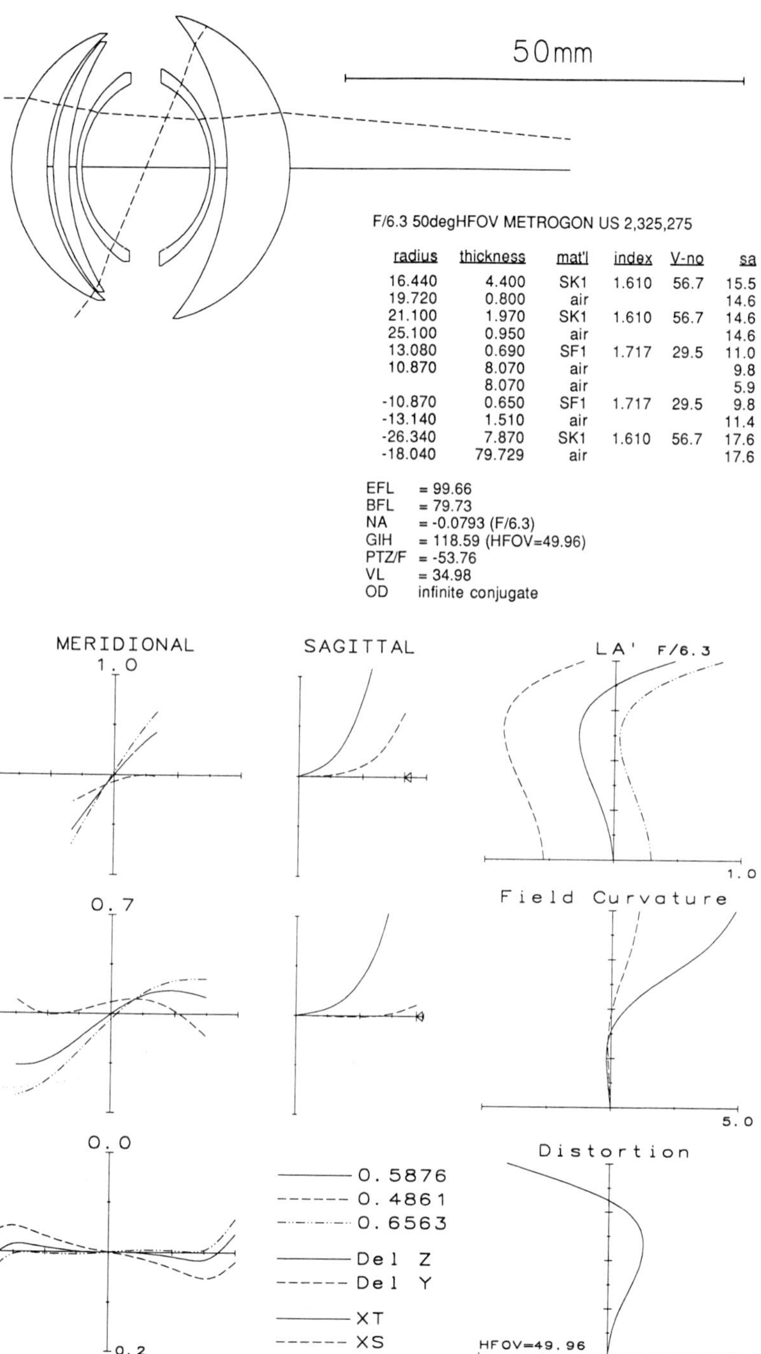

F/6.3 50degHFOV METROGON US 2,325,275

radius	thickness	mat'l	index	V-no	sa
16.440	4.400	SK1	1.610	56.7	15.5
19.720	0.800	air			14.6
21.100	1.970	SK1	1.610	56.7	14.6
25.100	0.950	air			14.6
13.080	0.690	SF1	1.717	29.5	11.0
10.870	8.070	air			9.8
	8.070	air			5.9
-10.870	0.650	SF1	1.717	29.5	9.8
-13.140	1.510	air			11.4
-26.340	7.870	SK1	1.610	56.7	17.6
-18.040	79.729	air			17.6

EFL = 99.66
BFL = 79.73
NA = -0.0793 (F/6.3)
GIH = 118.59 (HFOV=49.96)
PTZ/F = -53.76
VL = 34.98
OD infinite conjugate

Figure 11.3

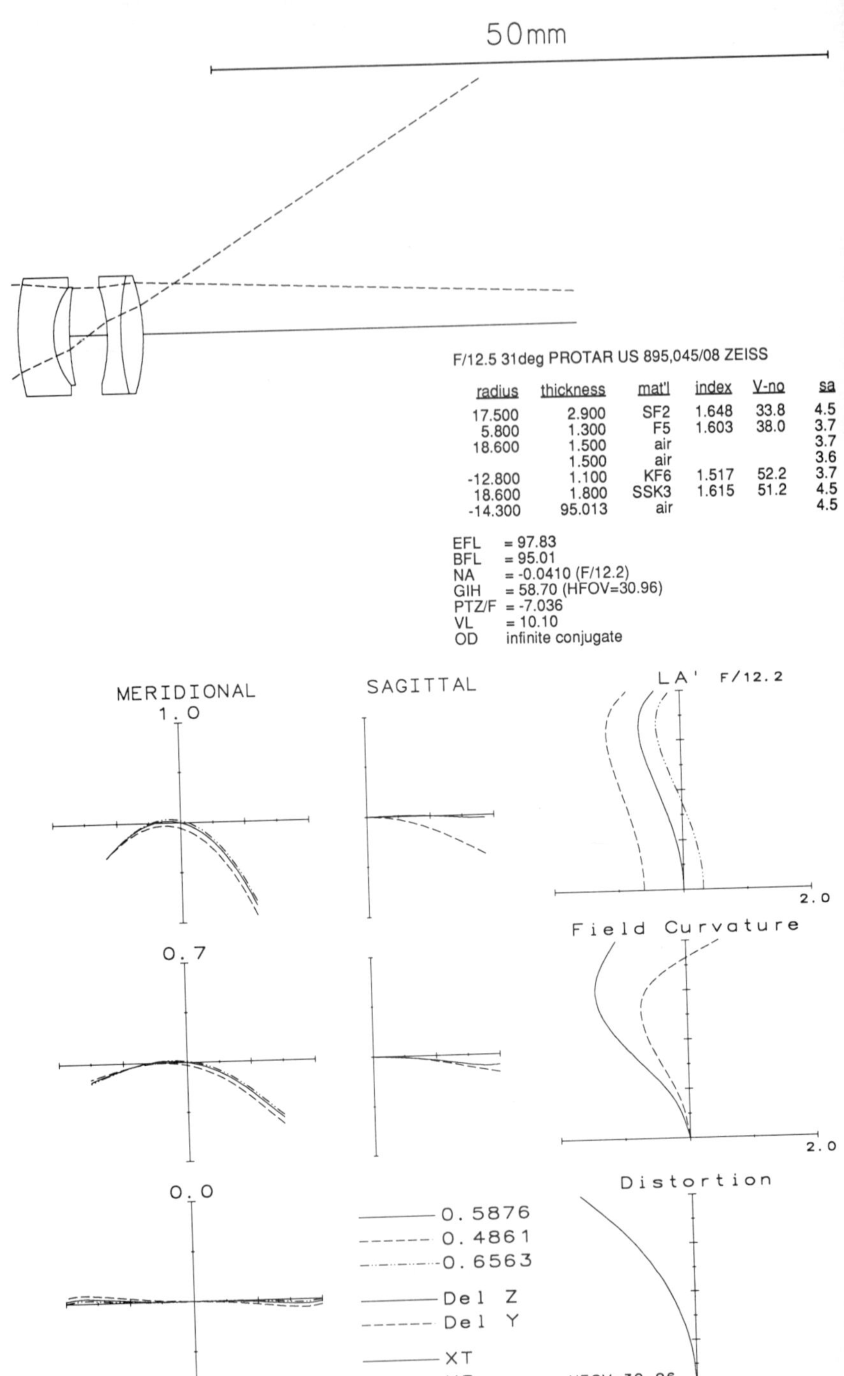

F/12.5 31deg PROTAR US 895,045/08 ZEISS

radius	thickness	mat'l	index	V-no	sa
17.500	2.900	SF2	1.648	33.8	4.5
5.800	1.300	F5	1.603	38.0	3.7
18.600	1.500	air			3.7
	1.500	air			3.6
-12.800	1.100	KF6	1.517	52.2	3.7
18.600	1.800	SSK3	1.615	51.2	4.5
-14.300	95.013	air			4.5

EFL = 97.83
BFL = 95.01
NA = -0.0410 (F/12.2)
GIH = 58.70 (HFOV=30.96)
PTZ/F = -7.036
VL = 10.10
OD infinite conjugate

Figure 11.4

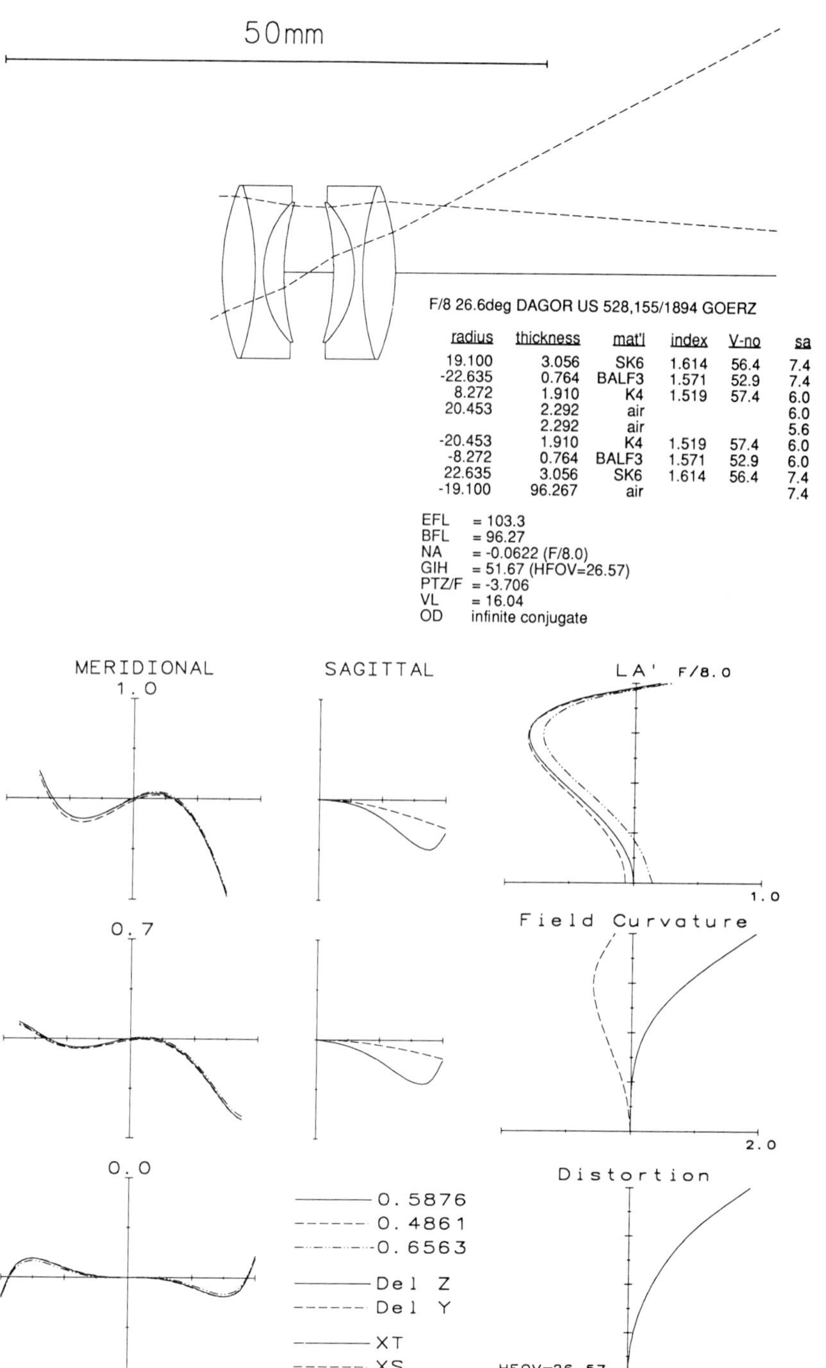

F/8 26.6deg DAGOR US 528,155/1894 GOERZ

radius	thickness	mat'l	index	V-no	sa
19.100	3.056	SK6	1.614	56.4	7.4
-22.635	0.764	BALF3	1.571	52.9	7.4
8.272	1.910	K4	1.519	57.4	6.0
20.453	2.292	air			6.0
	2.292	air			5.6
-20.453	1.910	K4	1.519	57.4	6.0
-8.272	0.764	BALF3	1.571	52.9	6.0
22.635	3.056	SK6	1.614	56.4	7.4
-19.100	96.267	air			7.4

EFL = 103.3
BFL = 96.27
NA = -0.0622 (F/8.0)
GIH = 51.67 (HFOV=26.57)
PTZ/F = -3.706
VL = 16.04
OD infinite conjugate

Figure 11.5

five elements cemented together on each side of the stop, for a total of 10 elements. Little was gained by adding elements, and one cannot help but suspect that many elaborations were more for the purpose of evading patent coverage than for the improvement of image quality.

There are many designs which can be regarded as descendants of these early thick meniscus anastigmats. The *Tessar* (Chap. 12), although it looks like a modification of the Cooke triplet, was actually designed as a rear new achromat combined with an airspaced front pair replacing the old achromat of the Protar. The airspaced pair can be regarded as a thick meniscus triplet component with a center lens made of air.

The *Biotar* or double Gauss lens (Chap. 17) is also a (very distant) descendant of this general form, as indicated by the thick-meniscus inner doublets. The Biotar has evolved into what is arguably the most versatile and powerful of the standard design types.

11.4 The Split Dagor

In the *Orthometar, Plasmat, W. A. Express,* and *Euryplan* (Fig. 11.6) types, the inner meniscus elements of the Dagor were split off, a higher-index glass was used, and complete symmetry was abandoned. The freedom to independently bend these inner elements allowed the designer to correct the spherical aberration without being limited to the low-index glass of the original old achromat part of the Dagor in order to do so. Of course, this doubled the number of air-glass surfaces from four to eight, but this design was a significant improvement. A fully symmetrical version of this form has been widely utilized in xerographic copiers, working at near-unity magnifications. Another thick meniscus form is shown in Fig. 11.7, as a copy lens for use at unit magnification.

11.5 The Dogmar

The *Dogmar* lens form (Fig. 11.8) is also a member of the double-meniscus family, as can be realized if one considers each half to be a triplet with a center air lens. The Dogmar form is used as an excellent general-purpose camera lens, and its symmetry and stability of correction make it eminently suitable for an enlarger lens, as shown in Fig. 11.9 at a 1:0.25 magnification. The performance of the Dogmar is improved by the use of higher-index crowns, as shown for the enlarger lens in Fig. 11.10. For use as a process lens, glasses with unusual partial dispersions can be chosen (see Sec. 6.2) to reduce the secondary spectrum.

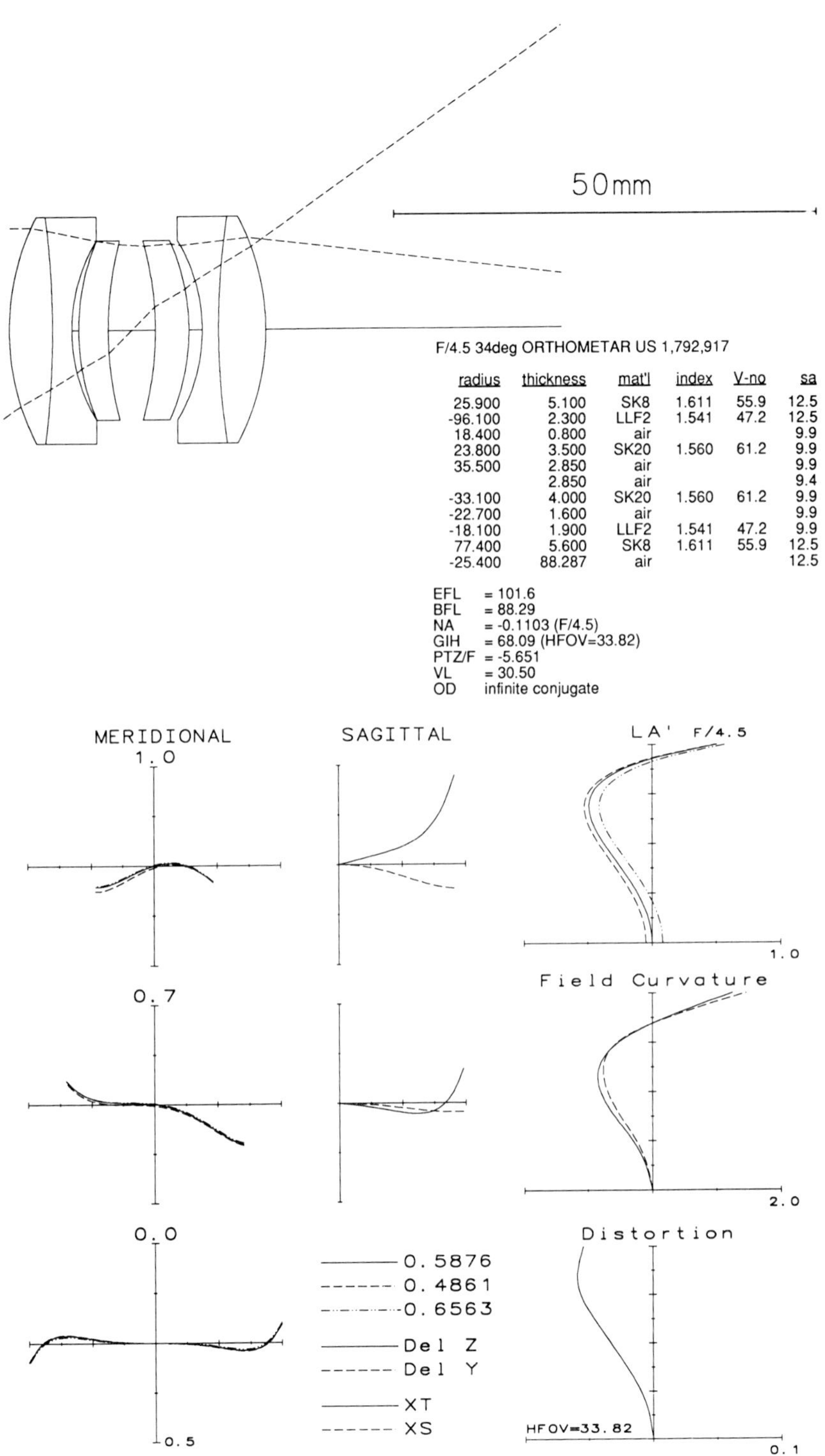

F/4.5 34deg ORTHOMETAR US 1,792,917

radius	thickness	mat'l	index	V-no	sa
25.900	5.100	SK8	1.611	55.9	12.5
-96.100	2.300	LLF2	1.541	47.2	12.5
18.400	0.800	air			9.9
23.800	3.500	SK20	1.560	61.2	9.9
35.500	2.850	air			9.9
	2.850	air			9.4
-33.100	4.000	SK20	1.560	61.2	9.9
-22.700	1.600	air			9.9
-18.100	1.900	LLF2	1.541	47.2	9.9
77.400	5.600	SK8	1.611	55.9	12.5
-25.400	88.287	air			12.5

EFL = 101.6
BFL = 88.29
NA = -0.1103 (F/4.5)
GIH = 68.09 (HFOV=33.82)
PTZ/F = -5.651
VL = 30.50
OD infinite conjugate

Figure 11.6

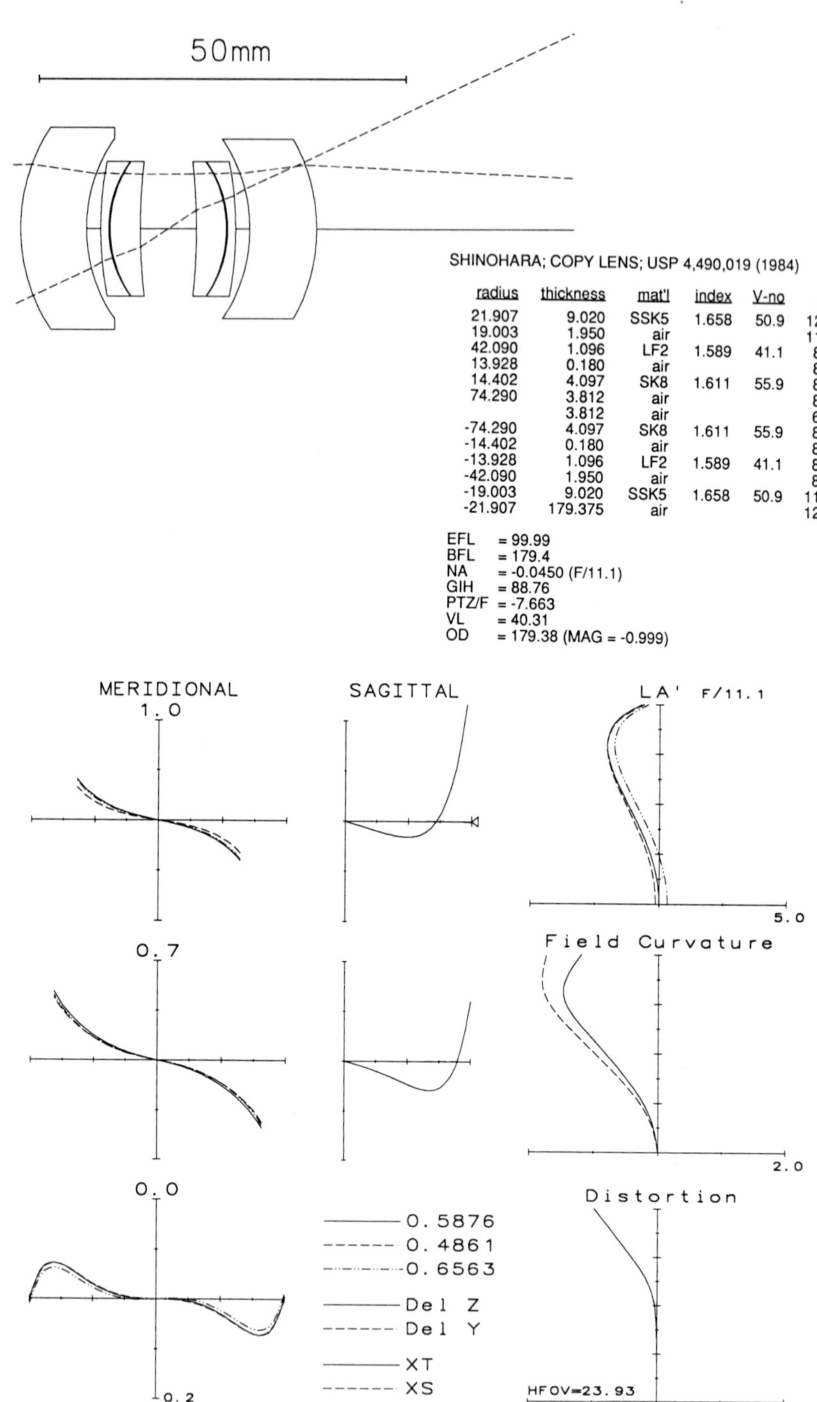

SHINOHARA; COPY LENS; USP 4,490,019 (1984)

radius	thickness	mat'l	index	V-no	sa
21.907	9.020	SSK5	1.658	50.9	12.8
19.003	1.950	air			11.4
42.090	1.096	LF2	1.589	41.1	8.6
13.928	0.180	air			8.6
14.402	4.097	SK8	1.611	55.9	8.6
74.290	3.812	air			8.6
	3.812	air			6.9
-74.290	4.097	SK8	1.611	55.9	8.6
-14.402	0.180	air			8.6
-13.928	1.096	LF2	1.589	41.1	8.6
-42.090	1.950	air			8.6
-19.003	9.020	SSK5	1.658	50.9	11.4
-21.907	179.375	air			12.8

EFL = 99.99
BFL = 179.4
NA = -0.0450 (F/11.1)
GIH = 88.76
PTZ/F = -7.663
VL = 40.31
OD = 179.38 (MAG = -0.999)

Figure 11.7

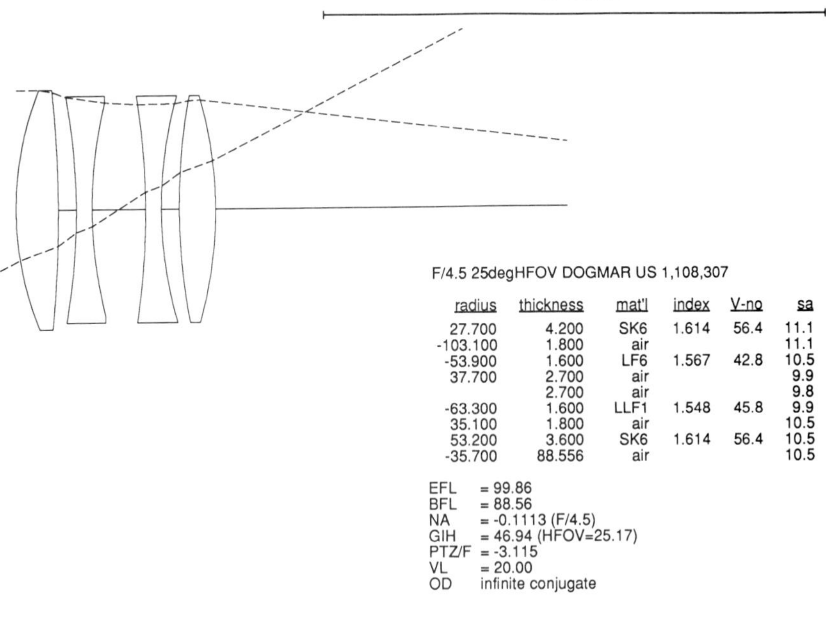

F/4.5 25degHFOV DOGMAR US 1,108,307

radius	thickness	mat'l	index	V-no	sa
27.700	4.200	SK6	1.614	56.4	11.1
-103.100	1.800	air			11.1
-53.900	1.600	LF6	1.567	42.8	10.5
37.700	2.700	air			9.9
	2.700	air			9.8
-63.300	1.600	LLF1	1.548	45.8	9.9
35.100	1.800	air			10.5
53.200	3.600	SK6	1.614	56.4	10.5
-35.700	88.556	air			10.5

EFL = 99.86
BFL = 88.56
NA = -0.1113 (F/4.5)
GIH = 46.94 (HFOV=25.17)
PTZ/F = -3.115
VL = 20.00
OD infinite conjugate

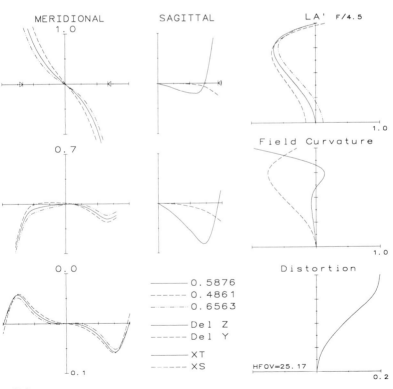

MERIDIONAL SAGITTAL LA' F/4.5

1.0

1.0

Field Curvature

0.7

1.0

0.0 Distortion

0.5876
0.4861
0.6563
Del Z
Del Y
XT
XS

0.1 HFOV=25.17 0.2

Figure 11.8

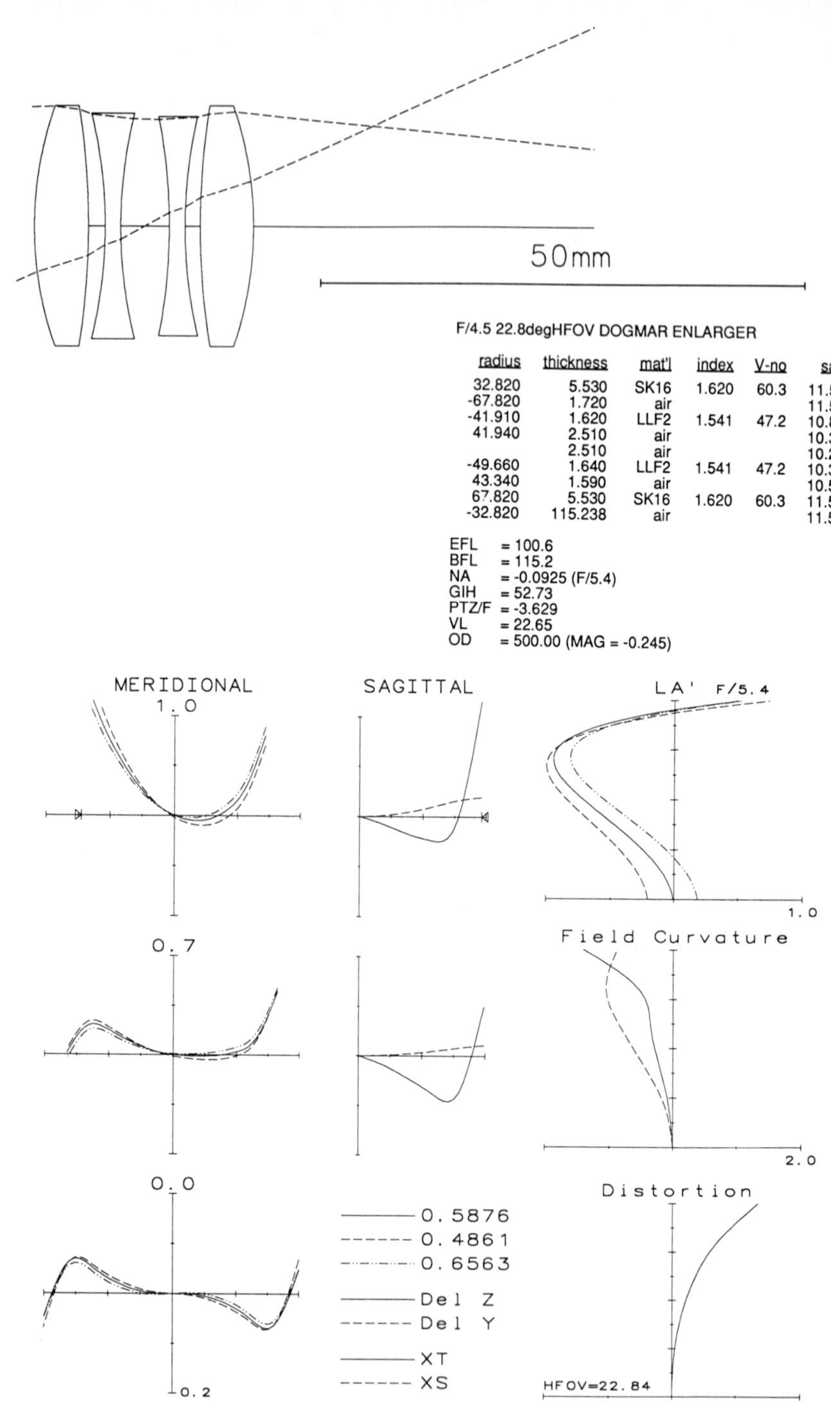

F/4.5 22.8degHFOV DOGMAR ENLARGER

radius	thickness	mat'l	index	V-no	sa
32.820	5.530	SK16	1.620	60.3	11.5
-67.820	1.720	air			11.5
-41.910	1.620	LLF2	1.541	47.2	10.8
41.940	2.510	air			10.3
	2.510	air			10.2
-49.660	1.640	LLF2	1.541	47.2	10.3
43.340	1.590	air			10.5
67.820	5.530	SK16	1.620	60.3	11.5
-32.820	115.238	air			11.5

EFL = 100.6
BFL = 115.2
NA = -0.0925 (F/5.4)
GIH = 52.73
PTZ/F = -3.629
VL = 22.65
OD = 500.00 (MAG = -0.245)

Figure 11.9

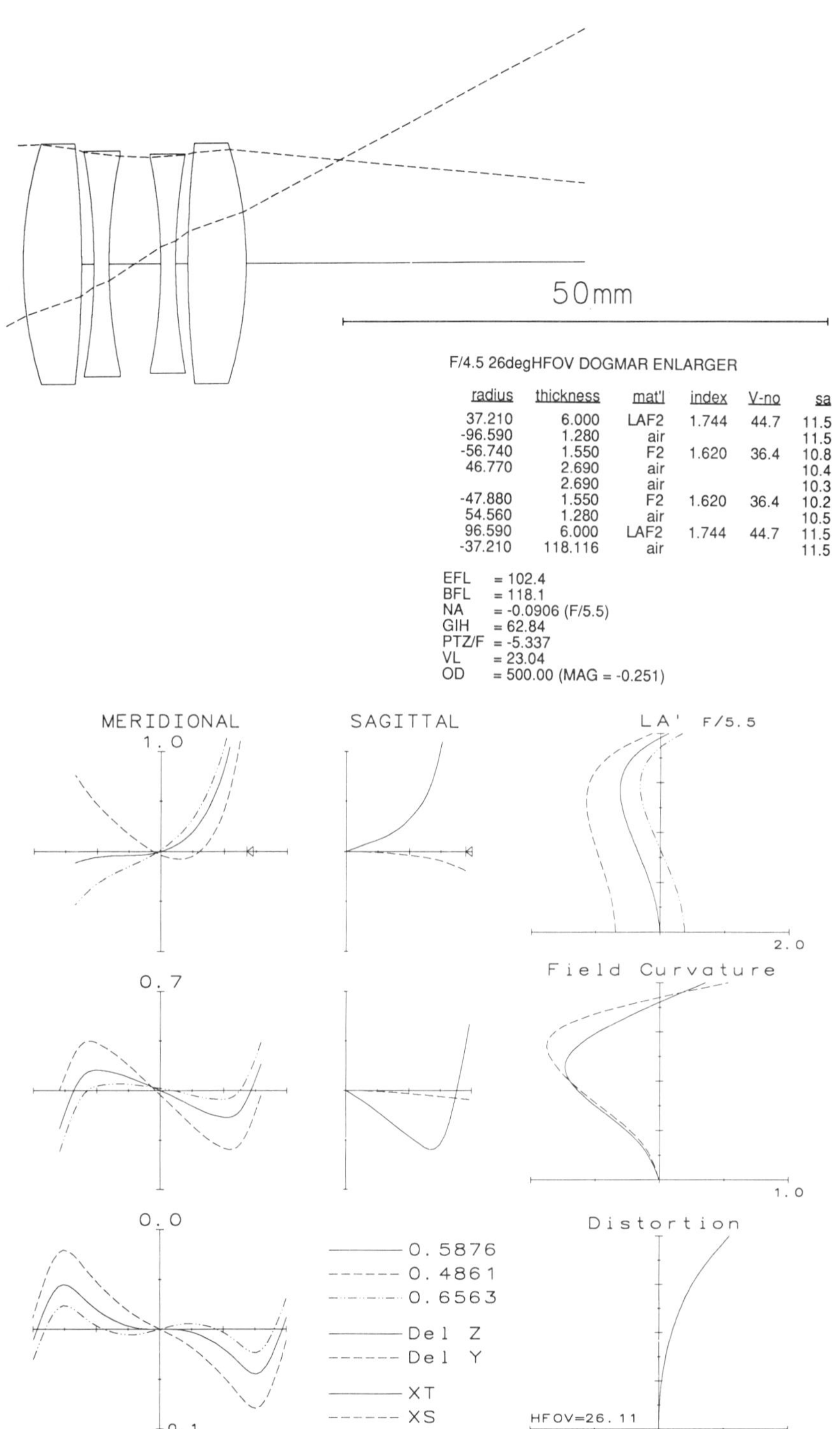

F/4.5 26degHFOV DOGMAR ENLARGER

radius	thickness	mat'l	index	V-no	sa
37.210	6.000	LAF2	1.744	44.7	11.5
-96.590	1.280	air			11.5
-56.740	1.550	F2	1.620	36.4	10.8
46.770	2.690	air			10.4
	2.690	air			10.3
-47.880	1.550	F2	1.620	36.4	10.2
54.560	1.280	air			10.5
96.590	6.000	LAF2	1.744	44.7	11.5
-37.210	118.116	air			11.5

EFL = 102.4
BFL = 118.1
NA = -0.0906 (F/5.5)
GIH = 62.84
PTZ/F = -5.337
VL = 23.04
OD = 500.00 (MAG = -0.251)

Figure 11.10

Chapter

12

The Tessar, Heliar, and Other Compounded Triplets

12.1 The Classic Tessar

Although the Tessar is actually a descendant of the double-meniscus anastigmat form, as described in Chap. 11, it can also be regarded as a modification of the Cooke triplet. As explained in more detail in Sec. 3.5, the substitution of a new achromat doublet for a crown in the triplet is the equivalent of using a high-index, high-V-value glass, and allows the possibility of utilizing the cemented surface of the doublet to control coma and oblique spherical aberration.

The classical Tessar form is illustrated in Figs. 12.1 to 12.12. Figure 12.1 is an early Tessar by Rudolph, using relatively standard glasses and covering a relatively standard (for its day) 60° field at a speed of *f*/6.3. Figures 12.2 and 12.3 each cover a 56° field at *f*/4.5; Fig. 12.3 achieves a modestly improved performance by using higher-index crown elements. Figure 12.4 is a lens similar to Fig. 12.2, but which has been corrected for use as an enlarger lens; it is shown at a magnification of 1:5.

Figures 12.5 and 12.6 are lenses of ordinary glass, at similar speeds (*f*/3.5) with angular fields of 44° and 56° respectively. Figure 12.7 covers the same 56° field as Fig. 12.6, but is balanced for an aperture of *f*/3.0 and has its rear doublet arranged in reversed order. According to Kingslake, this orientation is to be preferred when high-index glass is used.

A still higher speed (*f*/2.8) is illustrated in Figs. 12.8 and 12.9 at an angular field of 41° or 42°; Fig. 12.8 uses higher-index glasses all around. A third *f*/2.8 lens (Fig. 12.10) illustrates a design which is con-

(Text continues on page 210.)

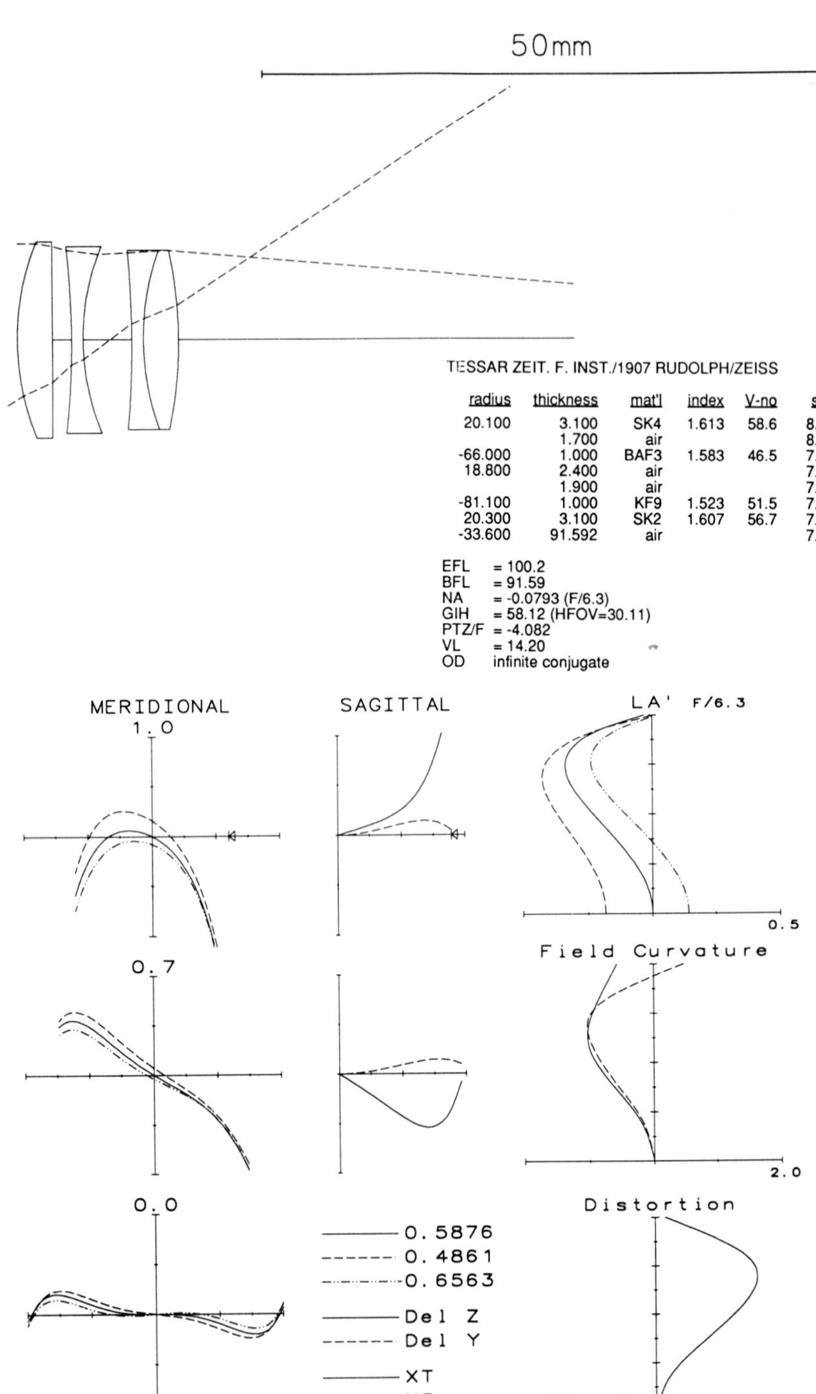

TESSAR ZEIT. F. INST./1907 RUDOLPH/ZEISS

radius	thickness	mat'l	index	V-no	sa
20.100	3.100	SK4	1.613	58.6	8.1
	1.700	air			8.1
-66.000	1.000	BAF3	1.583	46.5	7.7
18.800	2.400	air			7.7
	1.900	air			7.1
-81.100	1.000	KF9	1.523	51.5	7.4
20.300	3.100	SK2	1.607	56.7	7.4
-33.600	91.592	air			7.4

EFL = 100.2
BFL = 91.59
NA = -0.0793 (F/6.3)
GIH = 58.12 (HFOV=30.11)
PTZ/F = -4.082
VL = 14.20
OD infinite conjugate

Figure 12.1

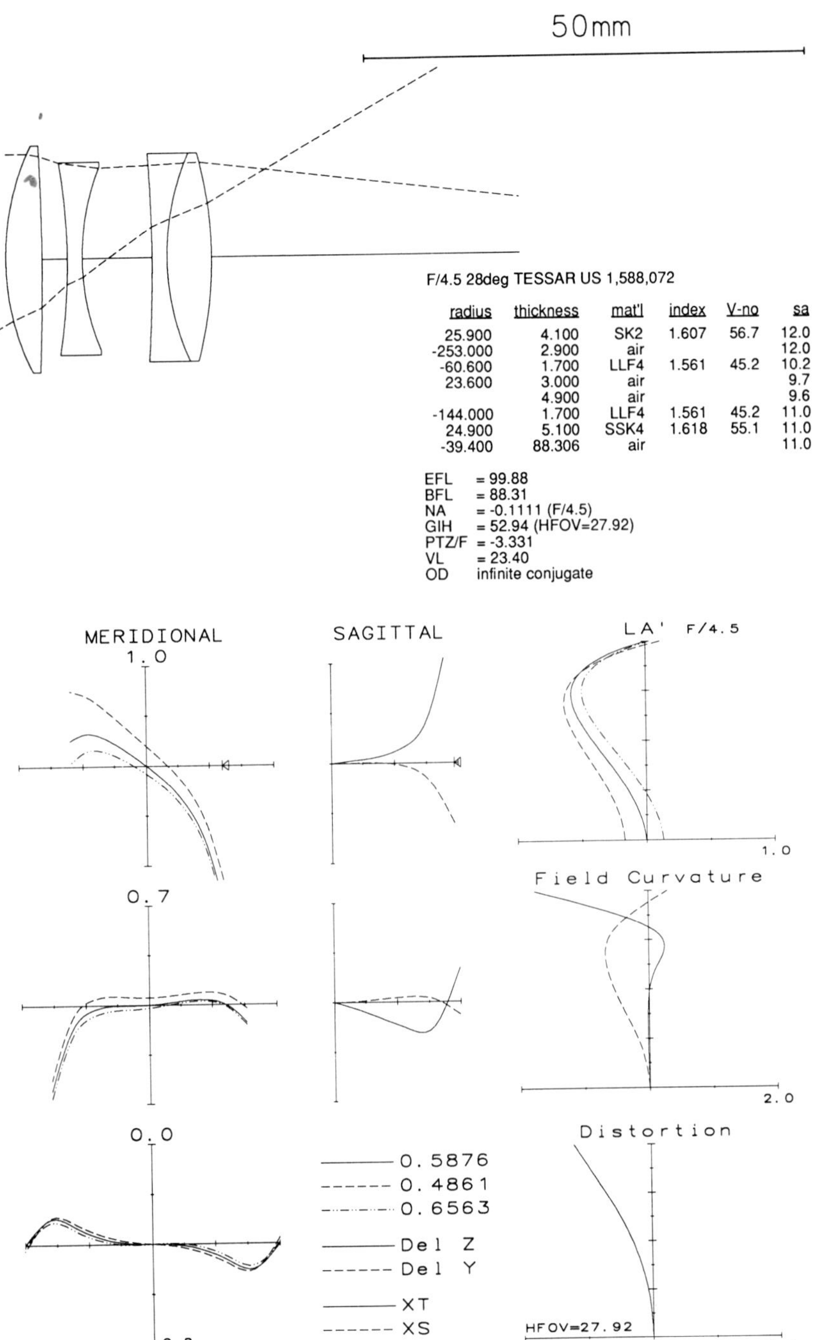

F/4.5 28deg TESSAR US 1,588,072

radius	thickness	mat'l	index	V-no	sa
25.900	4.100	SK2	1.607	56.7	12.0
-253.000	2.900	air			12.0
-60.600	1.700	LLF4	1.561	45.2	10.2
23.600	3.000	air			9.7
	4.900	air			9.6
-144.000	1.700	LLF4	1.561	45.2	11.0
24.900	5.100	SSK4	1.618	55.1	11.0
-39.400	88.306	air			11.0

EFL = 99.88
BFL = 88.31
NA = -0.1111 (F/4.5)
GIH = 52.94 (HFOV=27.92)
PTZ/F = -3.331
VL = 23.40
OD infinite conjugate

Figure 12.2

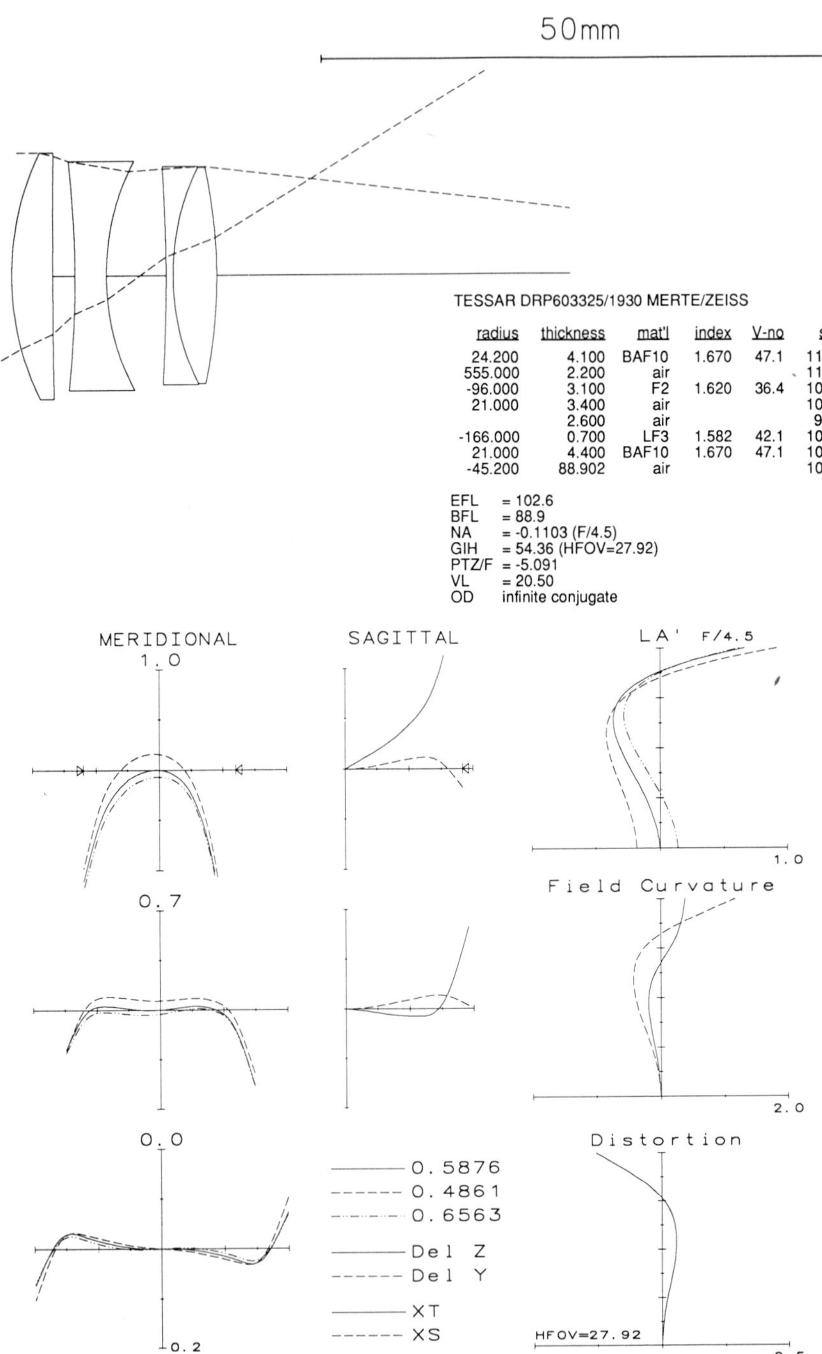

TESSAR DRP603325/1930 MERTE/ZEISS

radius	thickness	mat'l	index	V-no	sa
24.200	4.100	BAF10	1.670	47.1	11.4
555.000	2.200	air			11.4
-96.000	3.100	F2	1.620	36.4	10.6
21.000	3.400	air			10.6
	2.600	air			9.7
-166.000	0.700	LF3	1.582	42.1	10.1
21.000	4.400	BAF10	1.670	47.1	10.0
-45.200	88.902	air			10.1

EFL = 102.6
BFL = 88.9
NA = -0.1103 (F/4.5)
GIH = 54.36 (HFOV=27.92)
PTZ/F = -5.091
VL = 20.50
OD infinite conjugate

Figure 12.3

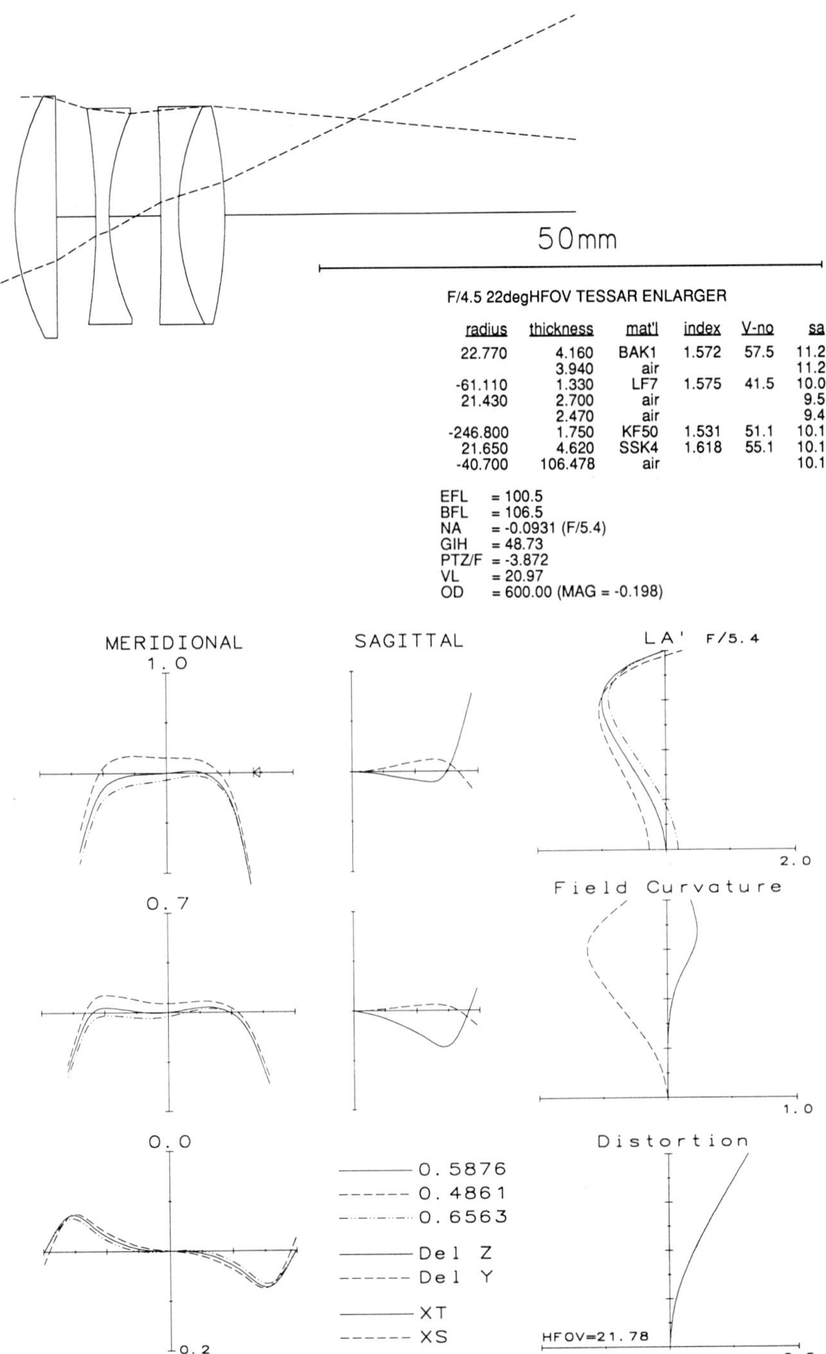

F/4.5 22degHFOV TESSAR ENLARGER

radius	thickness	mat'l	index	V-no	sa
22.770	4.160	BAK1	1.572	57.5	11.2
	3.940	air			11.2
-61.110	1.330	LF7	1.575	41.5	10.0
21.430	2.700	air			9.5
	2.470	air			9.4
-246.800	1.750	KF50	1.531	51.1	10.1
21.650	4.620	SSK4	1.618	55.1	10.1
-40.700	106.478	air			10.1

EFL = 100.5
BFL = 106.5
NA = -0.0931 (F/5.4)
GIH = 48.73
PTZ/F = -3.872
VL = 20.97
OD = 600.00 (MAG = -0.198)

Figure 12.4

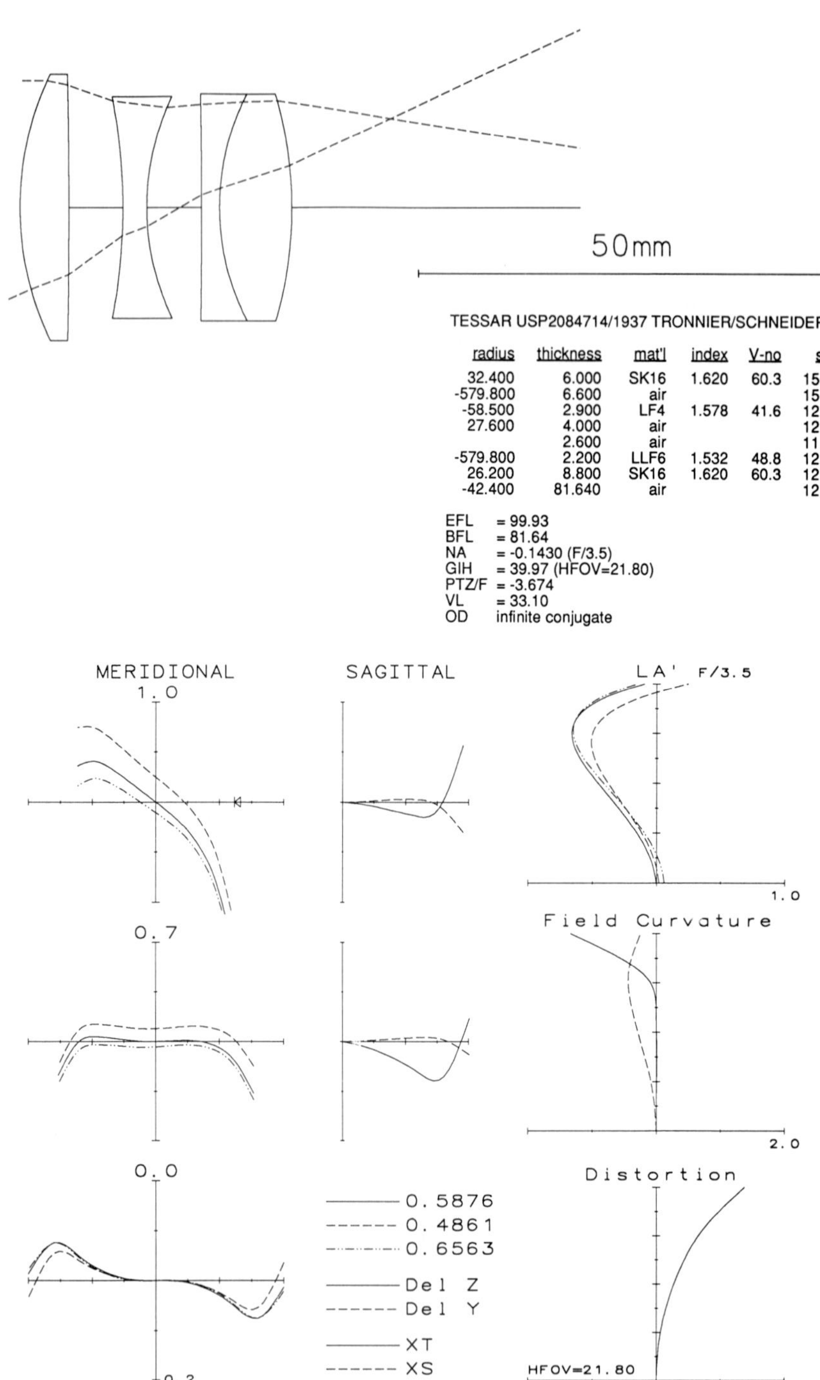

TESSAR USP2084714/1937 TRONNIER/SCHNEIDER

radius	thickness	mat'l	index	V-no	sa
32.400	6.000	SK16	1.620	60.3	15.0
-579.800	6.600	air			15.0
-58.500	2.900	LF4	1.578	41.6	12.5
27.600	4.000	air			12.5
	2.600	air			11.4
-579.800	2.200	LLF6	1.532	48.8	12.8
26.200	8.800	SK16	1.620	60.3	12.8
-42.400	81.640	air			12.8

EFL = 99.93
BFL = 81.64
NA = -0.1430 (F/3.5)
GIH = 39.97 (HFOV=21.80)
PTZ/F = -3.674
VL = 33.10
OD infinite conjugate

Figure 12.5

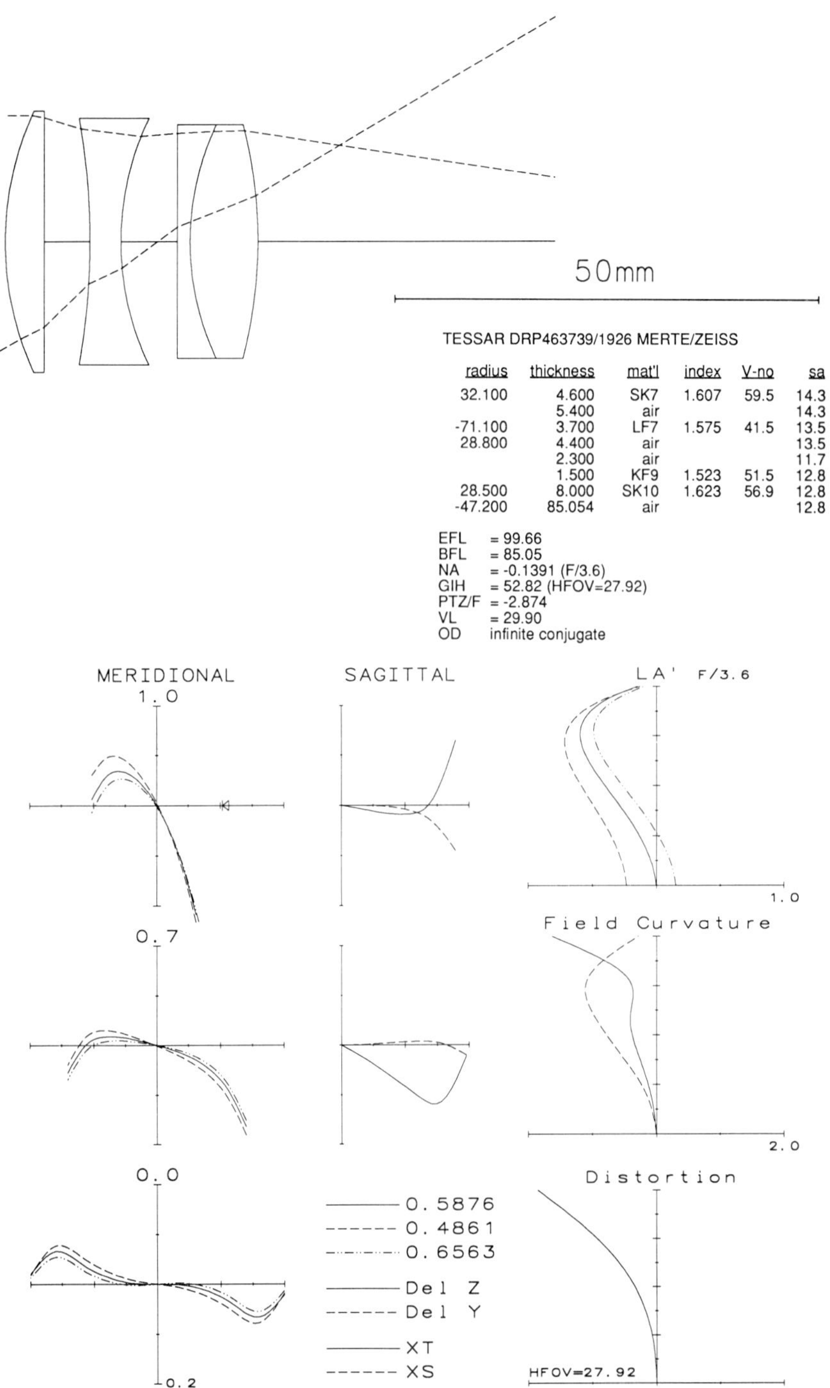

TESSAR DRP463739/1926 MERTE/ZEISS

radius	thickness	mat'l	index	V-no	sa
32.100	4.600	SK7	1.607	59.5	14.3
	5.400	air			14.3
-71.100	3.700	LF7	1.575	41.5	13.5
28.800	4.400	air			13.5
	2.300	air			11.7
	1.500	KF9	1.523	51.5	12.8
28.500	8.000	SK10	1.623	56.9	12.8
-47.200	85.054	air			12.8

EFL = 99.66
BFL = 85.05
NA = -0.1391 (F/3.6)
GIH = 52.82 (HFOV=27.92)
PTZ/F = -2.874
VL = 29.90
OD infinite conjugate

Figure 12.6

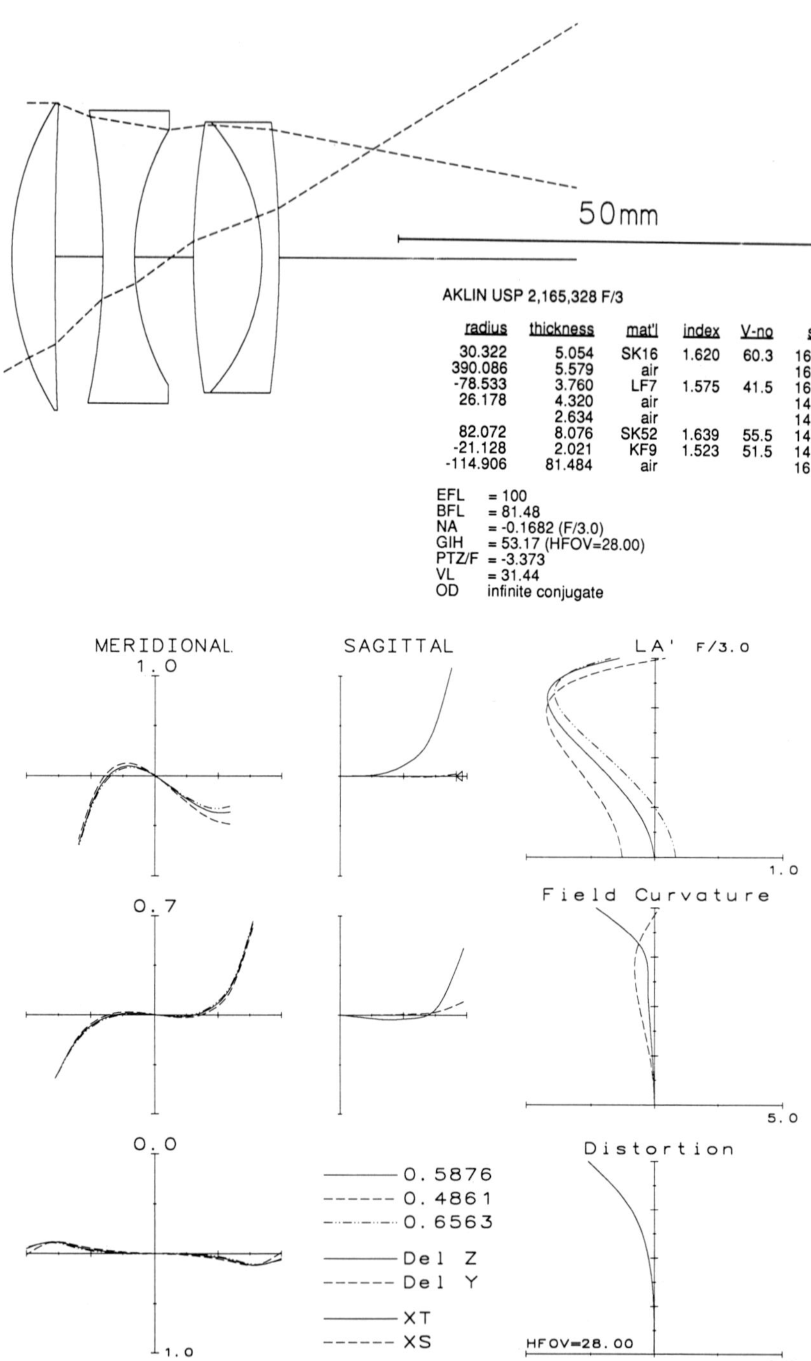

radius	thickness	mat'l	index	V-no	sa
30.322	5.054	SK16	1.620	60.3	16.8
390.086	5.579	air			16.8
-78.533	3.760	LF7	1.575	41.5	16.0
26.178	4.320	air			14.0
	2.634	air			14.0
82.072	8.076	SK52	1.639	55.5	14.8
-21.128	2.021	KF9	1.523	51.5	14.8
-114.906	81.484	air			16.0

Figure 12.7

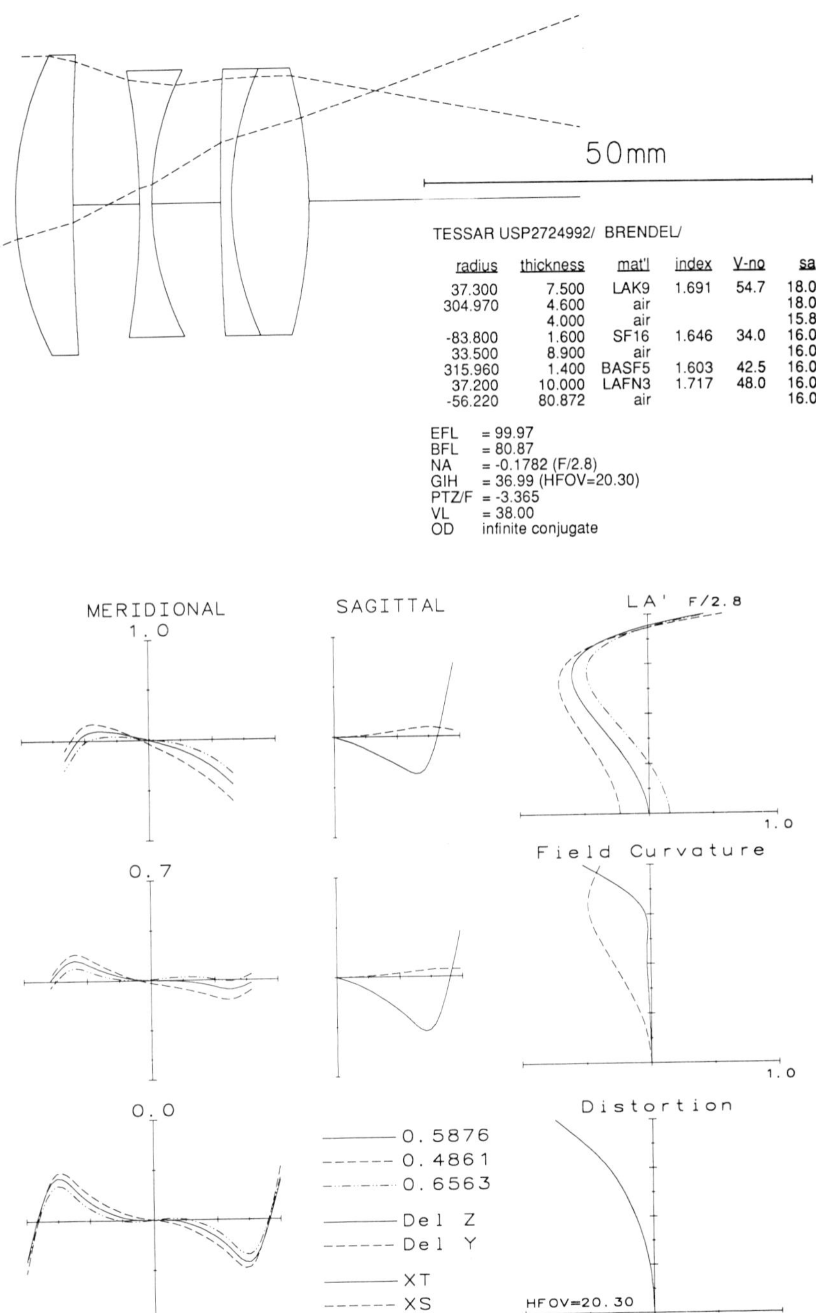

TESSAR USP2724992/ BRENDEL/

radius	thickness	mat'l	index	V-no	sa
37.300	7.500	LAK9	1.691	54.7	18.0
304.970	4.600	air			18.0
	4.000	air			15.8
-83.800	1.600	SF16	1.646	34.0	16.0
33.500	8.900	air			16.0
315.960	1.400	BASF5	1.603	42.5	16.0
37.200	10.000	LAFN3	1.717	48.0	16.0
-56.220	80.872	air			16.0

EFL = 99.97
BFL = 80.87
NA = -0.1782 (F/2.8)
GIH = 36.99 (HFOV=20.30)
PTZ/F = -3.365
VL = 38.00
OD infinite conjugate

Figure 12.8

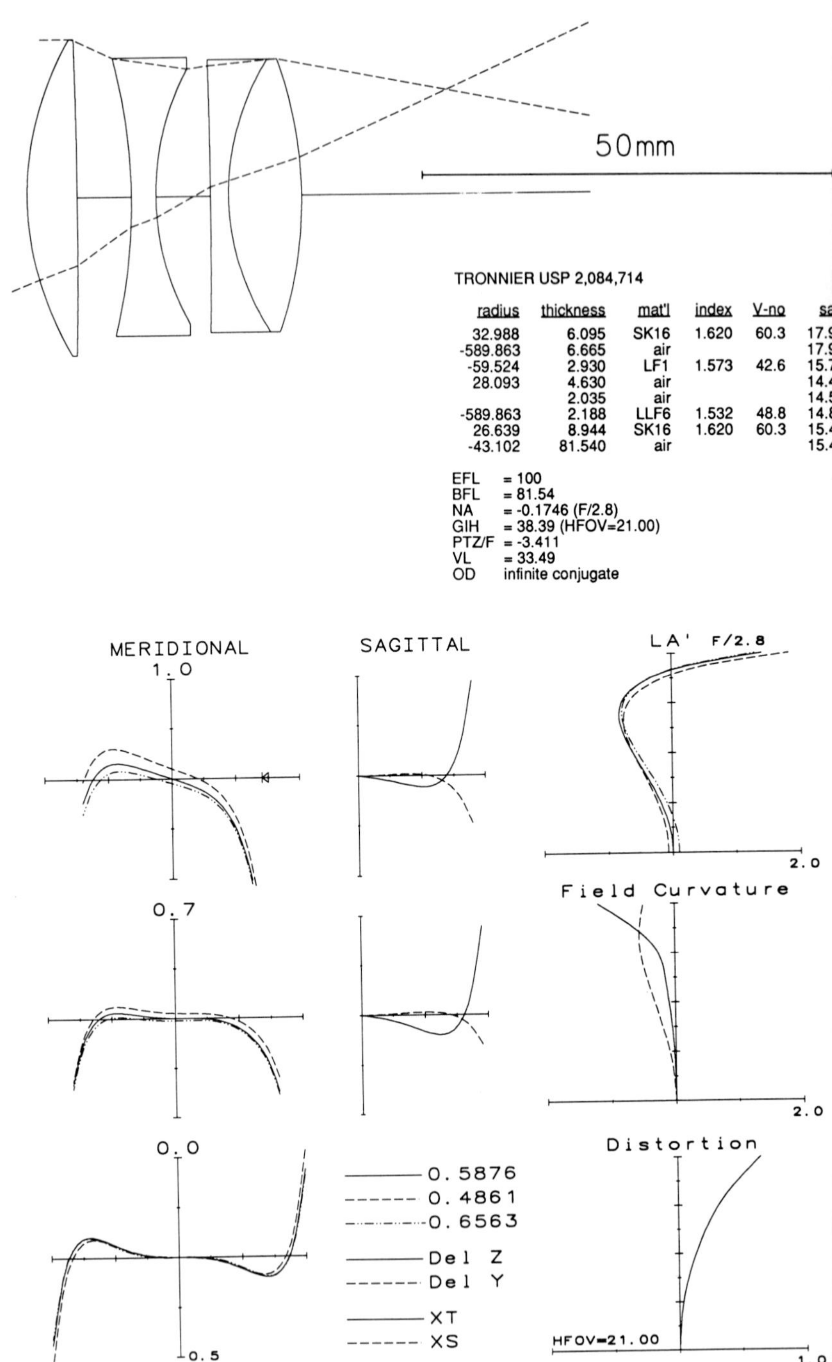

TRONNIER USP 2,084,714

radius	thickness	mat'l	index	V-no	sa
32.988	6.095	SK16	1.620	60.3	17.9
-589.863	6.665	air			17.9
-59.524	2.930	LF1	1.573	42.6	15.7
28.093	4.630	air			14.4
	2.035	air			14.5
-589.863	2.188	LLF6	1.532	48.8	14.8
26.639	8.944	SK16	1.620	60.3	15.4
-43.102	81.540	air			15.4

EFL = 100
BFL = 81.54
NA = -0.1746 (F/2.8)
GIH = 38.39 (HFOV=21.00)
PTZ/F = -3.411
VL = 33.49
OD infinite conjugate

Figure 12.9

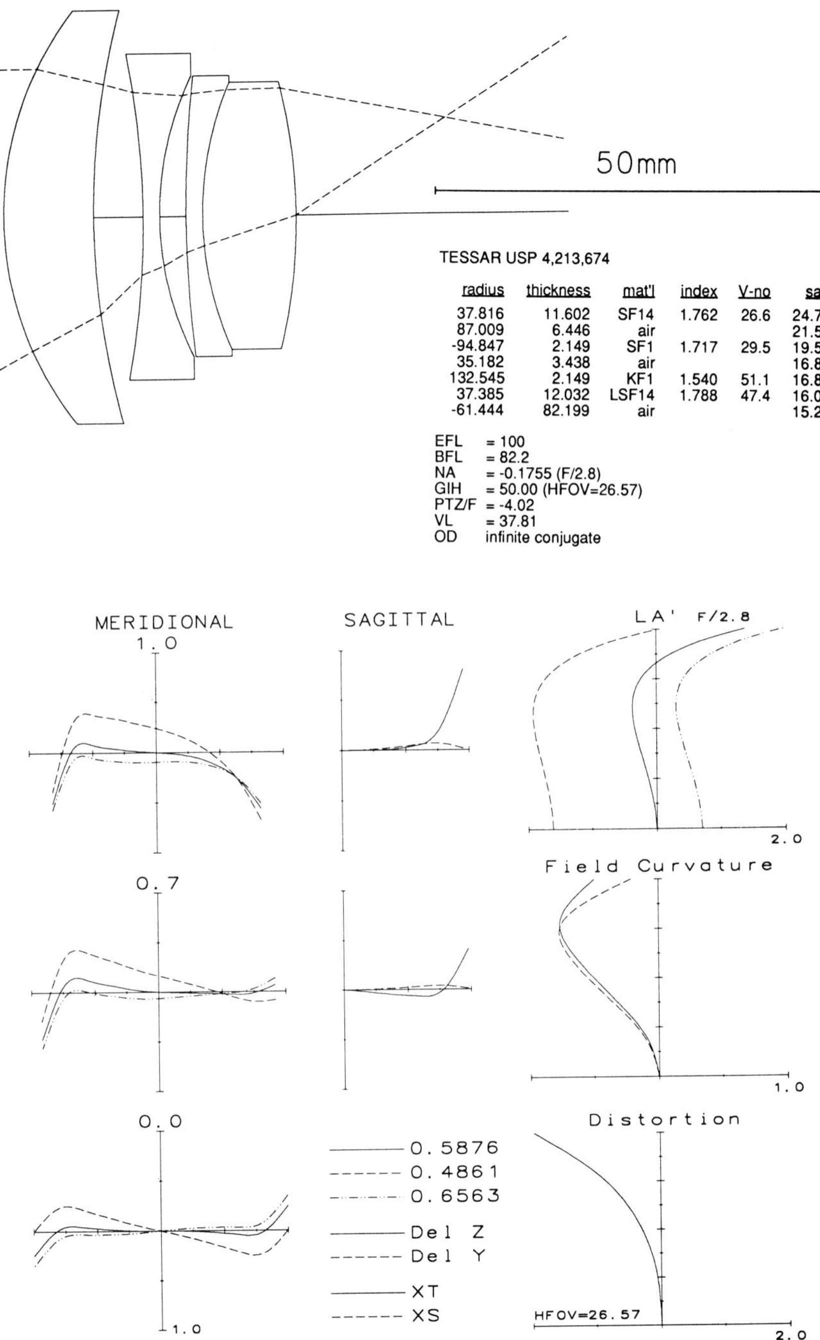

TESSAR USP 4,213,674

radius	thickness	mat'l	index	V-no	sa
37.816	11.602	SF14	1.762	26.6	24.7
87.009	6.446	air			21.5
-94.847	2.149	SF1	1.717	29.5	19.5
35.182	3.438	air			16.8
132.545	2.149	KF1	1.540	51.1	16.8
37.385	12.032	LSF14	1.788	47.4	16.0
-61.444	82.199	air			15.2

EFL = 100
BFL = 82.2
NA = -0.1755 (F/2.8)
GIH = 50.00 (HFOV=26.57)
PTZ/F = -4.02
VL = 37.81
OD infinite conjugate

Figure 12.10

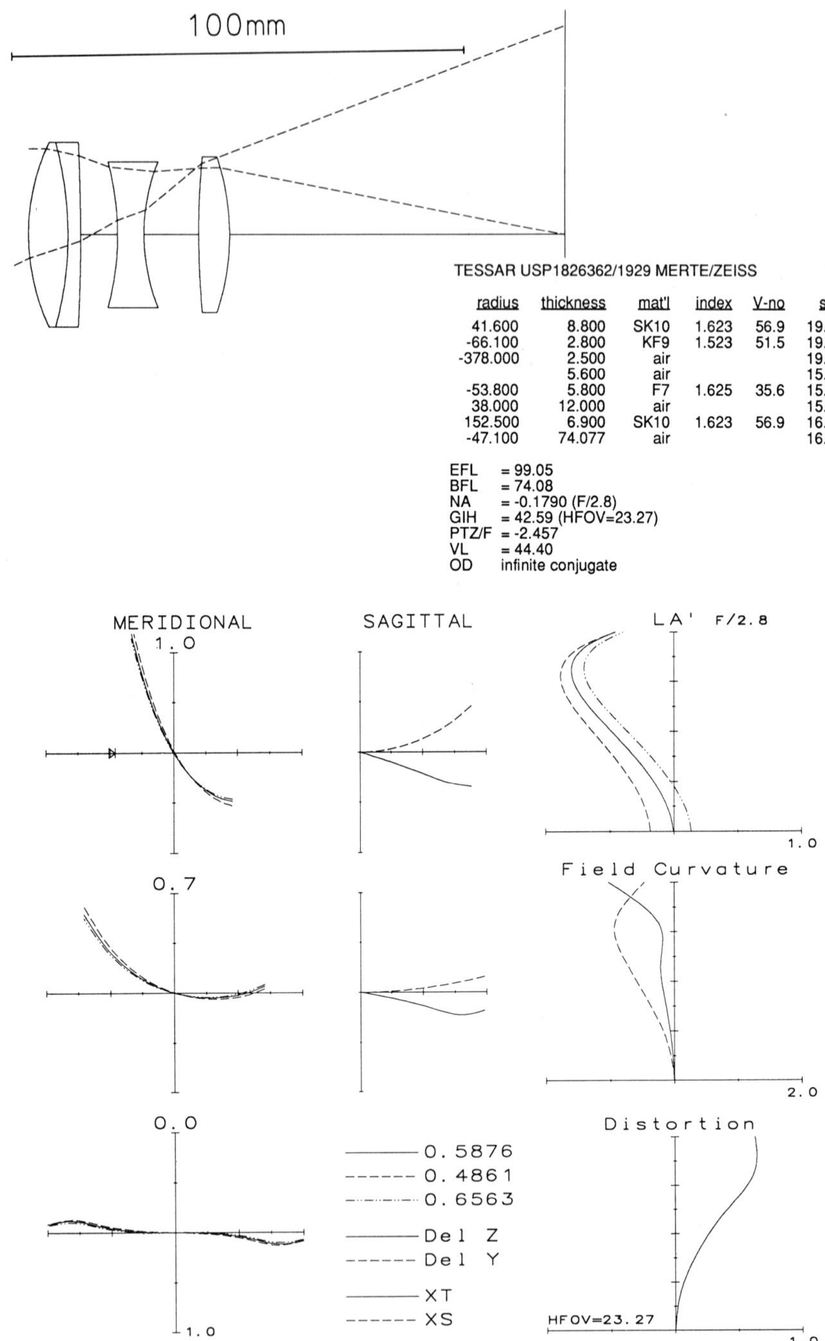

TESSAR USP1826362/1929 MERTE/ZEISS

radius	thickness	mat'l	index	V-no	sa
41.600	8.800	SK10	1.623	56.9	19.0
-66.100	2.800	KF9	1.523	51.5	19.0
-378.000	2.500	air			19.0
	5.600	air			15.2
-53.800	5.800	F7	1.625	35.6	15.0
38.000	12.000	air			15.0
152.500	6.900	SK10	1.623	56.9	16.0
-47.100	74.077	air			16.0

EFL = 99.05
BFL = 74.08
NA = -0.1790 (F/2.8)
GIH = 42.59 (HFOV=23.27)
PTZ/F = -2.457
VL = 44.40
OD infinite conjugate

Figure 12.11

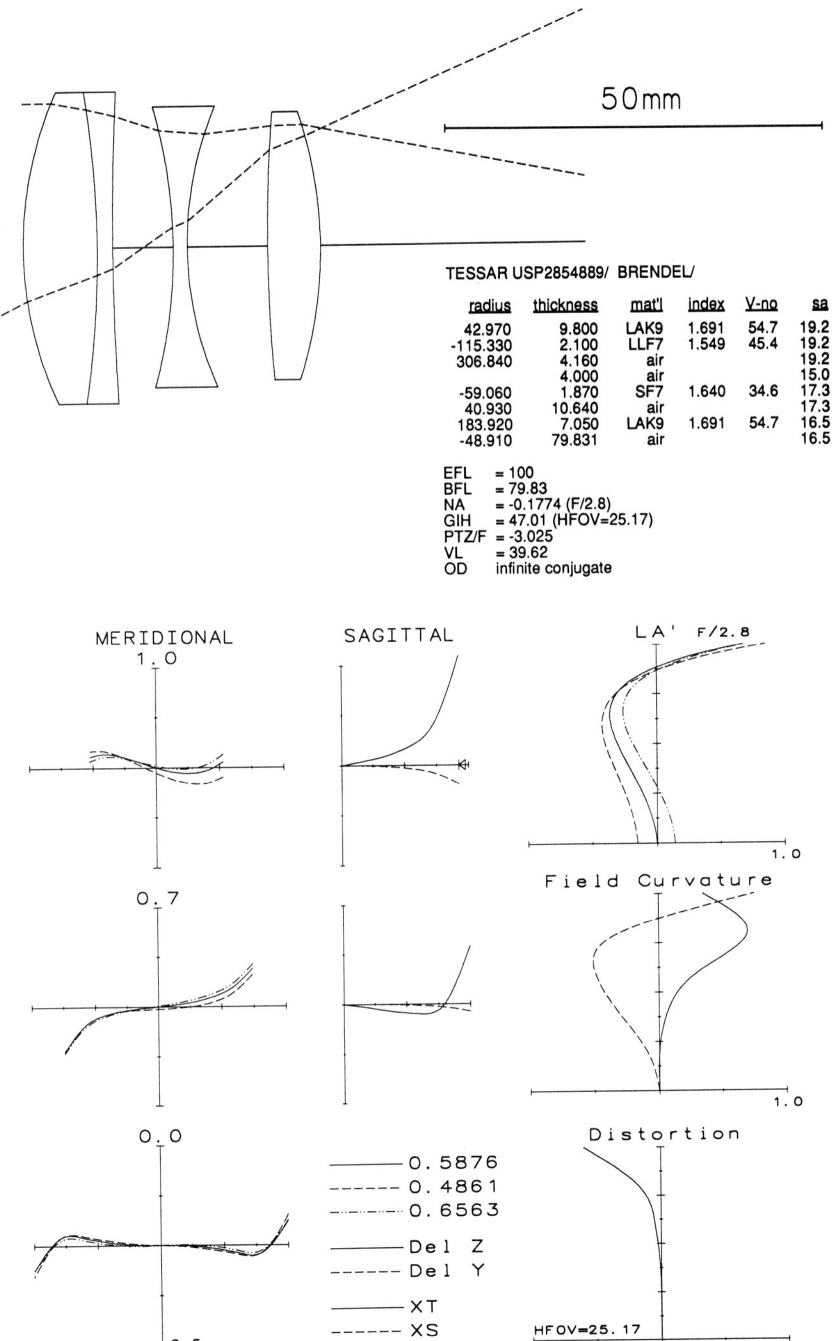

TESSAR USP2854889/ BRENDEL/

radius	thickness	mat'l	index	V-no	sa
42.970	9.800	LAK9	1.691	54.7	19.2
-115.330	2.100	LLF7	1.549	45.4	19.2
306.840	4.160	air			19.2
	4.000	air			15.0
-59.060	1.870	SF7	1.640	34.6	17.3
40.930	10.640	air			17.3
183.920	7.050	LAK9	1.691	54.7	16.5
-48.910	79.831	air			16.5

EFL = 100
BFL = 79.83
NA = -0.1774 (F/2.8)
GIH = 47.01 (HFOV=25.17)
PTZ/F = -3.025
VL = 39.62
OD infinite conjugate

Figure 12.12

figured for a rear, external aperture stop; again, high-index glasses are used.

Two Tessar-type lenses with the *front* member compounded (instead of the rear) are shown in Figs. 12.11 and 12.12. Both are *f*/2.8 lenses, with Fig. 12.12 having the higher indices. They can be compared to Figs. 12.8 and 12.9, although the fields of view are different. Although rarely encountered, it is also possible to compound the center element of the triplet.

12.2 The Heliar/Pentac

The Heliar (or Pentac) form compounds both outer elements of the triplet, and thus has the advantage of an approximately symmetrical form in addition to the double compounding. The result is an improvement over the Tessar. This design form has been executed with the doublets in the flint-out arrangement, as in Fig. 12.13 (in an early, fully symmetrical design), and in the (superior) crown-out arrangement, as in Figs. 12.14 to 12.16.

Two other orientations of the general Heliar format are possible. In Fig. 12.17 both doublets are arranged with their crown elements facing the object (long conjugate). In Fig. 12.18 both crowns face the short conjugate.

12.3 Other Compounded Triplets

Figure 12.19 has its rear element compounded to a triplet, although, because the three glasses have very similar *V* values, there is very little chromatic correction produced. The cemented triplet does, however, provide two buried convergent surfaces. By way of comparison, Figs. 12.2 and 12.3 are conventional Tessars of similar field and speed.

Both the central and rear elements are compounded in Fig. 12.20, which covers about 32° at a speed of *f*/2.7. This lens could work well as a long-focal-length 35-mm camera lens; when it is stopped down, the off-axis ray intercept plots are quite flat.

The *Hektor* of Fig. 12.21 is an example of what can be accomplished by compounding all three elements of the triplet. This high-speed (*f*/1.8) example also makes use of a strong cemented Merté surface (see Sec. 3.5) in the center doublet (as does Fig. 12.20). This surface introduces undercorrected seventh-order spherical aberration which has the effect of reducing the zonal spherical by causing the normally overcorrected spherical at the margin to reverse direction and go undercorrected.

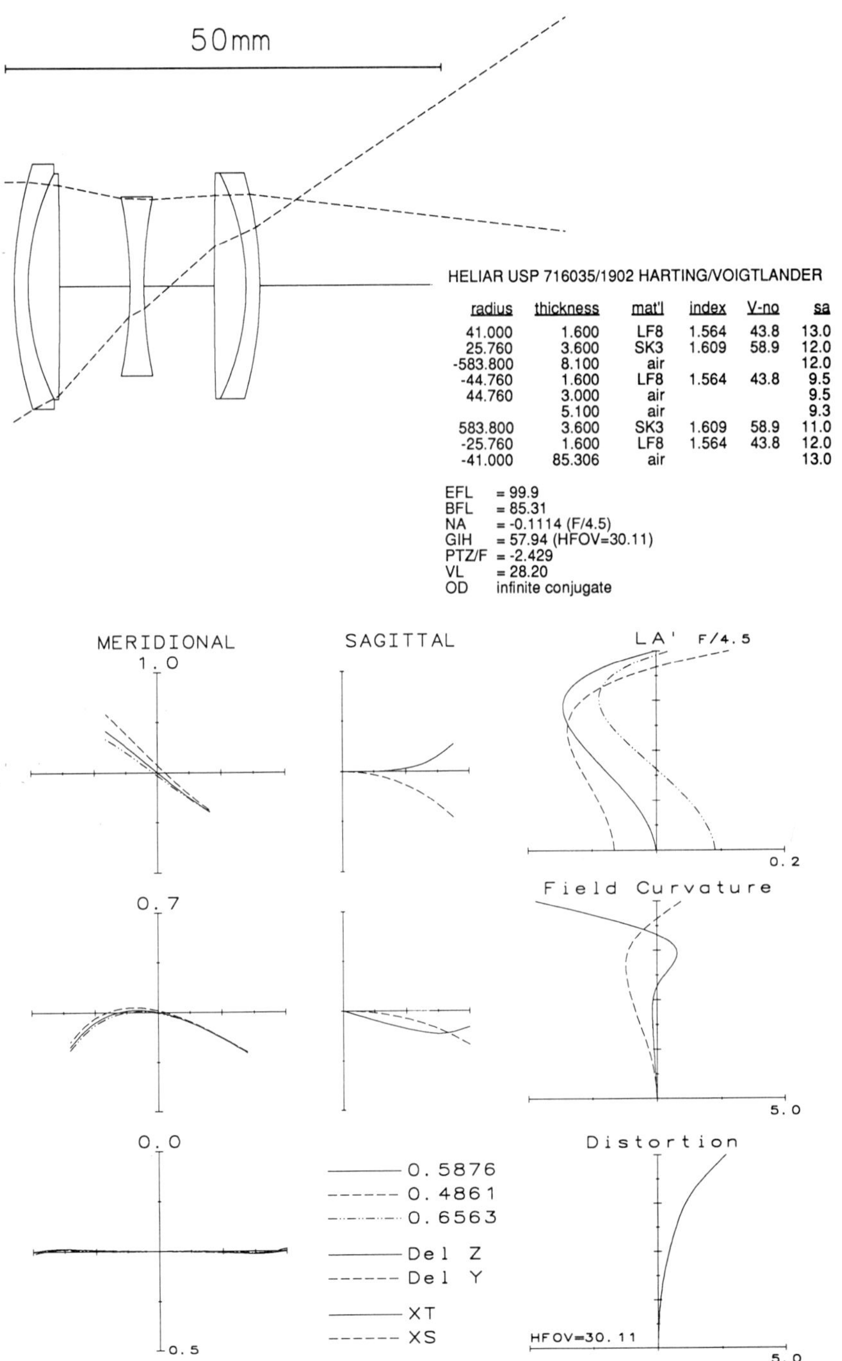

HELIAR USP 716035/1902 HARTING/VOIGTLANDER

radius	thickness	mat'l	index	V-no	sa
41.000	1.600	LF8	1.564	43.8	13.0
25.760	3.600	SK3	1.609	58.9	12.0
-583.800	8.100	air			12.0
-44.760	1.600	LF8	1.564	43.8	9.5
44.760	3.000	air			9.5
	5.100	air			9.3
583.800	3.600	SK3	1.609	58.9	11.0
-25.760	1.600	LF8	1.564	43.8	12.0
-41.000	85.306	air			13.0

EFL = 99.9
BFL = 85.31
NA = -0.1114 (F/4.5)
GIH = 57.94 (HFOV=30.11)
PTZ/F = -2.429
VL = 28.20
OD infinite conjugate

Figure 12.13

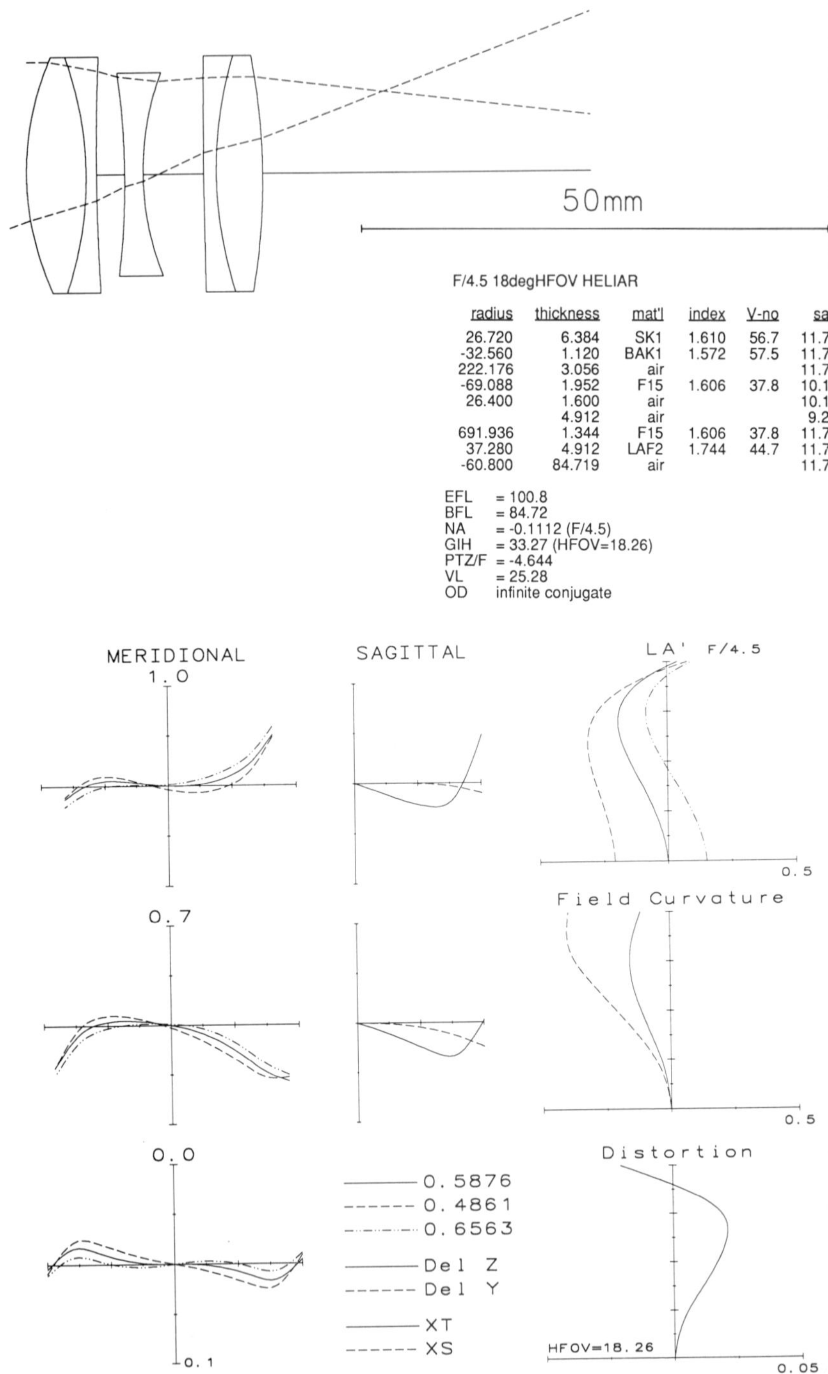

F/4.5 18degHFOV HELIAR

radius	thickness	mat'l	index	V-no	sa
26.720	6.384	SK1	1.610	56.7	11.7
-32.560	1.120	BAK1	1.572	57.5	11.7
222.176	3.056	air			11.7
-69.088	1.952	F15	1.606	37.8	10.1
26.400	1.600	air			10.1
	4.912	air			9.2
691.936	1.344	F15	1.606	37.8	11.7
37.280	4.912	LAF2	1.744	44.7	11.7
-60.800	84.719	air			11.7

EFL = 100.8
BFL = 84.72
NA = -0.1112 (F/4.5)
GIH = 33.27 (HFOV=18.26)
PTZ/F = -4.644
VL = 25.28
OD infinite conjugate

Figure 12.14

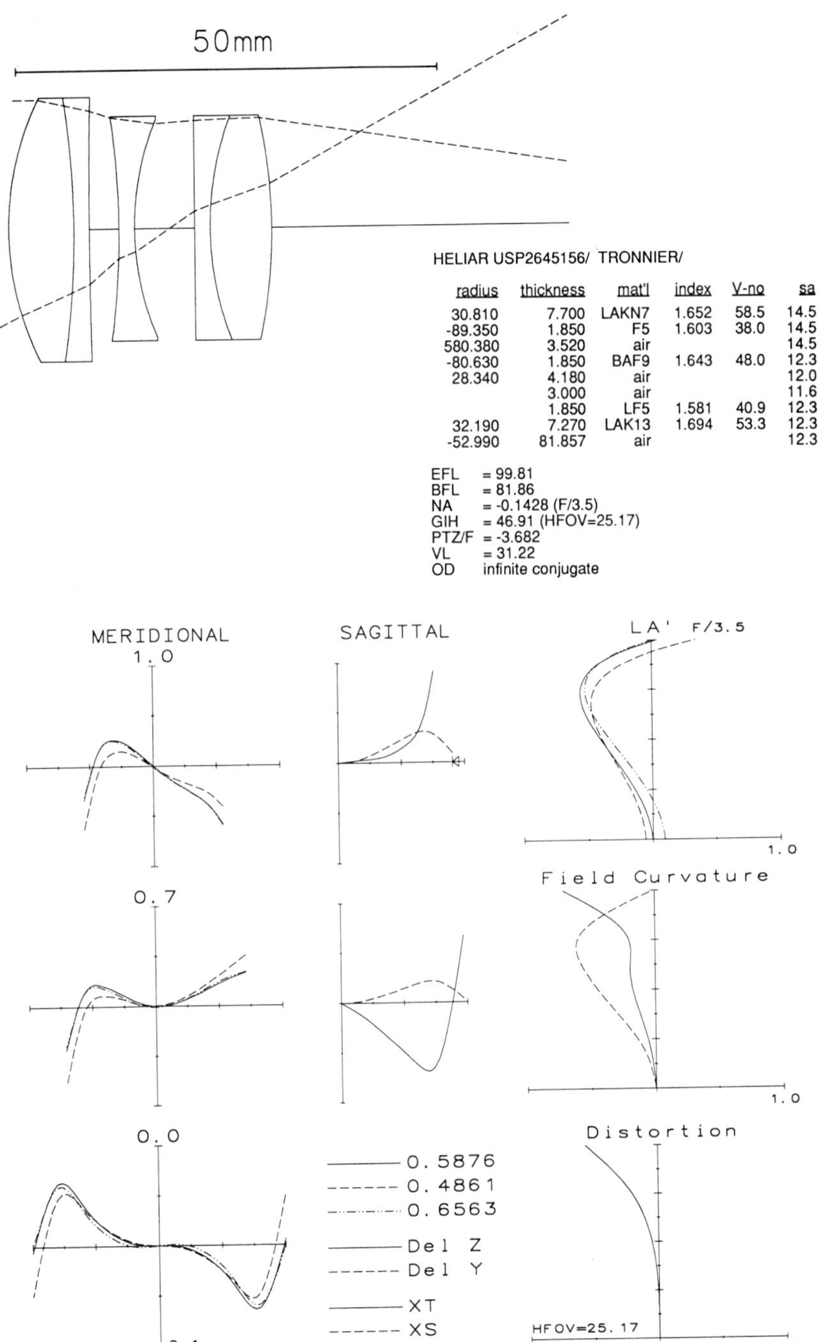

HELIAR USP2645156/ TRONNIER/

radius	thickness	mat'l	index	V-no	sa
30.810	7.700	LAKN7	1.652	58.5	14.5
-89.350	1.850	F5	1.603	38.0	14.5
580.380	3.520	air			14.5
-80.630	1.850	BAF9	1.643	48.0	12.3
28.340	4.180	air			12.0
	3.000	air			11.6
	1.850	LF5	1.581	40.9	12.3
32.190	7.270	LAK13	1.694	53.3	12.3
-52.990	81.857	air			12.3

EFL = 99.81
BFL = 81.86
NA = -0.1428 (F/3.5)
GIH = 46.91 (HFOV=25.17)
PTZ/F = -3.682
VL = 31.22
OD infinite conjugate

Figure 12.15

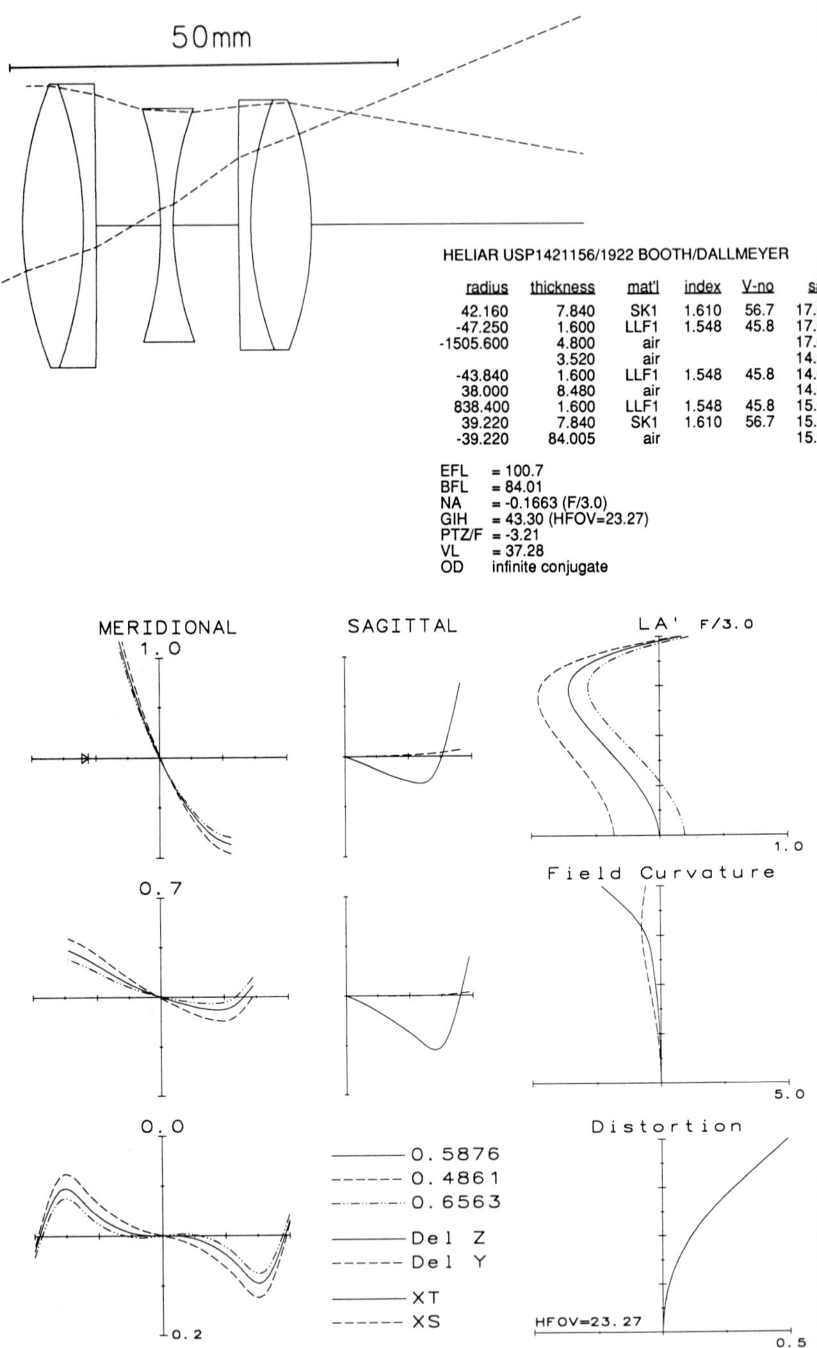

HELIAR USP1421156/1922 BOOTH/DALLMEYER

radius	thickness	mat'l	index	V-no	sa
42.160	7.840	SK1	1.610	56.7	17.0
-47.250	1.600	LLF1	1.548	45.8	17.0
-1505.600	4.800	air			17.0
	3.520	air			14.3
-43.840	1.600	LLF1	1.548	45.8	14.0
38.000	8.480	air			14.0
838.400	1.600	LLF1	1.548	45.8	15.0
39.220	7.840	SK1	1.610	56.7	15.0
-39.220	84.005	air			15.0

EFL = 100.7
BFL = 84.01
NA = -0.1663 (F/3.0)
GIH = 43.30 (HFOV=23.27)
PTZ/F = -3.21
VL = 37.28
OD infinite conjugate

Figure 12.16

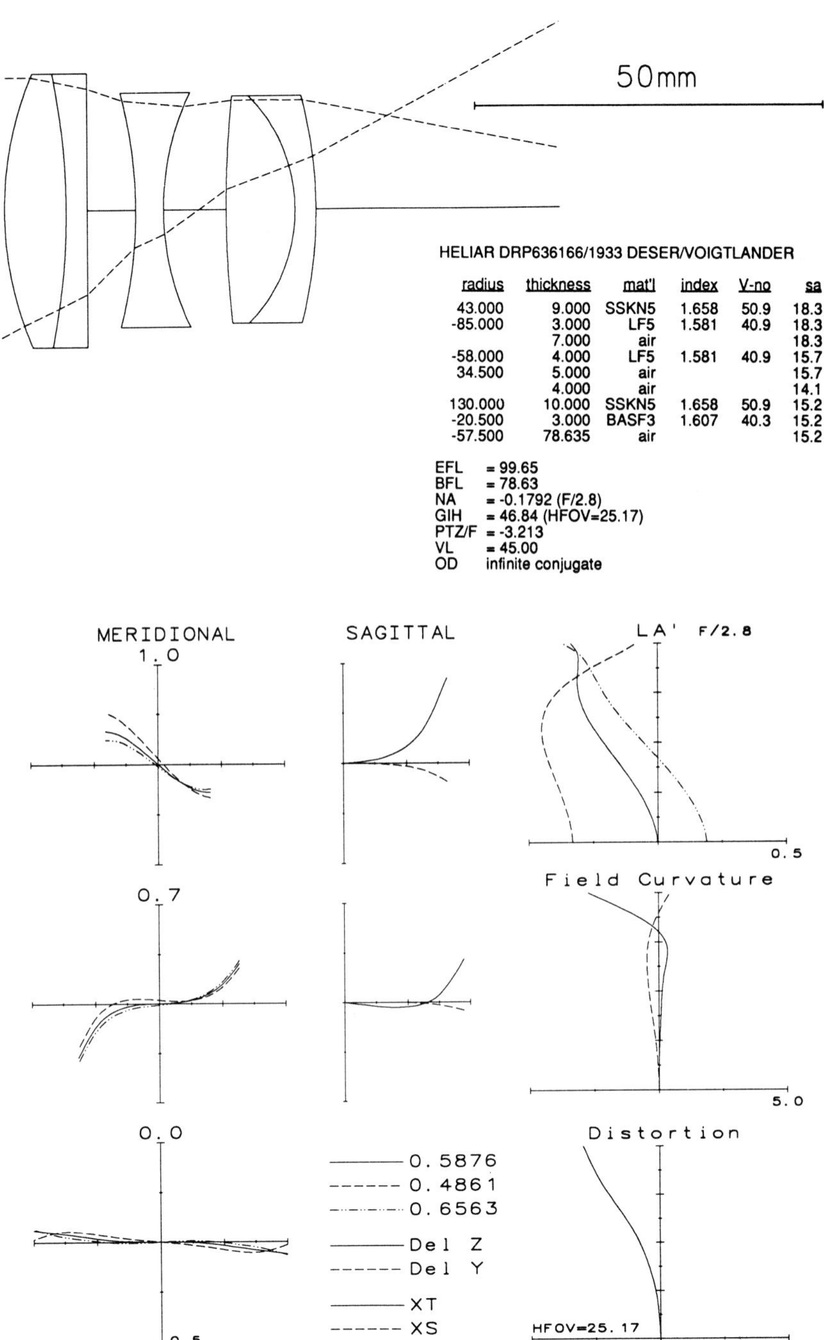

HELIAR DRP636166/1933 DESER/VOIGTLANDER

radius	thickness	mat'l	index	V-no	sa
43.000	9.000	SSKN5	1.658	50.9	18.3
-85.000	3.000	LF5	1.581	40.9	18.3
	7.000	air			18.3
-58.000	4.000	LF5	1.581	40.9	15.7
34.500	5.000	air			15.7
	4.000	air			14.1
130.000	10.000	SSKN5	1.658	50.9	15.2
-20.500	3.000	BASF3	1.607	40.3	15.2
-57.500	78.635	air			15.2

EFL = 99.65
BFL = 78.63
NA = -0.1792 (F/2.8)
GIH = 46.84 (HFOV=25.17)
PTZ/F = -3.213
VL = 45.00
OD infinite conjugate

Figure 12.17

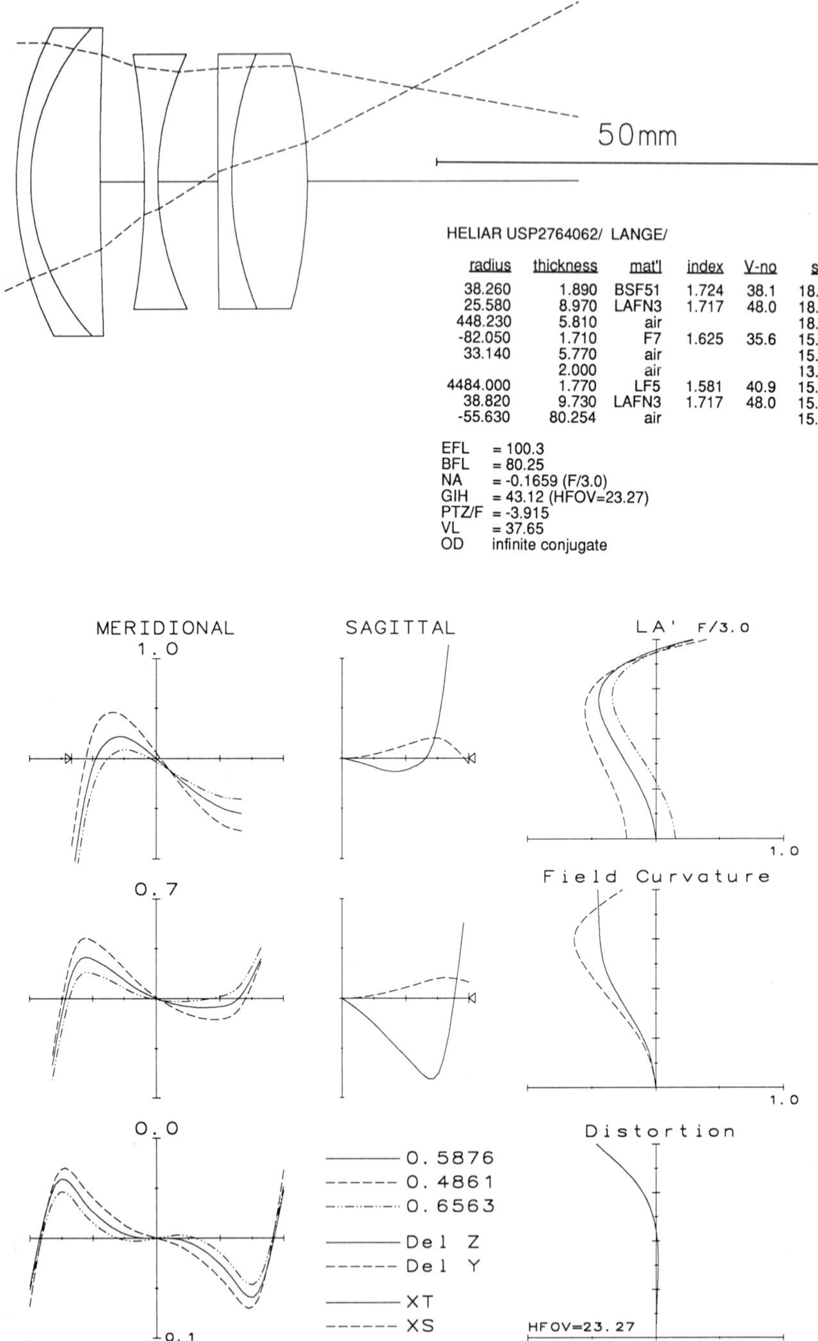

HELIAR USP2764062/ LANGE/

radius	thickness	mat'l	index	V-no	sa
38.260	1.890	BSF51	1.724	38.1	18.5
25.580	8.970	LAFN3	1.717	48.0	18.5
448.230	5.810	air			18.5
-82.050	1.710	F7	1.625	35.6	15.3
33.140	5.770	air			15.3
	2.000	air			13.4
4484.000	1.770	LF5	1.581	40.9	15.3
38.820	9.730	LAFN3	1.717	48.0	15.3
-55.630	80.254	air			15.3

EFL = 100.3
BFL = 80.25
NA = -0.1659 (F/3.0)
GIH = 43.12 (HFOV=23.27)
PTZ/F = -3.915
VL = 37.65
OD infinite conjugate

Figure 12.18

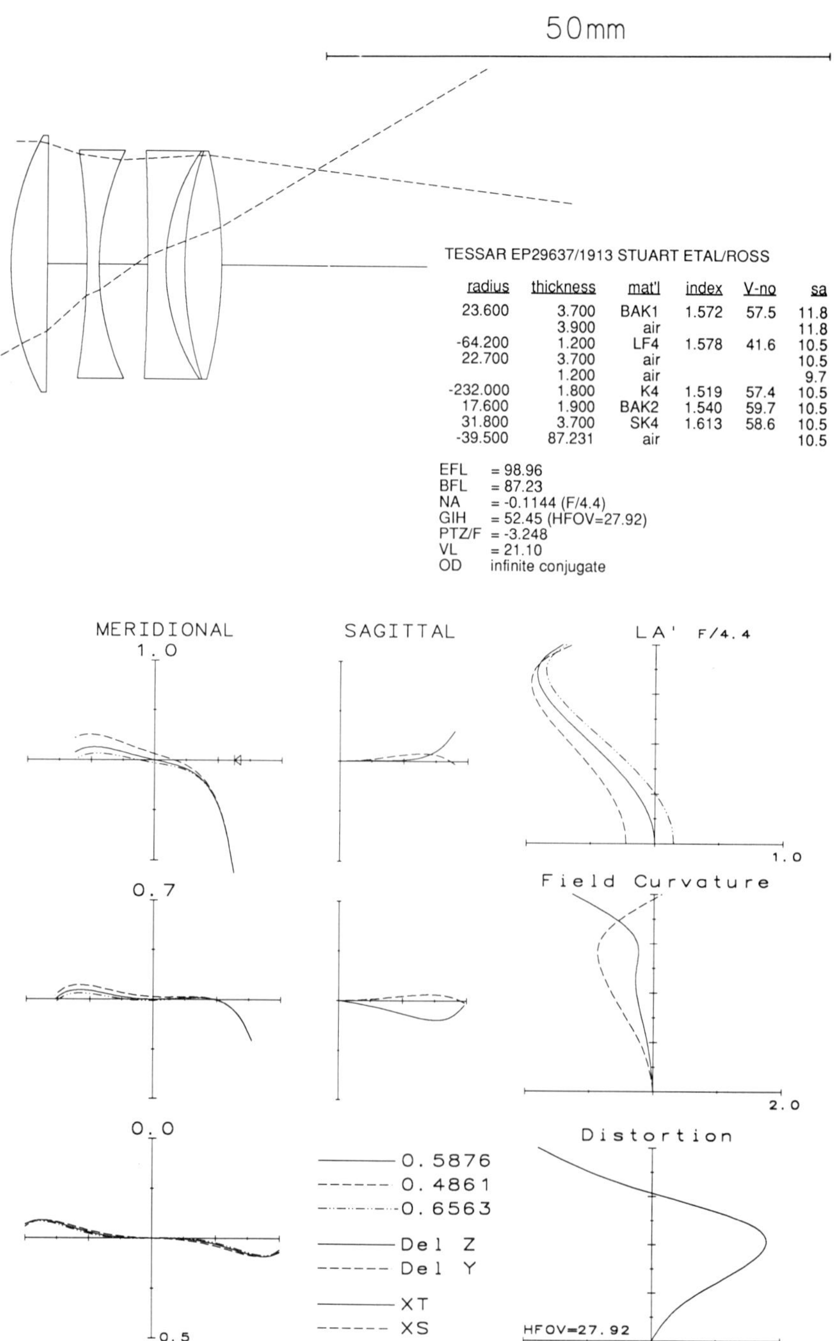

TESSAR EP29637/1913 STUART ETAL/ROSS

radius	thickness	mat'l	index	V-no	sa
23.600	3.700	BAK1	1.572	57.5	11.8
	3.900	air			11.8
-64.200	1.200	LF4	1.578	41.6	10.5
22.700	3.700	air			10.5
	1.200	air			9.7
-232.000	1.800	K4	1.519	57.4	10.5
17.600	1.900	BAK2	1.540	59.7	10.5
31.800	3.700	SK4	1.613	58.6	10.5
-39.500	87.231	air			10.5

EFL = 98.96
BFL = 87.23
NA = -0.1144 (F/4.4)
GIH = 52.45 (HFOV=27.92)
PTZ/F = -3.248
VL = 21.10
OD infinite conjugate

Figure 12.19

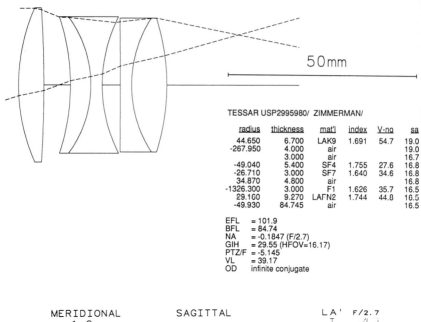

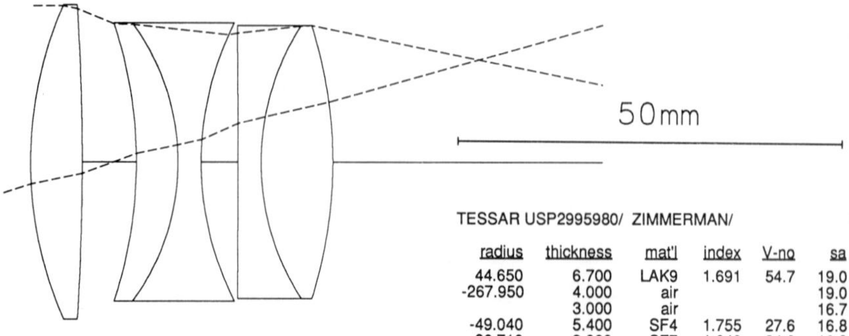

50mm

TESSAR USP2995980/ ZIMMERMAN/

radius	thickness	mat'l	index	V-no	sa
44.650	6.700	LAK9	1.691	54.7	19.0
-267.950	4.000	air			19.0
	3.000	air			16.7
-49.040	5.400	SF4	1.755	27.6	16.8
-26.710	3.000	SF7	1.640	34.6	16.8
34.870	4.800	air			16.8
-1326.300	3.000	F1	1.626	35.7	16.5
29.100	9.270	LAFN2	1.744	44.8	16.5
-49.930	84.745	air			16.5

EFL = 101.9
BFL = 84.74
NA = -0.1847 (F/2.7)
GIH = 29.55 (HFOV=16.17)
PTZ/F = -5.145
VL = 39.17
OD infinite conjugate

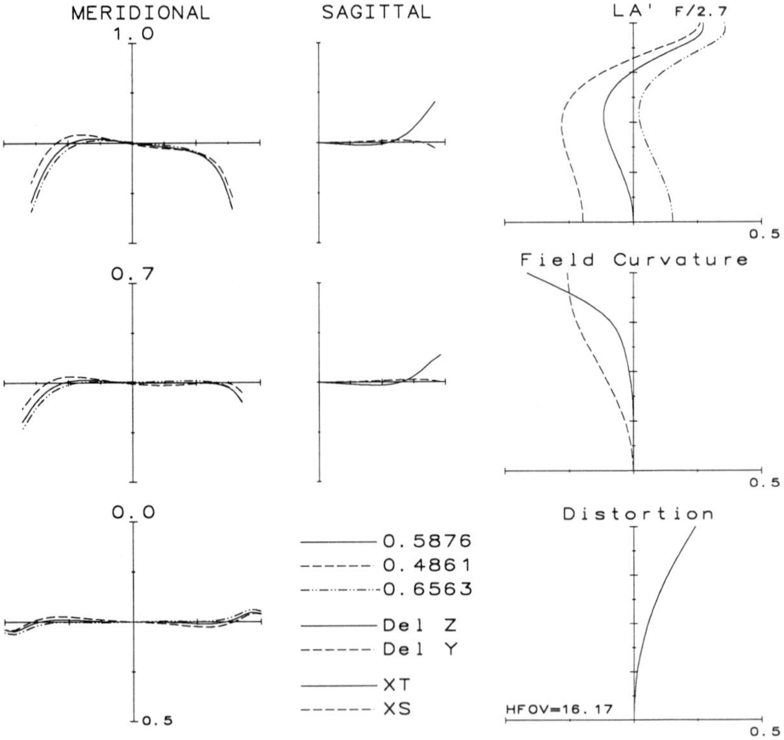

MERIDIONAL
SAGITTAL
1.0
0.7
0.0
0.5
LA' F/2.7
0.5
Field Curvature
0.5
Distortion
HFOV=16.17
0.5

0.5876
0.4861
0.6563
Del Z
Del Y
XT
XS

Figure 12.20

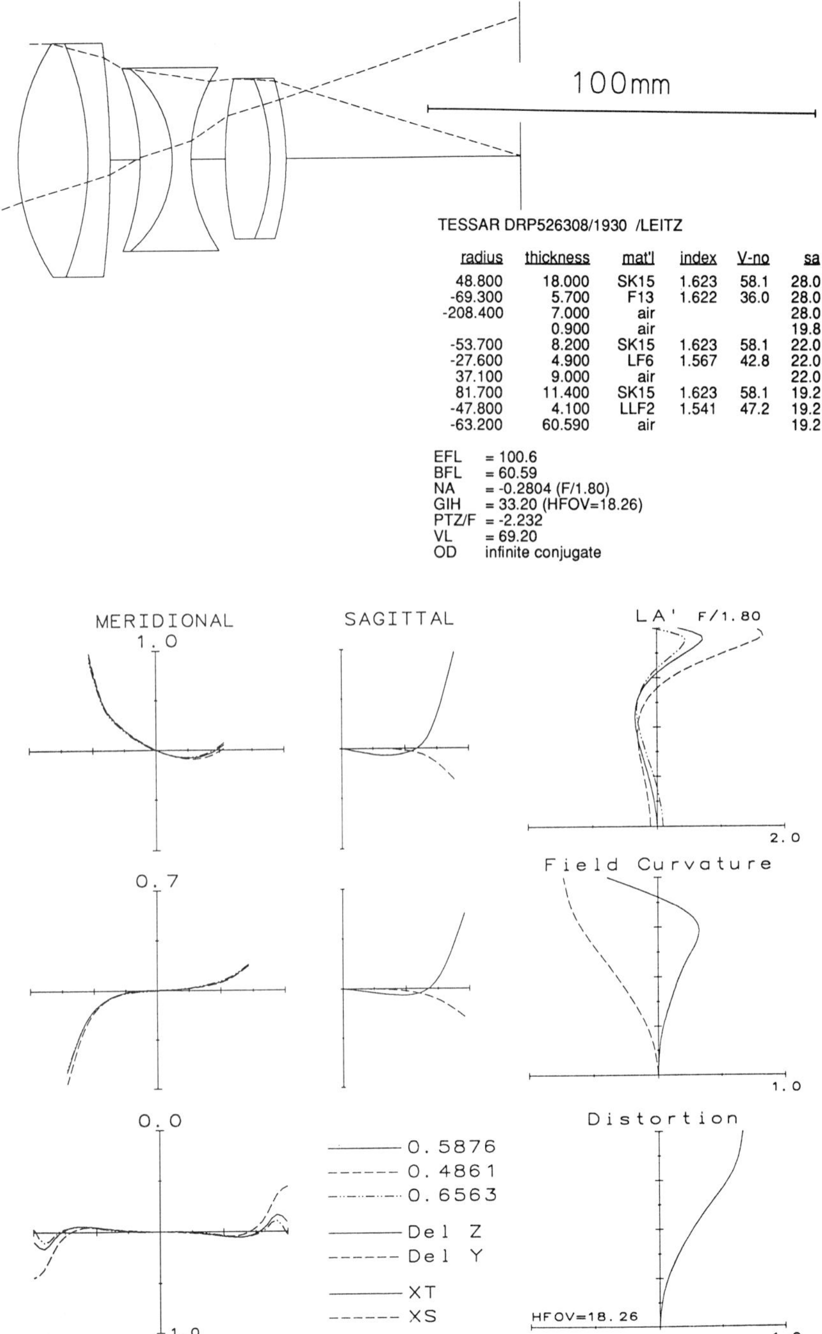

TESSAR DRP526308/1930 /LEITZ

radius	thickness	mat'l	index	V-no	sa
48.800	18.000	SK15	1.623	58.1	28.0
-69.300	5.700	F13	1.622	36.0	28.0
-208.400	7.000	air			28.0
	0.900	air			19.8
-53.700	8.200	SK15	1.623	58.1	22.0
-27.600	4.900	LF6	1.567	42.8	22.0
37.100	9.000	air			22.0
81.700	11.400	SK15	1.623	58.1	19.2
-47.800	4.100	LLF2	1.541	47.2	19.2
-63.200	60.590	air			19.2

EFL = 100.6
BFL = 60.59
NA = -0.2804 (F/1.80)
GIH = 33.20 (HFOV=18.26)
PTZ/F = -2.232
VL = 69.20
OD infinite conjugate

Figure 12.21

Chapter

13

The Petzval Lens; Head-up Display Lenses

13.1 The Petzval Portrait Lens

The basic Petzval lens consists of two positive components, spaced apart so that the astigmatism is controlled to be either zero or slightly positive. Usually the two components are achromats, typically doublets.

A significant exception to this is the infrared lens in the form of an airspaced doublet widely spaced from a positive singlet. This can be considered either as a triplet anastigmat or as a Petzval lens. This particular design configuration results from the high index and low dispersion of the infrared materials which are used—usually silicon and germanium for the mid-infrared, or germanium and zinc sulfide or selenide for the 8- to 12-μm region. See Chap. 21 for infrared lens designs.

The original Petzval lens (Fig. 13.1) was designed as a portrait lens for the Daguerrotype camera and, at a speed of about *f*/3.5, was (for its day) a very fast lens.

13.2 The Petzval Projection Lens

The more modern version, the Petzval projection lens (Fig. 13.2), is noted for covering a small field at high aperture with excellent image quality. The zonal spherical and spherochromatism are small, and the secondary spectrum is actually less than that of an achromatic doublet of the same glass. The classic arrangement is two achromats, each bending the axial ray toward the axis by the same amount, so that the work is equally divided—often a good principle to follow. For a system with a focal length of *f*, the front doublet has a focal length of 2*f*, the rear doublet has a focal length of *f*, and the (thin lens) spacing is equal to *f*. Then the (thin lens) back focus is equal to *f*/2.

The aperture stop of the Petzval projection lens is located at the

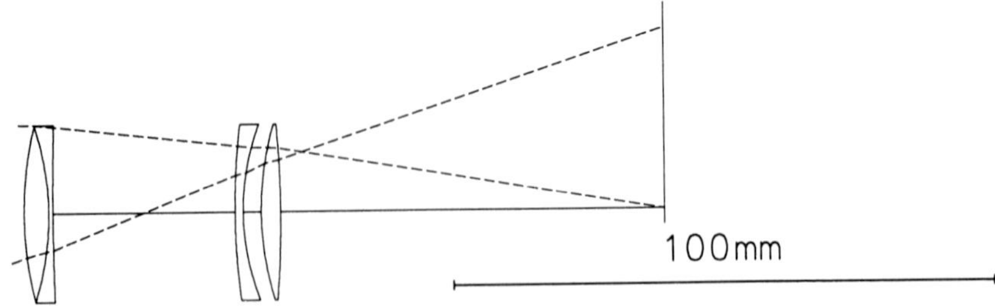

PETZVAL PORTRAIT LENS F/3.3 HFOV = 17DEG

radius	thickness	mat'l	index	V-no	sa
55.900	4.700	K3	1.518	59.0	15.1
-43.700	0.800	LF7	1.575	41.5	15.1
460.400	16.800	air			15.1
	16.800	air			13.1
110.600	1.500	LF7	1.575	41.5	15.0
38.900	3.300	air			15.0
48.000	3.600	BK7	1.517	64.2	15.0
-157.800	70.731	air			15.0

EFL = 99.52
BFL = 70.73
NA = -0.1517 (F/3.3)
GIH = 30.85 (HFOV=17.22)
PTZ/F = -1.268
VL = 47.50
OD infinite conjugate

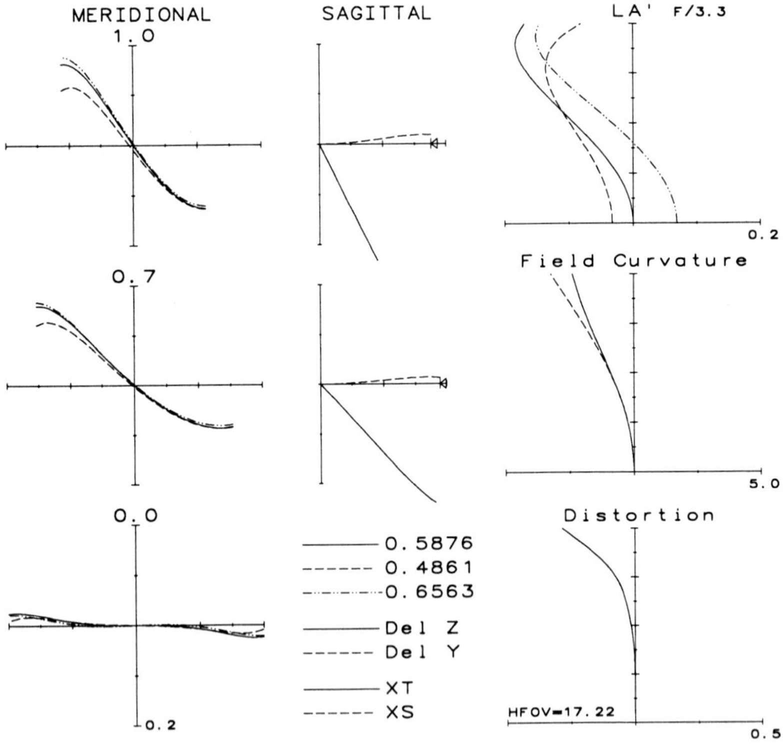

MERIDIONAL SAGITTAL LA' F/3.3
1.0
0.2
Field Curvature
0.7
5.0
0.0 Distortion
0.5876
0.4861
0.6563
Del Z
Del Y
XT
XS
0.2 HFOV=17.22 0.5

Figure 13.1

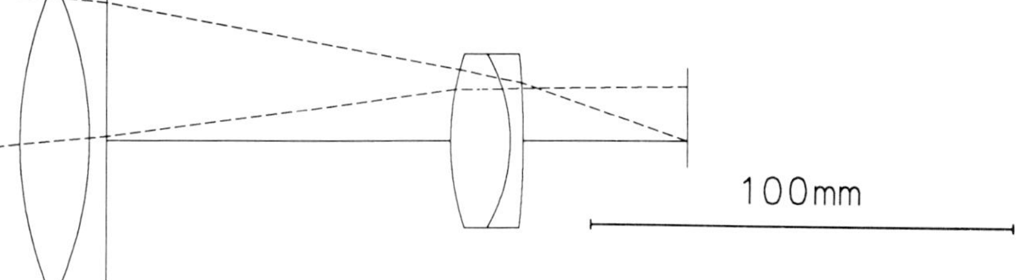

2" F/1.6 16MM PROJ PETZVAL - OPTIMIZED

radius	thickness	mat'l	index	V-no	sa
81.504	16.400	C1	1.523	58.6	31.8
-76.719	4.000	F4	1.617	36.6	31.8
	82.000	air			31.8
57.319	14.000	C1	1.523	58.6	19.0
-36.005	3.000	F4	1.617	36.6	19.0
-196.703	38.437	air			19.0

EFL = 101.3
BFL = 38.44
NA = -0.3128 (F/1.60)
GIH = 12.06 (HFOV=6.79)
PTZ/F = -0.9281
VL = 119.40
OD infinite conjugate

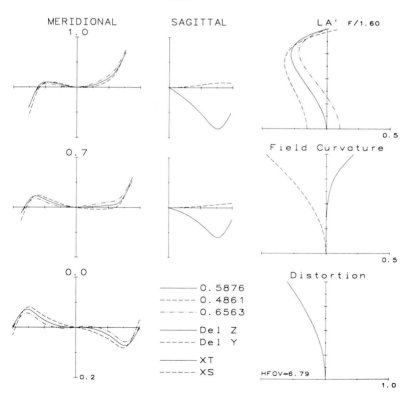

MERIDIONAL SAGITTAL LA' F/1.60
1.0
0.5
0.7 Field Curvature
0.5
0.0 Distortion
0.5876
0.4861
0.6563
Del Z
Del Y
XT
XS
0.2 HFOV=6.79
1.0

Figure 13.2

front component; stop shift theory tells us that a thin positive lens located at the stop can contribute only negative astigmatism. Therefore the rear component must contribute at least enough overcorrected astigmatism to offset this undercorrection. The astigmatism of the rear lens is, of course, a function of its shape and its distance from the stop, but it is also significantly affected by the cemented surface. If the index break $(n' - n)$ is not large enough, a broken contact at the rear doublet can be utilized to introduce the required overcorrected astigmatism.

In Fig. 13.3, the airspace between the doublets is made small to increase the working distance (bfl) of the lens. The contact is broken at both doublets in order to correct the aberrations. The spherical aberration residual is quite a bit bigger than in Fig. 13.2, despite the undercorrecting seventh-order which is apparent at the margin in the ray intercept plot. The shorter airspace reduces the Petzval field curvature at the same time it increases the bfl and makes the astigmatism correction more difficult.

Higher-index glasses are used in Fig. 13.4. Again, both doublets have been airspaced. The result is an improvement in both the zonal spherical and the field curvature, although the oblique spherical is increased somewhat.

13.3 The Petzval with a Field Flattener

Since the Petzval lens is afflicted with a very large Petzval curvature, a Piazzi-Smyth field flattener is often used to correct the situation. The field flattener is a strong negative lens placed close to the focal plane, where it has little effect on the focal length or on most aberrations, but it corrects the Petzval field curvature. The drawback to this arrangement is the short back focal length, or working distance, which results. Also, the fact that the field flattener is close to the focal plane means that defects or contamination on its surfaces may be apparent in the image. For this reason the field flattener is usually located some distance from the image plane in order to avoid these problems. When this is done both the field flattener and the basic lens must be made more powerful in order to maintain the focal length of the lens; this, of course, increases the residual spherical aberration and spherochromatism.

The improvement in the field curvature produced by a field flattener is apparent in Fig. 13.5. Note that contact has been broken in the front doublet, probably to correct the spherical; this was not necessary in the rear doublet because the combination of a sizable index break (1.611 to 1.720) at the cemented interface and a reasonably

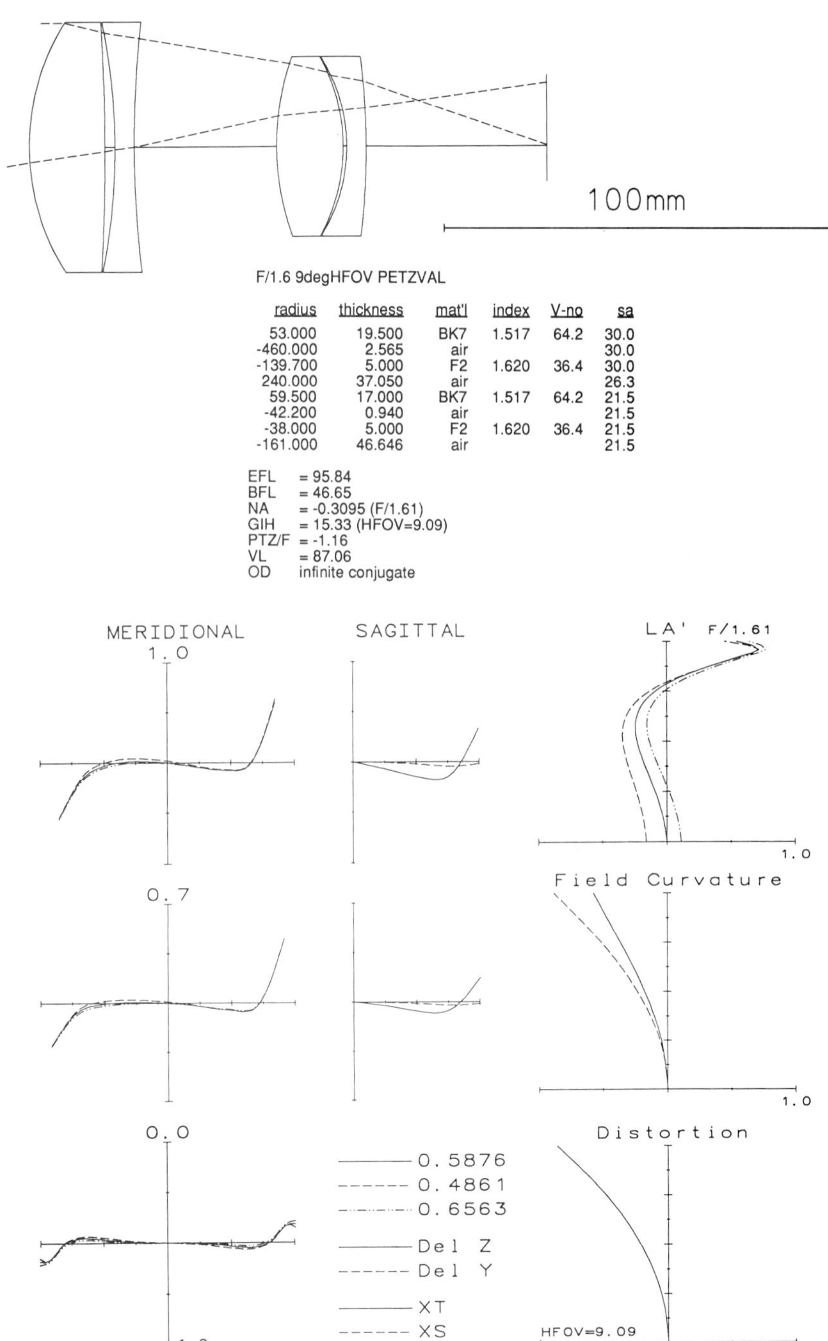

F/1.6 9degHFOV PETZVAL

radius	thickness	mat'l	index	V-no	sa
53.000	19.500	BK7	1.517	64.2	30.0
-460.000	2.565	air			30.0
-139.700	5.000	F2	1.620	36.4	30.0
240.000	37.050	air			26.3
59.500	17.000	BK7	1.517	64.2	21.5
-42.200	0.940	air			21.5
-38.000	5.000	F2	1.620	36.4	21.5
-161.000	46.646	air			21.5

EFL = 95.84
BFL = 46.65
NA = -0.3095 (F/1.61)
GIH = 15.33 (HFOV=9.09)
PTZ/F = -1.16
VL = 87.06
OD infinite conjugate

Figure 13.3

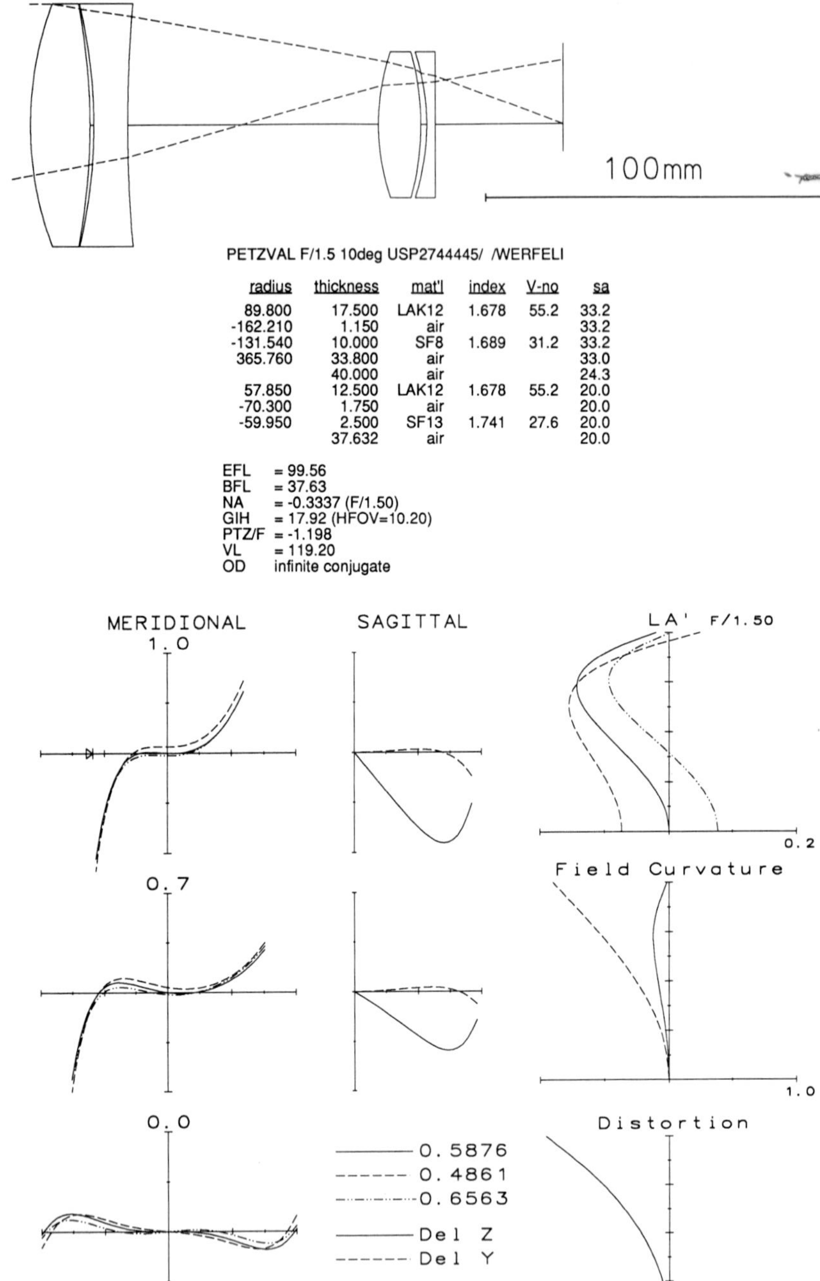

PETZVAL F/1.5 10deg USP2744445/ /WERFELI

radius	thickness	mat'l	index	V-no	sa
89.800	17.500	LAK12	1.678	55.2	33.2
-162.210	1.150	air			33.2
-131.540	10.000	SF8	1.689	31.2	33.2
365.760	33.800	air			33.0
	40.000	air			24.3
57.850	12.500	LAK12	1.678	55.2	20.0
-70.300	1.750	air			20.0
-59.950	2.500	SF13	1.741	27.6	20.0
	37.632	air			20.0

EFL = 99.56
BFL = 37.63
NA = -0.3337 (F/1.50)
GIH = 17.92 (HFOV=10.20)
PTZ/F = -1.198
VL = 119.20
OD infinite conjugate

Figure 13.4

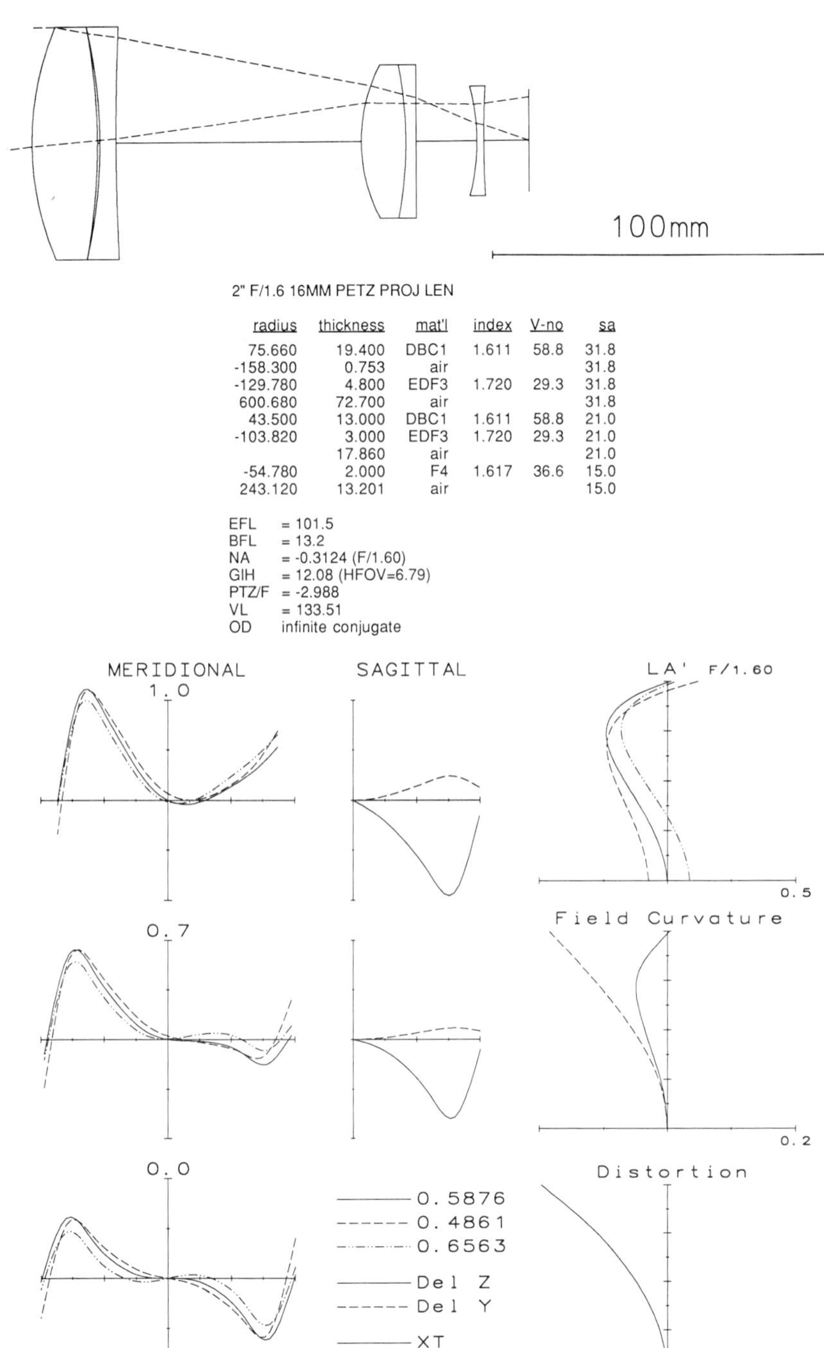

2" F/1.6 16MM PETZ PROJ LEN

radius	thickness	mat'l	index	V-no	sa
75.660	19.400	DBC1	1.611	58.8	31.8
-158.300	0.753	air			31.8
-129.780	4.800	EDF3	1.720	29.3	31.8
600.680	72.700	air			31.8
43.500	13.000	DBC1	1.611	58.8	21.0
-103.820	3.000	EDF3	1.720	29.3	21.0
	17.860	air			21.0
-54.780	2.000	F4	1.617	36.6	15.0
243.120	13.201	air			15.0

EFL = 101.5
BFL = 13.2
NA = -0.3124 (F/1.60)
GIH = 12.08 (HFOV=6.79)
PTZ/F = -2.988
VL = 133.51
OD infinite conjugate

Figure 13.5

large airspace between the doublets are sufficient to properly control the astigmatism.

In Fig. 13.6 the rear doublet is spaced so far apart that the flint element acts as a field flattener. Again the front doublet is split to correct the spherical aberration. The spherical zonal is relatively large in this construction. Note that all of the field flatteners in this chapter are made of high-dispersion flint glass. One might expect that the field flattener should be made of a low-index, high-V-value crown glass in order to reduce its chromatic effects and to increase its overcorrecting Petzval contribution. However, it turns out that a flint glass helps the axial chromatic correction greatly; this benefit is sufficient to make a low-V-value glass the best choice.

The lens of Fig. 13.7 is derived from Fig. 13.6 by splitting each positive element into two parts. The speed is increased to *f*/1.4. The improvement is quite obvious. A limiting aberration in this design is fifth-order coma; this shows up in the off-axis ray intercept plots with overcorrected (third-order) coma in the center of the aperture and undercorrected (fifth-order) at the margin of the aperture. This is often found in lenses where an edge contact airspace is used to correct the spherical aberration. A redesign which allowed the glasses to vary resulted in Fig. 13.8; the major improvement that this produced was the reduction of this coma.

Figures 13.9 and 13.10 are lenses of speed and angular coverage similar to those of Figs. 13.7 and 13.8. Both utilize high-index glasses to achieve a high level of correction. Figure 13.11, shown at a speed of *f*/1.25, uses high-index glasses but has a very short back focal length. All three illustrate the idea of splitting a positive element in two in order to improve the image quality. Notice that, in Fig. 13.10, in which the rear crown has apparently been split, the front doublet has only about half the power as in Figs. 13.9 and 13.11. Some of the positive power has been shifted from the front doublet to the rear; the front doublet almost functions as a low-power corrector. Another way to view this lens is as a three-doublet Petzval like Fig. 13.12 with the third doublet split and widely spaced to flatten the field, as in Fig. 13.6.

13.4 Very High Speed Petzval Lenses

The Petzval lens is the basis of many extremely fast lenses. With relatively modest modifications, projection lenses with speeds of *f*/1.2 have been made; more extreme modifications have been used to push the speed above *f*/1.0.

Doubling up on the rear doublet was used to increase the speed to *f*/1.3 in Fig. 13.12, although this format has not been especially suc-

(Text continues on page 236.)

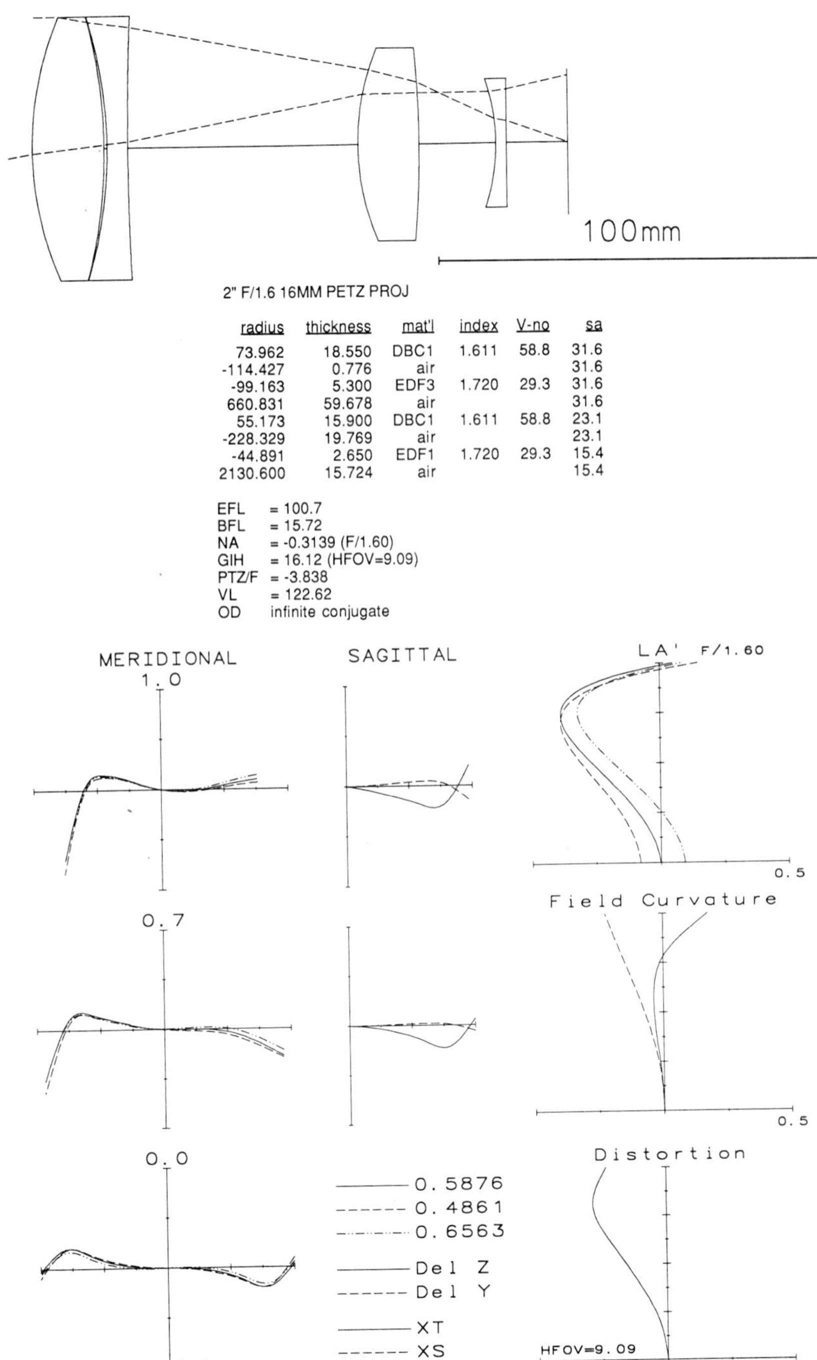

2" F/1.6 16MM PETZ PROJ

radius	thickness	mat'l	index	V-no	sa
73.962	18.550	DBC1	1.611	58.8	31.6
-114.427	0.776	air			31.6
-99.163	5.300	EDF3	1.720	29.3	31.6
660.831	59.678	air			31.6
55.173	15.900	DBC1	1.611	58.8	23.1
-228.329	19.769	air			23.1
-44.891	2.650	EDF1	1.720	29.3	15.4
2130.600	15.724	air			15.4

EFL = 100.7
BFL = 15.72
NA = -0.3139 (F/1.60)
GIH = 16.12 (HFOV=9.09)
PTZ/F = -3.838
VL = 122.62
OD infinite conjugate

Figure 13.6

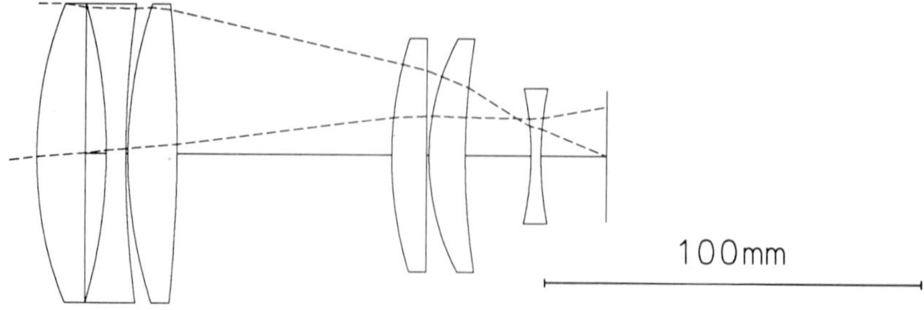

2" F/1.4 16MM PROJ USP3255664 1966

radius	thickness	mat'l	index	V-no	sa
91.500	13.000	DBC1	1.611	58.8	36.5
	5.570	air			36.5
-122.780	5.000	EDF3	1.720	29.3	36.5
263.200	0.600	air			36.5
108.560	13.200	DBC1	1.611	58.8	36.5
-340.740	56.720	air			36.5
90.320	9.200	DBC1	1.611	58.8	28.5
	0.600	air			28.5
58.110	9.600	DBC1	1.611	58.8	28.5
174.020	17.280	air			28.5
-73.400	2.600	EDF3	1.720	29.3	16.5
81.260	17.318	air			16.5

EFL = 102.1
BFL = 17.32
NA = -0.3576 (F/1.40)
GIH = 12.15 (HFOV=6.79)
PTZ/F = -6.801
VL = 133.37
OD infinite conjugate

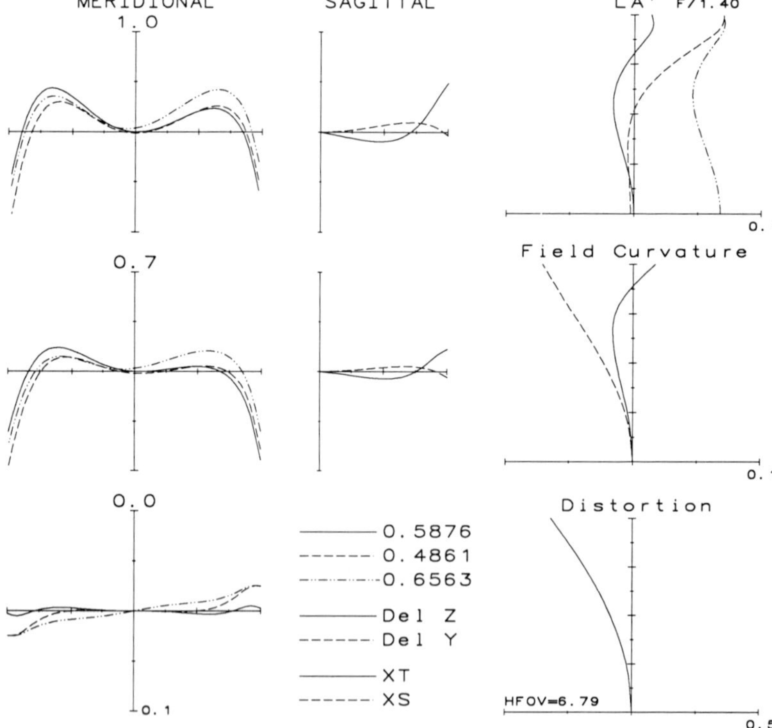

Figure 13.7

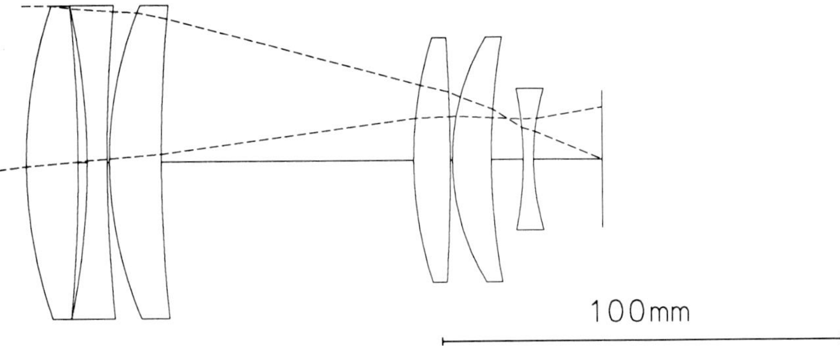

2" F/1.4 16MM PROJ USP3255664 1966 CHG'D GLASS

radius	thickness	mat'l	index	V-no	sa
108.061	13.000	LAK21	1.640	60.1	36.5
-345.695	2.315	air			36.5
-159.702	5.000	SF10	1.728	28.4	36.5
361.952	0.600	air			36.5
86.990	13.200	PSK52	1.603	65.4	36.5
360.984	63.402	air			36.5
89.189	9.200	PSK53	1.620	63.5	28.5
-419.257	0.600	air			28.5
53.078	9.600	PSK53	1.620	63.5	28.5
152.163	8.027	air			28.5
-78.525	2.600	SF5	1.673	32.2	16.5
55.142	17.322	air			16.5

EFL = 102.1
BFL = 17.32
NA = -0.3574 (F/1.40)
GIH = 12.15 (HFOV=6.79)
PTZ/F = -5.738
VL = 127.54
OD infinite conjugate

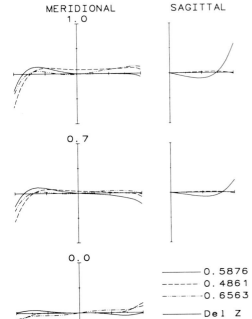

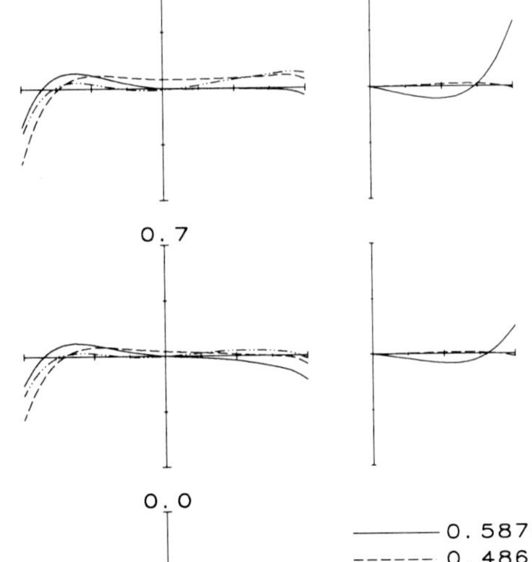

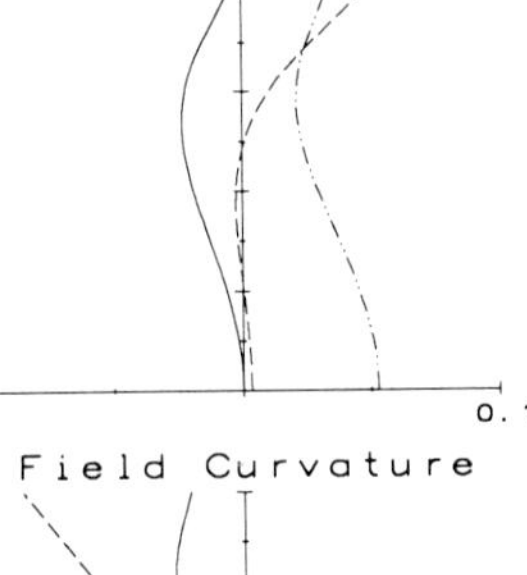

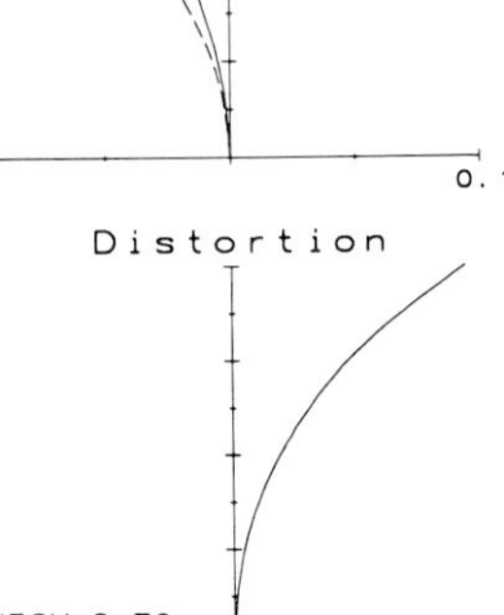

Figure 13.8

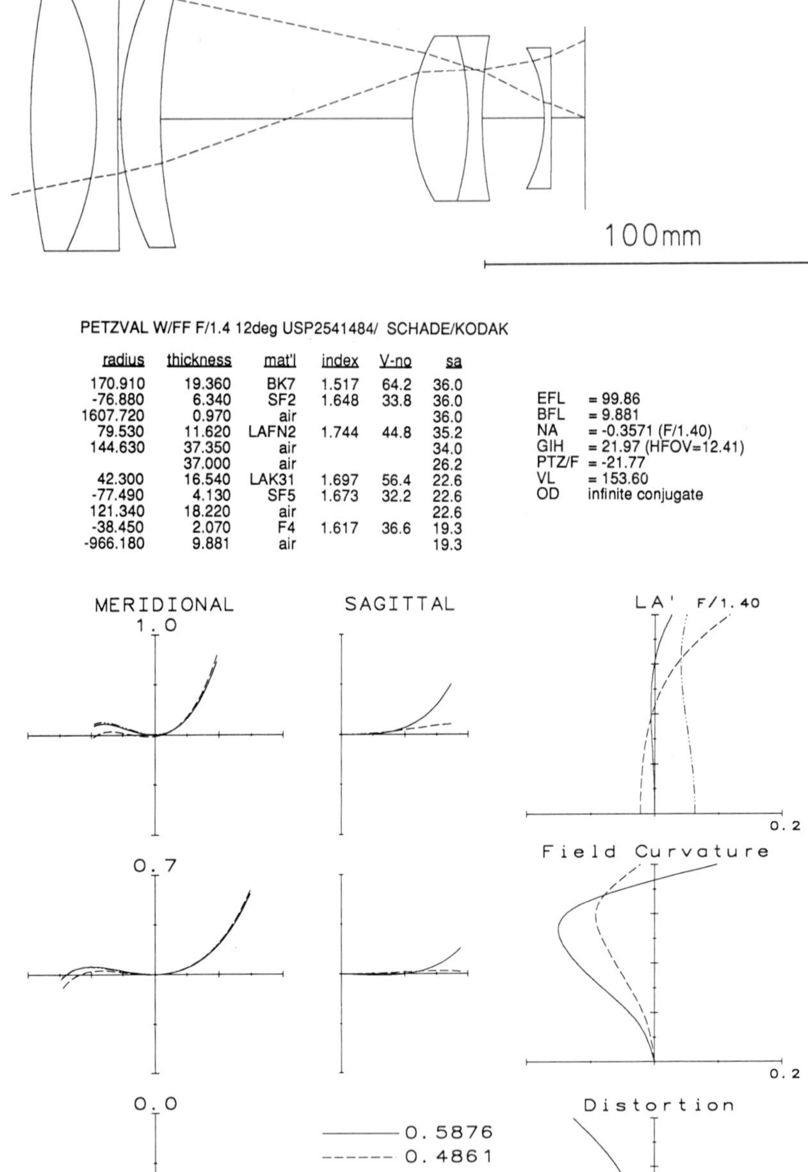

PETZVAL W/FF F/1.4 12deg USP2541484/ SCHADE/KODAK

radius	thickness	mat'l	index	V-no	sa
170.910	19.360	BK7	1.517	64.2	36.0
-76.880	6.340	SF2	1.648	33.8	36.0
1607.720	0.970	air			36.0
79.530	11.620	LAFN2	1.744	44.8	35.2
144.630	37.350	air			34.0
	37.000	air			26.2
42.300	16.540	LAK31	1.697	56.4	22.6
-77.490	4.130	SF5	1.673	32.2	22.6
121.340	18.220	air			22.6
-38.450	2.070	F4	1.617	36.6	19.3
-966.180	9.881	air			19.3

EFL	= 99.86
BFL	= 9.881
NA	= -0.3571 (F/1.40)
GIH	= 21.97 (HFOV=12.41)
PTZ/F	= -21.77
VL	= 153.60
OD	infinite conjugate

Figure 13.9

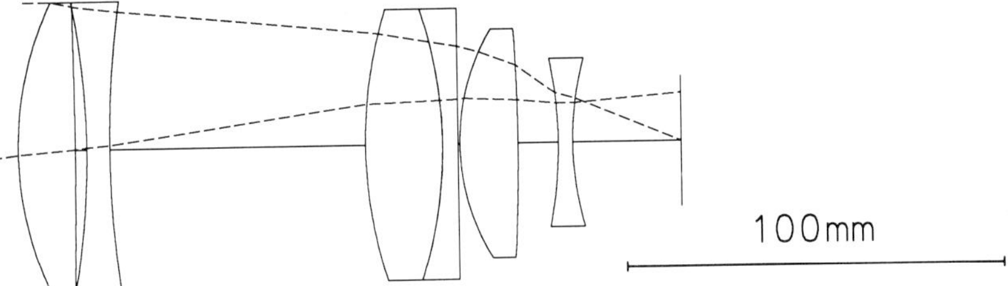

2" F/1.4 16MM PROJ USP2989895 1961

radius	thickness	mat'l	index	V-no	sa
78.232	15.240	DBC1	1.611	58.8	36.5
-1275.080	3.048	air			36.5
-181.407	6.096	EDF1	1.649	33.8	36.5
234.086	68.072	air			36.5
101.803	20.320	LAK22	1.651	55.9	33.4
-101.803	0.066	air			33.4
-100.686	4.064	EDF2	1.689	30.9	33.4
	0.508	air			33.4
51.562	15.240	SK16	1.620	60.3	28.2
-456.641	10.668	air			28.2
-97.790	3.759	SF5	1.673	32.2	20.8
71.222	28.562	air			20.8

EFL = 101.3
BFL = 28.56
NA = -0.3613 (F/1.39)
GIH = 12.06 (HFOV=6.79)
PTZ/F = -2.845
VL = 147.08
OD infinite conjugate

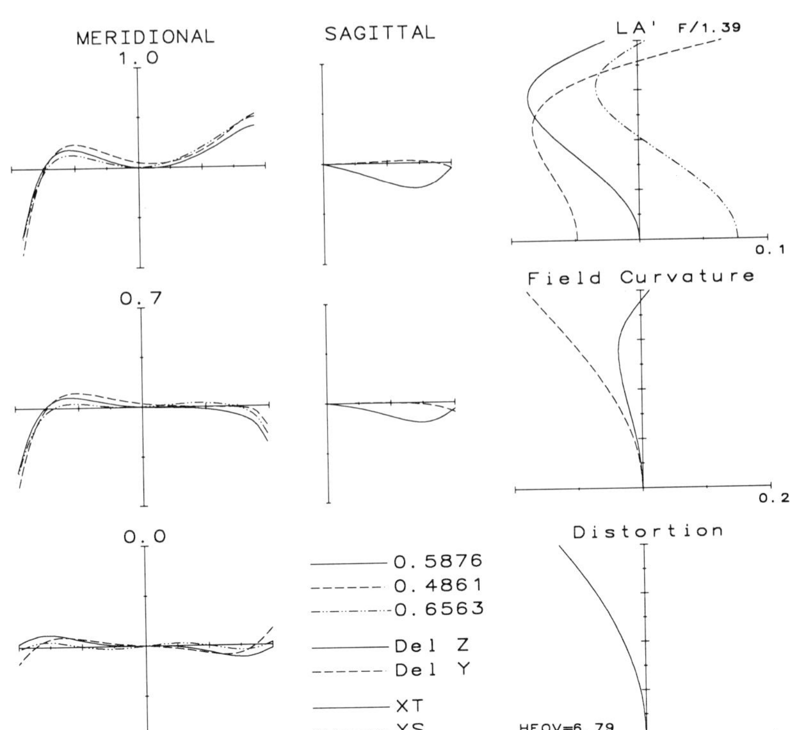

Figure 13.10

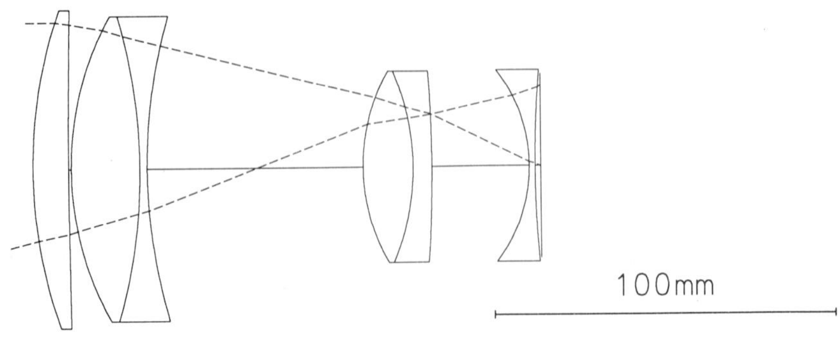

PETZVAL W/ FF F/1.25 12deg USP2649021/ ANGENIEUX/

radius	thickness	mat'l	index	V-no	sa
121.110	10.380	BK7	1.517	64.2	43.2
1600.000	0.860	air			43.2
81.320	19.900	BK7	1.517	64.2	41.4
-138.430	2.080	F1	1.626	35.7	41.4
138.430	32.150	air			41.4
	31.000	air			26.9
49.310	14.710	BK7	1.517	64.2	26.0
-60.560	5.540	F1	1.626	35.7	26.0
-332.230	28.300	air			26.0
-40.660	1.730	SF17	1.650	33.7	26.0
288.100	1.665	air			26.0

EFL = 99.5
BFL = 1.665
NA = -0.4008 (F/1.25)
GIH = 21.89 (HFOV=12.41)
PTZ/F = 337.3
VL = 146.65
OD infinite conjugate

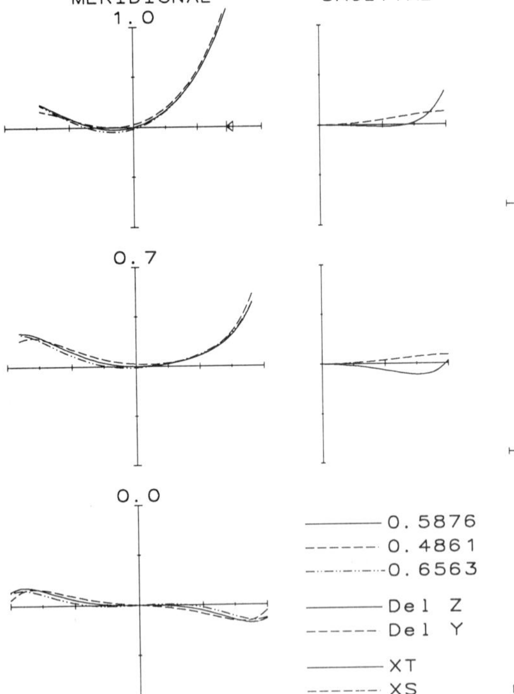

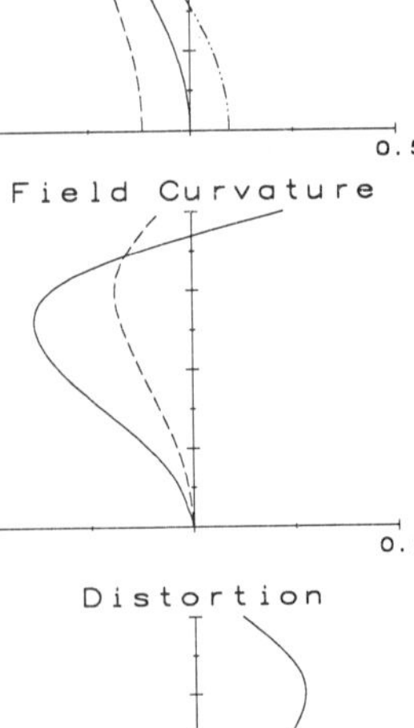

Figure 13.11

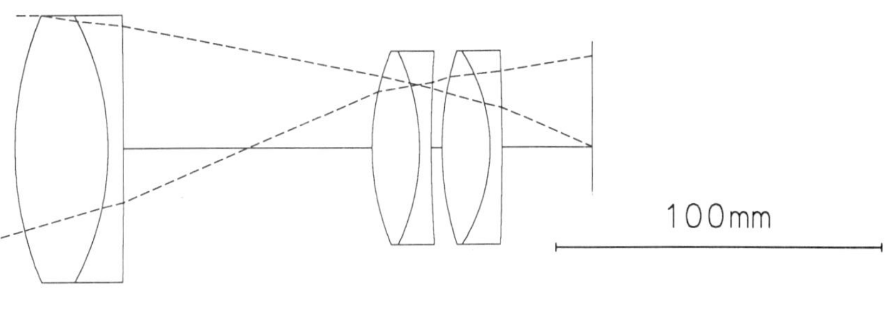

F/1.3 3-DOUB PETZ USP 2158202/1939(MODIF)

radius	thickness	mat'l	index	V-no	sa
93.200	28.500	K5	1.522	59.5	38.0
-75.100	4.800	F4	1.617	36.6	38.0
-2097.000	38.200	air			38.0
	38.100	air			27.9
67.300	14.600	BK7	1.517	64.2	27.5
-60.500	3.600	LF7	1.575	41.5	27.5
411.000	3.300	air			27.5
85.000	14.600	BK7	1.517	64.2	27.5
-48.900	3.600	LF7	1.575	41.5	27.5
-800.000	27.272	air			27.5

EFL = 100.4
BFL = 27.27
NA = -0.3788 (F/1.32)
GIH = 27.12 (HFOV=15.11)
PTZ/F = -0.8995
VL = 149.30
OD infinite conjugate

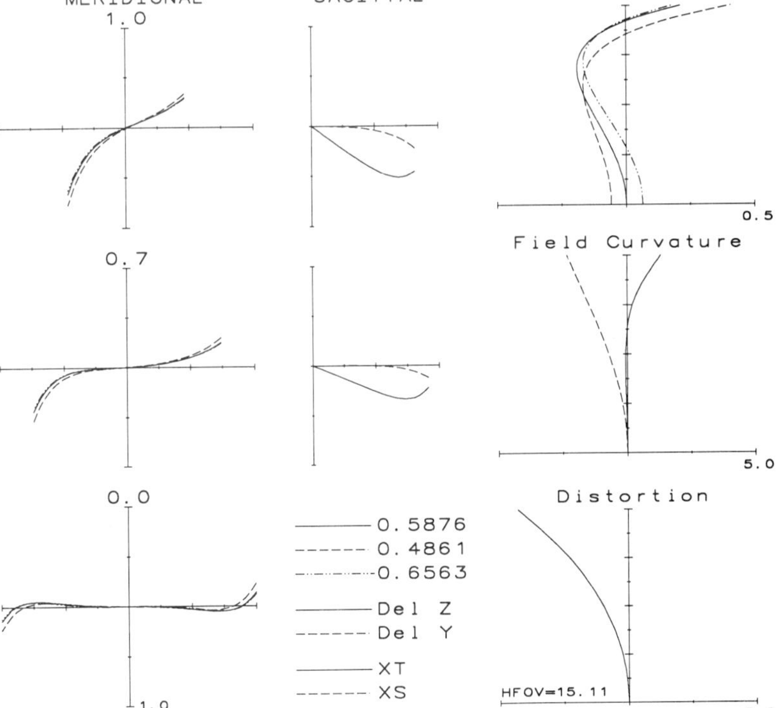

Figure 13.12

cessful. The *R-Biotar* of Fig. 13.13 uses a relatively simple construction to reach a speed of *f*/0.9; note that the slope of the marginal ray in the figure is increased by about the same amount at each component; this is usually helpful in reducing zonal spherical. The final high-speed Petzval lens is Fig. 13.14, in which both crowns have been split, and the image is formed on a curved film. (Note that the aberration plots in the figure are not with respect to a plane image surface, but on a curved surface, as shown in the lens drawing.)

The Petzval lens is also the basis of the classical microscope objective. The standard 10×, NA 0.25 objective is a Petzval projection lens much like Fig. 13.2, corrected for finite conjugates. The Amici and oil-immersion objectives are Petzval lenses with (more or less) aplanatic elements added to the short conjugate to increase the numerical aperture. See Chap. 15 for microscope objectives.

13.5 Head-up Display (HUD) Lenses; Biocular Lenses

The HUD lens is effectively a collimator which is used to present an infinitely distant image of a cathode-ray tube (CRT). This image is superimposed (by a semireflecting beam combiner) on a direct view of the scene, seen through the combiner. The system aperture is made large enough so both eyes can be used (hence, biocular). The large-aperture diameter requires a high-speed lens, typically to the order of *f*/1.0 or faster. Note that *binocular* means an optical system for each eye; *biocular* means a single optical system used simultaneously by both eyes. These systems are almost all based on the Petzval lens, usually with added elements, plus a field flattener. An aspheric surface is often used in the front component to control the spherical aberration; occasionally the concave surface of the field flattener is aspherized to help with the field aberrations. The analysis and evaluation of a biocular system is a bit tricky, since each eye uses only about a 5-mm-diameter part of the image beam. What counts most in this sort of system is the unpleasant "swimming" appearance of the image caused by spherical aberration, and the parallax between the directions at which the image is seen by the two eyes. This may take the form of convergence (of the eyes), which is quite tolerable, of divergence, or of dipvergence (a vertical difference of direction). HUD designers usually aim for a parallax of about a minute of arc or less over most of the field.

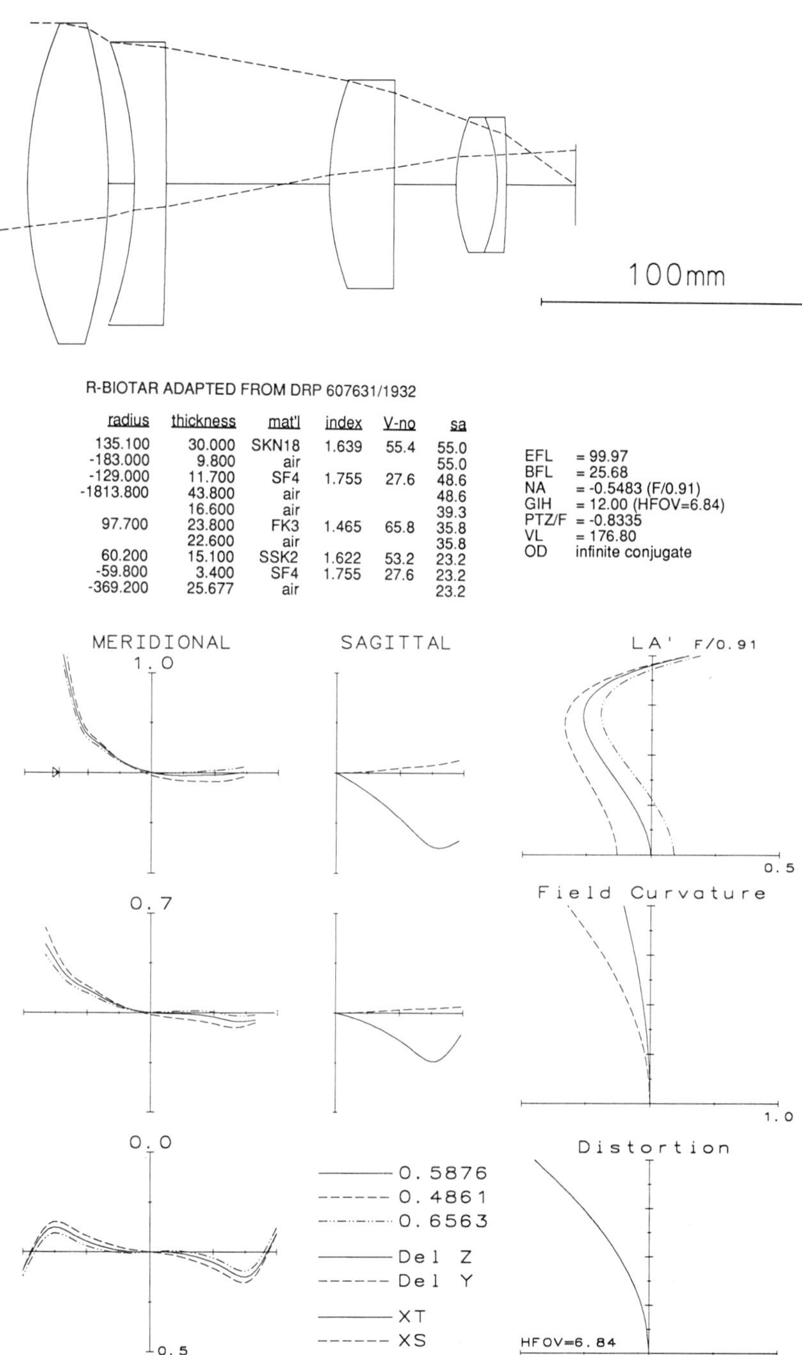

R-BIOTAR ADAPTED FROM DRP 607631/1932

radius	thickness	mat'l	index	V-no	sa
135.100	30.000	SKN18	1.639	55.4	55.0
-183.000	9.800	air			55.0
-129.000	11.700	SF4	1.755	27.6	48.6
-1813.800	43.800	air			48.6
	16.600	air			39.3
97.700	23.800	FK3	1.465	65.8	35.8
	22.600	air			35.8
60.200	15.100	SSK2	1.622	53.2	23.2
-59.800	3.400	SF4	1.755	27.6	23.2
-369.200	25.677	air			23.2
					23.2

EFL = 99.97
BFL = 25.68
NA = -0.5483 (F/0.91)
GIH = 12.00 (HFOV=6.84)
PTZ/F = -0.8335
VL = 176.80
OD infinite conjugate

Figure 13.13

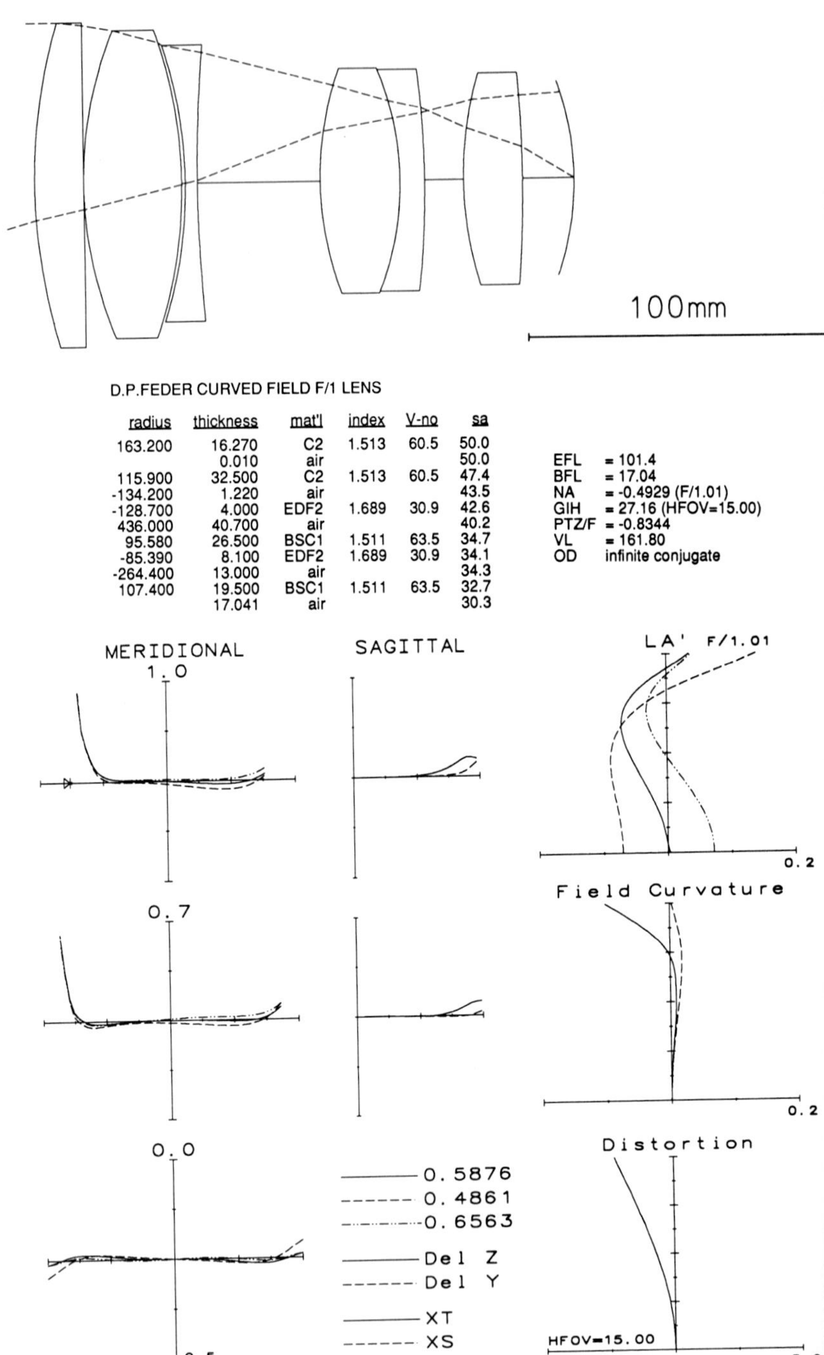

D.P.FEDER CURVED FIELD F/1 LENS

radius	thickness	mat'l	index	V-no	sa
163.200	16.270	C2	1.513	60.5	50.0
	0.010	air			50.0
115.900	32.500	C2	1.513	60.5	47.4
-134.200	1.220	air			43.5
-128.700	4.000	EDF2	1.689	30.9	42.6
436.000	40.700	air			40.2
95.580	26.500	BSC1	1.511	63.5	34.7
-85.390	8.100	EDF2	1.689	30.9	34.1
-264.400	13.000	air			34.3
107.400	19.500	BSC1	1.511	63.5	32.7
	17.041	air			30.3

EFL	= 101.4
BFL	= 17.04
NA	= -0.4929 (F/1.01)
GIH	= 27.16 (HFOV=15.00)
PTZ/F	= -0.8344
VL	= 161.80
OD	infinite conjugate

Figure 13.14

Chapter

14

Split Triplets

As discussed in Sec. 3.3, splitting an element into two elements allows the aberrations of the two to be substantially reduced from that of the original single element. If the balance of the system can be adjusted to take advantage of this reduction, a significant improvement in performance can be achieved.

In the Cooke triplet, the outer crowns are often split in order to reduce the zonal spherical aberration and to thus allow the speed of the lens to be increased. Examples of the split-rear crown type are shown in Figs. 14.1 and 14.2. Note that, as the speed is increased, the angular field coverage is reduced. The split-rear crown form is often used in camera lenses for 8-mm, 16-mm, small format TV, and charge-coupled device (CCD) cameras, typically at a speed of about *f*/2.

The split-front crown type has been widely used for (pseudo) telephoto camera lenses and for 35-mm slide projection lenses (Fig. 14.3). Figures 14.3 through 14.9 present an assortment of split-front triplets with glasses of low, medium, and high index, with thick and thin elements, with speeds from *f*/2.8 to *f*/1.5, and angular fields ranging from 10 to 45°.

The split-front triplet has also developed into two interesting and powerful, if rarely used, forms. The second element of the split front is meniscus in shape. If made thick, it can have a field-flattening effect of its own, and the *Sonnar* form, Figs. 14.10 through 14.13, makes use of this to eliminate the need for the center flint negative element. These lenses begin to look like the front half of a Biotar/double-Gauss combined with the rear of a triplet or Tessar. The tremendous variety of Sonnar configurations and characteristics is sampled here. They all cover reasonable angular fields, with speeds ranging from *f*/2.9 to *f*/1.2. Figure 14.11 is something of a curiosity, since it consists of only a cemented triplet and a singlet.

Another variation is the *Ernostar* type, Figs. 14.14 and 14.15, which uses the split triplet form as the basis for improvement by compounding the elements. Both the Sonnar and the Ernostar types had a brief vogue as 35-mm camera lenses of normal and long focal length, but they have been largely superseded by other constructions.

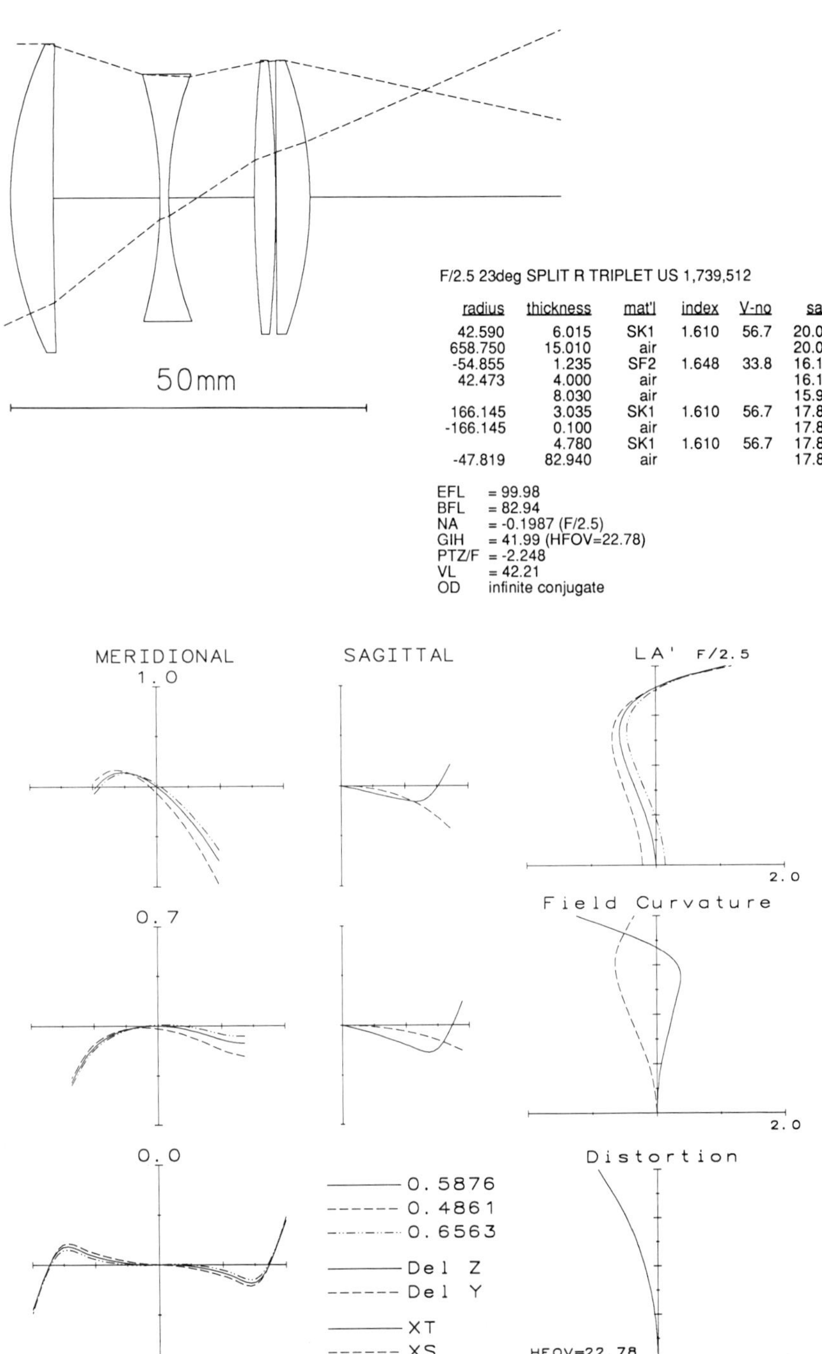

F/2.5 23deg SPLIT R TRIPLET US 1,739,512

radius	thickness	mat'l	index	V-no	sa
42.590	6.015	SK1	1.610	56.7	20.0
658.750	15.010	air			20.0
-54.855	1.235	SF2	1.648	33.8	16.1
42.473	4.000	air			16.1
	8.030	air			15.9
166.145	3.035	SK1	1.610	56.7	17.8
-166.145	0.100	air			17.8
	4.780	SK1	1.610	56.7	17.8
-47.819	82.940	air			17.8

EFL = 99.98
BFL = 82.94
NA = -0.1987 (F/2.5)
GIH = 41.99 (HFOV=22.78)
PTZ/F = -2.248
VL = 42.21
OD infinite conjugate

Figure 14.1

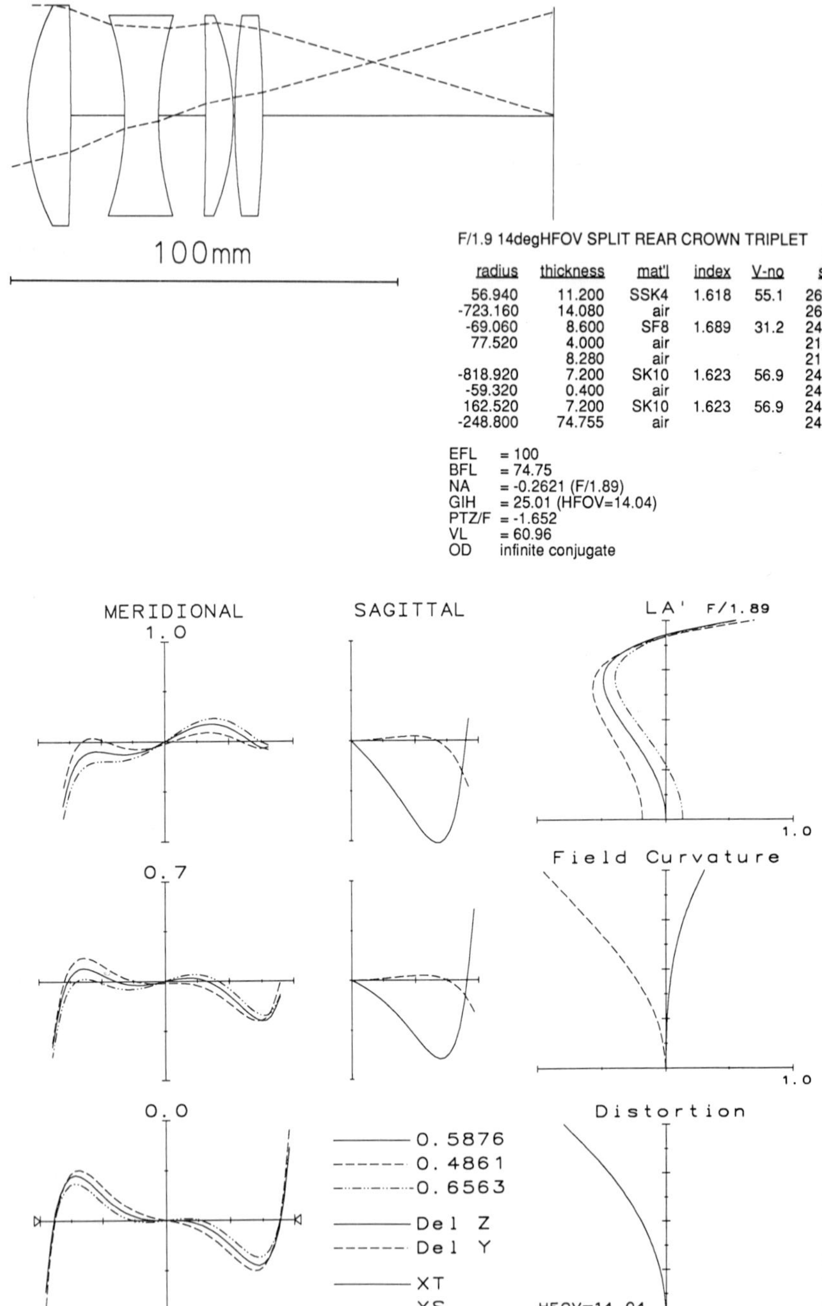

F/1.9 14degHFOV SPLIT REAR CROWN TRIPLET

radius	thickness	mat'l	index	V-no	sa
56.940	11.200	SSK4	1.618	55.1	26.4
-723.160	14.080	air			26.4
-69.060	8.600	SF8	1.689	31.2	24.0
77.520	4.000	air			21.0
	8.280	air			21.1
-818.920	7.200	SK10	1.623	56.9	24.0
-59.320	0.400	air			24.0
162.520	7.200	SK10	1.623	56.9	24.0
-248.800	74.755	air			24.0

EFL = 100
BFL = 74.75
NA = -0.2621 (F/1.89)
GIH = 25.01 (HFOV=14.04)
PTZ/F = -1.652
VL = 60.96
OD infinite conjugate

Figure 14.2

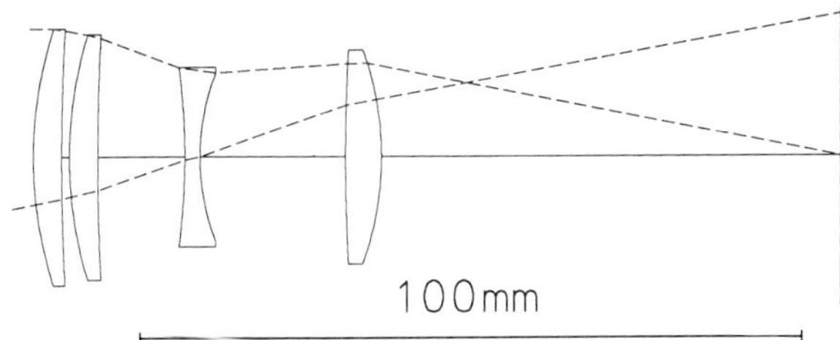

F/2.8 11deg SPLIT F TRIPLET US 2,767,614

radius	thickness	mat'l	index	V-no	sa
54.560	4.300	SK4	1.613	58.6	18.0
302.930	1.200	air			18.0
54.560	4.300	SK4	1.613	58.6	17.2
302.930	13.100	air			16.7
-89.240	2.300	SF8	1.689	31.2	12.6
29.610	22.300	air			11.8
252.780	5.500	SK4	1.613	58.6	15.0
-40.900	69.172	air			15.0

EFL = 99.36
BFL = 69.17
NA = -0.1811 (F/2.8)
GIH = 19.87 (HFOV=11.31)
PTZ/F = -2.604
VL = 53.00
OD infinite conjugate

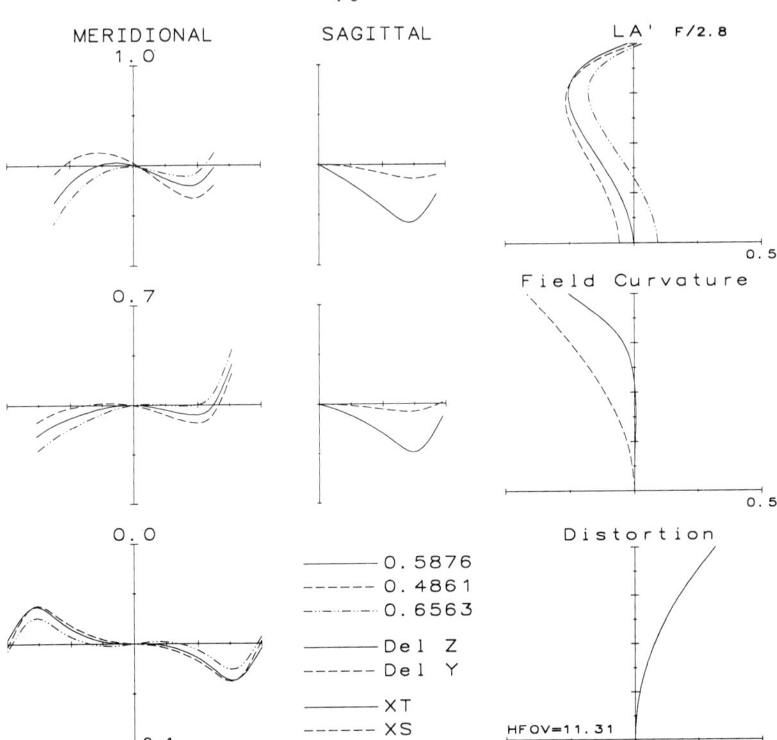

Figure 14.3

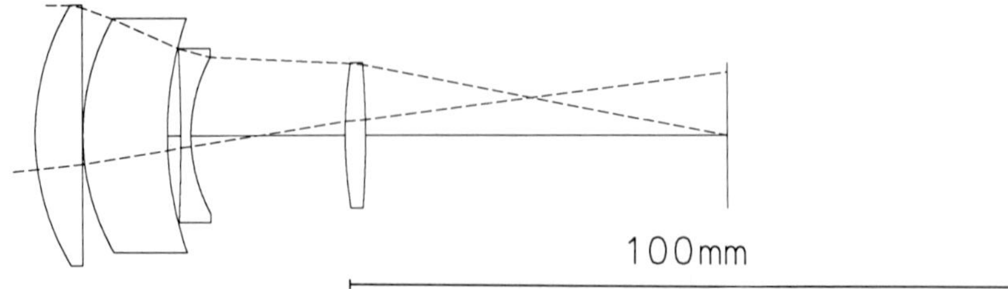

GARRY EDWARDS ET AL; USP 3649104; F/2.8 5 DEG. OBJECTIVE #14

radius	thickness	mat'l	index	V-no	sa
32.779	6.988	BK7	1.517	64.2	17.9
	0.100	air			17.6
31.057	12.443	BK7	1.517	64.2	16.1
46.253	1.996	air			12.0
-248.654	1.497	SF8	1.689	31.2	12.0
21.283	10.000	air			10.8
	12.980	air			0.0
62.076	2.995	LF3	1.582	42.1	10.0
-127.799	53.357	air			9.8

EFL = 100.1
BFL = 53.36
NA = -0.1789 (F/2.8)
GIH = 8.76
PTZ/F = -5.019
VL = 49.00
OD infinite conjugate

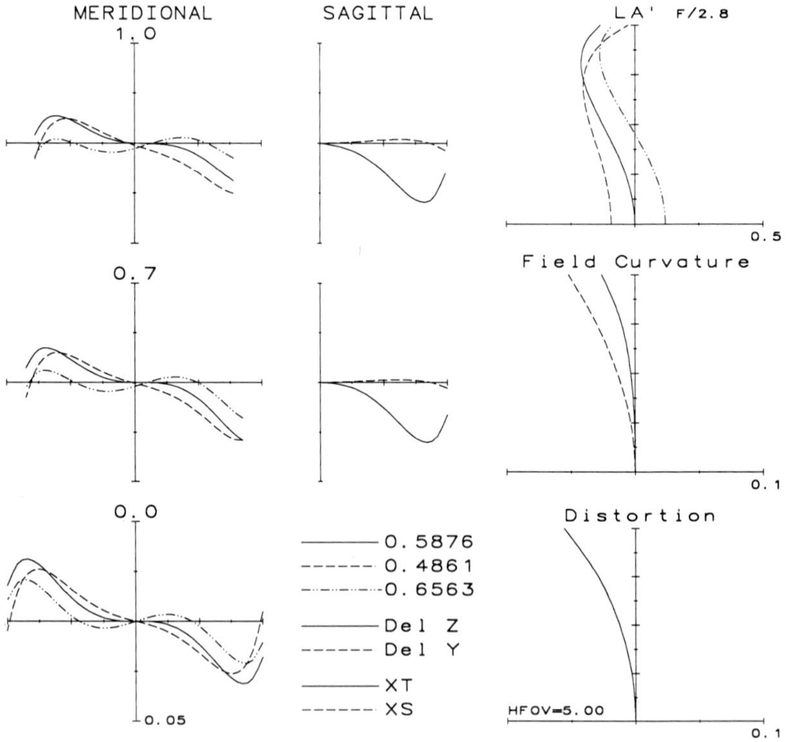

Figure 14.4

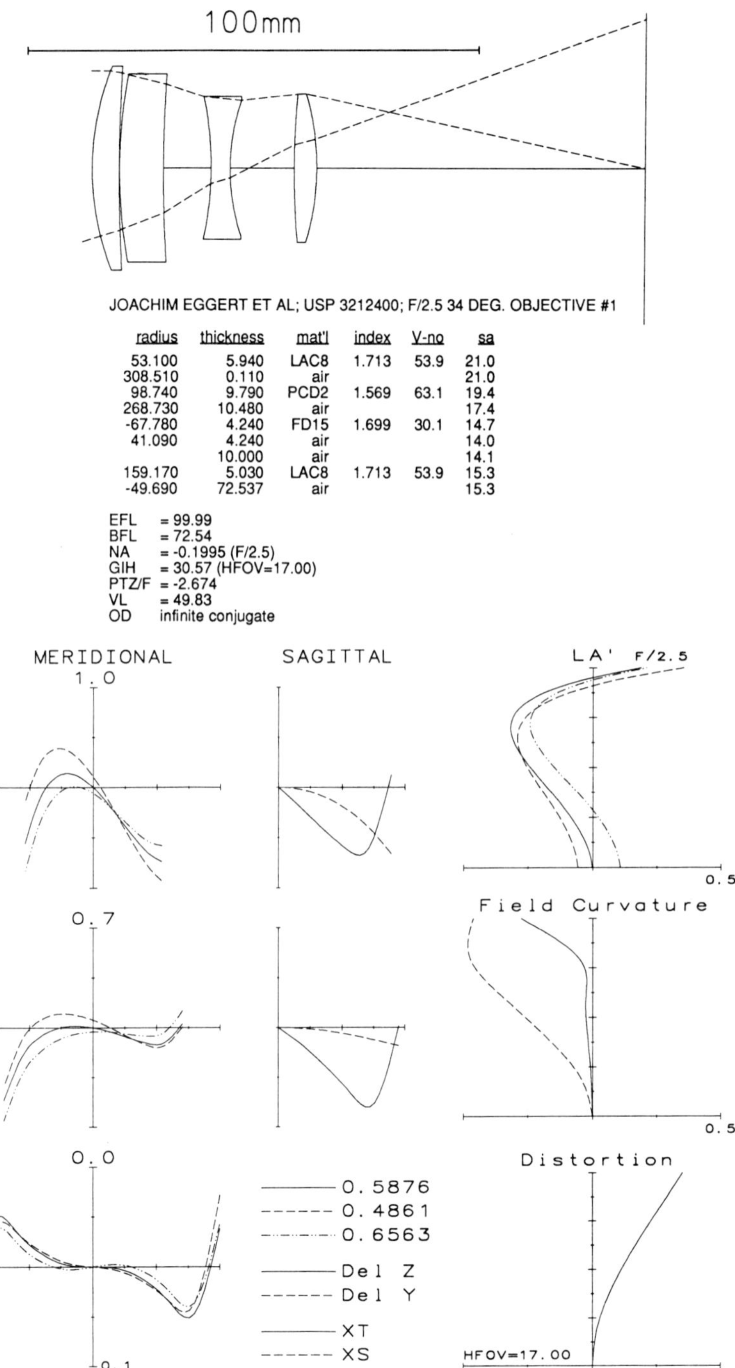

JOACHIM EGGERT ET AL; USP 3212400; F/2.5 34 DEG. OBJECTIVE #1

radius	thickness	mat'l	index	V-no	sa
53.100	5.940	LAC8	1.713	53.9	21.0
308.510	0.110	air			21.0
98.740	9.790	PCD2	1.569	63.1	19.4
268.730	10.480	air			17.4
-67.780	4.240	FD15	1.699	30.1	14.7
41.090	4.240	air			14.0
	10.000	air			14.1
159.170	5.030	LAC8	1.713	53.9	15.3
-49.690	72.537	air			15.3

EFL = 99.99
BFL = 72.54
NA = -0.1995 (F/2.5)
GIH = 30.57 (HFOV=17.00)
PTZ/F = -2.674
VL = 49.83
OD infinite conjugate

Figure 14.5

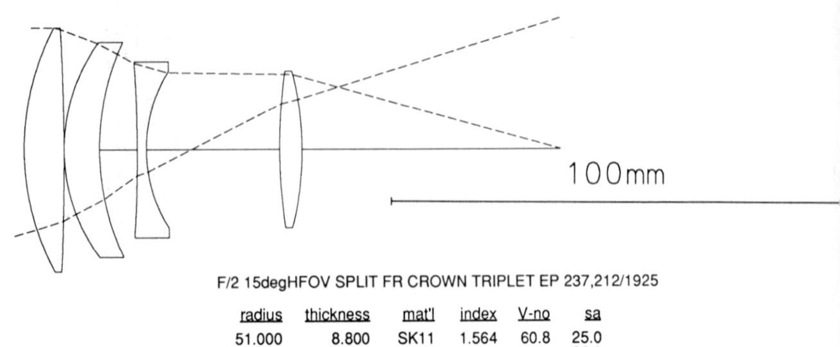

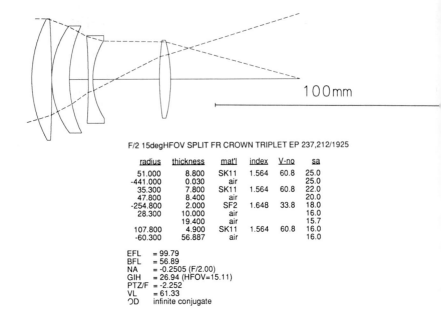

F/2 15degHFOV SPLIT FR CROWN TRIPLET EP 237,212/1925

radius	thickness	mat'l	index	V-no	sa
51.000	8.800	SK11	1.564	60.8	25.0
-441.000	0.030	air			25.0
35.300	7.800	SK11	1.564	60.8	22.0
47.800	8.400	air			20.0
-254.800	2.000	SF2	1.648	33.8	18.0
28.300	10.000	air			16.0
	19.400	air			15.7
107.800	4.900	SK11	1.564	60.8	16.0
-60.300	56.887	air			16.0

EFL = 99.79
BFL = 56.89
NA = -0.2505 (F/2.00)
GIH = 26.94 (HFOV=15.11)
PTZ/F = -2.252
VL = 61.33
OD infinite conjugate

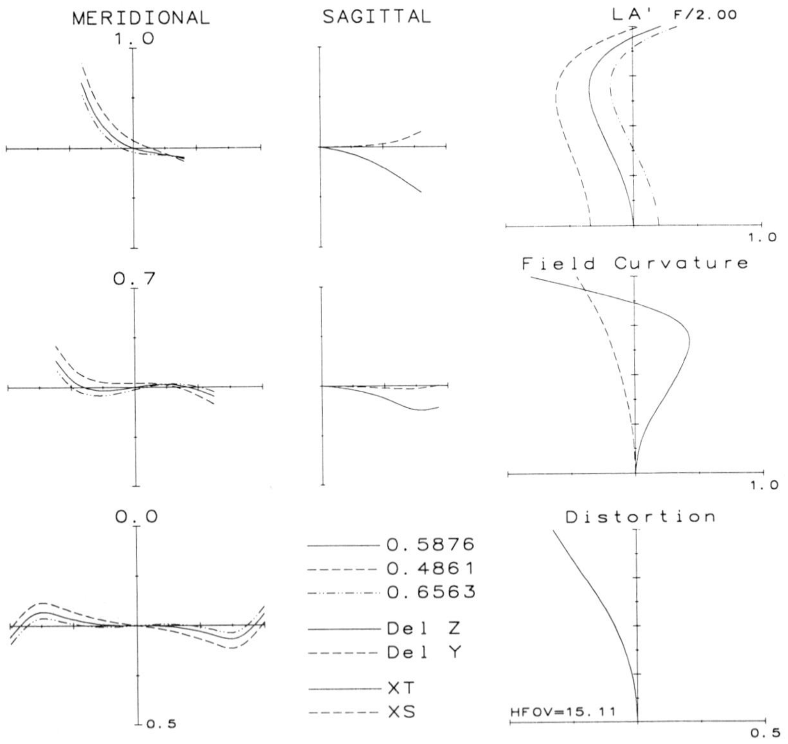

Figure 14.6

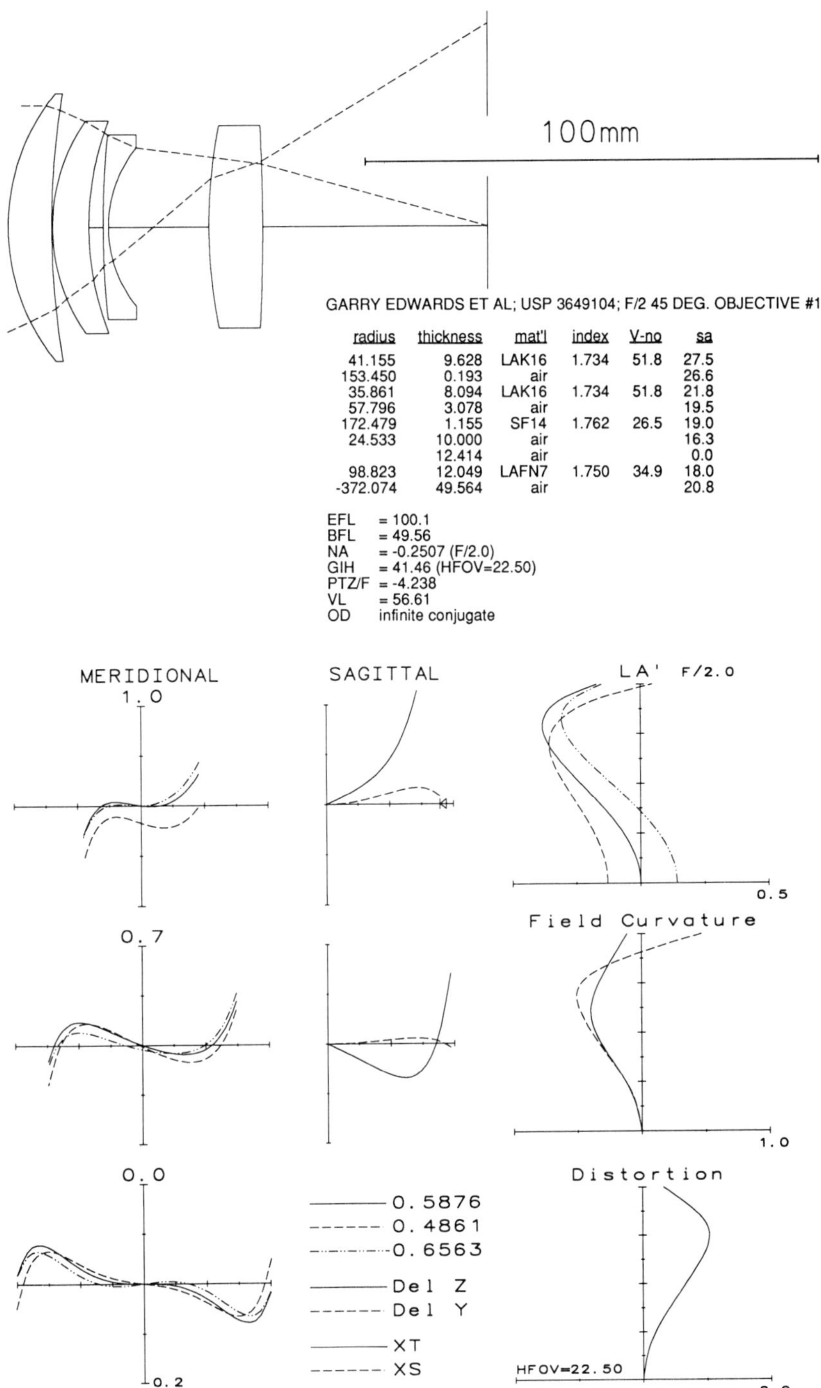

GARRY EDWARDS ET AL; USP 3649104; F/2 45 DEG. OBJECTIVE #1

radius	thickness	mat'l	index	V-no	sa
41.155	9.628	LAK16	1.734	51.8	27.5
153.450	0.193	air			26.6
35.861	8.094	LAK16	1.734	51.8	21.8
57.796	3.078	air			19.5
172.479	1.155	SF14	1.762	26.5	19.0
24.533	10.000	air			16.3
	12.414	air			0.0
98.823	12.049	LAFN7	1.750	34.9	18.0
-372.074	49.564	air			20.8

EFL = 100.1
BFL = 49.56
NA = -0.2507 (F/2.0)
GIH = 41.46 (HFOV=22.50)
PTZ/F = -4.238
VL = 56.61
OD infinite conjugate

Figure 14.7

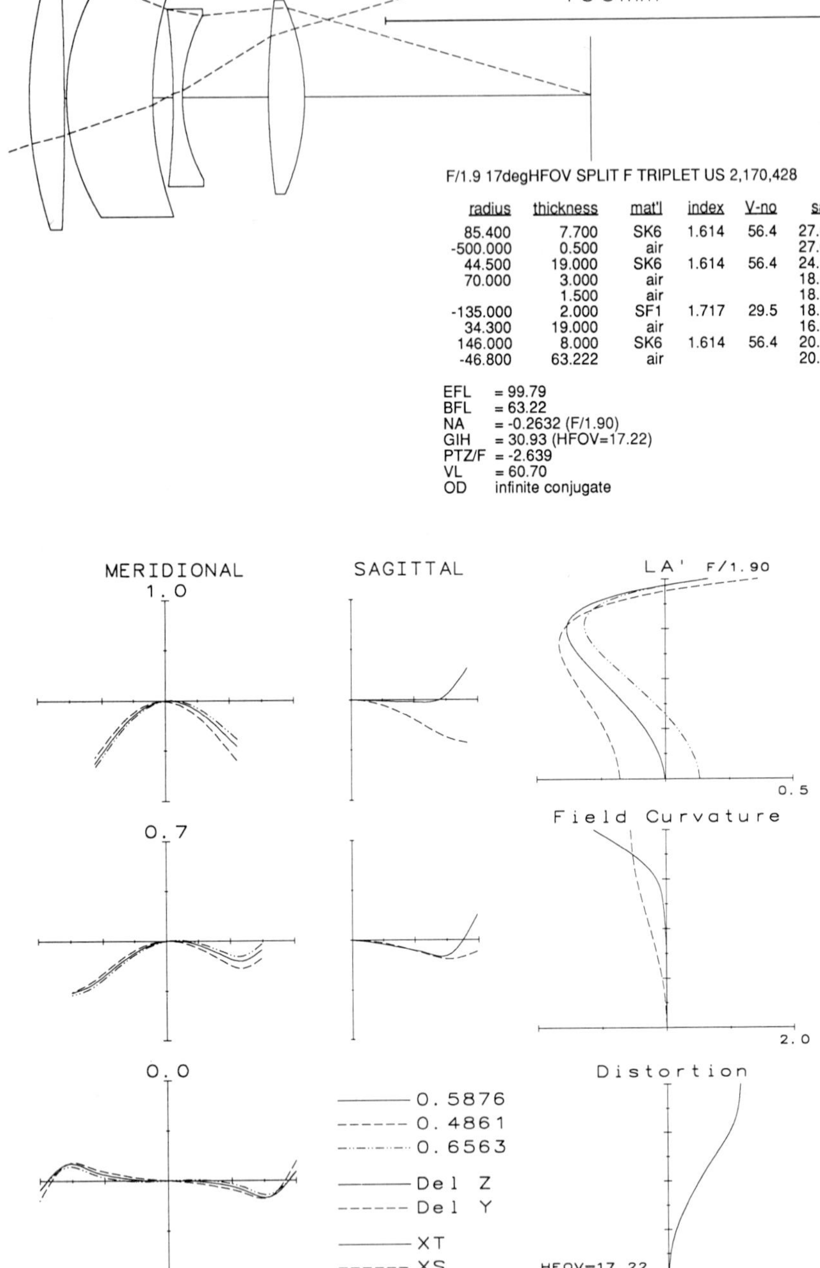

F/1.9 17degHFOV SPLIT F TRIPLET US 2,170,428

radius	thickness	mat'l	index	V-no	sa
85.400	7.700	SK6	1.614	56.4	27.0
-500.000	0.500	air			27.0
44.500	19.000	SK6	1.614	56.4	24.5
70.000	3.000	air			18.6
	1.500	air			18.3
-135.000	2.000	SF1	1.717	29.5	18.3
34.300	19.000	air			16.9
146.000	8.000	SK6	1.614	56.4	20.0
-46.800	63.222	air			20.0

EFL = 99.79
BFL = 63.22
NA = -0.2632 (F/1.90)
GIH = 30.93 (HFOV=17.22)
PTZ/F = -2.639
VL = 60.70
OD infinite conjugate

Figure 14.8

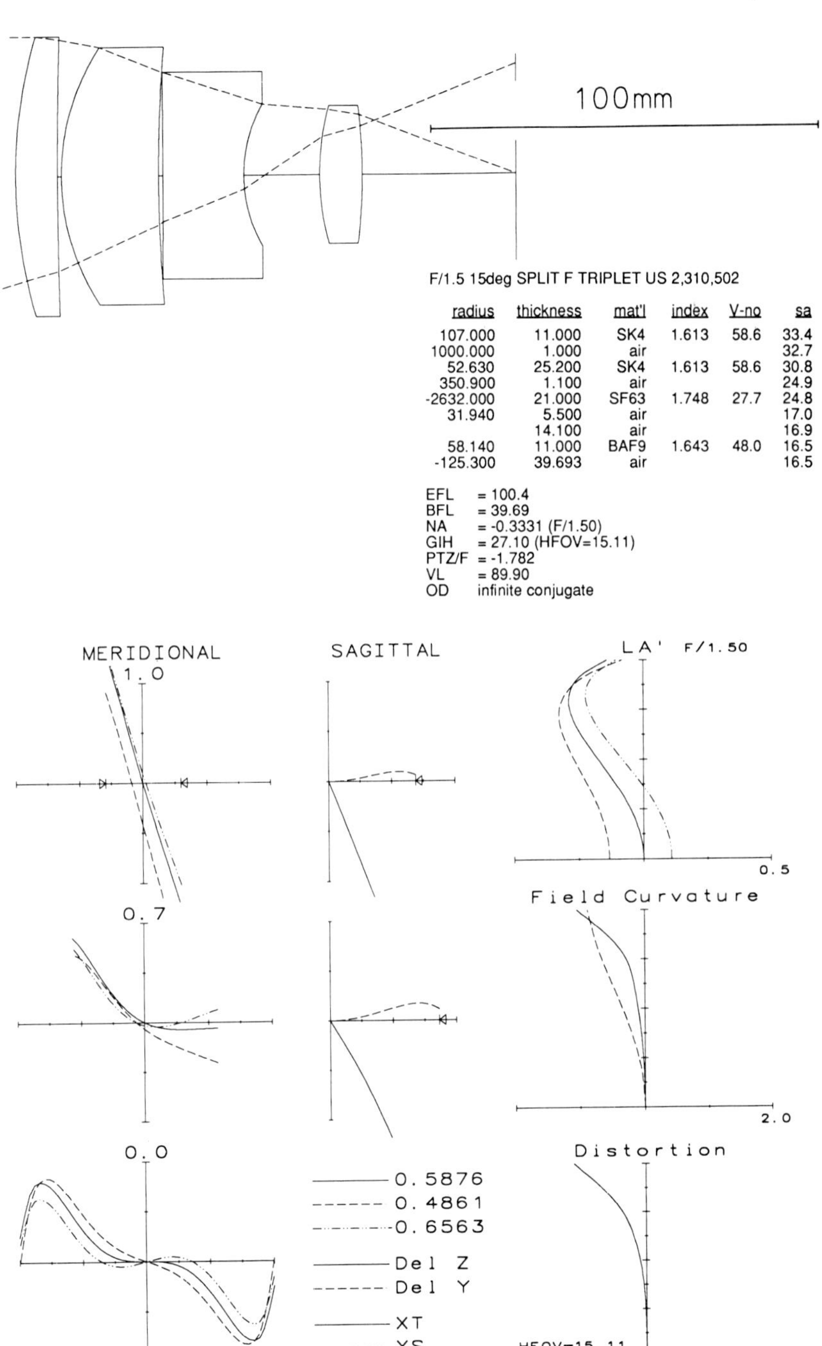

F/1.5 15deg SPLIT F TRIPLET US 2,310,502

radius	thickness	mat'l	index	V-no	sa
107.000	11.000	SK4	1.613	58.6	33.4
1000.000	1.000	air			32.7
52.630	25.200	SK4	1.613	58.6	30.8
350.900	1.100	air			24.9
-2632.000	21.000	SF63	1.748	27.7	24.8
31.940	5.500	air			17.0
	14.100	air			16.9
58.140	11.000	BAF9	1.643	48.0	16.5
-125.300	39.693	air			16.5

EFL = 100.4
BFL = 39.69
NA = -0.3331 (F/1.50)
GIH = 27.10 (HFOV=15.11)
PTZ/F = -1.782
VL = 89.90
OD infinite conjugate

Figure 14.9

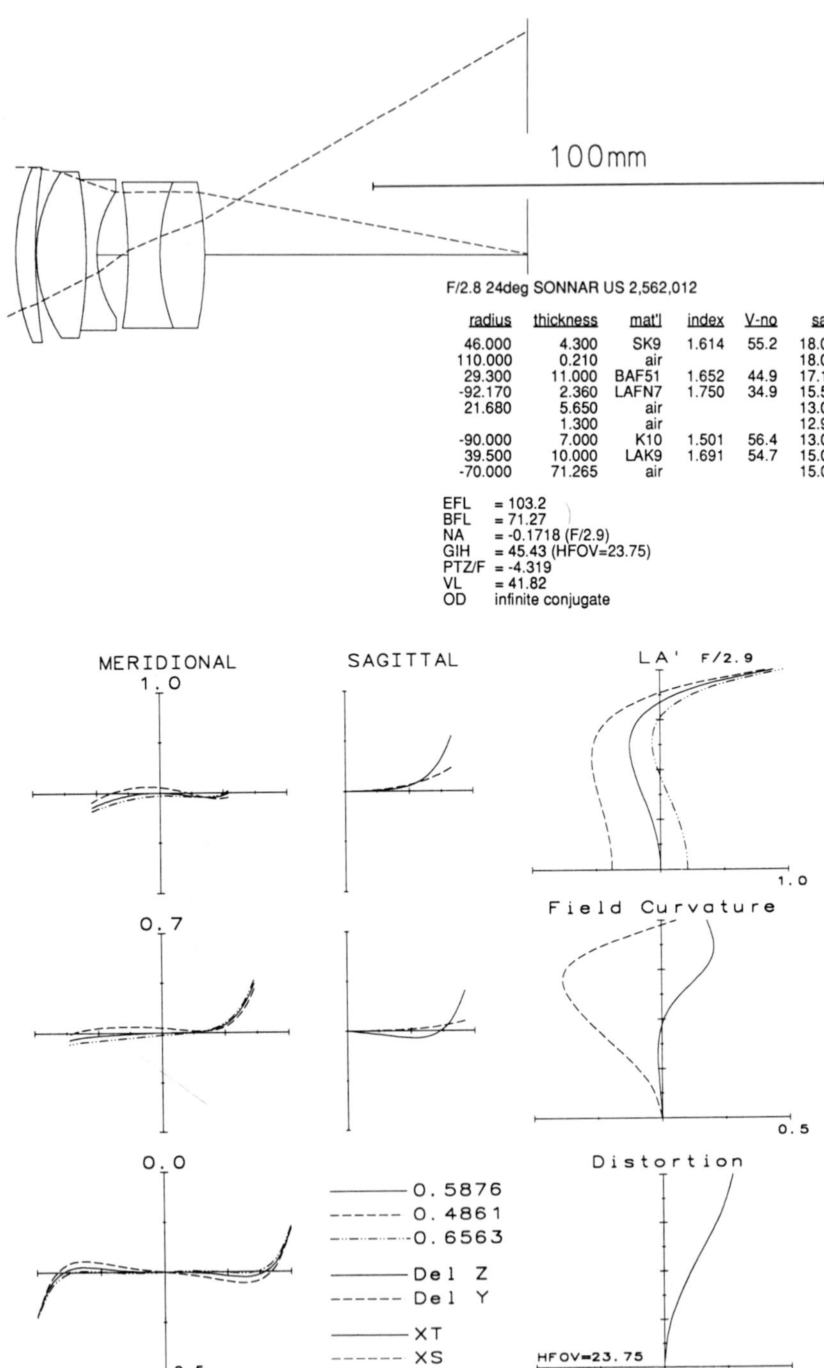

F/2.8 24deg SONNAR US 2,562,012

radius	thickness	mat'l	index	V-no	sa
46.000	4.300	SK9	1.614	55.2	18.0
110.000	0.210	air			18.0
29.300	11.000	BAF51	1.652	44.9	17.1
-92.170	2.360	LAFN7	1.750	34.9	15.5
21.680	5.650	air			13.0
	1.300	air			12.9
-90.000	7.000	K10	1.501	56.4	13.0
39.500	10.000	LAK9	1.691	54.7	15.0
-70.000	71.265	air			15.0

EFL = 103.2
BFL = 71.27
NA = -0.1718 (F/2.9)
GIH = 45.43 (HFOV=23.75)
PTZ/F = -4.319
VL = 41.82
OD infinite conjugate

Figure 14.10

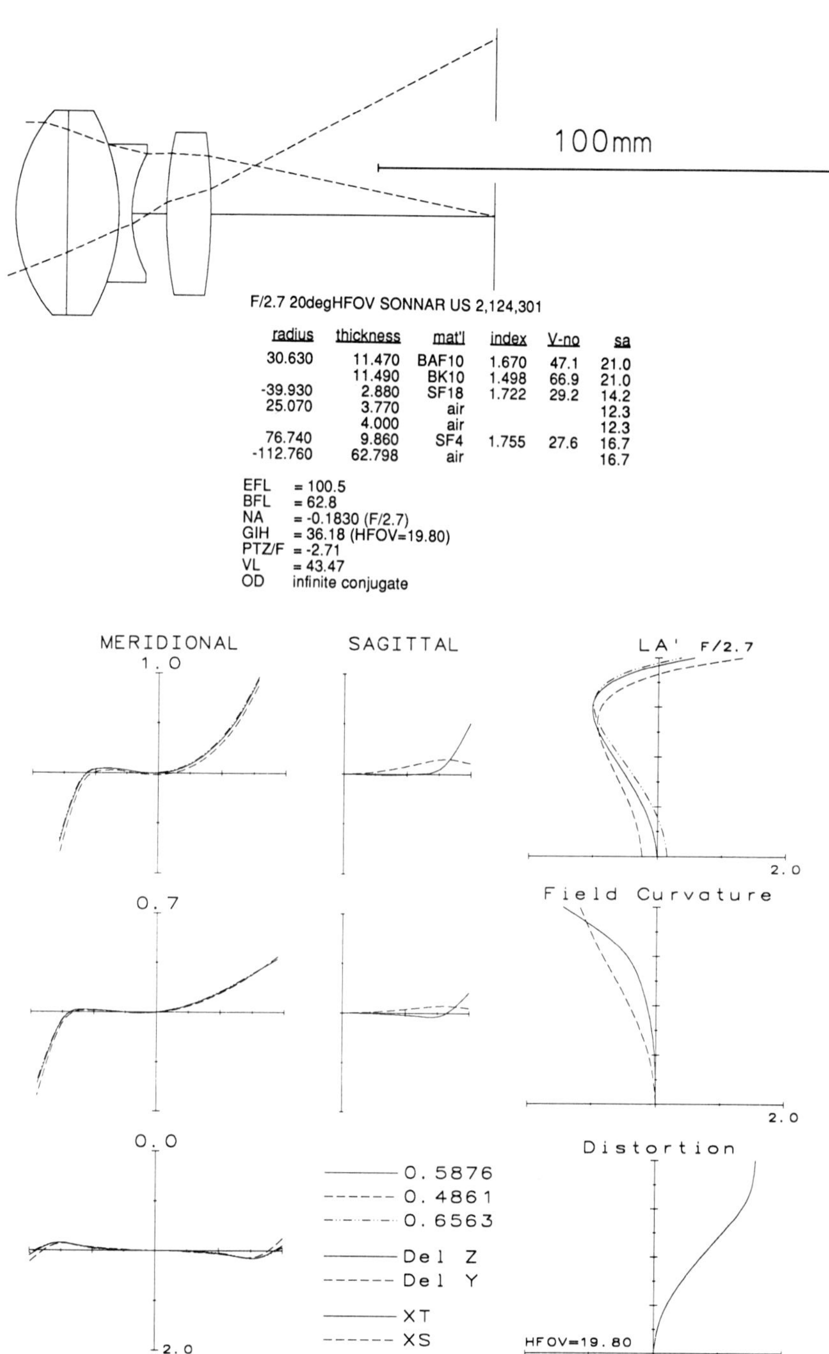

F/2.7 20degHFOV SONNAR US 2,124,301

radius	thickness	mat'l	index	V-no	sa
30.630	11.470	BAF10	1.670	47.1	21.0
	11.490	BK10	1.498	66.9	21.0
-39.930	2.880	SF18	1.722	29.2	14.2
25.070	3.770	air			12.3
	4.000	air			12.3
76.740	9.860	SF4	1.755	27.6	16.7
-112.760	62.798	air			16.7

EFL = 100.5
BFL = 62.8
NA = -0.1830 (F/2.7)
GIH = 36.18 (HFOV=19.80)
PTZ/F = -2.71
VL = 43.47
OD infinite conjugate

Figure 14.11

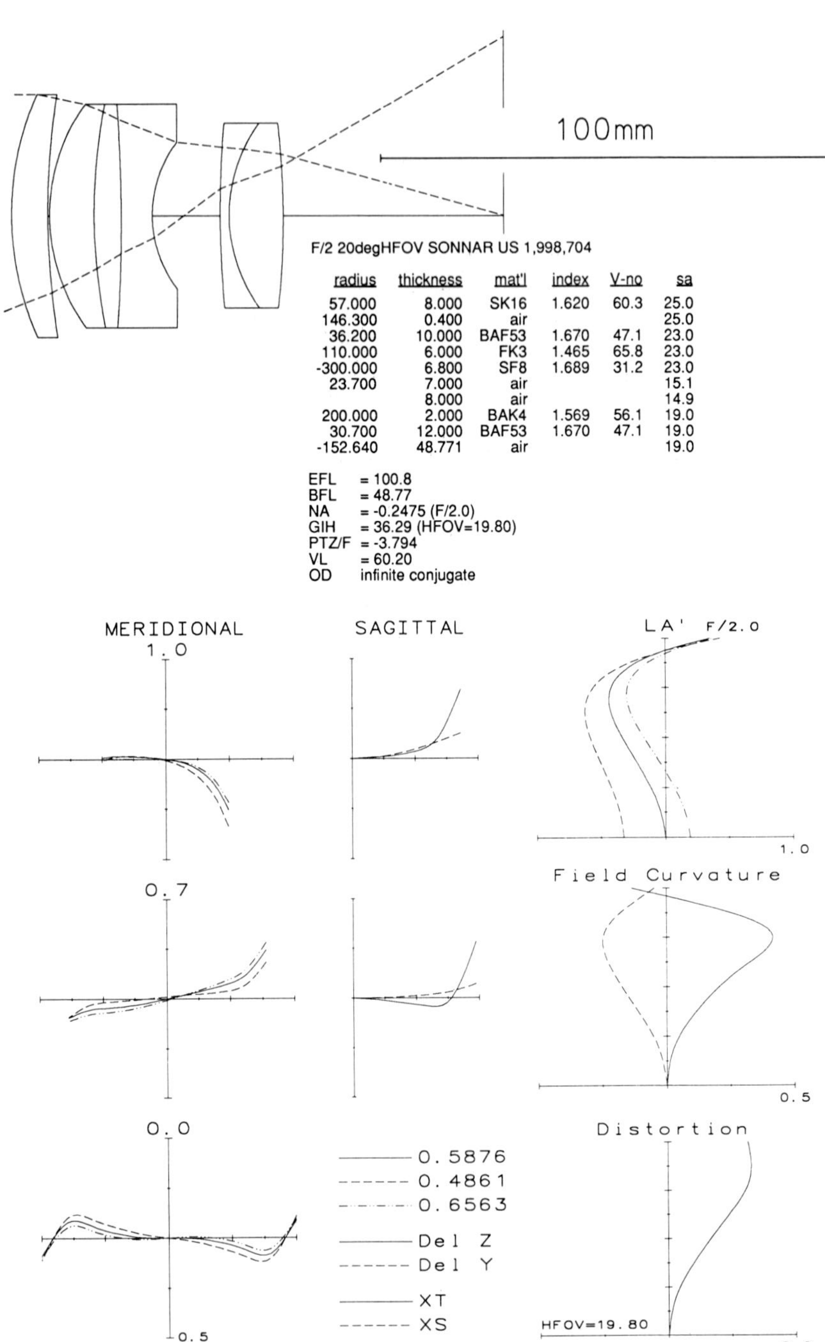

F/2 20degHFOV SONNAR US 1,998,704

radius	thickness	mat'l	index	V-no	sa
57.000	8.000	SK16	1.620	60.3	25.0
146.300	0.400	air			25.0
36.200	10.000	BAF53	1.670	47.1	23.0
110.000	6.000	FK3	1.465	65.8	23.0
-300.000	6.800	SF8	1.689	31.2	23.0
23.700	7.000	air			15.1
	8.000	air			14.9
200.000	2.000	BAK4	1.569	56.1	19.0
30.700	12.000	BAF53	1.670	47.1	19.0
-152.640	48.771	air			19.0

EFL = 100.8
BFL = 48.77
NA = -0.2475 (F/2.0)
GIH = 36.29 (HFOV=19.80)
PTZ/F = -3.794
VL = 60.20
OD infinite conjugate

Figure 14.12

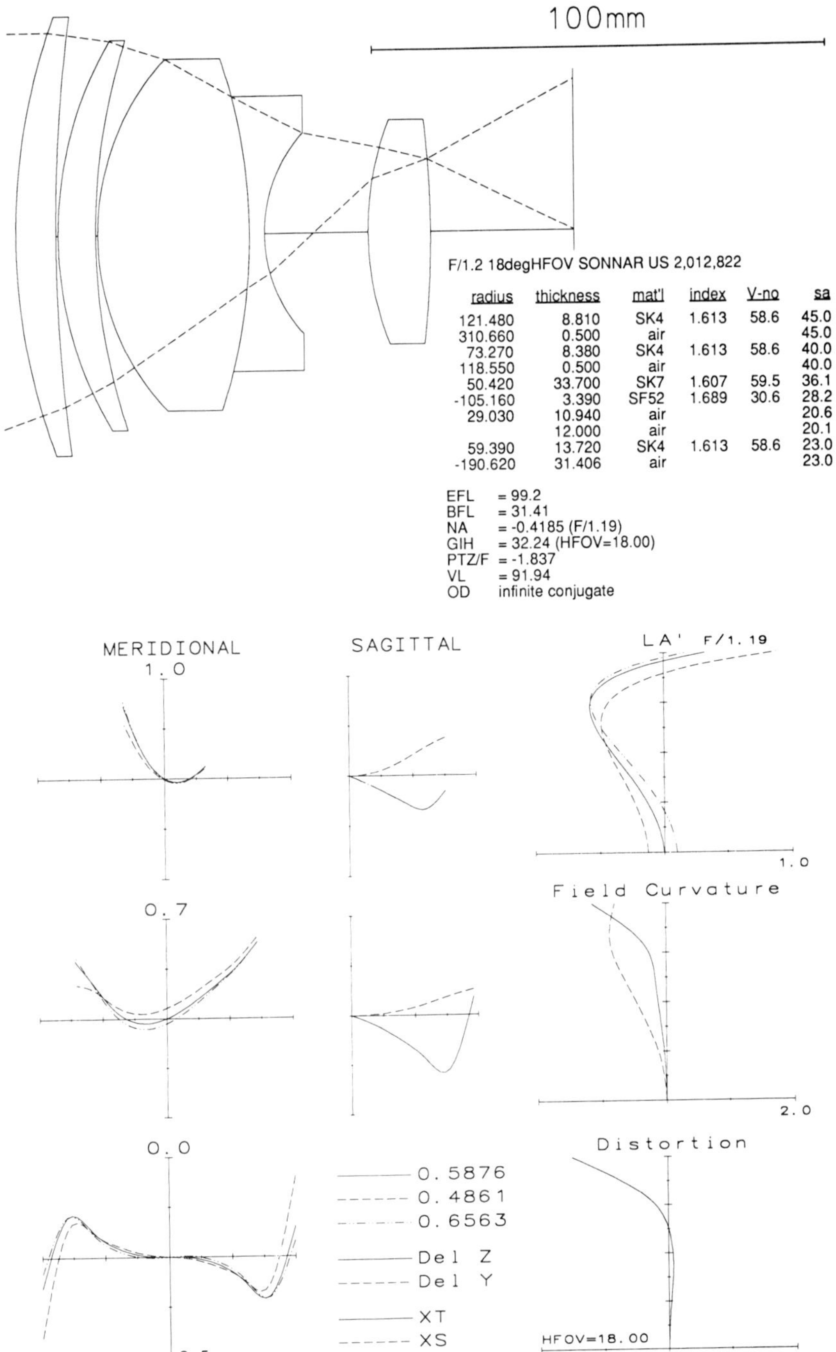

F/1.2 18degHFOV SONNAR US 2,012,822

radius	thickness	mat'l	index	V-no	sa
121.480	8.810	SK4	1.613	58.6	45.0
310.660	0.500	air			45.0
73.270	8.380	SK4	1.613	58.6	40.0
118.550	0.500	air			40.0
50.420	33.700	SK7	1.607	59.5	36.1
-105.160	3.390	SF52	1.689	30.6	28.2
29.030	10.940	air			20.6
	12.000	air			20.1
59.390	13.720	SK4	1.613	58.6	23.0
-190.620	31.406	air			23.0

EFL = 99.2
BFL = 31.41
NA = -0.4185 (F/1.19)
GIH = 32.24 (HFOV=18.00)
PTZ/F = -1.837
VL = 91.94
OD infinite conjugate

Figure 14.13

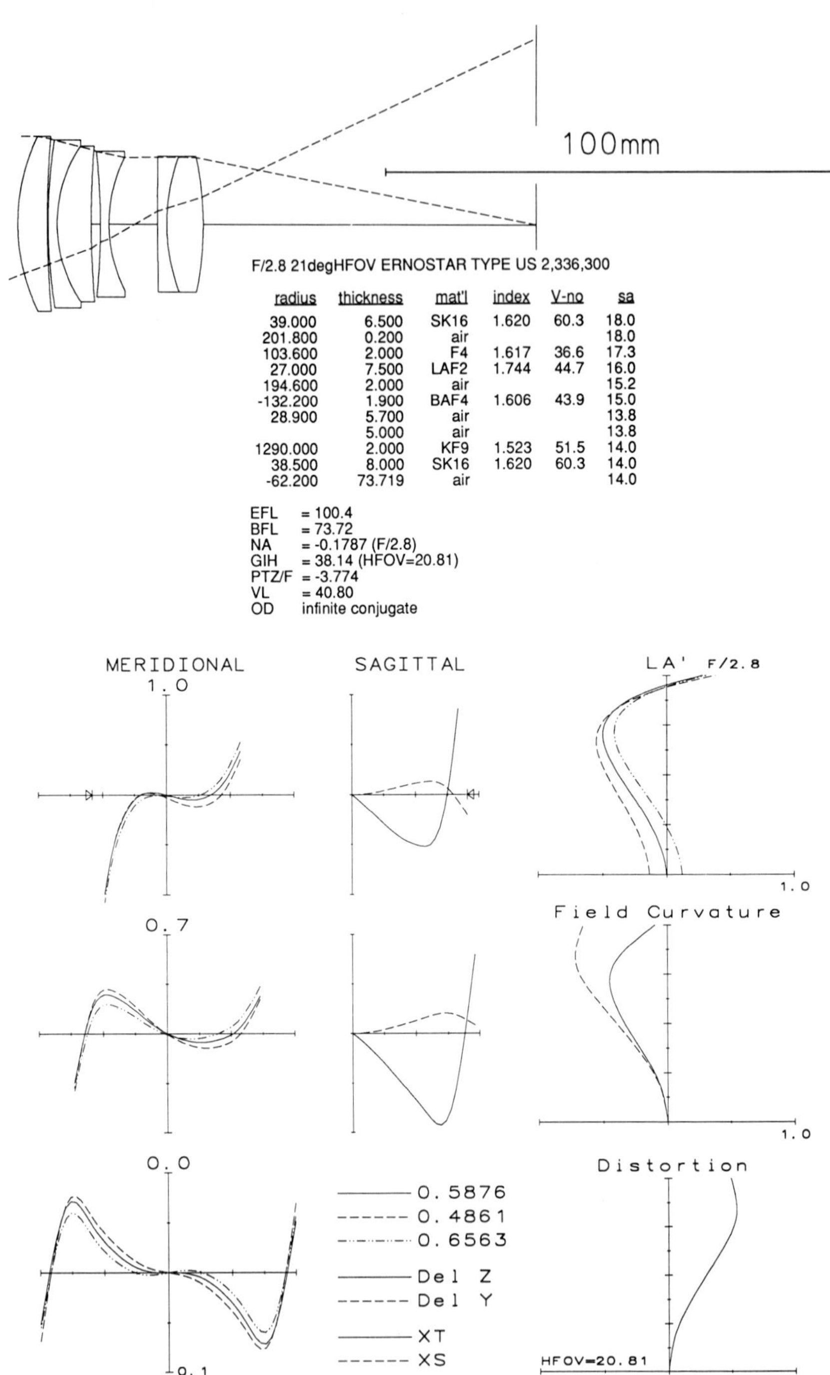

F/2.8 21degHFOV ERNOSTAR TYPE US 2,336,300

radius	thickness	mat'l	index	V-no	sa
39.000	6.500	SK16	1.620	60.3	18.0
201.800	0.200	air			18.0
103.600	2.000	F4	1.617	36.6	17.3
27.000	7.500	LAF2	1.744	44.7	16.0
194.600	2.000	air			15.2
-132.200	1.900	BAF4	1.606	43.9	15.0
28.900	5.700	air			13.8
	5.000	air			13.8
1290.000	2.000	KF9	1.523	51.5	14.0
38.500	8.000	SK16	1.620	60.3	14.0
-62.200	73.719	air			14.0

EFL = 100.4
BFL = 73.72
NA = -0.1787 (F/2.8)
GIH = 38.14 (HFOV=20.81)
PTZ/F = -3.774
VL = 40.80
OD infinite conjugate

Figure 14.14

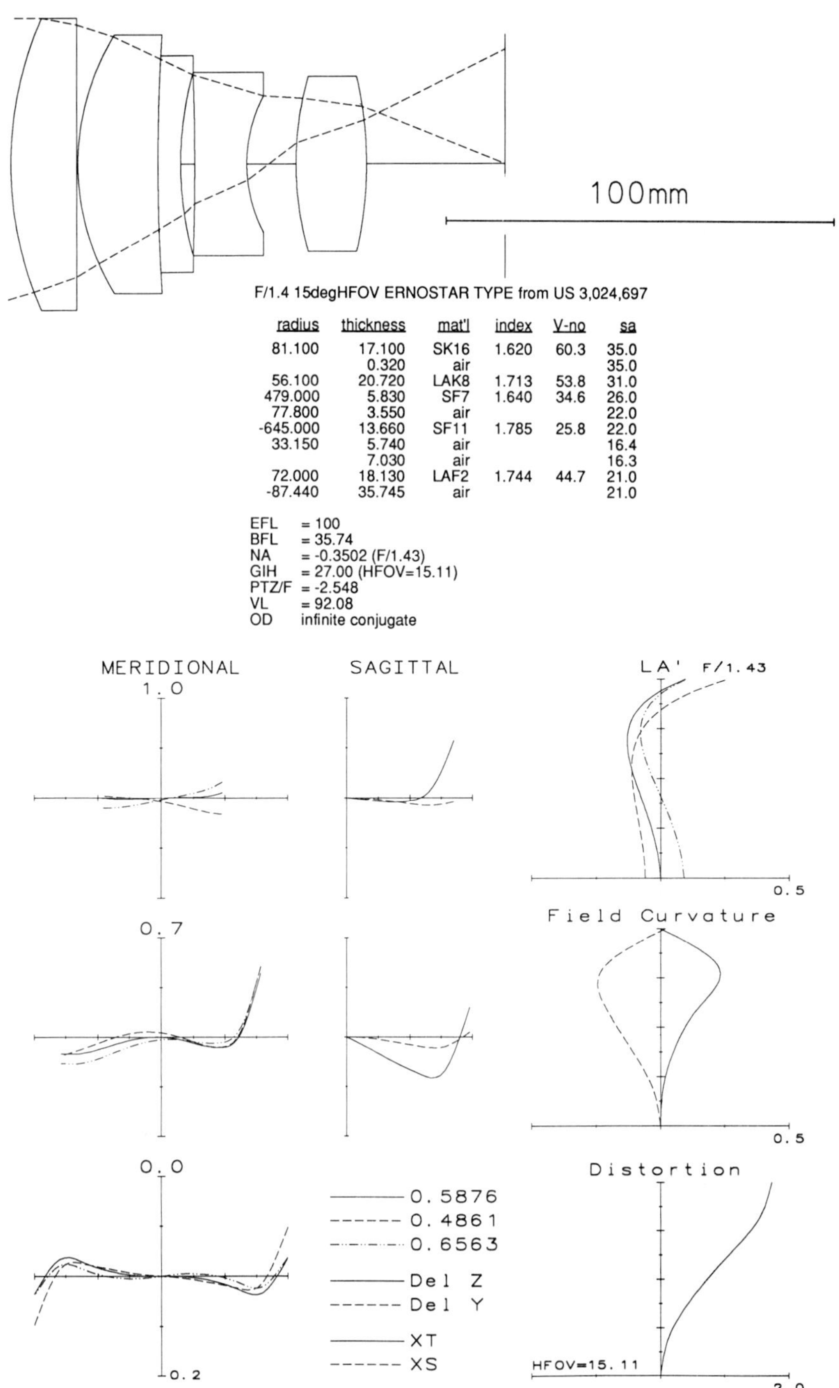

F/1.4 15degHFOV ERNOSTAR TYPE from US 3,024,697

radius	thickness	mat'l	index	V-no	sa
81.100	17.100	SK16	1.620	60.3	35.0
	0.320	air			35.0
56.100	20.720	LAK8	1.713	53.8	31.0
479.000	5.830	SF7	1.640	34.6	26.0
77.800	3.550	air			22.0
-645.000	13.660	SF11	1.785	25.8	22.0
33.150	5.740	air			16.4
	7.030	air			16.3
72.000	18.130	LAF2	1.744	44.7	21.0
-87.440	35.745	air			21.0

EFL = 100
BFL = 35.74
NA = -0.3502 (F/1.43)
GIH = 27.00 (HFOV=15.11)
PTZ/F = -2.548
VL = 92.08
OD infinite conjugate

Figure 14.15

Chapter

15

Microscope Objectives

15.1 General Considerations

An ordinary microscope objective has an image field with a diameter of about 20 mm; the image distance is about 160 mm. The total field of view is thus about 7°. The exit pupil diameter is typically about 8 mm; with the 160-mm image distance, the numerical aperture of the image cone is about 0.025. Therefore, the object-side numerical aperture is about $0.025M$, where M is the magnification of the objective.

A microscope objective may be designed to be used with or without a cover plate (or cover slip) over the object. The cover glass is nominally 0.18 mm (0.16- to 0.19-mm) thick with an index of 1.523 ± 0.005 and a V value of 56 ± 2 (close to Schott K4 or K5 glass). This thickness of glass in the strongly divergent cone of light at the object will contribute a significant amount of aberration and must always be included in the design calculations.

Most microscope objectives are designed to work at an image distance of about 160 mm. The aberration correction of the objective is significantly affected by the image distance. Often a microscope body tube will have a calibrated tube length adjustment; this adjustment can be used to fine-tune the spherical aberration correction of a less than perfectly made objective, or to compensate for a variation in cover plate thickness from the nominal value.

Some objectives are designed for an infinite image distance. These *infinity-corrected* objectives are used with what amounts to a telescope to view the image. One advantage of this arrangement is that a tilted-plate beam splitter can be placed in the collimated beam between the objective and the telescope without introducing the astigmatism which would result if the tilted plate were introduced into a convergent image cone.

It is generally much more convenient to design an optical system with the object at the long conjugate; this widespread practice is usually followed in designing and analyzing microscope objectives. The sample designs in this chapter are ray-traced from the long conjugate (image side) to the short.

15.2 Classical Objective Design Forms; The Aplanatic Front

The classical low-power microscope objective is a simple cemented achromatic doublet or triplet, essentially a telescope objective which has been designed for a finite object distance. Occasionally, a more complex arrangement of spaced-apart elements is used in order to obtain a long focal length in a compact package. The medium-power (10×, NA = 0.25) objective is the same design type as an ordinary *f*/2 Petzval projection lens (see Fig. 13.2, for example). Except for the finite-distance long conjugate, the design techniques for these lenses are the same as those outlined in Chaps. 6 and 13 respectively.

There are usually three possible combinations of shapes for which two cemented achromatic doublets are corrected for both coma and spherical aberration. One of these combinations is the Petzval projection lens configuration, where the astigmatism is controlled to reduce the field curvature. Another combination is the so-called divisible or separable Lister objective. In this form, both doublets are *independently* corrected for spherical and coma. Thus they can be used in combination as a 10× objective, or the small doublet nearest the object can be removed and the large doublet can be used alone as a low-power objective. The very severe drawback to this arrangement is that, since both components are corrected for spherical and coma, they both must contribute large amounts of negative astigmatism (as in Eqs. F.9.4 and F.10.5). The result is a badly inward-curving field and poor off-axis imagery.

The higher-power objectives make use of aplanatic elements located near the object. The function of these elements is to convert a large-numerical-aperture (NA = $n \sin U$), divergent cone of light from an object point into a cone with a lower numerical aperture, without introducing any coma or spherical aberration. The reduced numerical aperture is then within the capability of the (Petzval-type) "back" to handle. Figure 15.1 illustrates many of the standard configurations.

Aplanatic surfaces

There are three cases where a simple spherical surface is aplanatic, i.e., free of both coma and spherical aberration. One occurs when the object and image are both located at the surface. The second case is

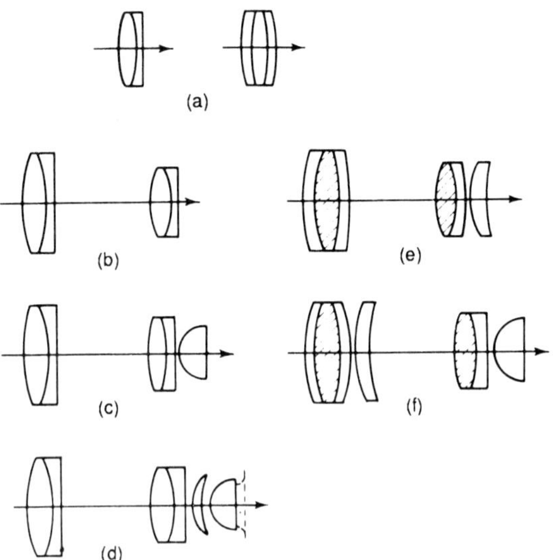

Figure 15.1 Microscope objectives. (*a*) Low-power doublet or triplet. (*b*) Medium-power objective; 10×, NA = 0.25. (*c*) Amici objective; 20×, NA = 0.5 to 40×, NA = 0.8. (*d*) Oil-immersion objective; 100×, NA = 1.2. (*e*) Apochromatic objective; 10×, NA = 0.3 (shading indicates a CaF_2 element). (*f*) Apochromatic objective; 50×, NA = 0.3. Note that the microscope object is to the right and the image is to the left; the arrow indicates the direction of the design raytrace rather than the direction of the light.

with both the object and image at the center of curvature of the surface. The third case occurs when the object and image distances are related by

$$l' = \frac{R(n' + n)}{n'} = \frac{nl}{n'} \tag{15.1}$$

where l = object distance
l' = image distance
n = index of object space
n' = index of the image space
R = radius of the surface

Obviously rays from an axial point pass radially through the second case undeviated; in the third case the slope of the ray (sin U) is changed by a factor equal to (n'/n).

The classic *full aplanatic front* is shown in Fig. 15.2. The object is immersed in a liquid whose index matches that of the front element; note that the higher-than-air immersion index also serves to increase

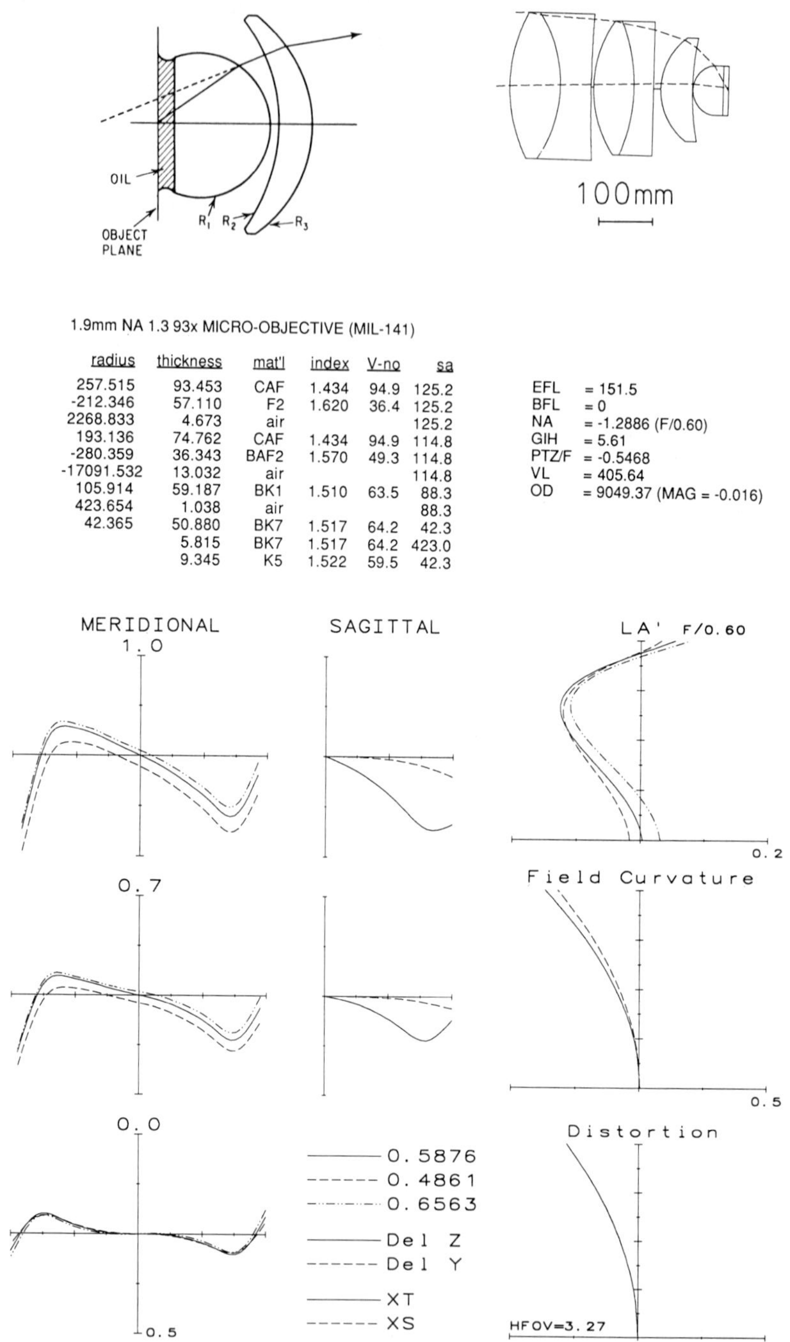

1.9mm NA 1.3 93x MICRO-OBJECTIVE (MIL-141)

radius	thickness	mat'l	index	V-no	sa
257.515	93.453	CAF	1.434	94.9	125.2
-212.346	57.110	F2	1.620	36.4	125.2
2268.833	4.673	air			125.2
193.136	74.762	CAF	1.434	94.9	114.8
-280.359	36.343	BAF2	1.570	49.3	114.8
-17091.532	13.032	air			114.8
105.914	59.187	BK1	1.510	63.5	88.3
423.654	1.038	air			88.3
42.365	50.880	BK7	1.517	64.2	42.3
	5.815	BK7	1.517	64.2	423.0
	9.345	K5	1.522	59.5	42.3

EFL = 151.5
BFL = 0
NA = -1.2886 (F/0.60)
GIH = 5.61
PTZ/F = -0.5468
VL = 405.64
OD = 9049.37 (MAG = -0.016)

Figure 15.2 The full aplanatic front of an oil-immersion objective. The object is immersed in a fluid whose index closely matches that of the hyperhemispheric first element. R_1 is an aplanatic surface of the third kind. The image formed by R_1 is located at the center of curvature of R_2, and R_3 is an aplanatic surface of the same type as R_1. All surfaces are thus aplanatic and the "front" has no coma or spherical aberration.

the NA and thus the resolution. Surface R_1 is the third aplanatic case and reduces the ray slope by a factor of the index of the lens. This first element must be hyperhemispheric to meet the requirements of the aplanatic case. Surface R_2 is the second case (with object and image at the center of curvature), and the rays pass through undeviated. Surface R_3 is again the third aplanatic case, so that the meniscus element also reduces the ray slope by a factor equal to its index. Actually, modest departures from the exact aplanatic cases are usually found in real designs; instead of zero spherical contributions, the element shapes are chosen to contribute some overcorrected spherical aberration and thus reduce the correction load on the Petzval style back.

These microscope objectives are usually very well corrected for the on-axis aberrations. However, they suffer from a strongly inward-curving Petzval field curvature. In addition, those with hyperhemispheric or full aplanatic fronts have lateral chromatic which results from their unsymmetrical construction, i.e., the chromatically undercorrected front and the overcorrected back. The lateral color can be balanced out by an equal but opposite amount of lateral color designed into a compensating eyepiece, but the Petzval curvature always severely limits the off-axis image quality. The other limiting aberration is secondary spectrum, which can be handled as outlined in Sec. 6.2. Elements of calcium fluoride (or of glass types which are composed primarily of calcium fluoride) are commonly used in apochromatic microscope objectives.

An unfortunate characteristic of the Amici objective (Fig. 15.1*c*) is a very short working distance. There is a direct correlation between the working distance and the amount of zonal spherical aberration in this type of objective. This limits the working distance of the higher-NA lenses. Note that occasionally the principle of the aplanatic front is incorporated near the image in photographic-type objectives to increase their speed without increasing the aberrations.

15.3 Flat-Field Objectives

The flat-field microscope objective is based on the usual principle of separating positive and negative power in order to correct the Petzval curvature. The general arrangement is the same as that found in reverse telephoto (or retrofocus) lenses, as described in Chap. 9. In the retrofocus lens, the angular field is usually much larger than in a microscope objective; thus in the retrofocus the general shape of the negative component is almost always a meniscus, concave toward the stop (which is at the rear positive component). However, the narrower field of the microscope objective does not require this shape, and an arrangement with a thick meniscus negative component oriented convex

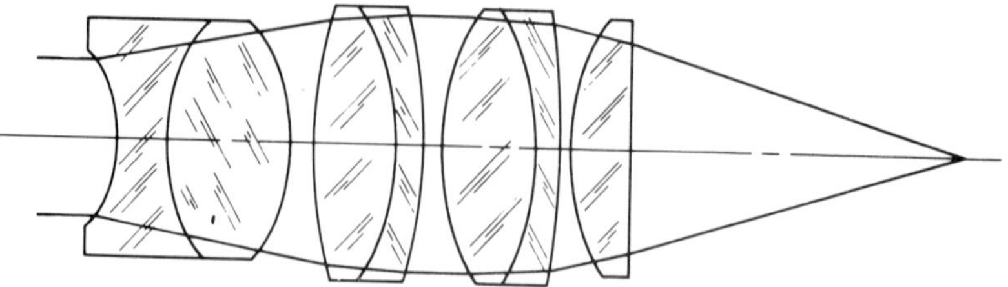

Figure 15.3 A flat-field microscope objective in which the Petzval curvature is corrected by a thick negative achromat placed above the positive components.

toward the positive component is common. A basic arrangement is shown in Fig. 15.3, with a thick negative achromatic component and a positive component consisting of two cemented doublets and a plano-convex singlet. In some designs, the negative component is a thin negative singlet spaced well away from the positive component. These objectives (like their retrofocus counterparts) have a long working distance; as a result the element diameters tend to be large.

15.4 Reflecting Objectives

The Schwarzschild mirror configuration described in Sec. 16.2 is the basis of the reflecting microscope objective. The most common version is the simple two-spherical-mirror arrangement; increased numerical apertures are possible either by utilizing aspheric surfaces or by incorporating refracting elements as shown in Fig. 15.4. An infinity-corrected two-mirror system can readily be derived using the relation-

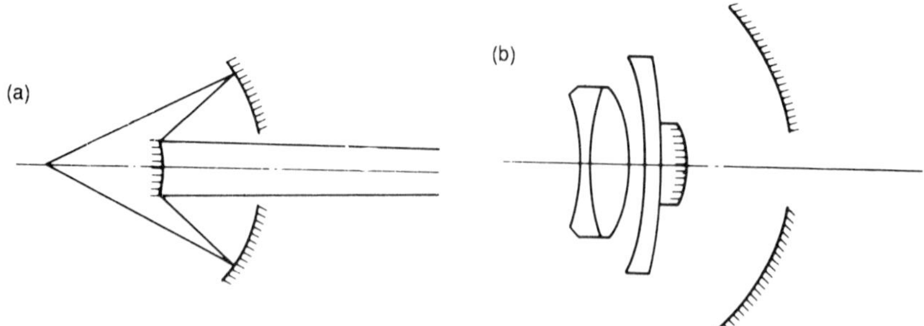

Figure 15.4 Reflecting microscope objectives. (*a*) The basic reflecting objective consists of simple concentric spherical surfaces in a Schwarzschild arrangement; 30×, NA = 0.5. (*b*) An ultraviolet modification at 50×, NA = 0.7 attributable to David Grey. The refracting elements are of SiO_2 and CaF_2.

ships given in Sec. 16.2; a modest adjustment will convert it to a design corrected for finite conjugates.

15.4 The Sample Lenses

Note that although the sample lens designs are shown at our standardized focal length of 100, they are, of course, intended for use at much shorter focal lengths, at which their aberrations are correspondingly smaller.

Figure 15.5 is an infinity-corrected lens which achieves a flat Petzval surface by utilizing thick menisci in a facing arrangement similar to that found in a double-Gauss or Biotar lens (Chap. 17). Note the incorporation of calcium fluoride (CaF_2) and FK51- and KzFS-type glasses in most of the sample designs. As noted in Sec. 15.2, such materials are often used in microscope objectives to reduce the secondary spectrum. Such lenses are variously referred to as *apochromatic, semiapochromatic,* or *fluorite* objectives, depending somewhat on the degree of correction achieved. Figure 15.6 is a simple construction using ordinary glass. Note the concave surface near the focal plane which helps to reduce the Petzval curvature somewhat. Note also the plano-surfaced cover plate incorporated as part of the design in this and several of the other lenses. Figures 15.7 and 15.10 are examples of *infinity-corrected* lenses which incorporate telescope objectives on the image side. Both use meniscus doublets to reduce field curvature and both have secondary spectrum reducing glass types. Figure 15.10 is a high-NA oil immersion objective. Figures 15.8 and 15.9 also incorporate unusual partial-dispersion materials as well as concave surfaces.

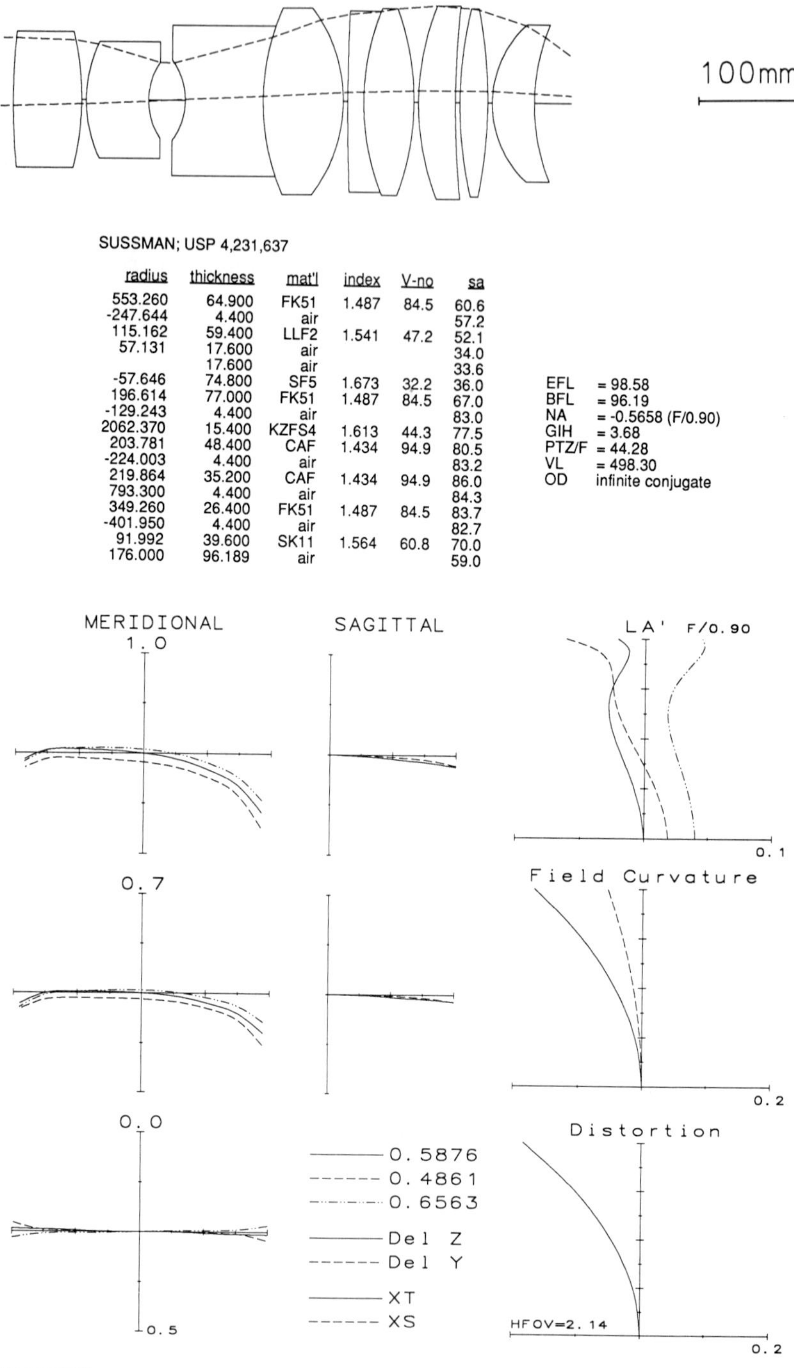

SUSSMAN; USP 4,231,637

radius	thickness	mat'l	index	V-no	sa
553.260	64.900	FK51	1.487	84.5	60.6
-247.644	4.400	air			57.2
115.162	59.400	LLF2	1.541	47.2	52.1
57.131	17.600	air			34.0
	17.600	air			33.6
-57.646	74.800	SF5	1.673	32.2	36.0
196.614	77.000	FK51	1.487	84.5	67.0
-129.243	4.400	air			83.0
2062.370	15.400	KZFS4	1.613	44.3	77.5
203.781	48.400	CAF	1.434	94.9	80.5
-224.003	4.400	air			83.2
219.864	35.200	CAF	1.434	94.9	86.0
793.300	4.400	air			84.3
349.260	26.400	FK51	1.487	84.5	83.7
-401.950	4.400	air			82.7
91.992	39.600	SK11	1.564	60.8	70.0
176.000	96.189	air			59.0

EFL = 98.58
BFL = 96.19
NA = -0.5658 (F/0.90)
GIH = 3.68
PTZ/F = 44.28
VL = 498.30
OD infinite conjugate

Figure 15.5

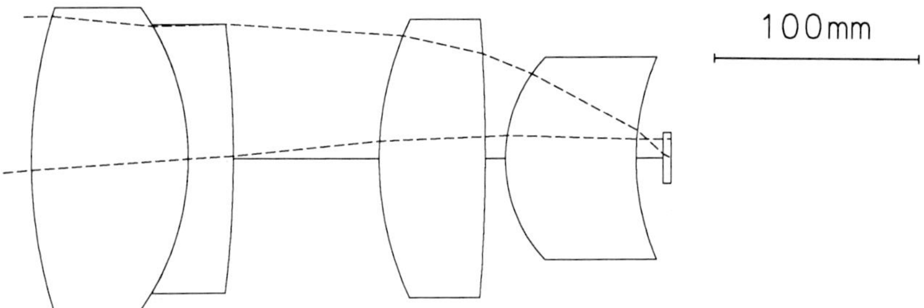

KARLHEINZ ESSWEIN; USP 4362365; .65NA 40X MICROSCOPE OBJ. #1

radius	thickness	mat'l	index	V-no	sa
216.912	77.028	K7	1.511	60.4	69.0
-111.847	21.965	SF56	1.785	26.1	60.5
-496.174	71.258	air			61.2
138.771	52.634	SK16	1.620	60.3	63.3
-803.477	9.919	air			57.0
64.780	63.161	SK2	1.607	56.7	46.0
110.227	13.158	air			20.2
	3.910	K3	1.518	59.0	11.5
	0.003	air			11.5

EFL = 101.2
BFL = 0.002956
NA = -0.6521 (F/0.77)
GIH = 8.63
PTZ/F = -1.386
VL = 313.03
OD = 3992.00 (MAG = -0.025)

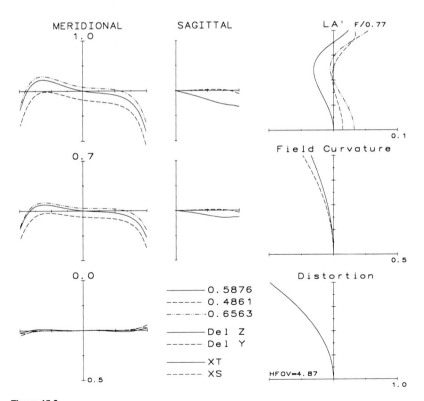

MERIDIONAL 1.0 SAGITTAL LA' F/0.77

0.1

0.7 Field Curvature

0.5

0.0 Distortion

0.5876
0.4861
0.6563
Del Z
Del Y
XT
XS

0.5 HFOV=4.87 1.0

Figure 15.6

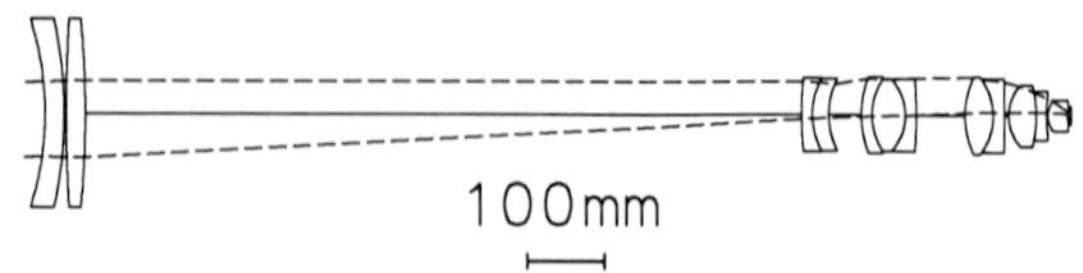

MILTON H. SUSSMAN; USP 4379623; 40X .8NA W/USP 3355234 TELESCOPE

radius	thickness	mat'l	index	V-no	sa
-352.265	24.702	BK1	1.510	63.5	112.8
-504.116	2.685	air			112.8
1729.446	24.702	SSK4	1.618	55.0	112.8
-1136.544	854.287	air			112.8
	69.578	air			0.0
402.742	13.962	KF9	1.523	51.5	44.0
81.612	21.480	SF11	1.785	25.8	43.0
90.085	41.885	air			39.7
117.064	13.962	SF15	1.699	30.1	45.1
72.032	42.959	FK51	1.487	84.5	43.0
-60.175	13.962	BK7	1.517	64.2	43.0
-605.756	63.365	air			43.5
96.819	36.515	FK51	1.487	84.5	46.2
-70.217	13.962	BK7	1.517	64.2	44.0
	6.122	air			40.8
51.551	35.441	FK51	1.487	84.5	36.5
-138.930	12.888	KF9	1.523	51.5	29.0
134.999	1.955	air			22.6
27.043	24.702	LSK02	1.786	50.0	18.6
23.166	3.211	air			8.6
	1.930	K3	1.518	59.0	10.7

EFL = 100
BFL = 0
NA = -0.8021 (F/0.62)
GIH = 2.49
PTZ/F = -2.359
VL = 1324.25
OD = 1876.20 (MAG = -0.025)

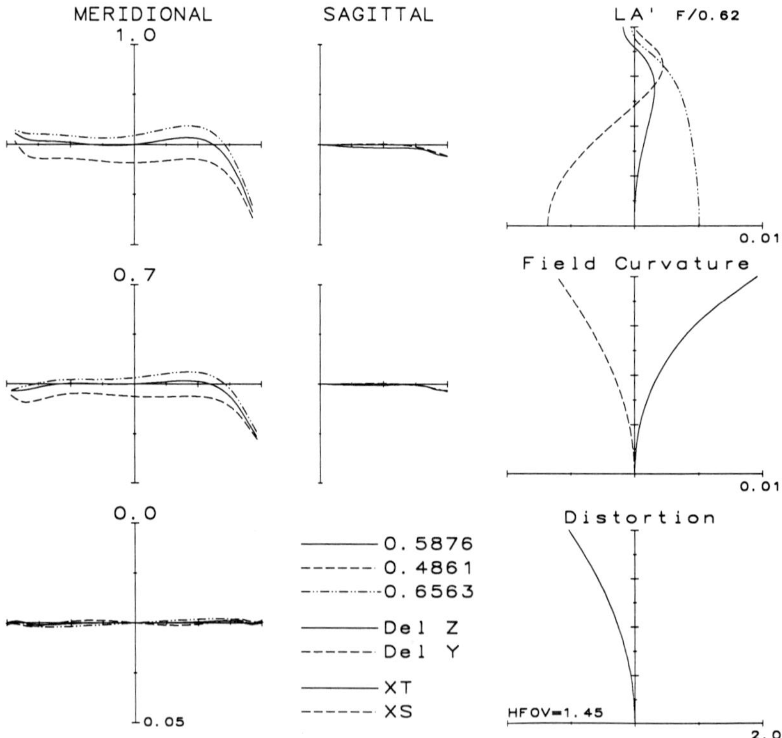

Figure 15.7

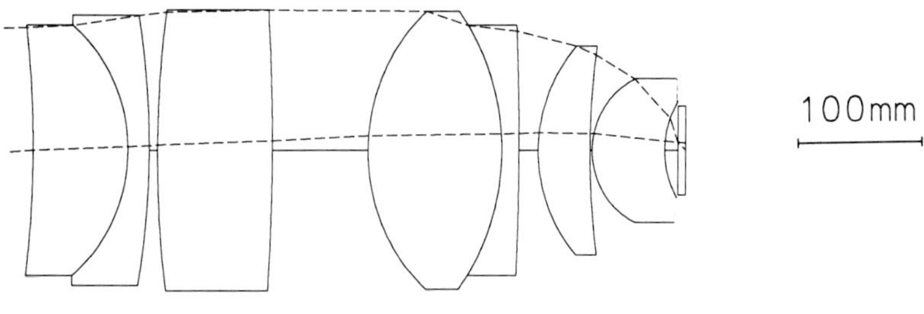

MASAKI MATSUBARA; USP 4037934; .95NA 60X MICROSCOPE OBJECTIVE #1

radius	thickness	mat'l	index	V-no	sa
-753.114	76.280	FK51	1.487	84.5	92.0
-121.010	17.373	PCD4	1.618	63.4	94.0
-577.791	6.889	air			101.3
808.826	93.153	PCD4	1.618	63.4	103.7
-1635.724	77.878	air			105.3
139.381	107.531	CAF	1.434	94.9	104.3
-175.224	13.878	SF3	1.740	28.3	94.0
-2129.653	15.576	air			89.7
116.217	41.635	CAF	1.434	94.9	78.3
571.301	1.697	air			71.7
59.007	58.907	LAF28	1.773	49.6	54.0
70.289	10.667	air			30.0
	6.000	K3	1.518	59.0	33.3
	0.002	air			33.3

EFL = 100
BFL = -0.001501
NA = -0.9472 (F/0.53)
GIH = 5.56
PTZ/F = -1.853
VL = 527.46
OD = 5834.39 (MAG = -0.016)

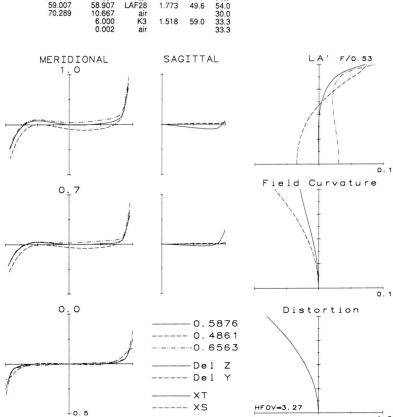

MERIDIONAL SAGITTAL
1.0
0.7
0.0
0.5

LA' F/0.53
0.1

Field Curvature
0.1

Distortion
HFOV=3.27
1.0

0.5876
0.4861
0.6563
Del Z
Del Y
XT
XS

Figure 15.8

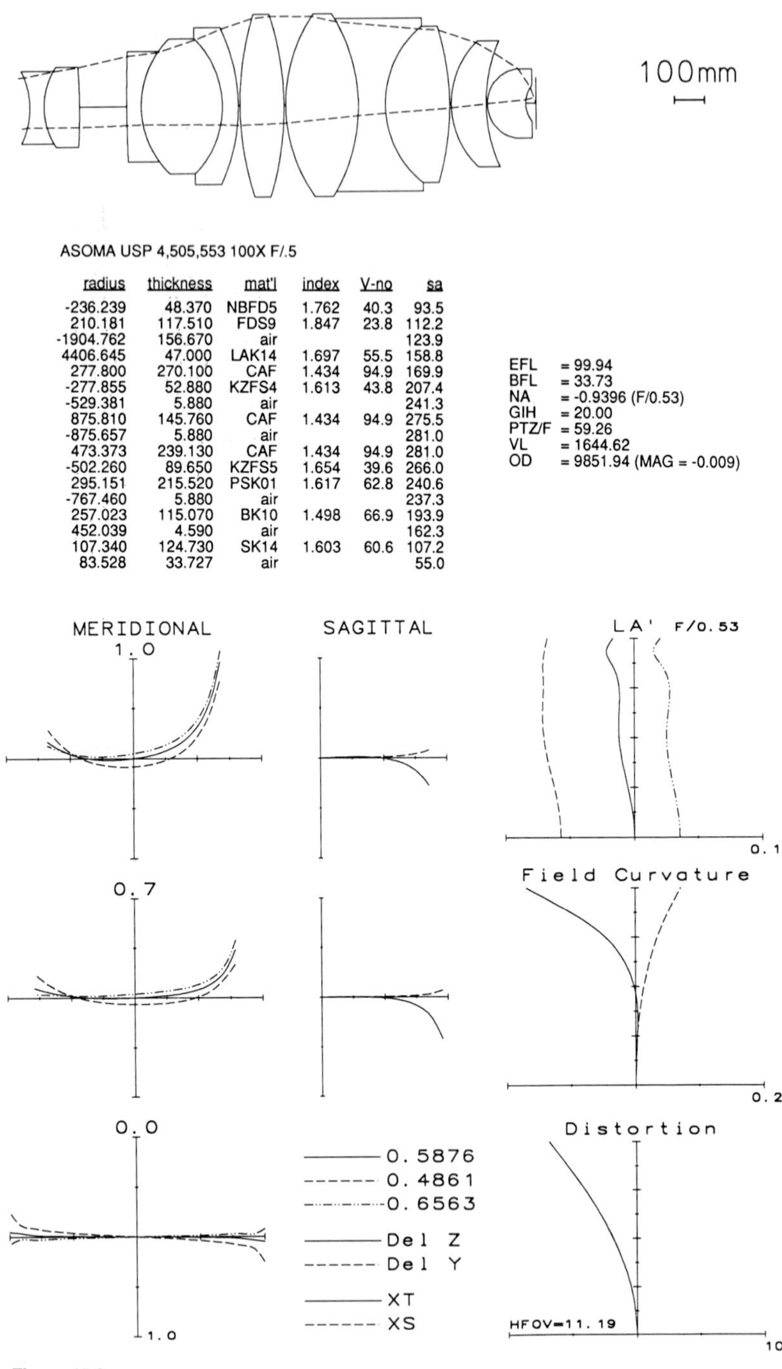

ASOMA USP 4,505,553 100X F/.5

radius	thickness	mat'l	index	V-no	sa
-236.239	48.370	NBFD5	1.762	40.3	93.5
210.181	117.510	FDS9	1.847	23.8	112.2
-1904.762	156.670	air			123.9
4406.645	47.000	LAK14	1.697	55.5	158.8
277.800	270.100	CAF	1.434	94.9	169.9
-277.855	52.880	KZFS4	1.613	43.8	207.4
-529.381	5.880	air			241.3
875.810	145.760	CAF	1.434	94.9	275.5
-875.657	5.880	air			281.0
473.373	239.130	CAF	1.434	94.9	281.0
-502.260	89.650	KZFS5	1.654	39.6	266.0
295.151	215.520	PSK01	1.617	62.8	240.6
-767.460	5.880	air			237.3
257.023	115.070	BK10	1.498	66.9	193.9
452.039	4.590	air			162.3
107.340	124.730	SK14	1.603	60.6	107.2
83.528	33.727	air			55.0

EFL = 99.94
BFL = 33.73
NA = -0.9396 (F/0.53)
GIH = 20.00
PTZ/F = 59.26
VL = 1644.62
OD = 9851.94 (MAG = -0.009)

Figure 15.9

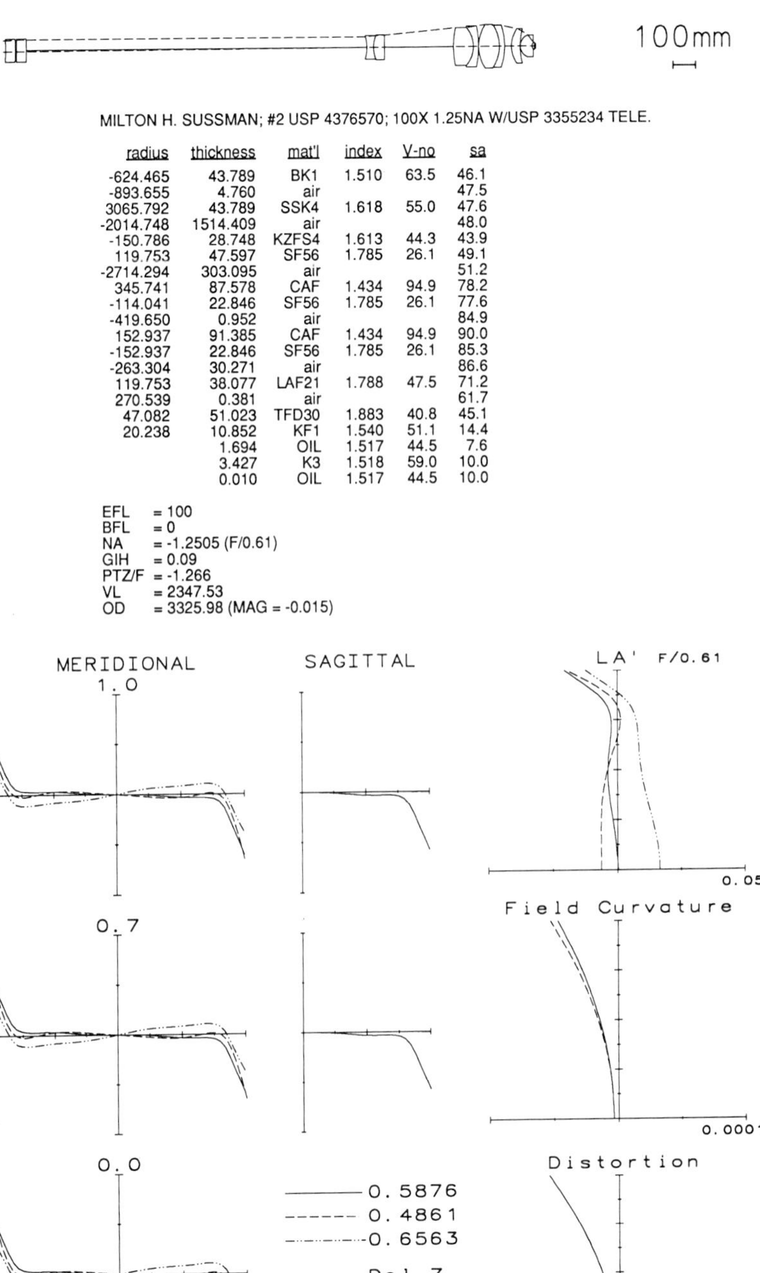

MILTON H. SUSSMAN; #2 USP 4376570; 100X 1.25NA W/USP 3355234 TELE.

radius	thickness	mat'l	index	V-no	sa
-624.465	43.789	BK1	1.510	63.5	46.1
-893.655	4.760	air			47.5
3065.792	43.789	SSK4	1.618	55.0	47.6
-2014.748	1514.409	air			48.0
-150.786	28.748	KZFS4	1.613	44.3	43.9
119.753	47.597	SF56	1.785	26.1	49.1
-2714.294	303.095	air			51.2
345.741	87.578	CAF	1.434	94.9	78.2
-114.041	22.846	SF56	1.785	26.1	77.6
-419.650	0.952	air			84.9
152.937	91.385	CAF	1.434	94.9	90.0
-152.937	22.846	SF56	1.785	26.1	85.3
-263.304	30.271	air			86.6
119.753	38.077	LAF21	1.788	47.5	71.2
270.539	0.381	air			61.7
47.082	51.023	TFD30	1.883	40.8	45.1
20.238	10.852	KF1	1.540	51.1	14.4
	1.694	OIL	1.517	44.5	7.6
	3.427	K3	1.518	59.0	10.0
	0.010	OIL	1.517	44.5	10.0

EFL = 100
BFL = 0
NA = -1.2505 (F/0.61)
GIH = 0.09
PTZ/F = -1.266
VL = 2347.53
OD = 3325.98 (MAG = -0.015)

Figure 15.10

Chapter

16

Mirror and Catadioptric Systems

16.1 The Good and Bad Points of Mirrors

A mirror element has several advantages over a refracting element. It is completely achromatic, having neither axial nor lateral color, nor chromatic variation of the aberrations (e.g., spherochromatism). A mirror can be used in any spectral region for which a reflecting coating is available (and the available coatings make this a very broad region), whereas a refractor is quite severely limited by the transmission characteristics of its material. Yet a third advantage is the fact that the aberrations of a spherical mirror are inherently smaller than those of a comparable lens. For example, the spherical aberration of a mirror is only one-eighth that of an equivalent lens of index 1.5, even with the lens bent for minimum spherical. Another unique feature of mirrors is that the Petzval field curvature is backward-curving for a converging (concave) mirror; this is the reverse of a refracting element.

The drawback to a single mirror is that its image is located in, and thus obscures, the incoming beam of light. Also, in a centered multiple-mirror system, the secondary mirror obscures the incoming beam. The obscuration not only reduces the illumination in the image, but, as indicated in Fig. 16.1, can drastically reduce the contrast in the image, especially at the lower spatial frequencies. The obscuration can be avoided by decentering the system aperture and/or tilting the mirrors of the system. Figures 16.20 and 16.21 are examples of unobscured systems in which the aperture has been decentered and the axial symmetry of the surfaces has been maintained.

16.2 The Classical Two-Mirror Systems

The third-order aberrations of *any* two-mirror system can readily be determined from the following equations. Given the focal length F, the

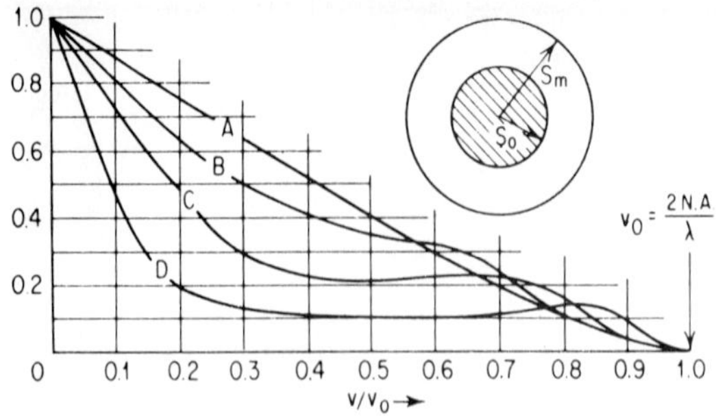

Figure 16.1 The effect of a central obscuration on the MTF of an aberration-free system. (*A*) $S_o/S_m = 0.00$. (*B*) $S_o/S_m = 0.25$. (*C*) $S_o/S_m = 0.50$. (*D*) $S_o/S_m = 0.75$.

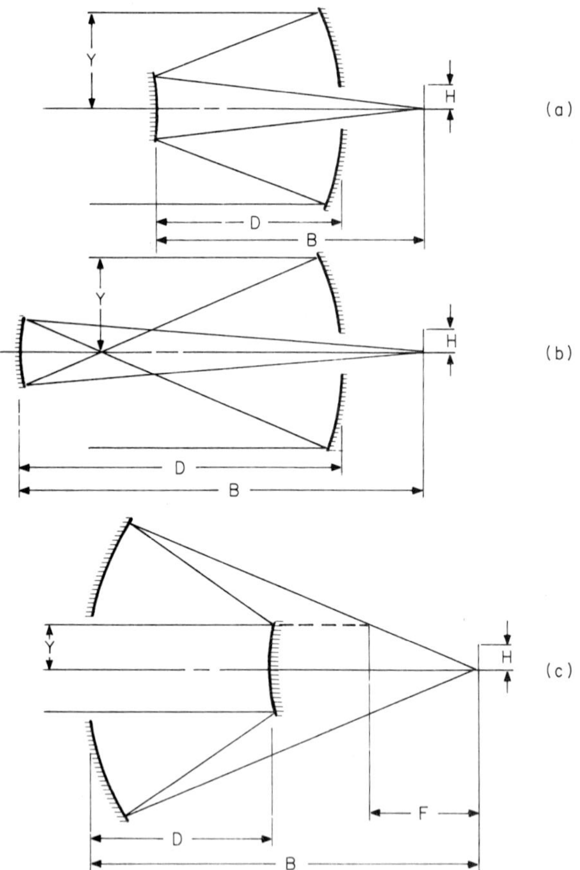

Figure 16.2 Three common two-mirror configurations. (*a*) The Cassegrain arrangement. (*b*) The Gregorian arrangement. (*c*) The Schwarzschild arrangement.

back focus B, and the mirror spacing D, one can determine the required mirror curvatures for any configuration from Eqs. 16.1 and 16.2. For example, if F is longer than B or D, and B and D are similar, the result is a Cassegrain system, Fig. 16.2*a*. If the focal length were chosen negative, the result would be a Gregorian objective, Fig. 16.2*b*. If the focal length is short compared to the back focus, the result can be the Schwarzschild system. Fig 16.2*c*.

$$C_1 = \frac{(B - F)}{2DF} \tag{16.1}$$

$$C_2 = \frac{(B + D - F)}{2DB} \tag{16.2}$$

Equations 16.3 through 16.6 give the third-order aberrations of any two-mirror system with the object located at infinity.

$$\Sigma\text{TSC} = \frac{Y^3[F(B - F)^3 + 64D^3F^4K_1 + B(F - D - B)(F + D - B)^2 - 64B^4D^3K_2]}{8D^3F^3} \tag{16.3}$$

$$\Sigma\text{CC} = \frac{HY^2[2F(B - F)^2 + (F - D - B)(F + D - B)(D - F - B) - 64B^3D^3K_2]}{8D^2F^3} \tag{16.4}$$

$$\Sigma\text{TAC} = \frac{H^2Y[4BF(B - F) + (F - D - B)(D - F - B)^2 - 64B^3D^3K_2]}{8BDF^3} \tag{16.5}$$

$$\Sigma\text{TPC} = \frac{H^2Y[DF - (B - F)^2]}{2BDF^2} \tag{16.6}$$

where Y = semiaperture of the system
H = image height
B = distance from mirror 2 to image (i.e., the back focal length)
F = system focal length
D = spacing (use positive sign)
ΣTSC = transverse third-order spherical aberration sum
ΣCC = third-order sagittal coma sum
ΣTAC = transverse third-order astigmatism sum
ΣTPC = transverse Petzval curvature sum

and where K_1 and K_2 are the equivalent fourth-order deformation coefficients for the primary and secondary mirrors. For a conic section, K is equal to the conic constant κ divided by 8 times the cube of the surface radius. Thus $K = \kappa/8R^3$.

Equations 16.7 through 16.11 describe the case when both mirrors are *individually* corrected for spherical aberration (and are thus easy to test). This case includes the classical *Cassegrain* (paraboloid primary and hyperboloid secondary) and the classical *Gregorian* (paraboloid primary and ellipsoid secondary). Note that the coma (per Eq. 16.10) is exactly the same for *any* arrangement of the mirrors when each mirror is individually free of spherical aberration.

$$K_1 = \frac{(F - B)^3}{64D^3F^3} \tag{16.7}$$

$$K_2 = \frac{(F - D - B)(F + D - B)^2}{64B^3D^3} \tag{16.8}$$

$$\Sigma\text{TSC} = 0.0 \tag{16.9}$$

$$\Sigma\text{CC} = \frac{HY^2}{4F^2} \tag{16.10}$$

$$\Sigma\text{TAC} = \frac{H^2Y(D - F)}{2BF^2} \tag{16.11}$$

Equations 16.12 through 16.16 cover the case when the aspherics are chosen so that both the spherical and coma are simultaneously corrected. The *Ritchey-Chretien* design (both mirrors hyperboloidal) falls in this category.

$$K_1 = \frac{[2BD^2 - (B - F)^3]}{64D^3F^3} \tag{16.12}$$

$$K_2 = \frac{[2F(B - F)^2 + (F - D - B)(F + D - B)(D - F - B)]}{64B^3D^3} \tag{16.13}$$

$$\Sigma\text{TSC} = 0.0 \tag{16.14}$$

$$\Sigma\text{CC} = 0.0 \tag{16.15}$$

$$\Sigma\text{TAC} = \frac{H^2Y(D - 2F)}{4BF^2} \tag{16.16}$$

Equations 16.17 through 16.21 describe the *Dall-Kirkham* system (with a spherical secondary mirror that is easier to fabricate and test).

$$K_1 = \frac{[F(F - B)^3 - B(F - D - B)(F + D - B)^2]}{64D^3F^4} \tag{16.17}$$

$$K_2 = 0.0 \tag{16.18}$$

$$\Sigma\text{TSC} = 0.0 \tag{16.19}$$

$$\Sigma\text{CC} = \frac{HY^2[2F(B - F)^2 + (F - D - B)(F + D - B)(D - F - B)]}{8D^2F^3} \tag{16.20}$$

$$\Sigma\text{TAC} = \frac{H^2Y[4BF(B - F) + (F - D - B)(D - F - B)^2]}{8DBF^3} \tag{16.21}$$

While these equations are exact only for the third-order aberrations, they are remarkably accurate for systems of modest aperture, and can provide good starting points for high-speed designs.

Figures 16.3 through 16.10 are examples of the classical two-mirror forms. So that the reader can compare the various configurations, all of them (except Figs. 16.6 and 16.7) have been designed at a speed of *f*/2 and a total angular field of two degrees. This speed is, of course, much, much faster than one would find in an astronomical telescope, but it is not atypical of the speeds used for many other applications (although the Gregorian example *is* rather extreme!). The back focus has been set at 30 percent of the focal length to hold the obscuration by the secondary mirror to a reasonable value. The mirror spacing has been set at 20 percent of the focal length so that (in combination with the 30 percent back focus) the final image is behind the primary mirror by 10 percent of the focal length. Obviously these choices significantly affect the aberrations, and other arrangements will differ from these particular examples.

The Cassegrain configuration at *f*/2.0 is sketched in Fig. 16.3; this lens drawing also applies for Figs. 16.4 and 16.5. The prescription and aberration plots for the classical version with a paraboloidal primary and hyperboloidal secondary are also given in Fig. 16.3. Note the large overcorrected coma. Figure 16.4 is a conic Ritchey-Chretien, where the conic constants have been modified slightly from the values of Eqs. 16.12 and 16.13 (which correct the *third-order* spherical and coma) so as to correct the marginal aberrations. The Dall-Kirkham,

(*Text continues on page 284.*)

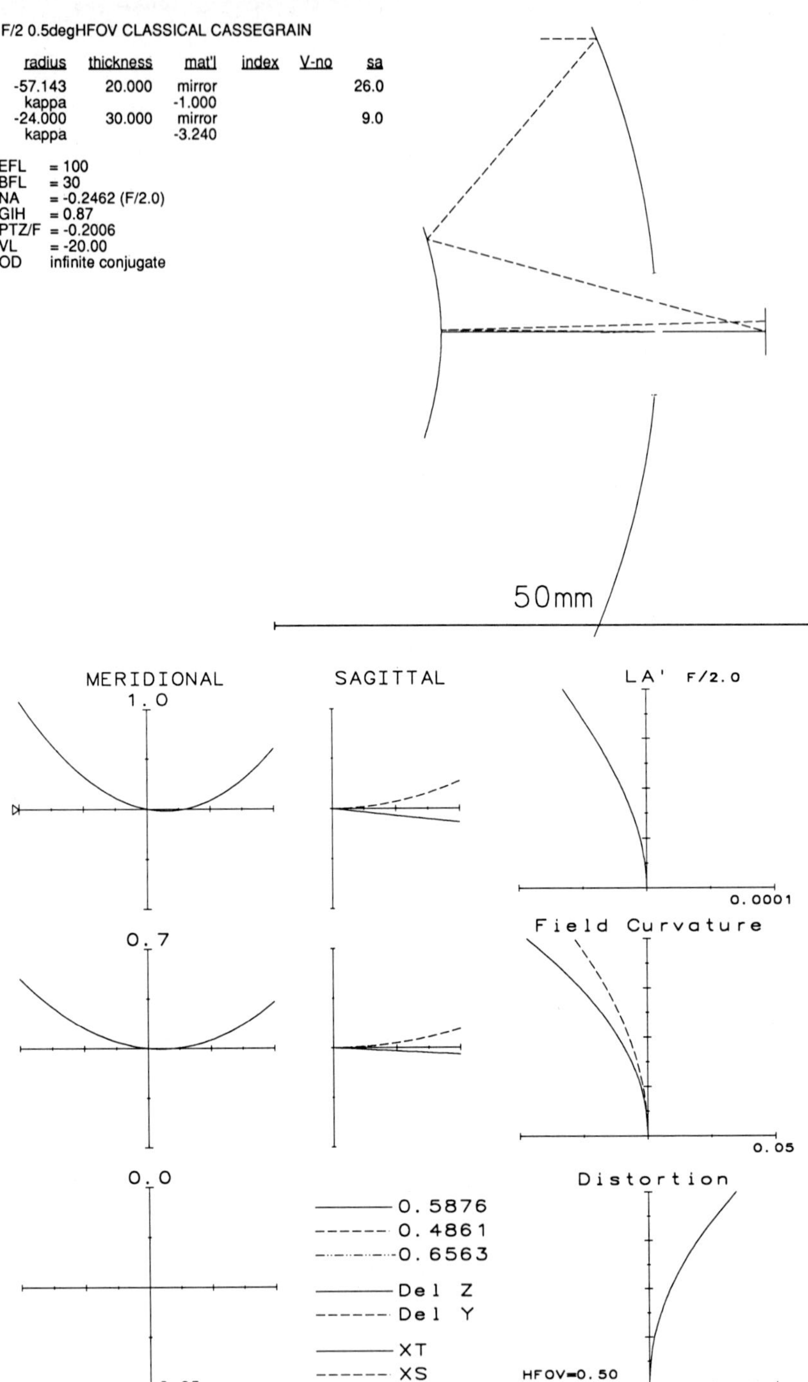

F/2 0.5degHFOV CLASSICAL CASSEGRAIN

radius	thickness	mat'l	index	V-no	sa
-57.143	20.000	mirror			26.0
kappa		-1.000			
-24.000	30.000	mirror			9.0
kappa		-3.240			

EFL = 100
BFL = 30
NA = -0.2462 (F/2.0)
GIH = 0.87
PTZ/F = -0.2006
VL = -20.00
OD infinite conjugate

Figure 16.3

F/2 0.5degHFOV CONIC RITCHEY-CHRETIEN

radius	thickness	mat'l	index	V-no	sa
-57.143	20.000	mirror			26.0
kappa		-1.082			
-24.000	30.000	mirror			9.0
kappa		-4.001			

EFL = 100
BFL = 30
NA = -0.2503 (F/2.0)
GIH = 0.87
PTZ/F = -0.2074
VL = -20.00
OD infinite conjugate

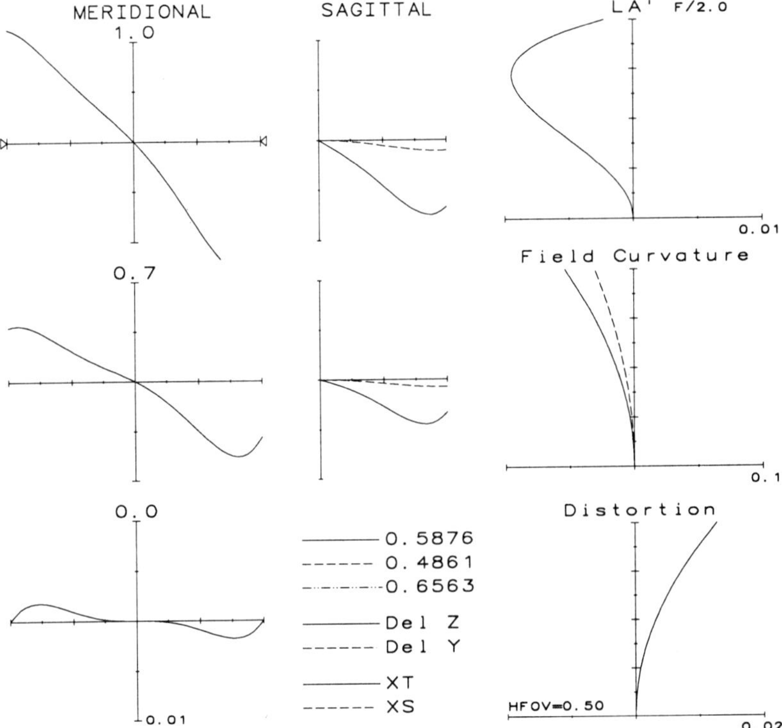

Figure 16.4

F/2 0.5degHFOV DALL-KIRKHAM

radius	thickness	mat'l	index	V-no	sa
-57.143	20.000	mirror			26.0
kappa		-0.646			
ae		1.883E-11			
af		1.821E-15			
-24.000	30.000	mirror			9.0

EFL = 100
BFL = 30
NA = -0.2284 (F/2.0)
GIH = 0.87
PTZ/F = -0.1727
VL = -20.00
OD infinite conjugate

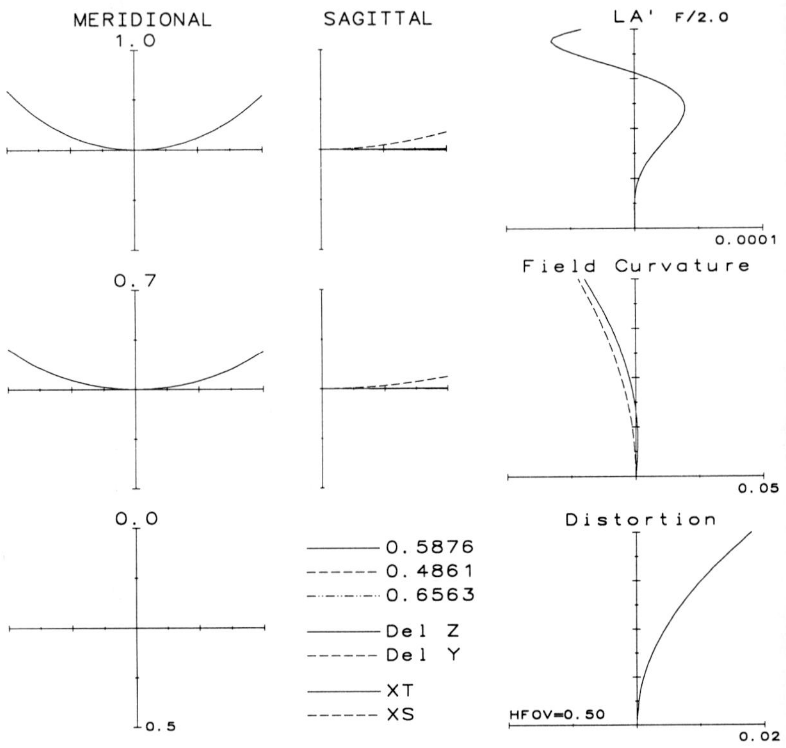

Figure 16.5

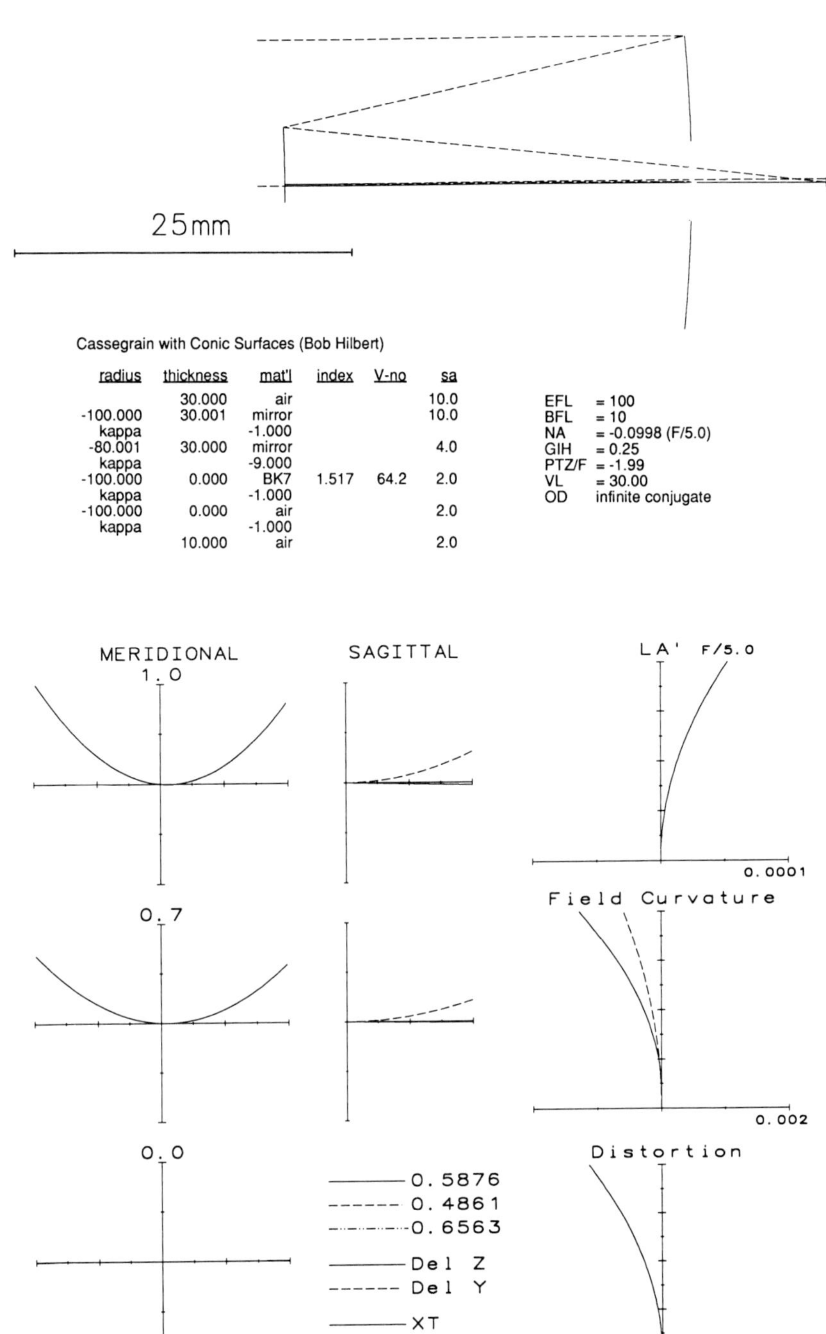

Cassegrain with Conic Surfaces (Bob Hilbert)

radius	thickness	mat'l	index	V-no	sa
	30.000	air			10.0
-100.000	30.001	mirror			10.0
kappa		-1.000			
-80.001	30.000	mirror			4.0
kappa		-9.000			
-100.000	0.000	BK7	1.517	64.2	2.0
kappa		-1.000			
-100.000	0.000	air			2.0
kappa		-1.000			
	10.000	air			2.0

EFL = 100
BFL = 10
NA = -0.0998 (F/5.0)
GIH = 0.25
PTZ/F = -1.99
VL = 30.00
OD infinite conjugate

Figure 16.6

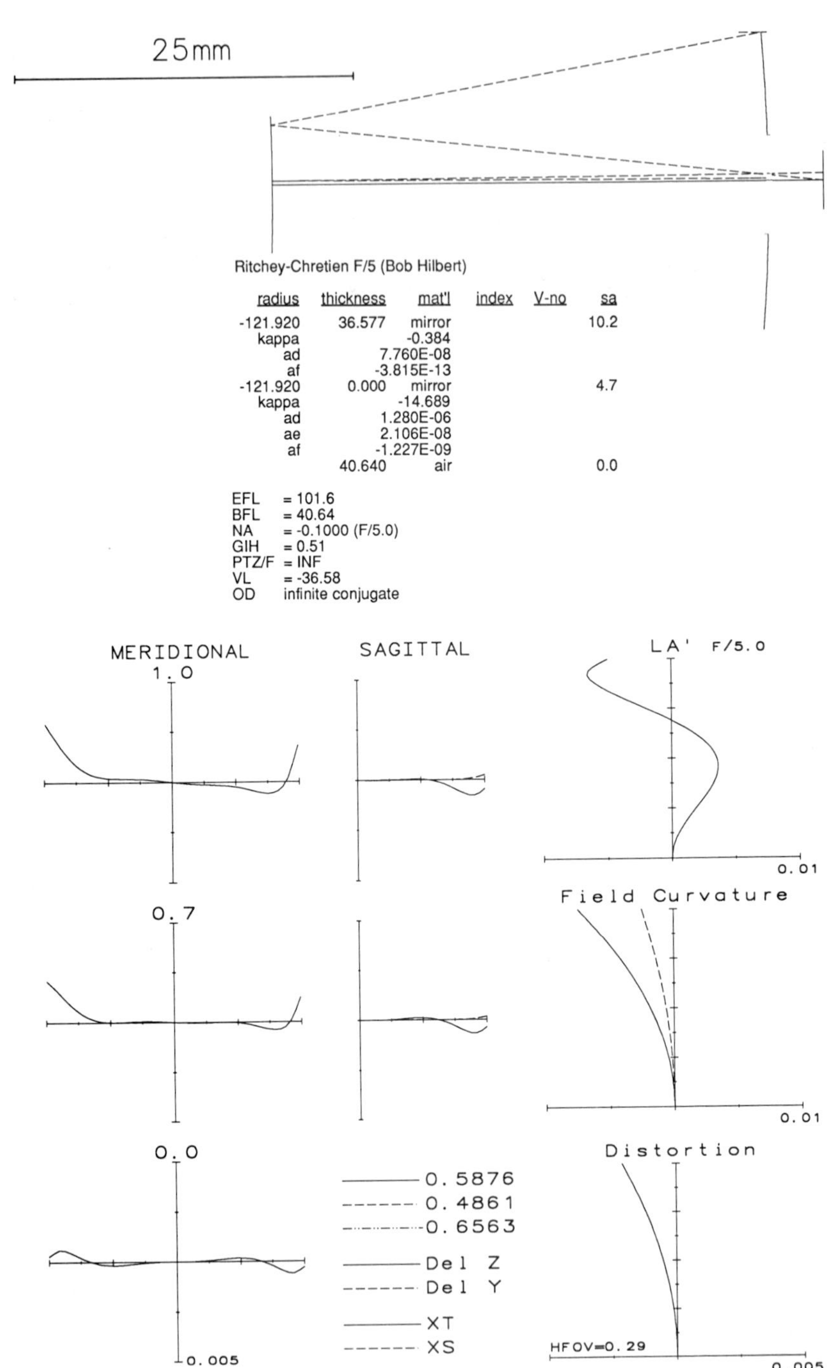

Ritchey-Chretien F/5 (Bob Hilbert)

radius	thickness	mat'l	index	V-no	sa
-121.920	36.577	mirror			10.2
kappa		-0.384			
ad		7.760E-08			
af		-3.815E-13			
-121.920	0.000	mirror			4.7
kappa		-14.689			
ad		1.280E-06			
ae		2.106E-08			
af		-1.227E-09			
	40.640	air			0.0

EFL = 101.6
BFL = 40.64
NA = -0.1000 (F/5.0)
GIH = 0.51
PTZ/F = INF
VL = -36.58
OD infinite conjugate

Figure 16.7

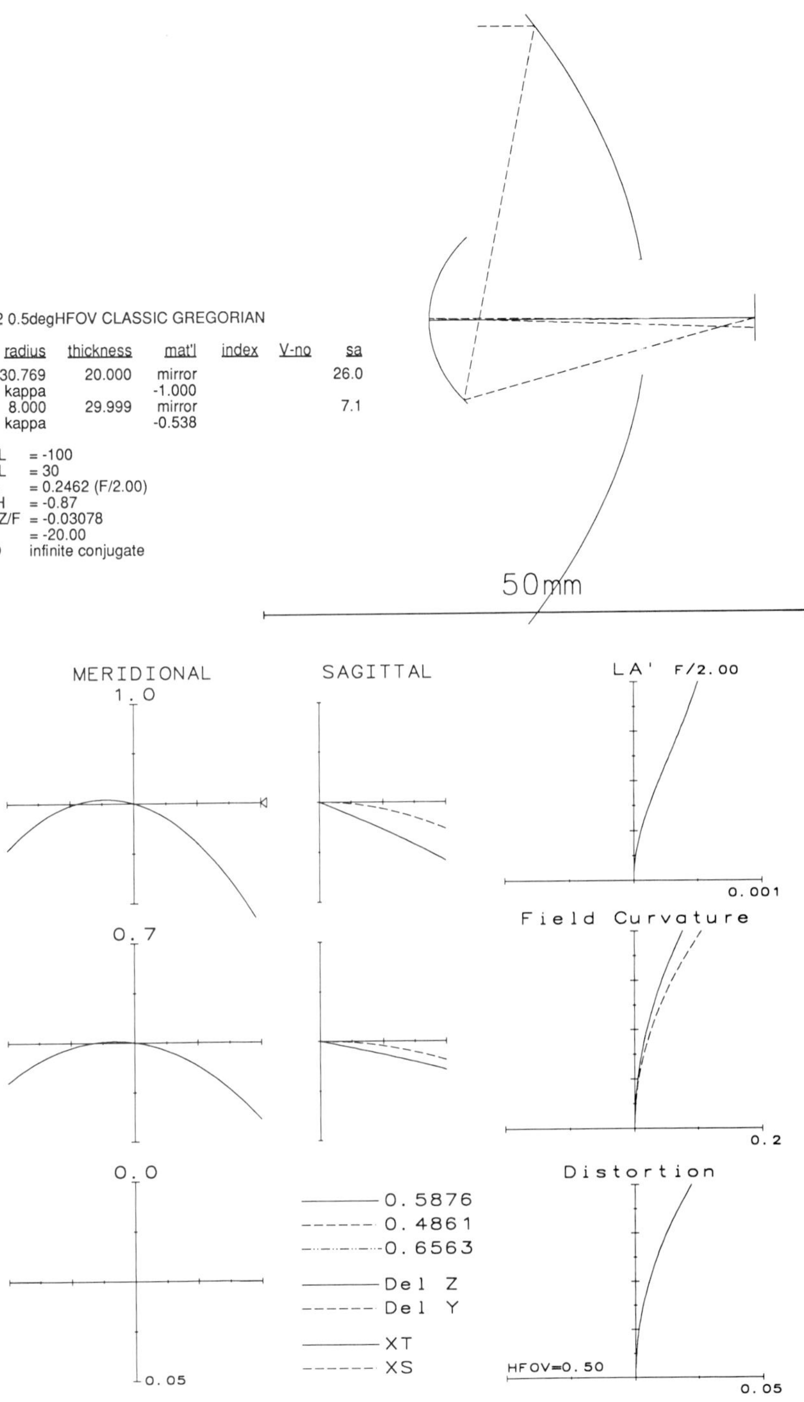
F/2 0.5degHFOV CLASSIC GREGORIAN
radius thickness mat'l index V-no sa
-30.769 20.000 mirror 26.0
kappa -1.000
8.000 29.999 mirror 7.1
kappa -0.538
EFL = -100
BFL = 30
NA = 0.2462 (F/2.00)
GIH = -0.87
PTZ/F = -0.03078
VL = -20.00
OD infinite conjugate
50mm
MERIDIONAL
1.0
SAGITTAL
LA' F/2.00
0.001
Field Curvature
0.7
0.2
0.0
Distortion
0.5876
0.4861
0.6563
Del Z
Del Y
XT
XS
0.05
HFOV=0.50
0.05

Figure 16.8

F/2 0.5degHFOV ADJUSTED GREGORIAN

radius	thickness	mat'l	index	V-no	sa
-30.769	20.000	mirror			26.0
kappa		-0.989			
8.000	29.999	mirror			7.2
kappa		-0.561			

EFL = -100
BFL = 30
NA = 0.2500 (F/2.00)
GIH = -0.87
PTZ/F = -0.03174
VL = -20.00
OD infinite conjugate

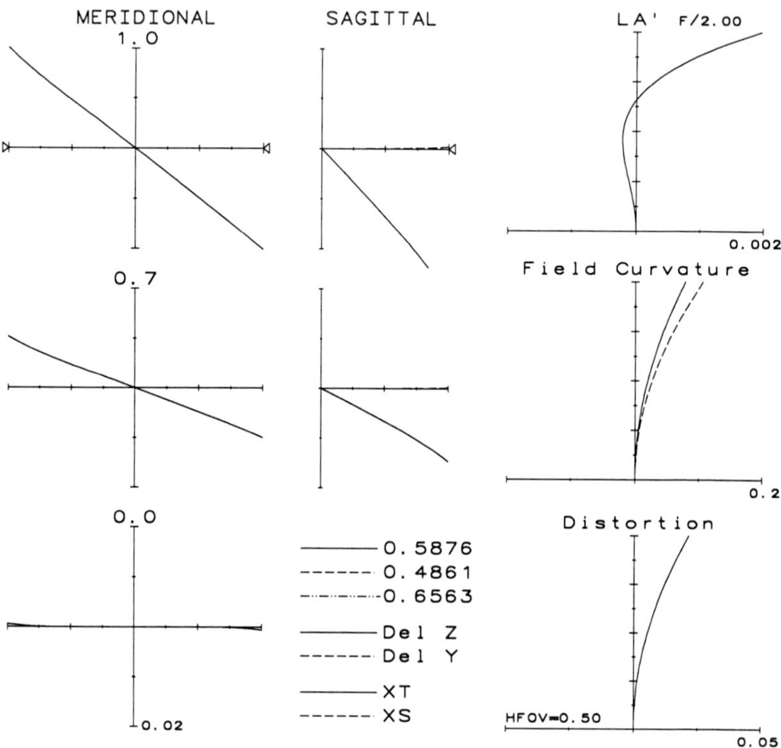

Figure 16.9

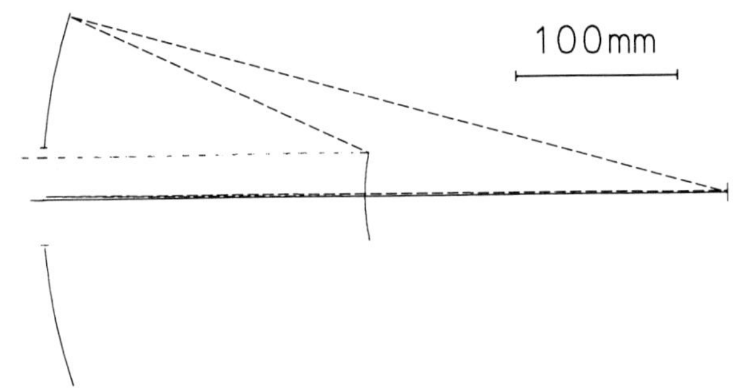

F/2.0 0.5degHFOV SCHWARZSCHILD

radius	thickness	mat'l	index	V-no	sa
123.510	200.000	mirror			25.1
323.644	423.862	mirror			106.7

EFL = 100
BFL = 423.9
NA = -0.2500 (F/2.00)
GIH = 0.87
PTZ/F = -0.9986
VL = -200.00
OD infinite conjugate

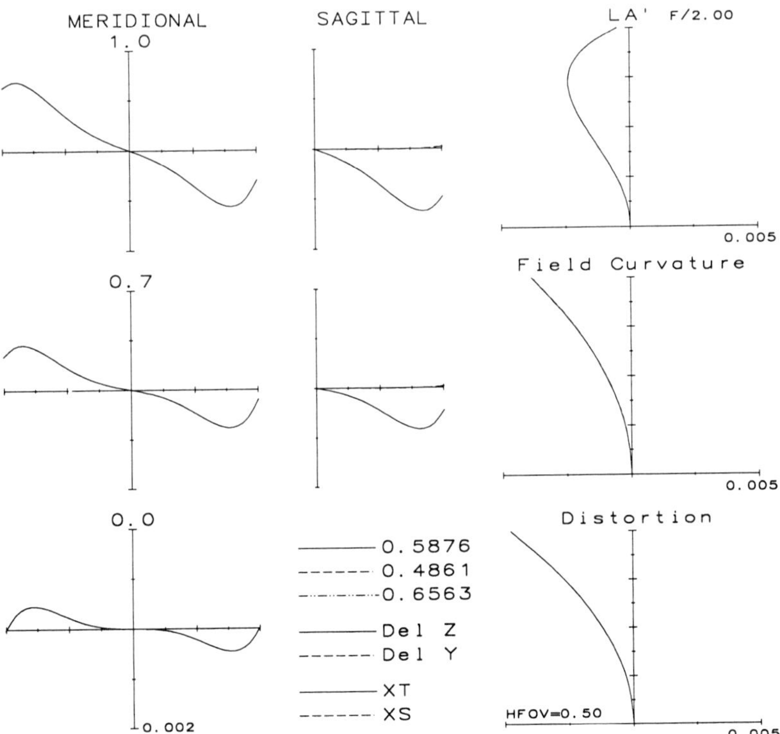

Figure 16.10

with a spherical secondary, is shown in Fig. 16.5; while easier to fabricate, its coma is even worse than that of the classic Cassegrain.

Figures 16.6 and 16.7 show Cassegrains at a somewhat more modest speed of *f*/5. The first is a classic Cassegrain and the second is a general aspheric Ritchey-Chretien, with radii chosen equal in order to obtain a flat Petzval surface.

The Gregorian form is infrequently encountered for two primary reasons. With all else equal, it is a longer system than the Cassegrain. In addition, the Cassegrain is preferred because it has a flatter field. The classic parabola-ellipse Gregorian is shown at an extreme speed of *f*/2 in Fig. 16.8. Figure 16.9 (same lens drawing) shows a sort of Ritchey-Chretien-Gregorian combination with the conic constants selected to correct both the spherical and coma.

The Schwarzschild arrangement (Fig. 16.10) suffers from the fact that the concave mirror must be several times as large as the system aperture and its length is more than 4 times its focal length. However, it can be corrected for spherical, coma, and astigmatism using only spherical surfaces. For an infinite object distance, a mirror spacing of twice the focal length f, a convex radius of $(\sqrt{5} - 1)f$ and a concave radius of $(\sqrt{5} + 1)f$ will produce a system corrected for spherical aberration. Note that the mirrors have a common center of curvature. In a *monocentric* system of this type, if the aperture stop is located at the common center, the system is free of coma and astigmatism, and the image surface is a sphere of radius f. There are several systems of this type which are close to the construction given above but which have somewhat differing aberration characteristics; for example, there is a nearby form with very low high-order spherical.

An all-spherical-surface Schwarzschild system is shown in Fig. 16.10. The radii have been adjusted (very slightly) from the values which would result from the expressions in the preceding paragraph, so that the marginal aberrations are corrected.

The Schwarzschild configuration is often used as a microscope objective, where its relatively large-diameter mirror is not a problem because the short focal length means that the actual size is not inconveniently large. The mirror construction allows the system to be used in the infrared or in the ultraviolet. Its simple all-spherical construction makes it an economical design to manufacture, and its long working distance is an added convenience for many applications.

The two degrees of freedom represented by the choice of the conic constants in a two-mirror system allow one to correct only two aberrations (assuming that the mirror curvatures have been determined by requirements for f, B, and D). If there is a free choice of mirror curvatures, the Petzval curvature can be corrected in a Cassegrain configuration with identical radii on both primary and secondary, as in

Fig. 16.7. Control of astigmatism requires one additional effective degree of freedom, which can be provided by another optical component. There are a few arrangements where the choice of *f, B,* and *D* is used to correct the astigmatism.

16.3 Catadioptric Systems

The fundamental philosophy behind most catadioptric systems is to use refracting elements to correct the aberrations of a system of spherical mirrors, but to do it without introducing any new aberrations. Thus the typical system uses mirrors for most of the focusing power and uses very weak refracting elements to provide the aberration correction. The thick meniscus corrector elements of the well-known Bouwers and Maksutov systems are examples of weak refractor elements which are strongly bent in order to increase their overcorrecting spherical aberration. (See Figs. 16.22 and 16.23.) Where more than one refracting corrector is used, the approach is usually to have them closely spaced and to use approximately equal amounts of positive and negative power. The same glass type is often used for all the elements. The effect of this is that the net chromatic aberration is zero; both ordinary chromatic and secondary spectrum can be eliminated. For example, in a two-element corrector located at the entrance pupil, the positive element is shaped to minimize its spherical aberration contribution and the negative element is shaped to provide enough overcorrected spherical to balance out the undercorrected spherical aberration from the positive element and the spherical mirrors. Both can be bent to simultaneously correct the coma and spherical aberration. Correctors can also be located in the convergent beam in front of the image. In this location, the corrector's function is more likely to be the correction of the field aberrations, coma, astigmatism, and, occasionally, the Petzval curvature.

Figures 16.11 and 16.12 incorporate three BK7 refracting correctors near the stop. The axial color is well-corrected, but in these two systems the spherochromatism is rather large. These systems both cover a field of 2.5°, are fast—*f*/1.5 and *f*/1.0 respectively—and use no aspheric surfaces. Figure 16.12 also incorporates a small meniscus field corrector near the focal plane.

Figures 16.13 through 16.16 show a series of moderately complex catadioptric designs. They all have several features in common. They all incorporate at least one Mangin-type mirror (with the reflector on the second surface of a lens element). This provides the effect of a lens plus a mirror, in a component which is probably at least partially corrected, for about the same cost as the mirror alone. An ordinary Mangin mirror is corrected for spherical aberration and also has a re-

(Text continues on page 292.)

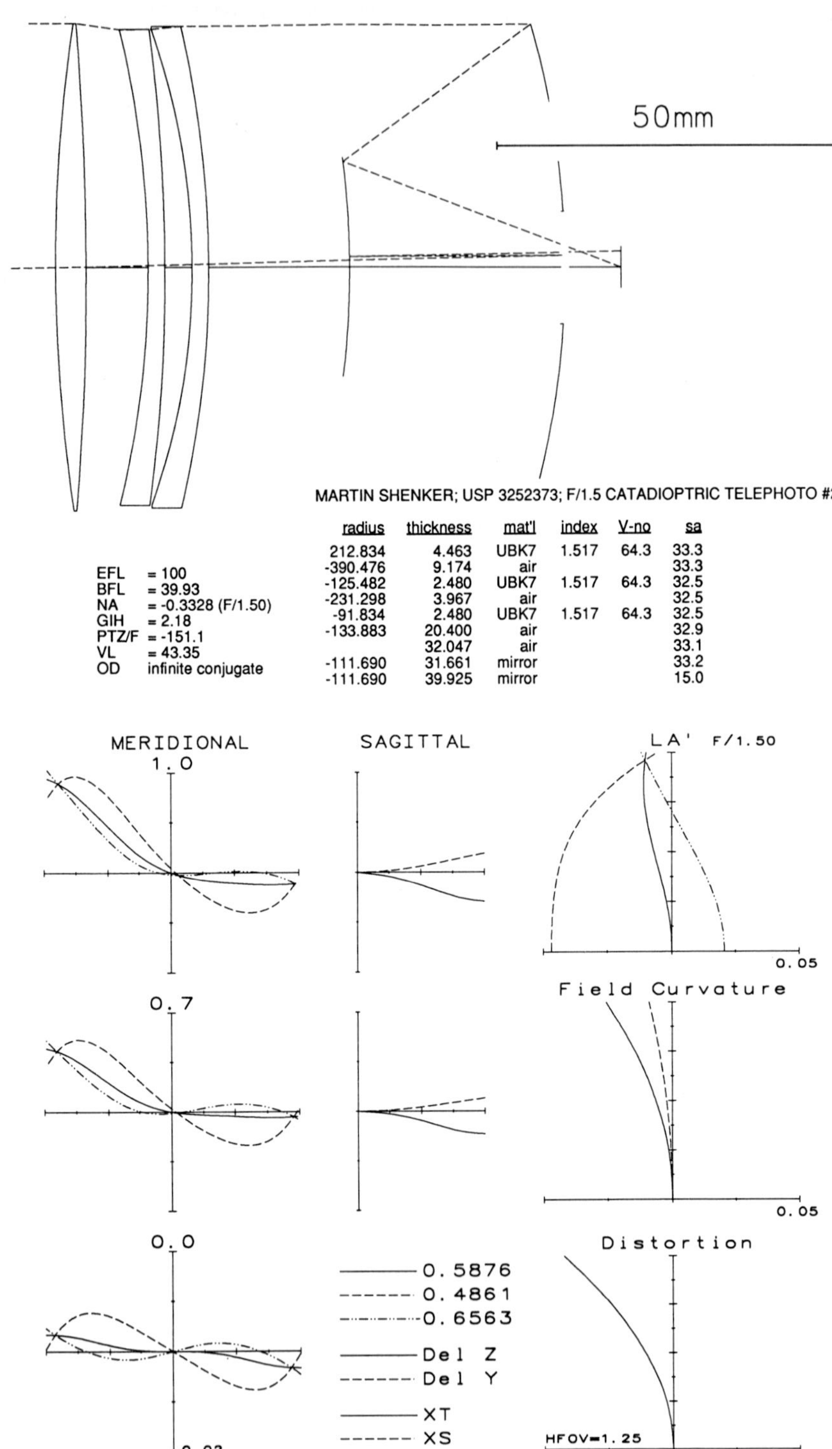

MARTIN SHENKER; USP 3252373; F/1.5 CATADIOPTRIC TELEPHOTO #2

radius	thickness	mat'l	index	V-no	sa
212.834	4.463	UBK7	1.517	64.3	33.3
-390.476	9.174	air			33.3
-125.482	2.480	UBK7	1.517	64.3	32.5
-231.298	3.967	air			32.5
-91.834	2.480	UBK7	1.517	64.3	32.5
-133.883	20.400	air			32.9
	32.047	air			33.1
-111.690	31.661	mirror			33.2
-111.690	39.925	mirror			15.0

EFL = 100
BFL = 39.93
NA = -0.3328 (F/1.50)
GIH = 2.18
PTZ/F = -151.1
VL = 43.35
OD infinite conjugate

Figure 16.11

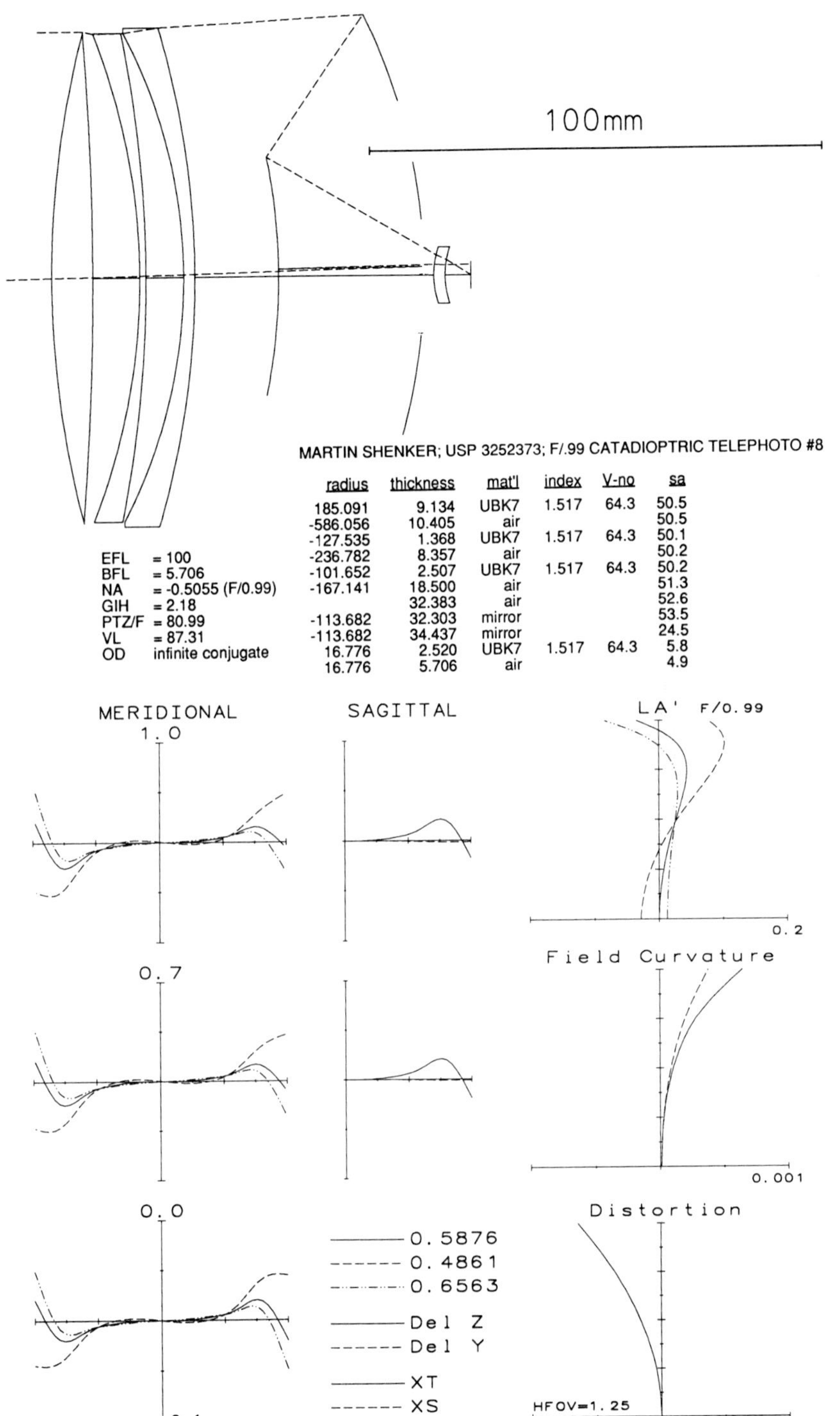

MARTIN SHENKER; USP 3252373; F/.99 CATADIOPTRIC TELEPHOTO #8

radius	thickness	mat'l	index	V-no	sa
185.091	9.134	UBK7	1.517	64.3	50.5
-586.056	10.405	air			50.5
-127.535	1.368	UBK7	1.517	64.3	50.1
-236.782	8.357	air			50.2
-101.652	2.507	UBK7	1.517	64.3	50.2
-167.141	18.500	air			51.3
	32.383	air			52.6
-113.682	32.303	mirror			53.5
-113.682	34.437	mirror			24.5
16.776	2.520	UBK7	1.517	64.3	5.8
16.776	5.706	air			4.9

EFL = 100
BFL = 5.706
NA = -0.5055 (F/0.99)
GIH = 2.18
PTZ/F = 80.99
VL = 87.31
OD infinite conjugate

Figure 16.12

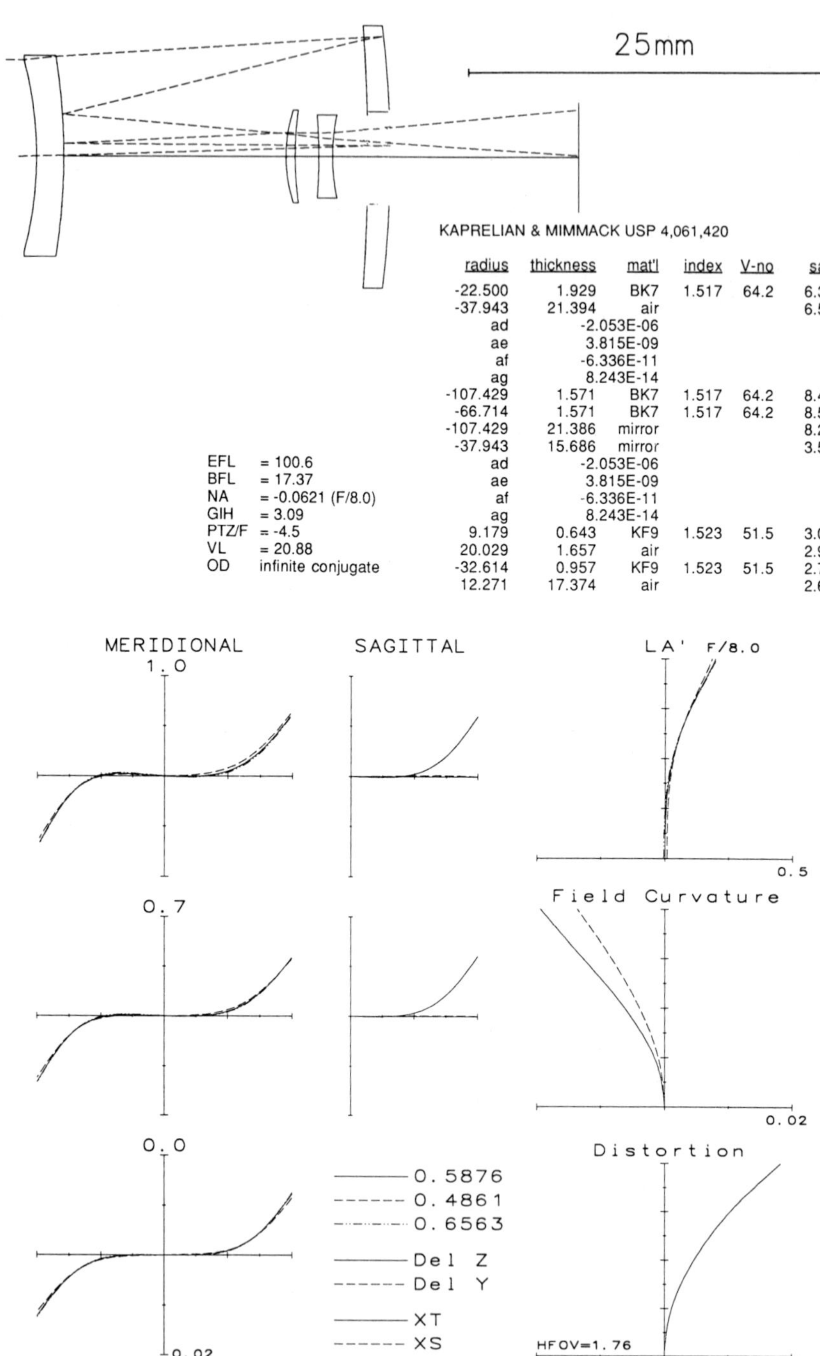

KAPRELIAN & MIMMACK USP 4,061,420

radius	thickness	mat'l	index	V-no	sa
-22.500	1.929	BK7	1.517	64.2	6.3
-37.943	21.394	air			6.5
ad	-2.053E-06				
ae	3.815E-09				
af	-6.336E-11				
ag	8.243E-14				
-107.429	1.571	BK7	1.517	64.2	8.4
-66.714	1.571	BK7	1.517	64.2	8.5
-107.429	21.386	mirror			8.2
-37.943	15.686	mirror			3.5
ad	-2.053E-06				
ae	3.815E-09				
af	-6.336E-11				
ag	8.243E-14				
9.179	0.643	KF9	1.523	51.5	3.0
20.029	1.657	air			2.9
-32.614	0.957	KF9	1.523	51.5	2.7
12.271	17.374	air			2.6

EFL = 100.6
BFL = 17.37
NA = -0.0621 (F/8.0)
GIH = 3.09
PTZ/F = -4.5
VL = 20.88
OD infinite conjugate

Figure 16.13

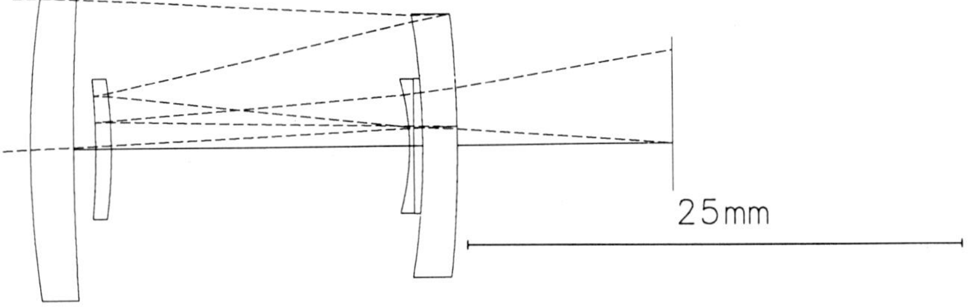

YOSHIYUKI SHIMIZU; USP 3632190; F/7 CATADIOPTRIC TELEPHOTO #2

radius	thickness	mat'l	index	V-no	sa
45.730	2.200	K3	1.518	59.0	7.1
122.760	1.120	air			7.1
	16.620	air			7.1
-36.856	1.760	LLF2	1.541	47.2	6.2
-57.800	1.760	LLF2	1.541	47.2	6.2
-36.856	15.840	mirror			6.2
-23.166	0.780	LLF1	1.548	45.8	3.3
-34.500	0.780	LLF1	1.548	45.8	3.2
-23.166	15.180	mirror			3.3
-10.640	0.220	K5	1.522	59.5	3.1
	0.440	SF10	1.728	28.4	3.2
-36.856	1.760	LLF2	1.541	47.2	3.2
-57.800	10.820	air			3.3

EFL = 99.77
BFL = 10.82
NA = -0.0714 (F/7.0)
GIH = 4.36
PTZ/F = -1.448
VL = 21.70
OD infinite conjugate

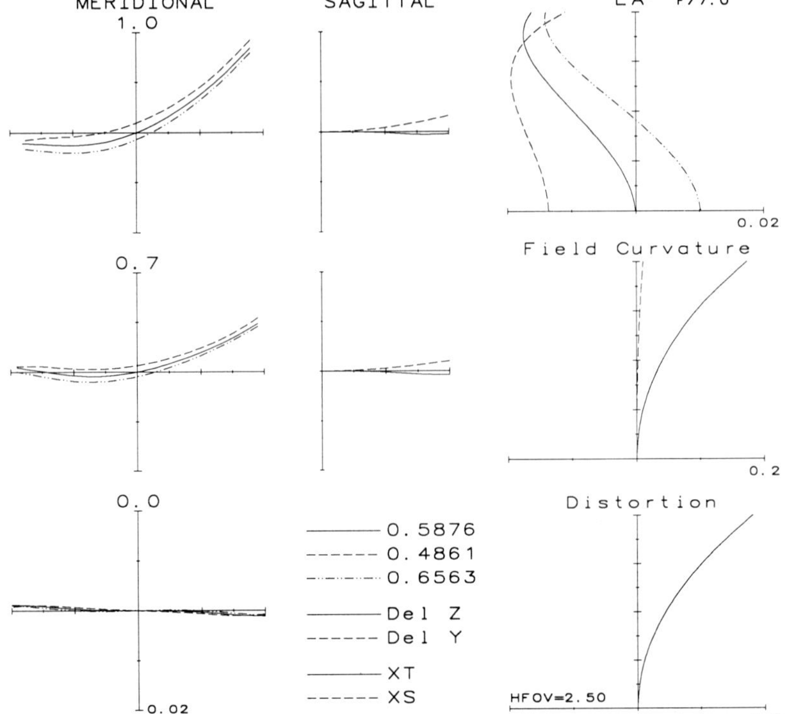

Figure 16.14

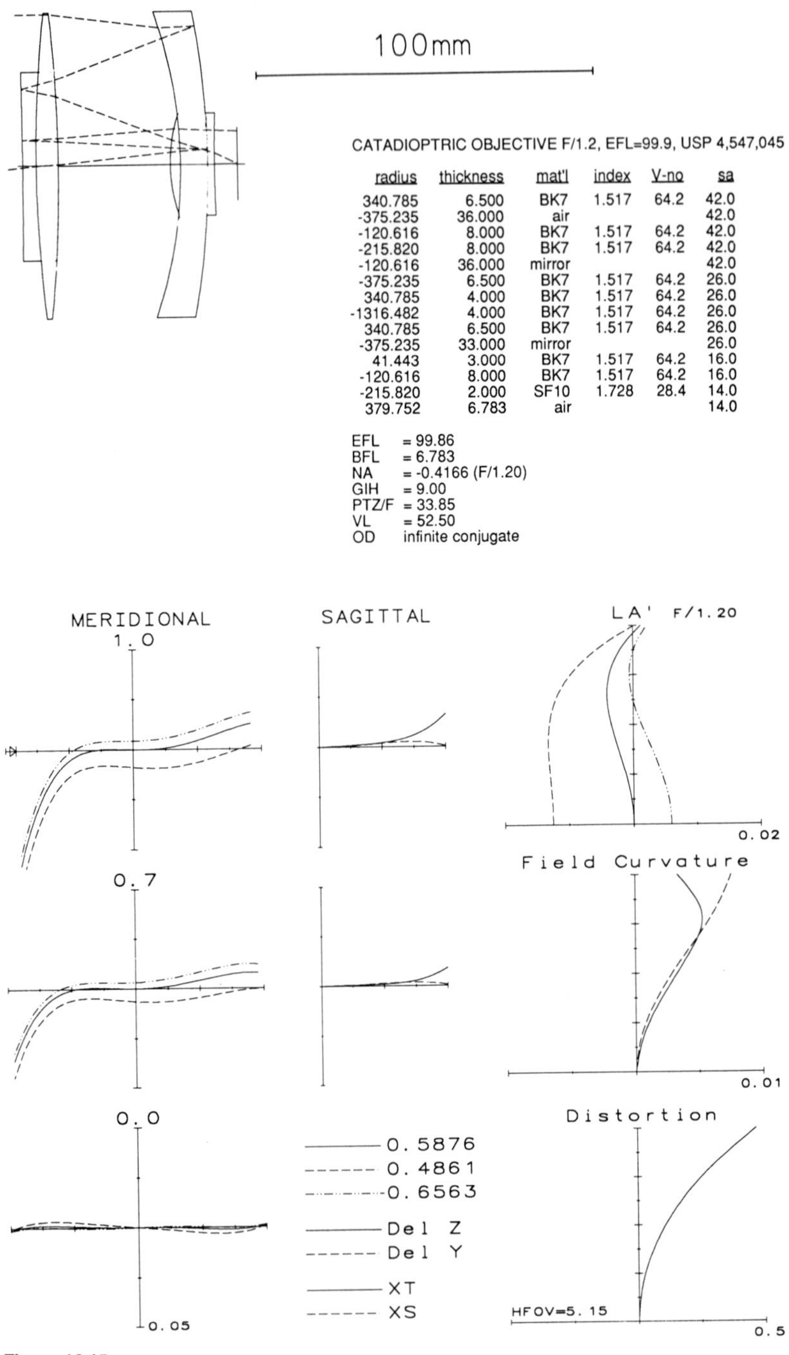

CATADIOPTRIC OBJECTIVE F/1.2, EFL=99.9, USP 4,547,045

radius	thickness	mat'l	index	V-no	sa
340.785	6.500	BK7	1.517	64.2	42.0
-375.235	36.000	air			42.0
-120.616	8.000	BK7	1.517	64.2	42.0
-215.820	8.000	BK7	1.517	64.2	42.0
-120.616	36.000	mirror			42.0
-375.235	6.500	BK7	1.517	64.2	26.0
340.785	4.000	BK7	1.517	64.2	26.0
-1316.482	4.000	BK7	1.517	64.2	26.0
340.785	6.500	BK7	1.517	64.2	26.0
-375.235	33.000	mirror			26.0
41.443	3.000	BK7	1.517	64.2	16.0
-120.616	8.000	BK7	1.517	64.2	16.0
-215.820	2.000	SF10	1.728	28.4	14.0
379.752	6.783	air			14.0

EFL = 99.86
BFL = 6.783
NA = -0.4166 (F/1.20)
GIH = 9.00
PTZ/F = 33.85
VL = 52.50
OD infinite conjugate

Figure 16.15

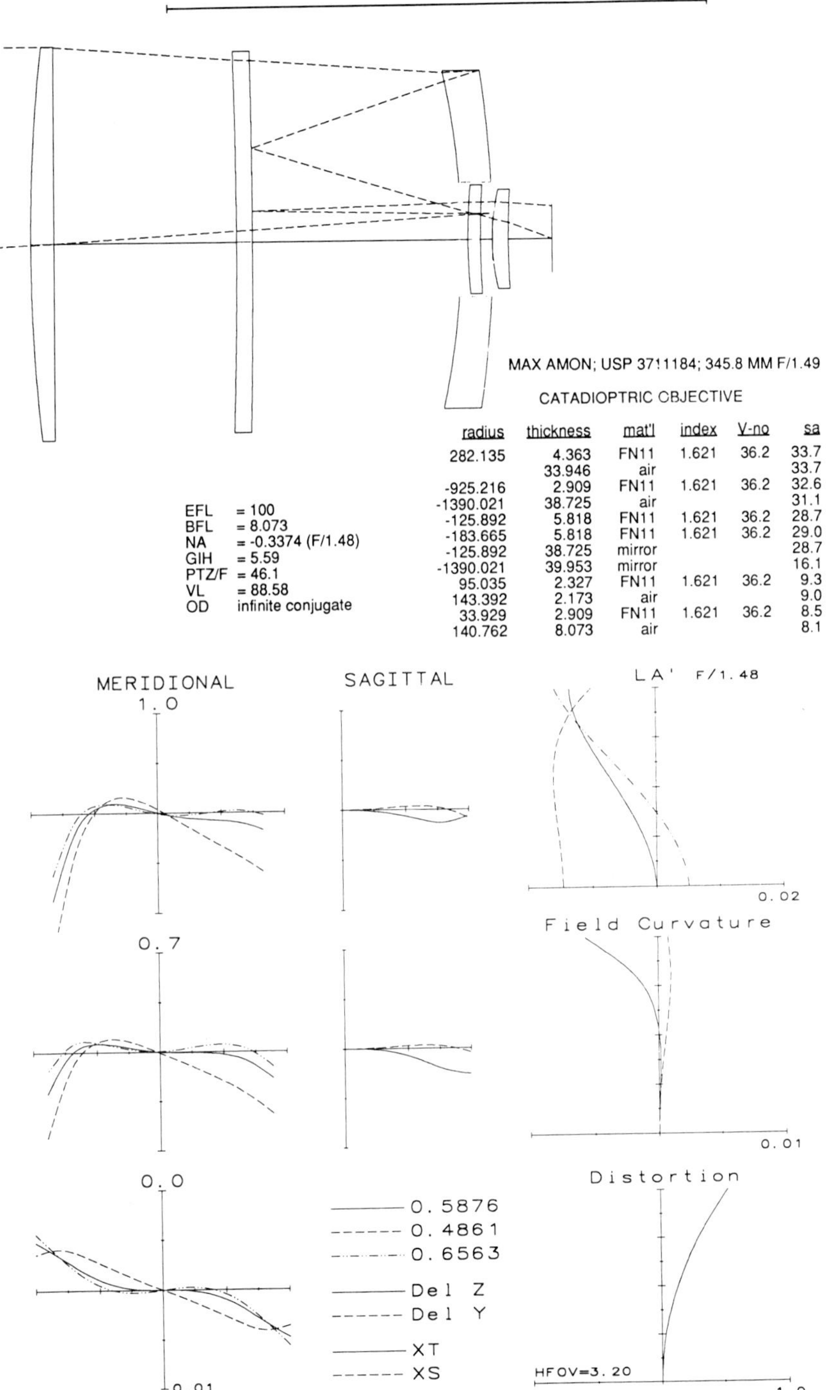

MAX AMON; USP 3711184; 345.8 MM F/1.49

CATADIOPTRIC OBJECTIVE

radius	thickness	mat'l	index	V-no	sa
282.135	4.363	FN11	1.621	36.2	33.7
	33.946	air			33.7
-925.216	2.909	FN11	1.621	36.2	32.6
-1390.021	38.725	air			31.1
-125.892	5.818	FN11	1.621	36.2	28.7
-183.665	5.818	FN11	1.621	36.2	29.0
-125.892	38.725	mirror			28.7
-1390.021	39.953	mirror			16.1
95.035	2.327	FN11	1.621	36.2	9.3
143.392	2.173	air			9.0
33.929	2.909	FN11	1.621	36.2	8.5
140.762	8.073	air			8.1

EFL = 100
BFL = 8.073
NA = -0.3374 (F/1.48)
GIH = 5.59
PTZ/F = 46.1
VL = 88.58
OD infinite conjugate

Figure 16.16

duced coma contribution, but it is afflicted with severe chromatic aberration. All the systems shown have a full-aperture element on the object side which can be used as a protection or sealing window for the system. The secondary mirror is on a surface which does double duty; it is either a surface of a corrector element, or in one case, a Mangin secondary. All have field correctors in the convergent cone near the image. In two systems, the primary Mangin is also used in transmission as a refracting corrector in the convergent cone near the image.

Figures 16.17 and 16.18 are what are called *solid cats*. The volume of the lens is almost completely filled with glass. A modest number of lenses similar to Fig. 16.17 were sold as extremely compact telephoto lenses for 35-mm cameras. The system of Fig. 16.18 is much faster than that of Fig. 16.17 and is modestly more complex in construction (although the aberration plots make it clear that this design might profitably be limited to a speed of about *f*/2).

An infrared catadioptric telescope is shown in Fig. 21.12 of Chap. 21.

Aspheric correctors and Schmidt systems

An aspheric refracting corrector can be used to correct an aberration in a mirror system. Added to a Ritchey-Chretien system, for example, such a corrector can be used in combination with the two aspheric mirrors to correct the astigmatism as well as the spherical and coma. Such a corrector can be located in either the incoming beam or in the converging beam in front of the image. Often an aspheric corrector is used at the entrance pupil with an all-spherical-mirror system (usually in a Cassegrain configuration) to correct the spherical aberration. This Schmidt-Cassegrain is a popular form for medium-sized commercial telescopes.

In the *Schmidt system,* as shown in Fig. 16.19, a spherical mirror with the aperture stop at its center of curvature (and thus free of coma and astigmatism) is corrected for spherical aberration by an aspheric corrector located at the stop (where it affects only the spherical—see Eqs. F.11.1 through F.11.7).

If a refracting aspheric corrector is used to correct the spherical aberration, it can exactly and completely correct the spherical at the nominal wavelength. However, at shorter wavelengths its higher index overcorrects the spherical, and at longer wavelengths its lower index leaves the spherical undercorrected. This is ordinary spherochromatism. It can be balanced (but not eliminated) by introducing undercorrected axial chromatic, using some weak positive power to do so. This is usually combined with the aspheric surface so that the central part of the corrector is convex (converging) and the margin of the aspheric is diverging. The neutral, nondiverging zone is usually lo-

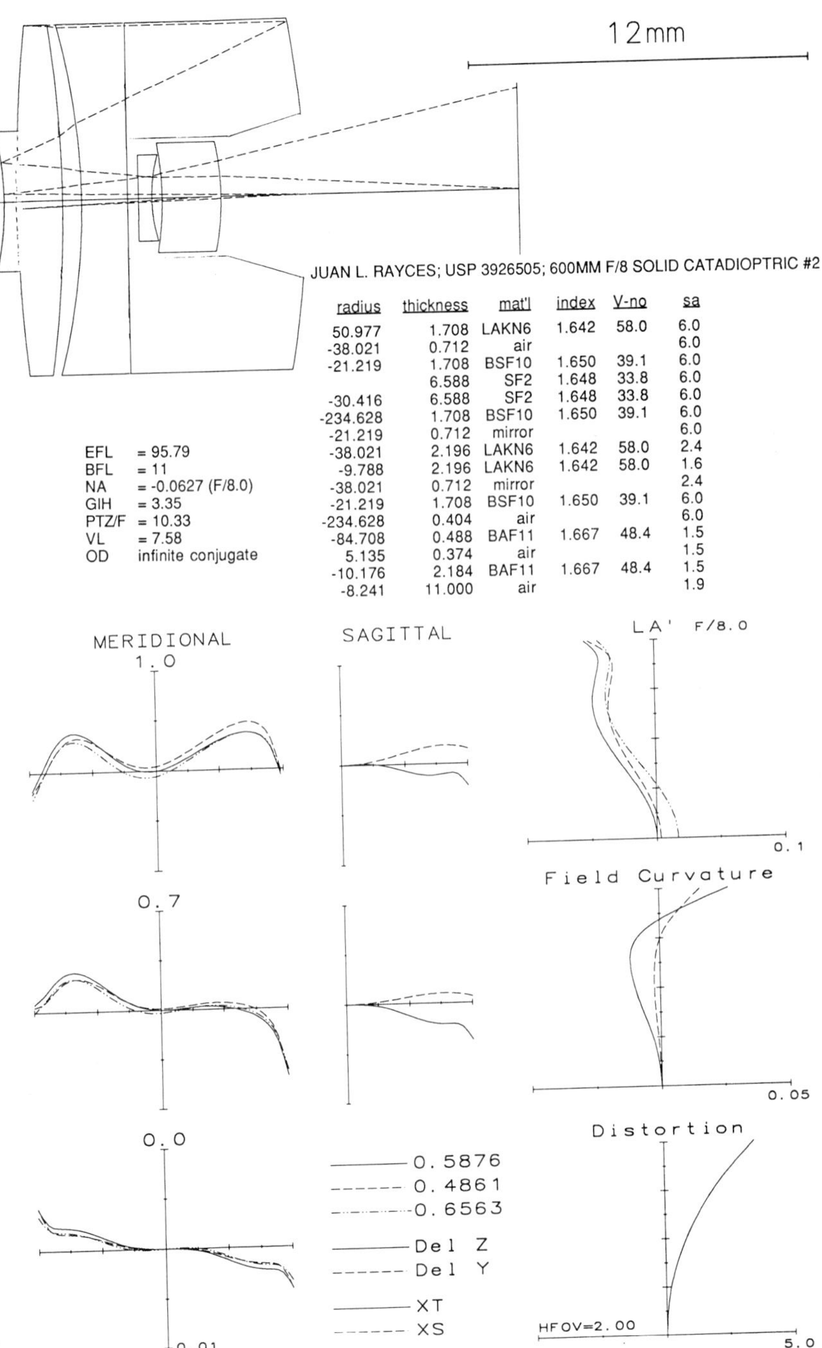

JUAN L. RAYCES; USP 3926505; 600MM F/8 SOLID CATADIOPTRIC #2

radius	thickness	mat'l	index	V-no	sa
50.977	1.708	LAKN6	1.642	58.0	6.0
-38.021	0.712	air			6.0
-21.219	1.708	BSF10	1.650	39.1	6.0
	6.588	SF2	1.648	33.8	6.0
-30.416	6.588	SF2	1.648	33.8	6.0
-234.628	1.708	BSF10	1.650	39.1	6.0
-21.219	0.712	mirror			6.0
-38.021	2.196	LAKN6	1.642	58.0	2.4
-9.788	2.196	LAKN6	1.642	58.0	1.6
-38.021	0.712	mirror			2.4
-21.219	1.708	BSF10	1.650	39.1	6.0
-234.628	0.404	air			6.0
-84.708	0.488	BAF11	1.667	48.4	1.5
5.135	0.374	air			1.5
-10.176	2.184	BAF11	1.667	48.4	1.5
-8.241	11.000	air			1.9

EFL = 95.79
BFL = 11
NA = -0.0627 (F/8.0)
GIH = 3.35
PTZ/F = 10.33
VL = 7.58
OD infinite conjugate

Figure 16.17

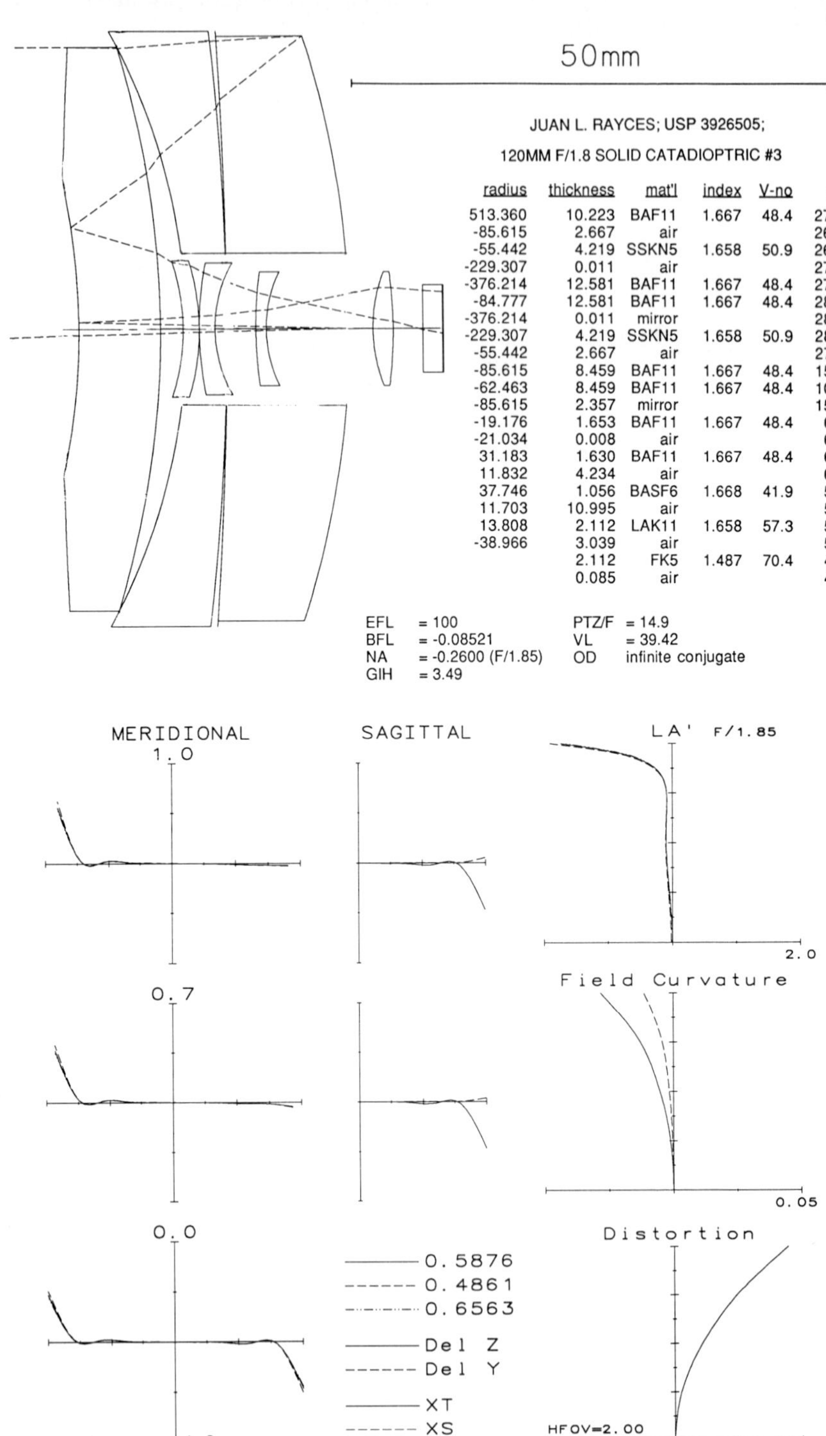

radius	thickness	mat'l	index	V-no	sa
513.360	10.223	BAF11	1.667	48.4	27.0
-85.615	2.667	air			26.9
-55.442	4.219	SSKN5	1.658	50.9	26.9
-229.307	0.011	air			27.6
-376.214	12.581	BAF11	1.667	48.4	27.6
-84.777	12.581	BAF11	1.667	48.4	28.0
-376.214	0.011	mirror			28.5
-229.307	4.219	SSKN5	1.658	50.9	28.5
-55.442	2.667	air			27.9
-85.615	8.459	BAF11	1.667	48.4	15.8
-62.463	8.459	BAF11	1.667	48.4	10.6
-85.615	2.357	mirror			15.8
-19.176	1.653	BAF11	1.667	48.4	6.5
-21.034	0.008	air			6.5
31.183	1.630	BAF11	1.667	48.4	6.3
11.832	4.234	air			6.3
37.746	1.056	BASF6	1.668	41.9	5.5
11.703	10.995	air			5.5
13.808	2.112	LAK11	1.658	57.3	5.5
-38.966	3.039	air			5.5
	2.112	FK5	1.487	70.4	4.2
	0.085	air			4.2

Figure 16.18

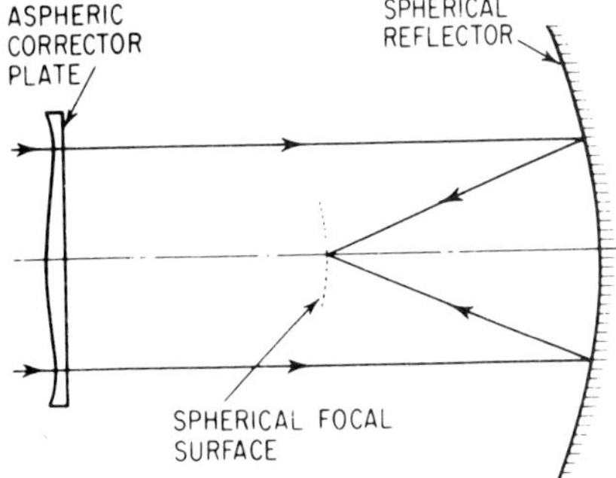

Figure 16.19 The Schmidt system consists of a spherical reflector with an aspheric corrector plate (and the aperture stop) located at its center of curvature. The aspheric surface of the f/1 system shown here is greatly exaggerated.

cated at about 0.866 of the marginal ray height in order to get the optimum spherochromatic balance. In the case of the classical Schmidt system, the equation for a near-optimum aspheric surface for the corrector plate is

$$x = 0.5Cy^2 + Ky^4 + Ly^6$$

where

$$C = \frac{3}{128(n - 1)f(f/\#)^2}$$

$$K = \frac{\left[1 - \dfrac{3}{64\,(f/\#)^2}\right]^2}{32(1 - n)f^3}$$

$$L = \frac{1}{85.8(1 - n)f^5}$$

and x = sag of surface
f = focal length
n = index of corrector material
$(f/\#)$ = relative aperture of Schmidt system

16.4 Confocal Paraboloids

If two paraboloids are arranged with their focal points coincident, the result is an afocal system with rather interesting properties. It is free of spherical, coma, and astigmatism, and is, of course, free of any chromatic aberrations. Both paraboloids may be concave, or one may be concave and the other convex (provided that the convex radius is shorter than the concave). The Petzval curvature is not corrected. For the two-concave-mirror arrangement, the Petzval field is strongly backward-curving; the concave-convex arrangement has a lesser, inward curvature. Confocal paraboloids are often used as afocal attachments to increase the aperture and focal length, or conversely, to shorten the focal length and increase the field of view.

16.5 Unobscured Systems

Any of the standard mirror systems can be used with a decentered aperture stop in order to produce a system with an unobscured aperture. The classical Herschel mount is simply a paraboloid with the aperture stop displaced laterally so that the aperture is completely off the optical axis and the image/focal point is outside the beam defined by the aperture. Note that, although this arrangement is often called an *off-axis parabola,* the paraboloid is not used off-axis; the *aperture* is off the axis, but the object and image are both on the optical axis. Since the paraboloid is completely free of spherical aberration for an infinitely distant object, the stop shift equations (specifically Eq. F.10.4) indicate that the coma is unchanged by moving the stop off-axis.

Two three-mirror systems with axially symmetrical mirrors which are the basis of many of the more complex, tilted or decentered, unobscured systems are illustrated in this section. Figure 16.20 shows a (slightly modified) Baker system which, in its basic form, consists of a confocal paraboloid front followed by a concave spherical reflector. The secondary mirror of the confocal pair is at the aperture stop, and is placed at the center of curvature of the tertiary sphere; the secondary is modified by aspheric deformation terms so that it acts as a reflecting Schmidt corrector for the spherical tertiary. (In this design the primary mirror has been modified from its original paraboloid form.) The confocal paraboloid pair is free of spherical, coma, and astigmatism. The tertiary sphere with the stop at its center of curvature is free of coma and astigmatism, and the aspheric deformation of the secondary corrects its spherical aberration. The Petzval curvature can be corrected if the curvature of the convex mirror is made equal to the sum of the concave curvatures. Note that the confocal front may be used in either orientation; with the concave as the primary, the focal length of the system is long, and with the convex as primary, the focal length is short.

The second system (Fig. 16.21) is afocal, with a magnification of $5\times$. It consists of a classical Cassegrain (with a paraboloid primary and hyperboloid secondary) as the "objective," plus a concave paraboloid as the "eyepiece" of the system to collimate the output beam. This arrangement has properties similar to those of confocal paraboloids, being free of spherical, coma, and astigmatism. Again, if the convex curvature is made equal to the sum of the concave curvatures, the Petzval will be corrected. This type of system (i.e., afocal) is often used as a device to reduce

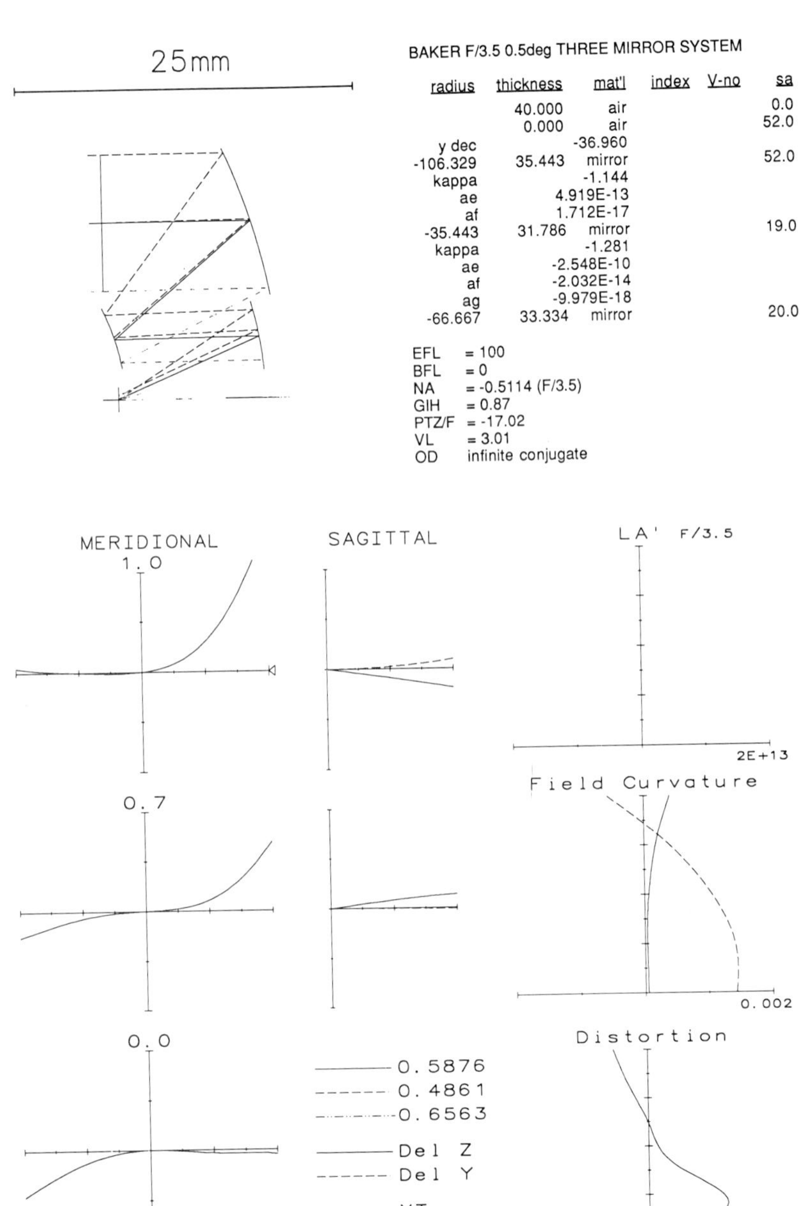
25mm
BAKER F/3.5 0.5deg THREE MIRROR SYSTEM
radius thickness mat'l index V-no sa
40.000 air 0.0
0.000 air 52.0
y dec -36.960
-106.329 35.443 mirror 52.0
kappa -1.144
ae 4.919E-13
af 1.712E-17
-35.443 31.786 mirror 19.0
kappa -1.281
ae -2.548E-10
af -2.032E-14
ag -9.979E-18
-66.667 33.334 mirror 20.0
EFL = 100
BFL = 0
NA = -0.5114 (F/3.5)
GIH = 0.87
PTZ/F = -17.02
VL = 3.01
OD infinite conjugate
MERIDIONAL
SAGITTAL
1.0
0.7
0.0
0.0005
LA' F/3.5
2E+13
Field Curvature
0.002
Distortion
HFOV=0.50
0.5
0.5876
0.4861
0.6563
Del Z
Del Y
XT
XS

Figure 16.20

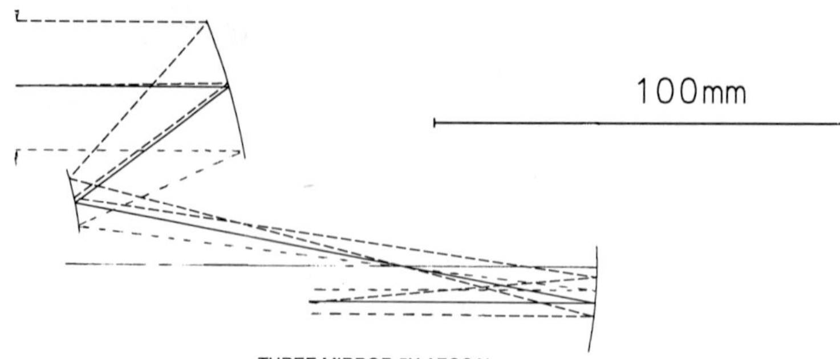

THREE MIRROR 5X AFOCAL

radius	thickness	mat'l	index	V-no	sa
	43.840	air			12.5
	0.000	air			44.0
y dec		-32.000			
-100.725	33.514	mirror			44.0
kappa		-1.000			
-46.109	100.000	mirror			17.0
kappa		-3.016			
-74.819	55.560	mirror			15.0
kappa		-1.000			
Afo	1000				

EFL = -5000
BFL = 1000
NA = 0.0087 (F/213.7)
VL = 54.77
OD infinite conjugate
MAG = 5

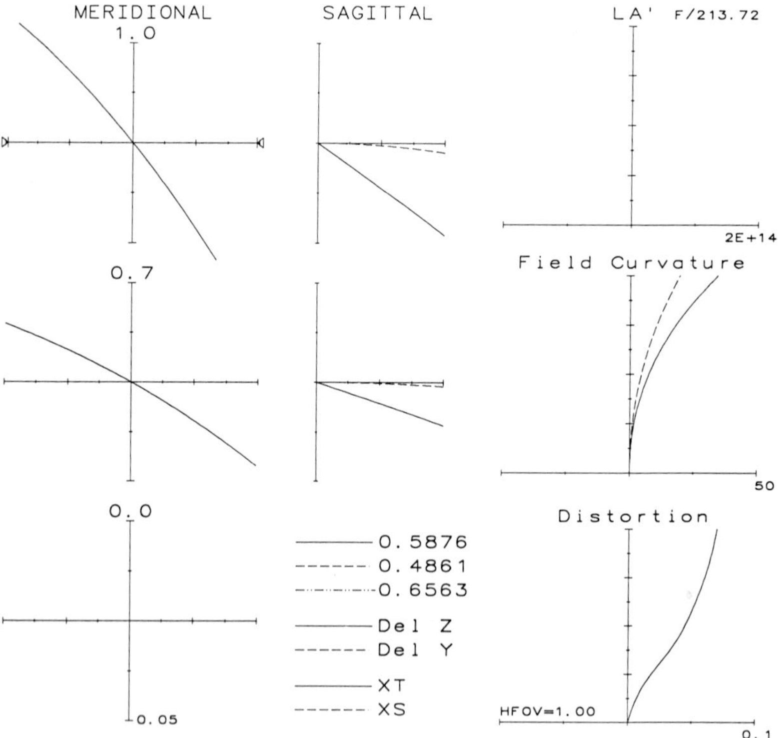

Figure 16.21

a large beam diameter to a smaller, more easily handled beam for the balance of the optical train. Note the external exit pupil, which can be located at the entrance pupil of the rest of the system.

Figures 16.22 and 16.23 are examples of systems which incorporate the Gabor-Bouwers-Maksutov principle of a corrector which is a weak, but strongly bent, meniscus element. Each also incorporates a field flattener element near the focal plane—a positive element in one case, negative in the other. A drawback to these systems is that the full aperture meniscus corrector tends to be thick and heavy and requires a large piece of good-quality glass. The basic Bouwers system is monocentric, with all radii having a common center which is located at the aperture stop, so that the system is thereby free of coma and astigmatism. The Maksutov system corrector has its radii chosen so that the corrector is free of axial chromatic; since it departs slightly from monocentricity, the freedom from coma and astigmatism is given up in exchange for chromatic correction.

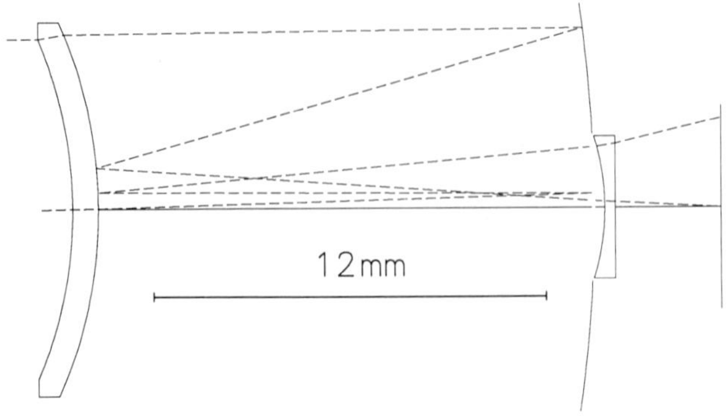

MAKSUTOV CASSEGRAIN F/10 1.5deg HFOV

radius	thickness	mat'l	index	V-no	sa
-11.796	0.831	BK8	1.520	63.7	5.0
-12.528	15.763	air			5.5
-39.711	15.763	mirror			6.0
-12.528	16.096	mirror			2.0
-6.651	0.333	BK8	1.520	63.7	2.0
	3.314	air			2.1

EFL	= 99.89
BFL	= 3.314
NA	= -0.0496 (F/10.0)
GIH	= 2.60
PTZ/F	= -0.1751
VL	= 17.26
OD	infinite conjugate

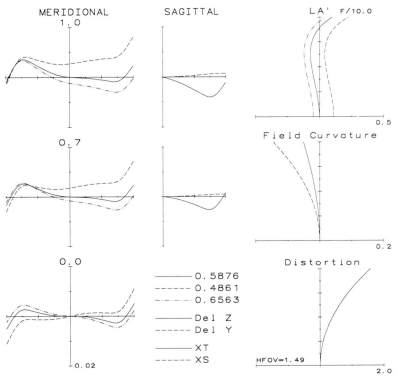

MERIDIONAL SAGITTAL LA' F/10.0

1.0

0.5

Field Curvature

0.7

0.2

0.0

Distortion

0.5876
0.4861
0.6563
Del Z
Del Y
XT
XS

0.02

HFOV=1.49

2.0

Figure 16.22

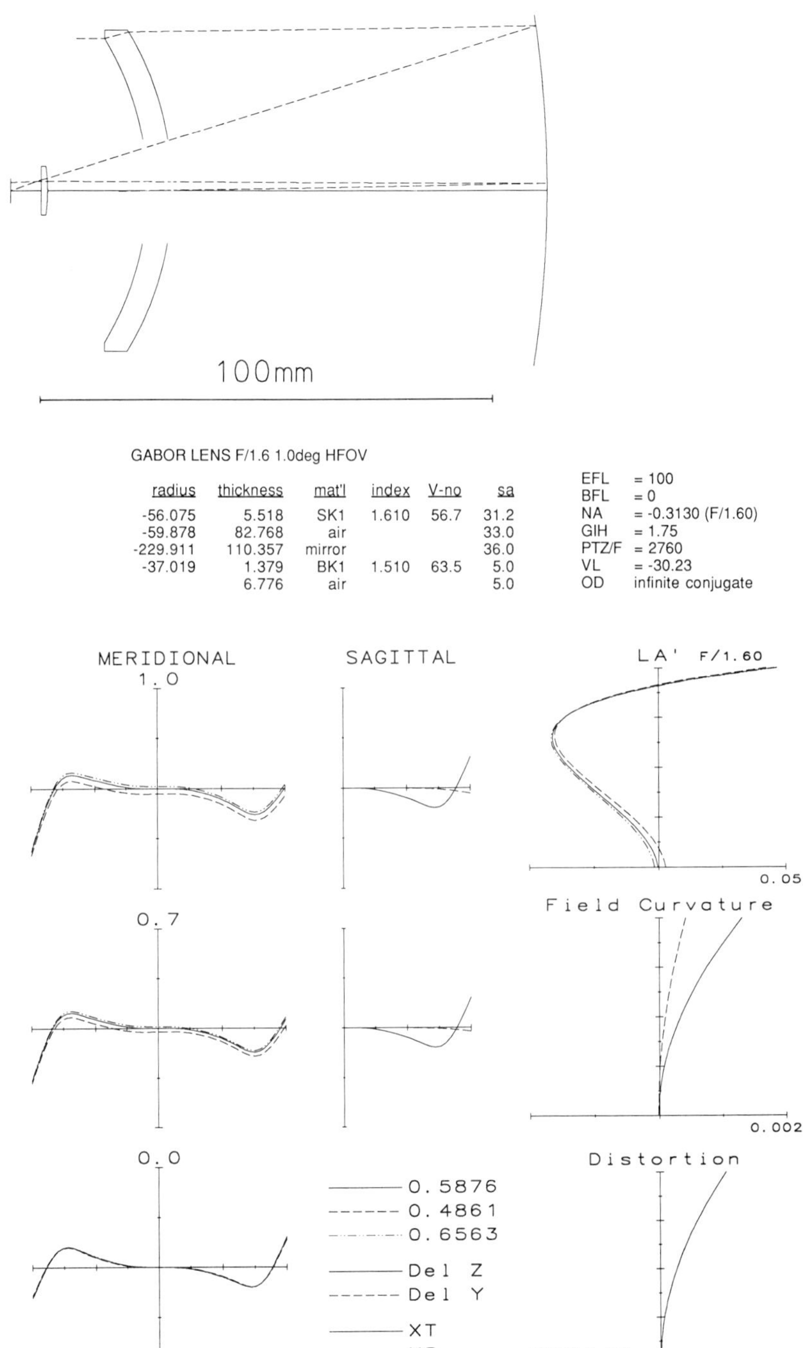

GABOR LENS F/1.6 1.0deg HFOV

radius	thickness	mat'l	index	V-no	sa
-56.075	5.518	SK1	1.610	56.7	31.2
-59.878	82.768	air			33.0
-229.911	110.357	mirror			36.0
-37.019	1.379	BK1	1.510	63.5	5.0
	6.776	air			5.0

EFL = 100
BFL = 0
NA = -0.3130 (F/1.60)
GIH = 1.75
PTZ/F = 2760
VL = -30.23
OD infinite conjugate

Figure 16.23

Chapter

17

The Biotar or Double-Gauss Lens

17.1 The Basic Six-Element Version

The Biotar or double-Gauss type is a descendant of the double-meniscus anastigmat lenses discussed in Chap. 11. One of the many variants of the double-meniscus form consisted of outer positive singlets and inner cemented negative doublets, in a symmetrical construction. These lenses had the speed and angular coverage typical of their genre, i.e., good angular coverage at a quite modest aperture. However, a departure from symmetry allowed the speed of the lens to be increased (initially) to *f*/2, and a tremendously useful and powerful design form was born.

This design type is the basis of most normal-focal-length 35-mm camera lenses, having supplanted the Sonnar and Ernostar types which evolved about the same time. It is found in many applications where extremely high performance is required of a lens. It can be made into a moderately wide-angle lens, an enlarger lens, a high-resolution objective, or a lens of extremely high speed. It has been subject to almost every imaginable modification, including splits, compoundings, doublings, inserted components, and even complete doubling or duplication of all components.

The basic six-element version is shown in Figs. 17.1 through 17.14, arranged in order of speed, which ranges from *f*/1.25 to *f*/8.0 in these examples. Typically, the positive front element is meniscus in shape, usually of a lanthanum flint glass with an index of about 1.7 and a *V* value of about 48. Dense barium crowns and barium flints are also used. The second element tends to be meniscus, although a plano-convex or a mild biconvex form is not unusual; its index is typically a bit lower and its *V* value higher than the first element. The third el-

(Text continues on page 318.)

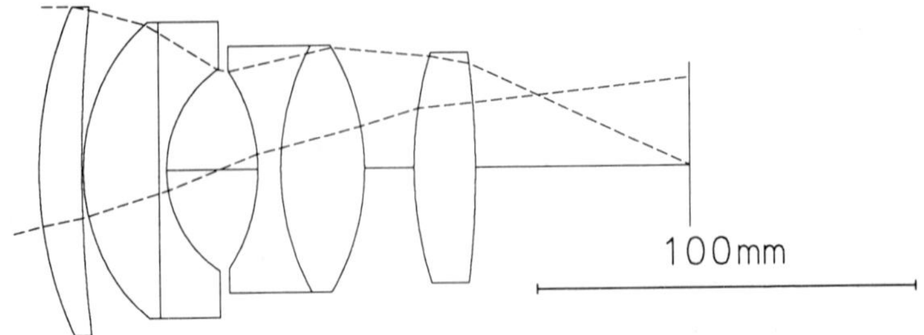

D-GAUSS F/1.25 12deg USP2771006/ WERFELI/

radius	thickness	mat'l	index	V-no	sa
93.320	11.320	LAF3	1.717	48.0	40.0
358.290	0.400	air			40.0
46.320	20.000	BAF9	1.643	48.0	36.0
	2.000	LF2	1.589	40.9	36.0
28.680	14.000	air			24.5
	10.000	air			24.3
-41.320	6.000	SF14	1.762	26.5	24.0
60.800	22.000	LAF2	1.744	44.7	30.0
-55.000	13.000	air			30.0
90.200	16.000	LAF3	1.717	48.0	28.0
-212.580	56.424	air			28.0

EFL = 100.1
BFL = 56.42
NA = -0.3992 (F/1.25)
GIH = 22.03 (HFOV=12.41)
PTZ/F = -3.801
VL = 114.72
OD infinite conjugate

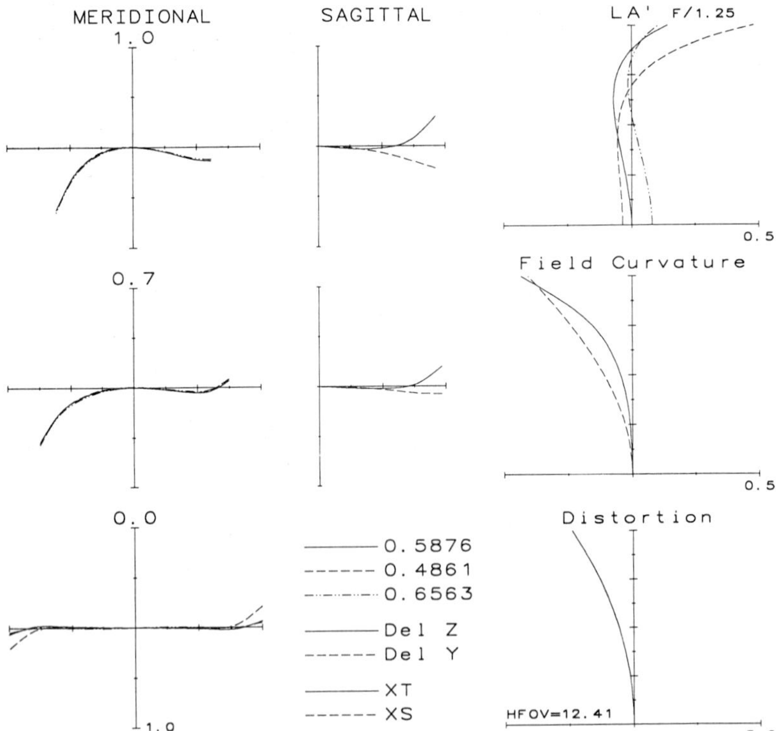

Figure 17.1

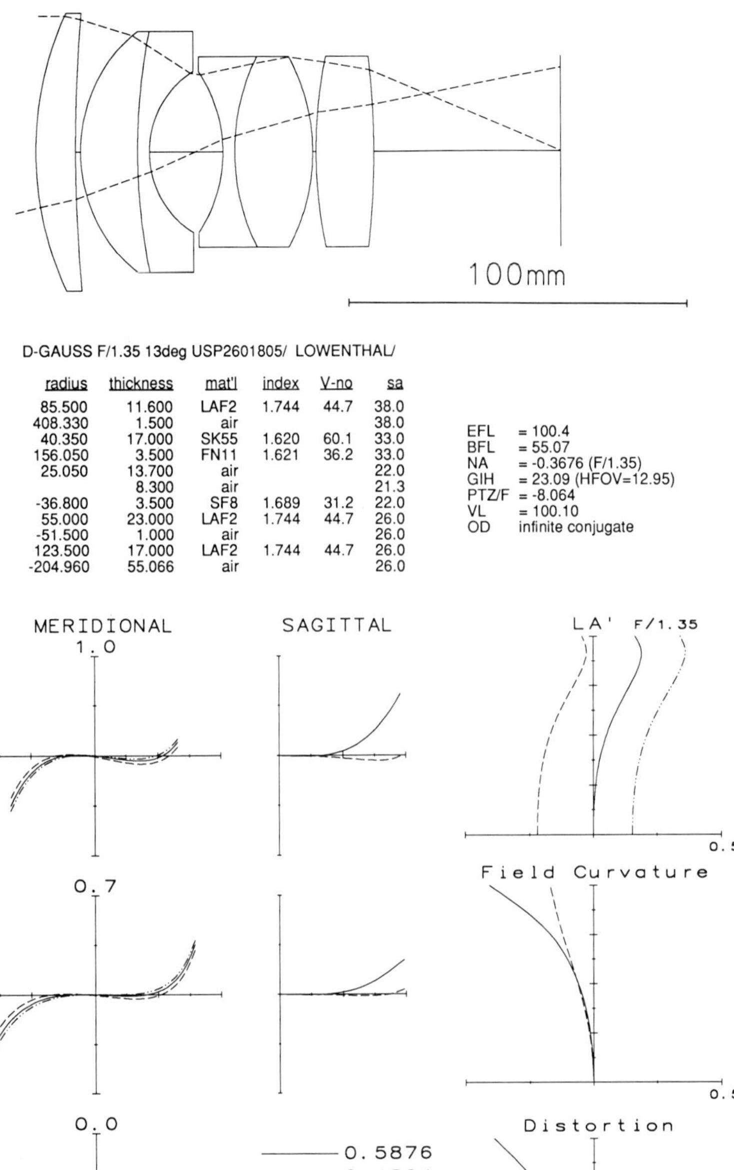

D-GAUSS F/1.35 13deg USP2601805/ LOWENTHAL/

radius	thickness	mat'l	index	V-no	sa
85.500	11.600	LAF2	1.744	44.7	38.0
408.330	1.500	air			38.0
40.350	17.000	SK55	1.620	60.1	33.0
156.050	3.500	FN11	1.621	36.2	33.0
25.050	13.700	air			22.0
	8.300	air			21.3
-36.800	3.500	SF8	1.689	31.2	22.0
55.000	23.000	LAF2	1.744	44.7	26.0
-51.500	1.000	air			26.0
123.500	17.000	LAF2	1.744	44.7	26.0
-204.960	55.066	air			26.0

EFL = 100.4
BFL = 55.07
NA = -0.3676 (F/1.35)
GIH = 23.09 (HFOV=12.95)
PTZ/F = -8.064
VL = 100.10
OD infinite conjugate

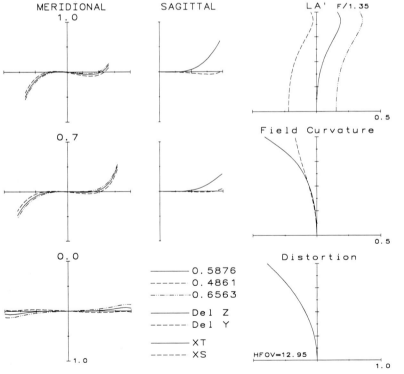

Figure 17.2

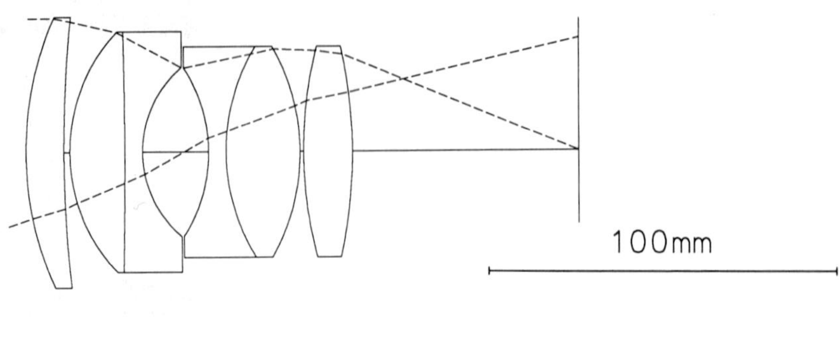

D-GAUSS F/1.4 17deg DRP485798/1927 MERTE/ZEISS

radius	thickness	mat'l	index	V-no	sa
83.600	10.800	BAF9	1.643	48.0	36.0
321.000	1.700	air			36.0
44.800	15.600	SK10	1.623	56.9	32.0
-1150.000	5.100	LF7	1.575	41.5	32.0
28.300	11.900	air			22.5
	7.000	air			22.3
-38.500	5.100	SF5	1.673	32.2	22.4
50.500	21.200	BAF9	1.643	48.0	28.0
-53.200	1.000	air			28.0
106.000	13.900	BAF9	1.643	48.0	28.0
-120.000	64.690	air			28.0

EFL = 99.66
BFL = 64.69
NA = -0.3577 (F/1.40)
GIH = 30.89 (HFOV=17.22)
PTZ/F = -3.594
VL = 93.30
OD infinite conjugate

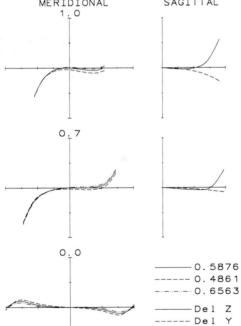

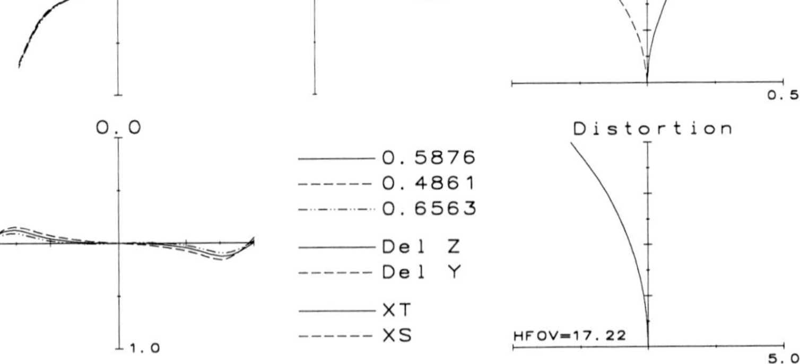

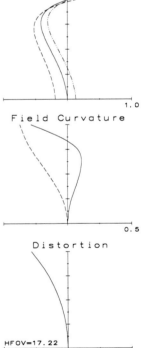

Figure 17.3

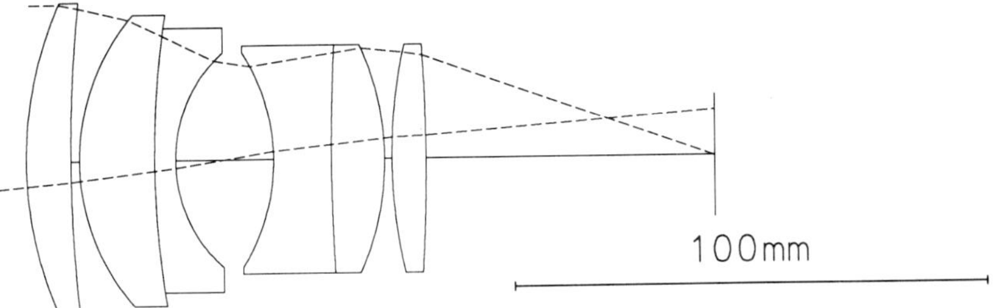

F/1.6 5degHFOV DOUBLE GAUSS

radius	thickness	mat'l	index	V-no	sa
70.525	9.490	SK4	1.613	58.6	31.2
274.170	1.820	air			31.2
41.860	15.860	SK4	1.613	58.6	28.6
166.920	4.550	F5	1.603	38.0	26.0
26.910	10.400	air			21.1
	10.400	air			19.2
-37.245	12.220	SF1	1.717	29.5	21.1
357.500	11.310	LAK16	1.734	51.8	22.4
-50.557	1.560	air			22.4
92.040	7.280	LAK16	1.734	51.8	22.4
-309.712	60.762	air			22.4

EFL = 98.42
BFL = 60.76
NA = -0.3129 (F/1.60)
GIH = 9.05 (HFOV=5.26)
PTZ/F = -4.622
VL = 84.89
OD infinite conjugate

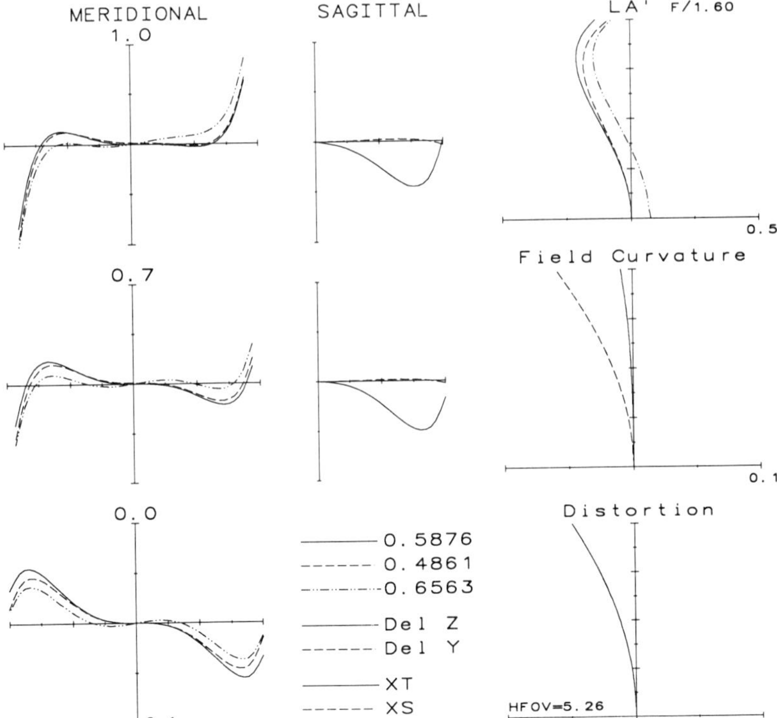

Figure 17.4

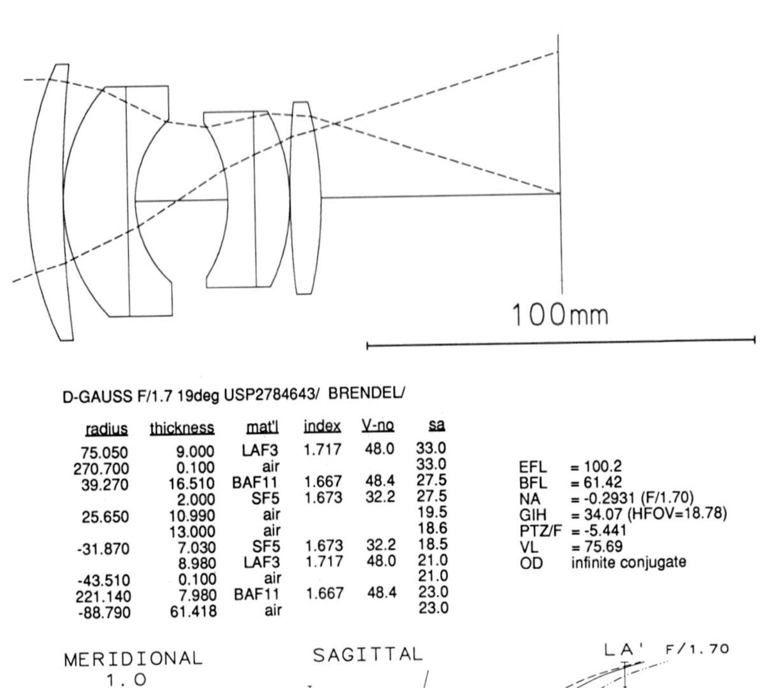

D-GAUSS F/1.7 19deg USP2784643/ BRENDEL/

radius	thickness	mat'l	index	V-no	sa
75.050	9.000	LAF3	1.717	48.0	33.0
270.700	0.100	air			33.0
39.270	16.510	BAF11	1.667	48.4	27.5
	2.000	SF5	1.673	32.2	27.5
25.650	10.990	air			19.5
	13.000	air			18.6
-31.870	7.030	SF5	1.673	32.2	18.5
	8.980	LAF3	1.717	48.0	21.0
-43.510	0.100	air			21.0
221.140	7.980	BAF11	1.667	48.4	23.0
-88.790	61.418	air			23.0

EFL = 100.2
BFL = 61.42
NA = -0.2931 (F/1.70)
GIH = 34.07 (HFOV=18.78)
PTZ/F = -5.441
VL = 75.69
OD infinite conjugate

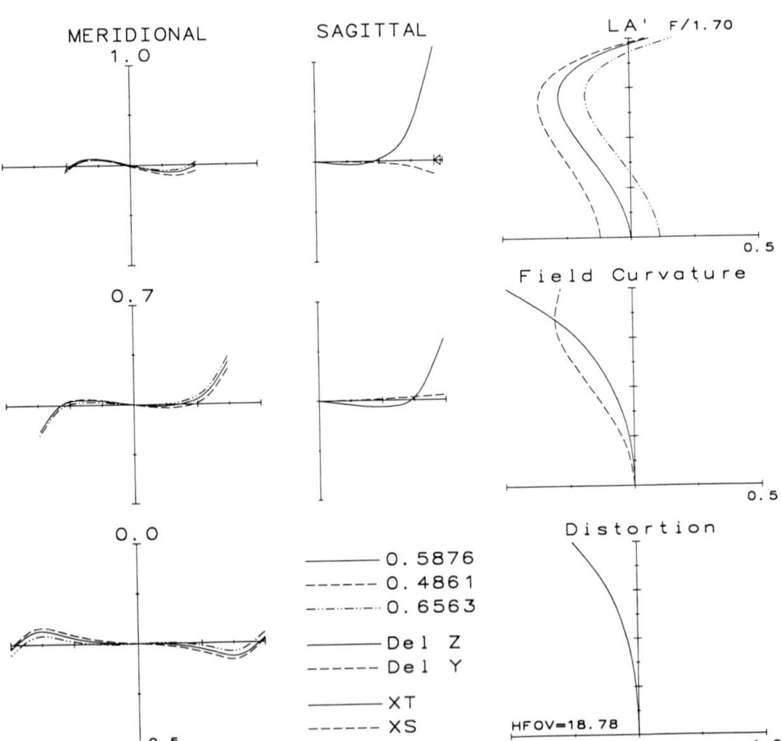

Figure 17.5

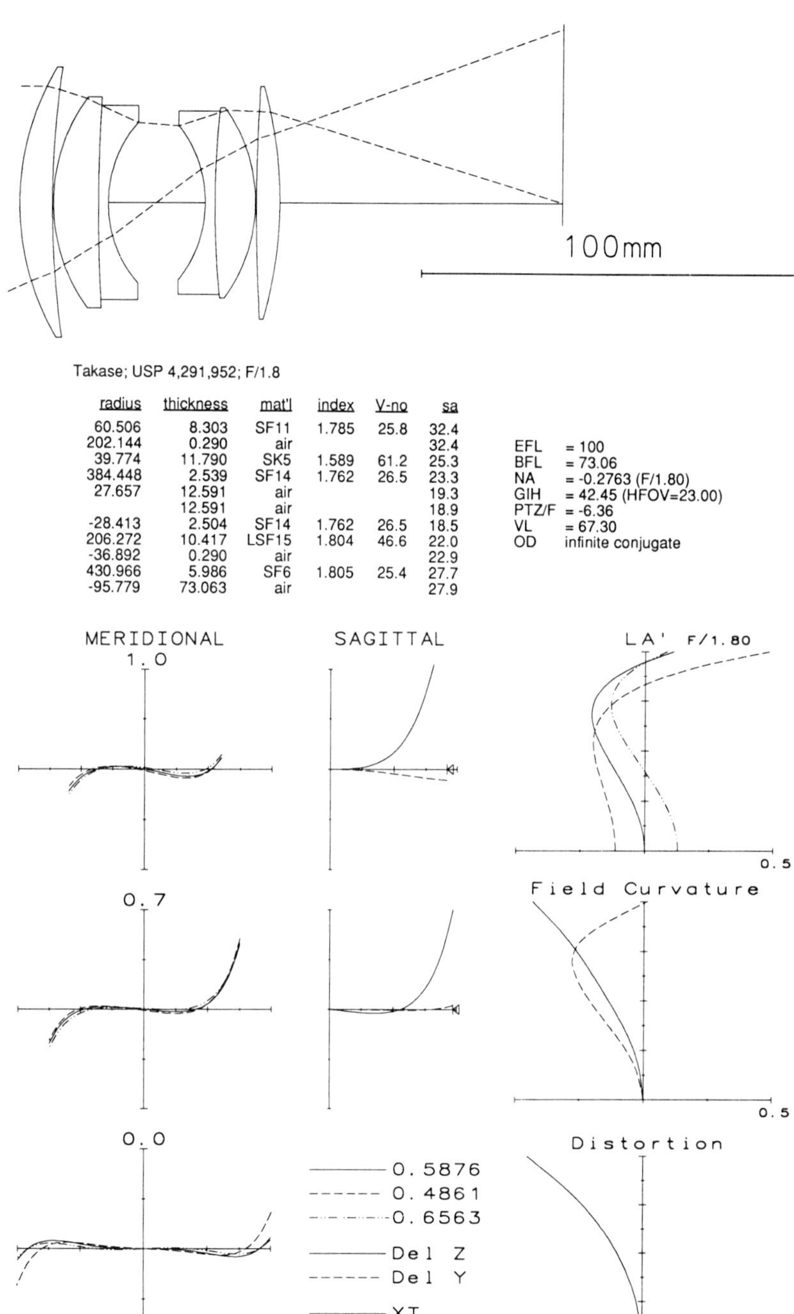

Takase; USP 4,291,952; F/1.8

radius	thickness	mat'l	index	V-no	sa
60.506	8.303	SF11	1.785	25.8	32.4
202.144	0.290	air			32.4
39.774	11.790	SK5	1.589	61.2	25.3
384.448	2.539	SF14	1.762	26.5	23.3
27.657	12.591	air			19.3
	12.591	air			18.9
-28.413	2.504	SF14	1.762	26.5	18.5
206.272	10.417	LSF15	1.804	46.6	22.0
-36.892	0.290	air			22.9
430.966	5.986	SF6	1.805	25.4	27.7
-95.779	73.063	air			27.9

EFL = 100
BFL = 73.06
NA = -0.2763 (F/1.80)
GIH = 42.45 (HFOV=23.00)
PTZ/F = -6.36
VL = 67.30
OD infinite conjugate

Figure 17.6

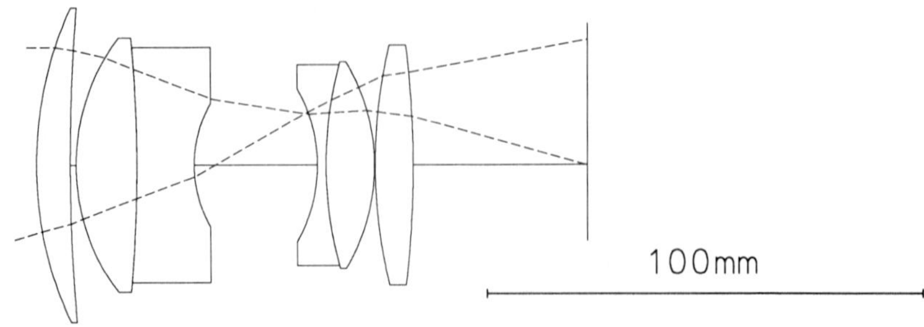

BRIXNER DOUBLE GAUSS - FROM PLANOS

radius	thickness	mat'l	index	V-no	sa
73.878	7.861	BSF51	1.724	38.1	33.4
404.472	1.294	air			33.4
42.314	13.931	LAKN7	1.652	58.5	27.0
-271.759	13.135	SF1	1.717	29.5	25.0
24.961	4.976	air			13.1
	23.285	air			13.9
-31.056	1.990	BASF2	1.664	35.8	16.3
77.345	10.946	LAFN2	1.744	44.8	21.3
-40.486	0.199	air			21.9
103.569	8.657	LAFN2	1.744	44.8	25.5
-192.555	39.913	air			25.5

EFL = 100
BFL = 39.91
NA = -0.2506 (F/2.0)
GIH = 26.83 (HFOV=15.02)
PTZ/F = -6.611
VL = 86.27
OD infinite conjugate

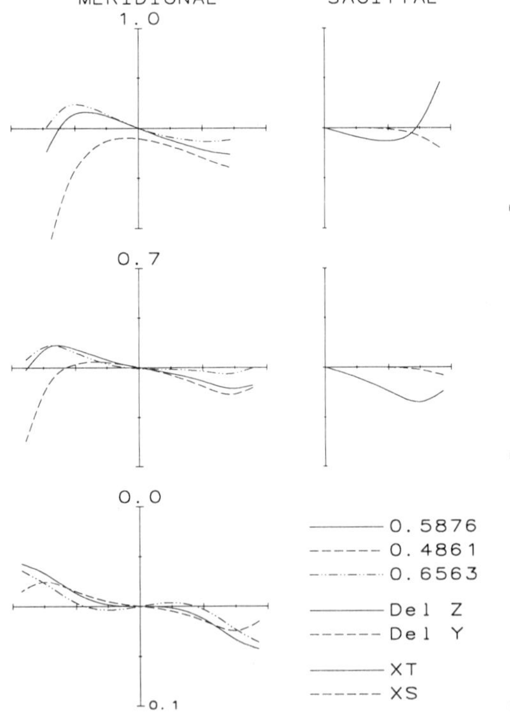

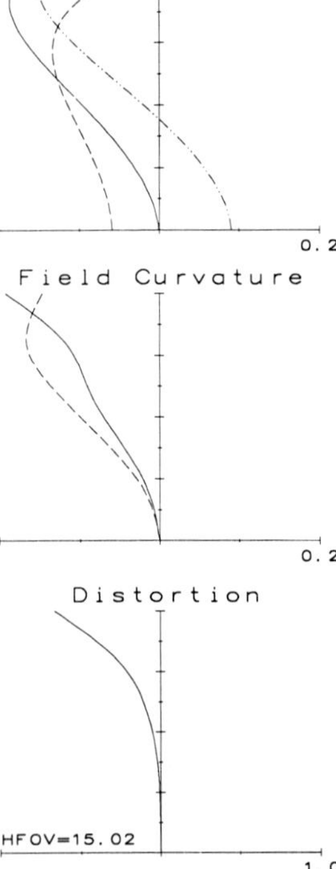

Figure 17.7

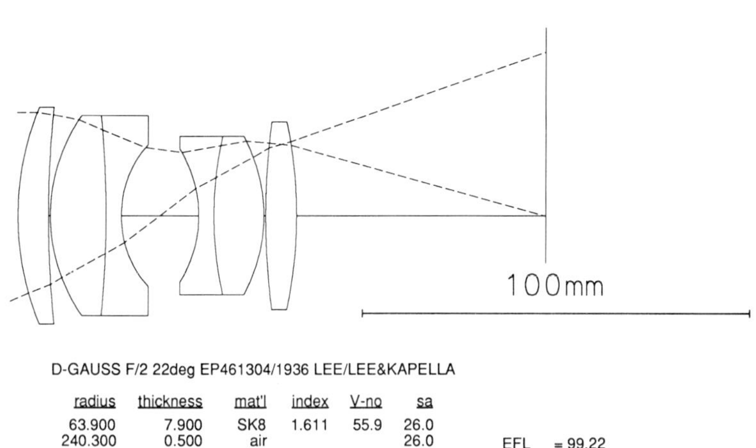

D-GAUSS F/2 22deg EP461304/1936 LEE/LEE&KAPELLA

radius	thickness	mat'l	index	V-no	sa
63.900	7.900	SK8	1.611	55.9	26.0
240.300	0.500	air			26.0
39.500	14.500	SK10	1.623	56.9	24.0
-220.500	4.000	F15	1.606	37.8	24.0
24.500	9.950	air			17.0
	9.950	air			15.9
-28.700	4.000	F15	1.606	37.8	16.0
78.800	12.900	SK10	1.623	56.9	19.0
-37.900	0.500	air			19.0
161.900	8.000	SK10	1.623	56.9	22.5
-103.200	64.005	air			22.5

EFL = 99.22
BFL = 64
NA = -0.2505 (F/2.0)
GIH = 39.69 (HFOV=21.80)
PTZ/F = -5.429
VL = 72.20
OD infinite conjugate

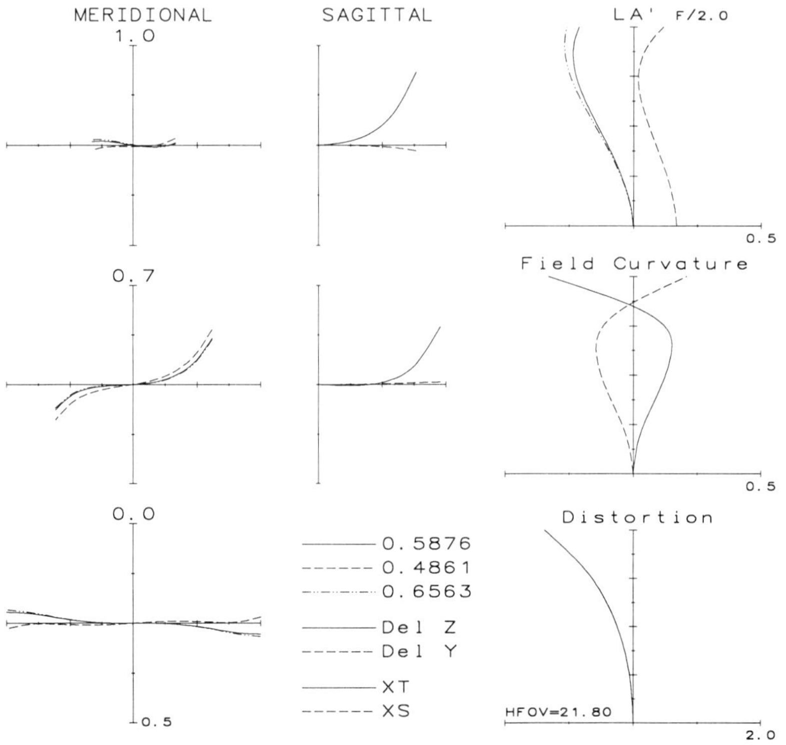

MERIDIONAL
1.0
SAGITTAL
LA' F/2.0
0.5
0.7
Field Curvature
0.5
0.0
Distortion
0.5
0.5876
0.4861
0.6563
Del Z
Del Y
XT
XS
HFOV=21.80
2.0

Figure 17.8

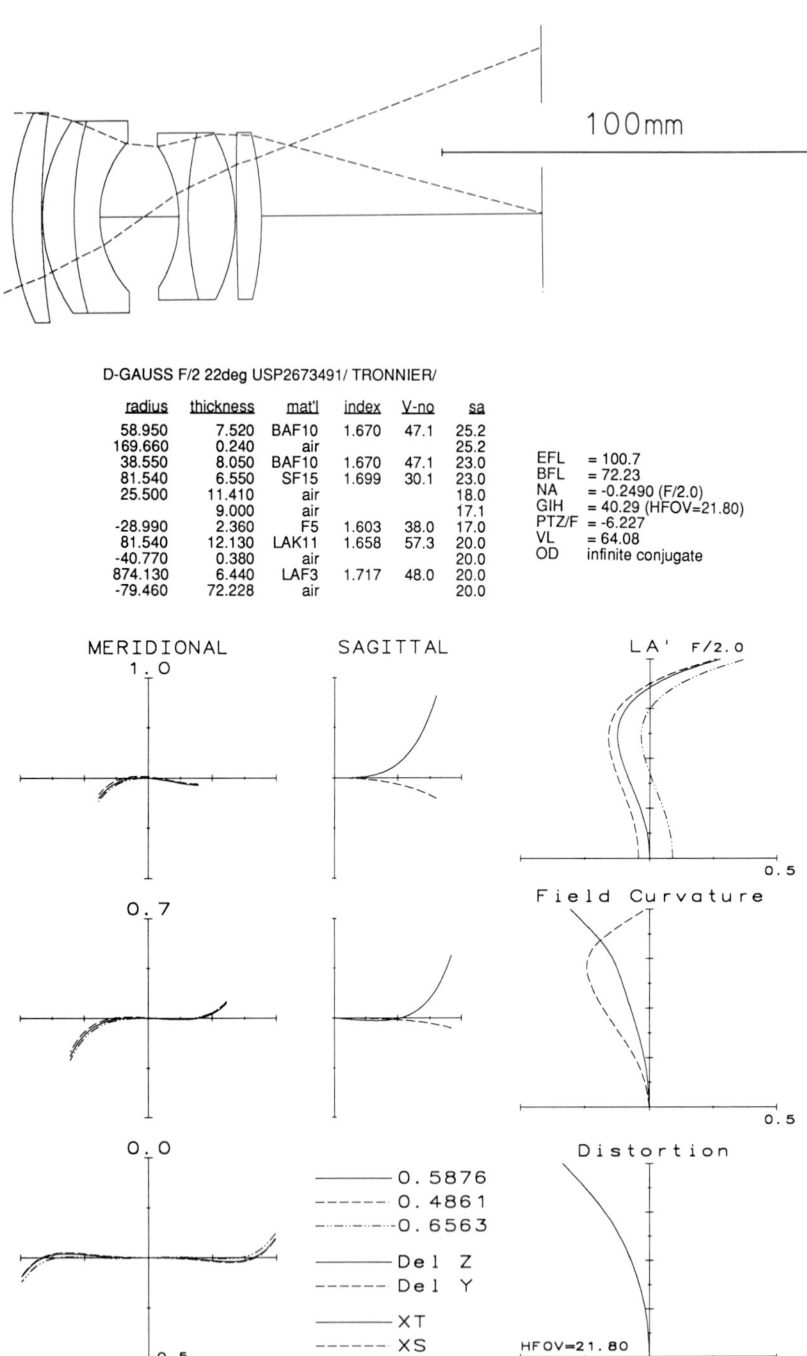

D-GAUSS F/2 22deg USP2673491/ TRONNIER/

radius	thickness	mat'l	index	V-no	sa
58.950	7.520	BAF10	1.670	47.1	25.2
169.660	0.240	air			25.2
38.550	8.050	BAF10	1.670	47.1	23.0
81.540	6.550	SF15	1.699	30.1	23.0
25.500	11.410	air			18.0
	9.000	air			17.1
-28.990	2.360	F5	1.603	38.0	17.0
81.540	12.130	LAK11	1.658	57.3	20.0
-40.770	0.380	air			20.0
874.130	6.440	LAF3	1.717	48.0	20.0
-79.460	72.228	air			20.0

EFL = 100.7
BFL = 72.23
NA = -0.2490 (F/2.0)
GIH = 40.29 (HFOV=21.80)
PTZ/F = -6.227
VL = 64.08
OD infinite conjugate

Figure 17.9

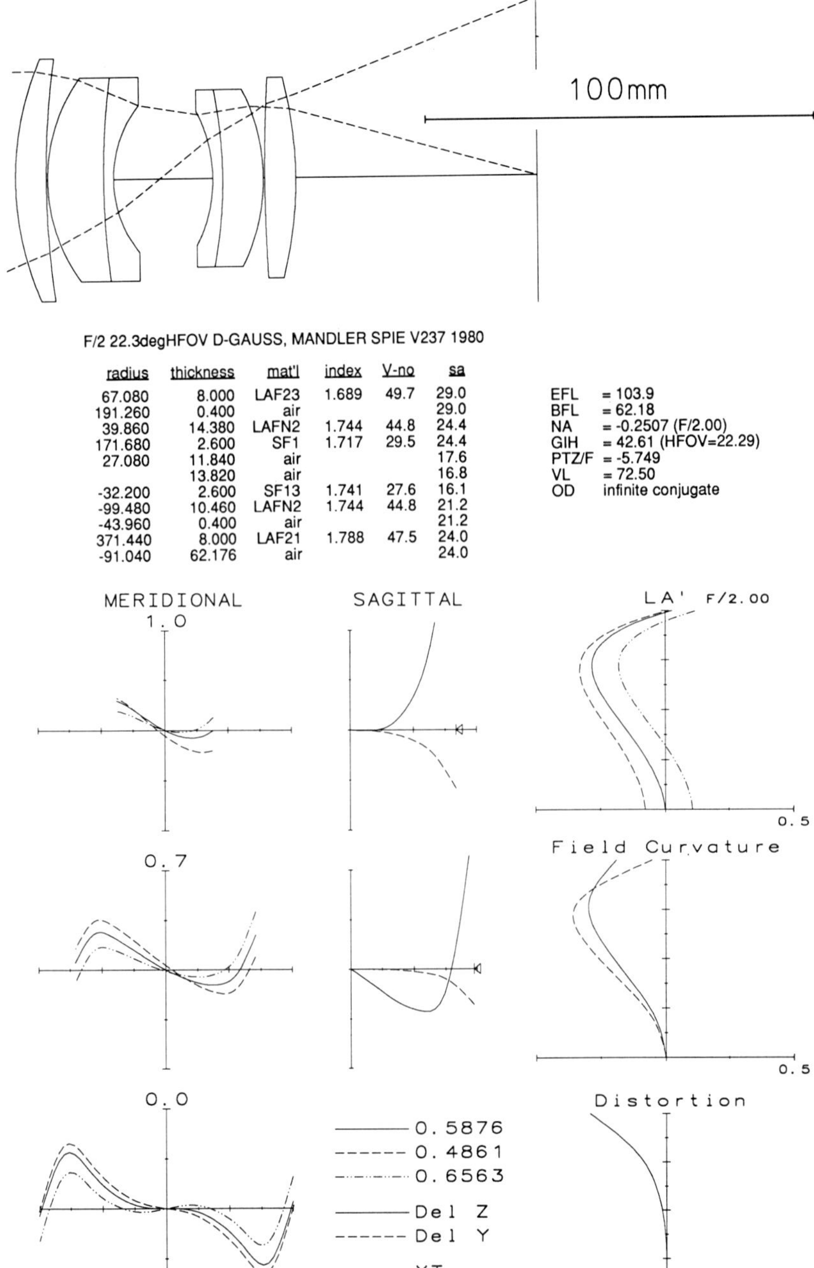

F/2 22.3degHFOV D-GAUSS, MANDLER SPIE V237 1980

radius	thickness	mat'l	index	V-no	sa
67.080	8.000	LAF23	1.689	49.7	29.0
191.260	0.400	air			29.0
39.860	14.380	LAFN2	1.744	44.8	24.4
171.680	2.600	SF1	1.717	29.5	24.4
27.080	11.840	air			17.6
	13.820	air			16.8
-32.200	2.600	SF13	1.741	27.6	16.1
-99.480	10.460	LAFN2	1.744	44.8	21.2
-43.960	0.400	air			21.2
371.440	8.000	LAF21	1.788	47.5	24.0
-91.040	62.176	air			24.0

EFL = 103.9
BFL = 62.18
NA = -0.2507 (F/2.00)
GIH = 42.61 (HFOV=22.29)
PTZ/F = -5.749
VL = 72.50
OD infinite conjugate

Figure 17.10

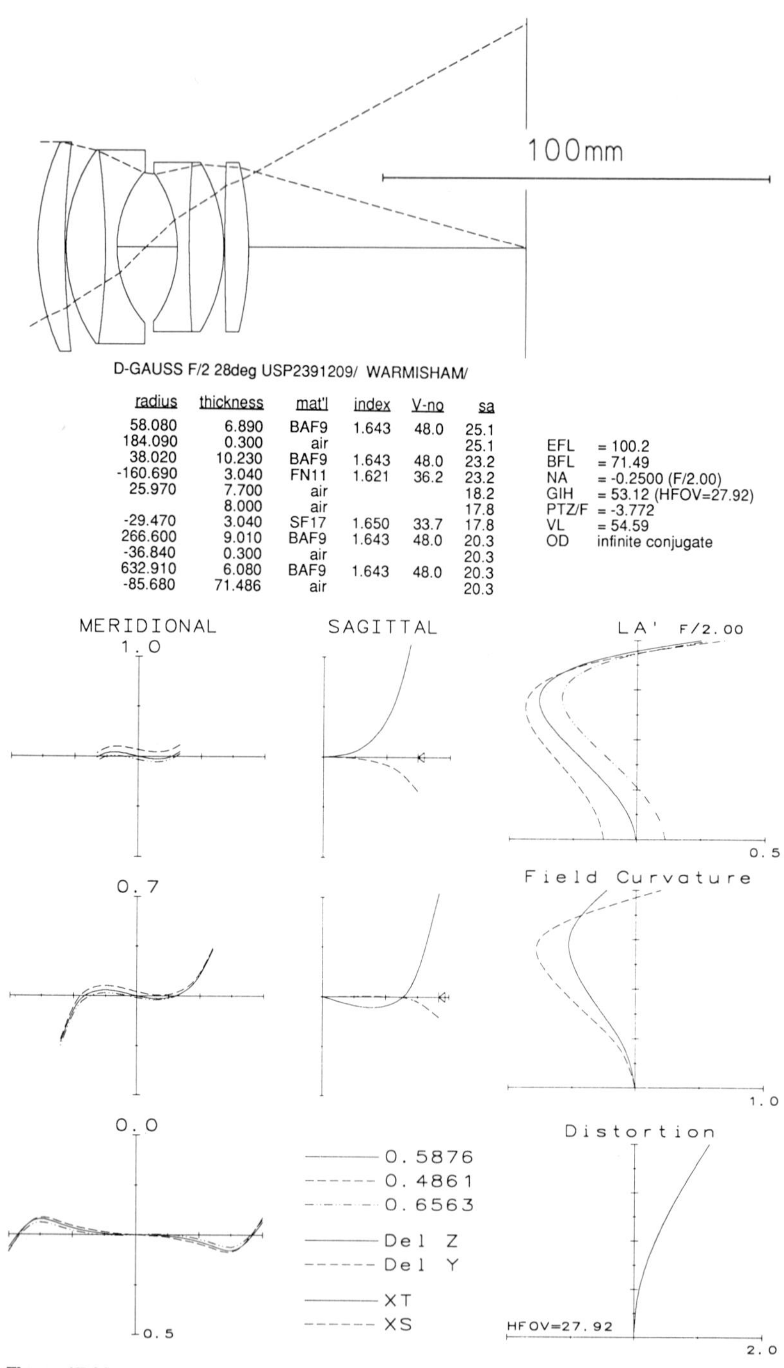

D-GAUSS F/2 28deg USP2391209/ WARMISHAM/

radius	thickness	mat'l	index	V-no	sa
58.080	6.890	BAF9	1.643	48.0	25.1
184.090	0.300	air			25.1
38.020	10.230	BAF9	1.643	48.0	23.2
-160.690	3.040	FN11	1.621	36.2	23.2
25.970	7.700	air			18.2
	8.000	air			17.8
-29.470	3.040	SF17	1.650	33.7	17.8
266.600	9.010	BAF9	1.643	48.0	20.3
-36.840	0.300	air			20.3
632.910	6.080	BAF9	1.643	48.0	20.3
-85.680	71.486	air			20.3

EFL = 100.2
BFL = 71.49
NA = -0.2500 (F/2.00)
GIH = 53.12 (HFOV=27.92)
PTZ/F = -3.772
VL = 54.59
OD infinite conjugate

Figure 17.11

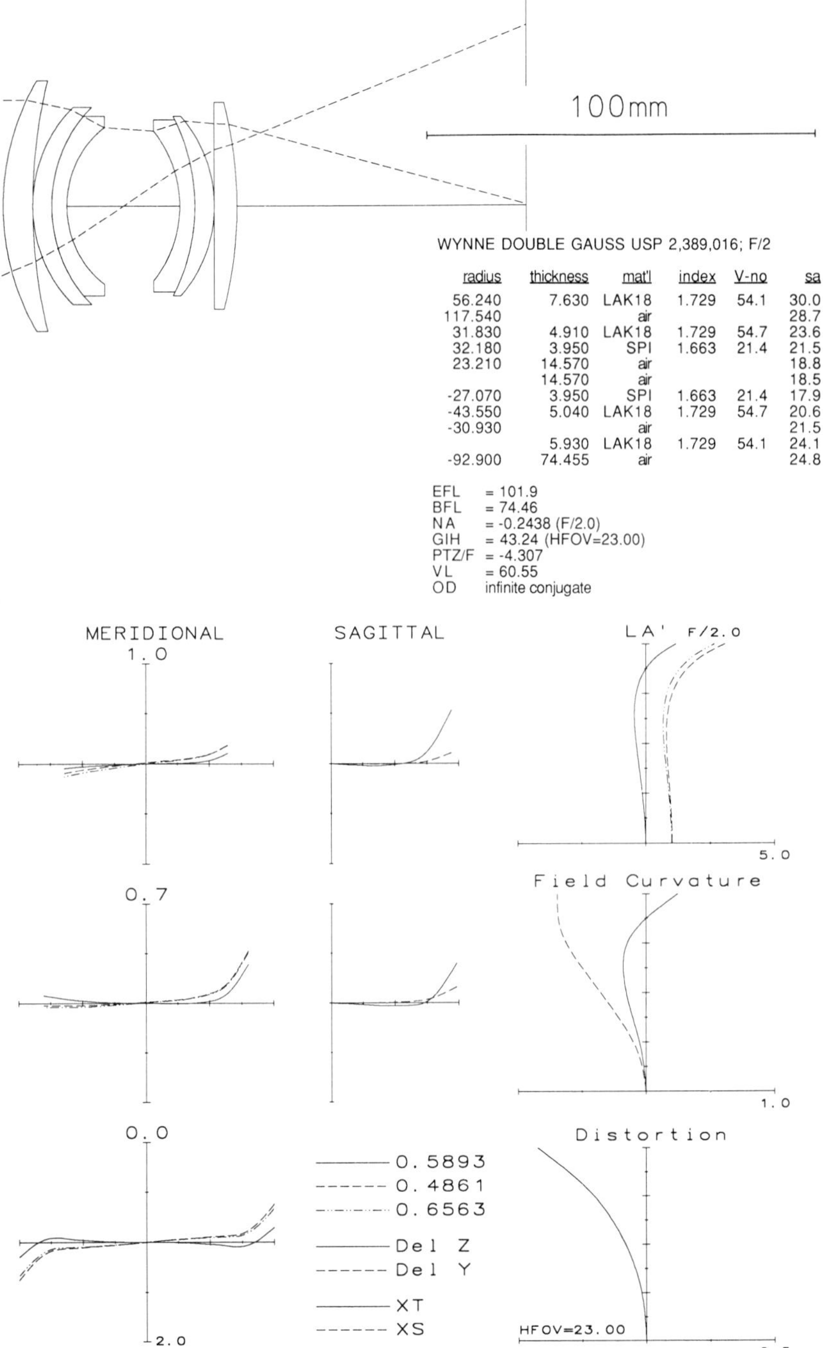

WYNNE DOUBLE GAUSS USP 2,389,016; F/2

radius	thickness	mat'l	index	V-no	sa
56.240	7.630	LAK18	1.729	54.1	30.0
117.540		air			28.7
31.830	4.910	LAK18	1.729	54.7	23.6
32.180	3.950	SPI	1.663	21.4	21.5
23.210	14.570	air			18.8
	14.570	air			18.5
-27.070	3.950	SPI	1.663	21.4	17.9
-43.550	5.040	LAK18	1.729	54.7	20.6
-30.930		air			21.5
	5.930	LAK18	1.729	54.1	24.1
-92.900	74.455	air			24.8

EFL = 101.9
BFL = 74.46
NA = -0.2438 (F/2.0)
GIH = 43.24 (HFOV=23.00)
PTZ/F = -4.307
VL = 60.55
OD infinite conjugate

Figure 17.12

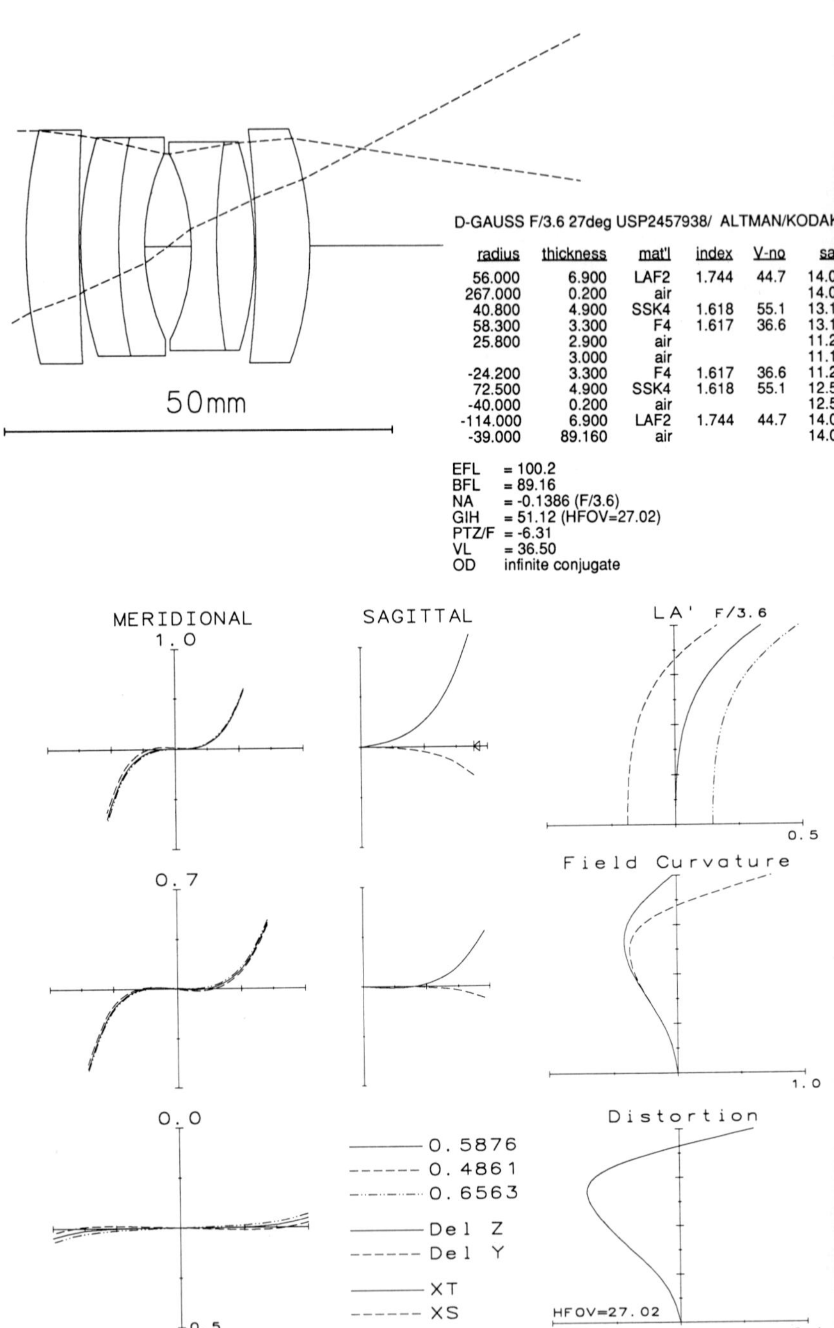

D-GAUSS F/3.6 27deg USP2457938/ ALTMAN/KODAK

radius	thickness	mat'l	index	V-no	sa
56.000	6.900	LAF2	1.744	44.7	14.0
267.000	0.200	air			14.0
40.800	4.900	SSK4	1.618	55.1	13.1
58.300	3.300	F4	1.617	36.6	13.1
25.800	2.900	air			11.2
	3.000	air			11.1
-24.200	3.300	F4	1.617	36.6	11.2
72.500	4.900	SSK4	1.618	55.1	12.5
-40.000	0.200	air			12.5
-114.000	6.900	LAF2	1.744	44.7	14.0
-39.000	89.160	air			14.0

EFL = 100.2
BFL = 89.16
NA = -0.1386 (F/3.6)
GIH = 51.12 (HFOV=27.02)
PTZ/F = -6.31
VL = 36.50
OD infinite conjugate

Figure 17.13

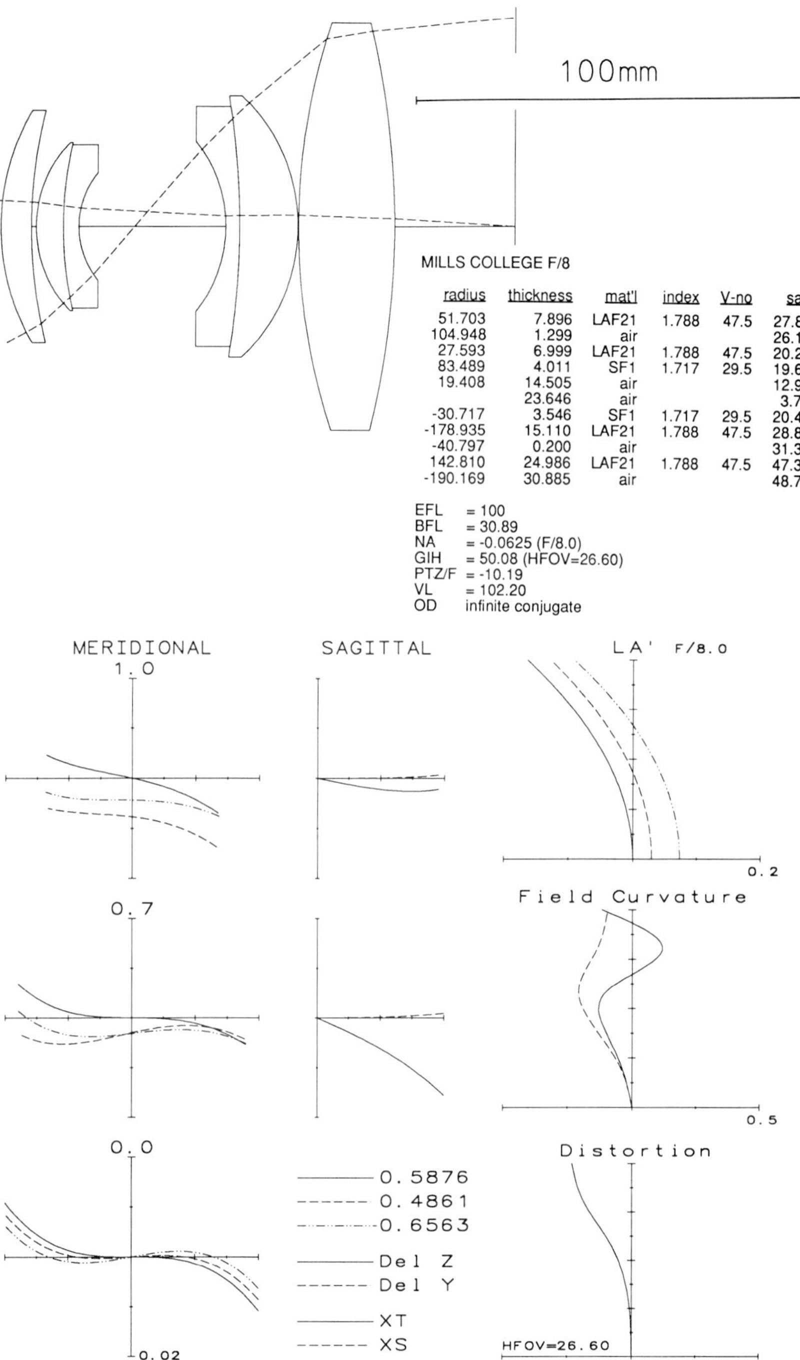

MILLS COLLEGE F/8

radius	thickness	mat'l	index	V-no	sa
51.703	7.896	LAF21	1.788	47.5	27.8
104.948	1.299	air			26.1
27.593	6.999	LAF21	1.788	47.5	20.2
83.489	4.011	SF1	1.717	29.5	19.6
19.408	14.505	air			12.9
	23.646	air			3.7
-30.717	3.546	SF1	1.717	29.5	20.4
-178.935	15.110	LAF21	1.788	47.5	28.8
-40.797	0.200	air			31.3
142.810	24.986	LAF21	1.788	47.5	47.3
-190.169	30.885	air			48.7

EFL = 100
BFL = 30.89
NA = -0.0625 (F/8.0)
GIH = 50.08 (HFOV=26.60)
PTZ/F = -10.19
VL = 102.20
OD infinite conjugate

Figure 17.14

ement usually has an index close to that of the second element, and it is a dense flint from along the glass line. The fourth element is usually biconcave and made of a glass-line flint with a slightly higher index than the third element. The fifth and sixth elements are ordinarily both biconvex and of lanthanum flint glass.

Often the rear positive elements are of higher-index glass than those in front. Since these rear elements are more powerful than those in front, the higher index is more effective there; if cost limits the amount of high-index glass which may be used, the smaller rear elements are the logical (and economical) place to use it. Occasionally, the front element is made of dense flint glass; Fig. 17.6 shows a design in which both outer elements are made of SF-type glass. The dense flint glass is a less expensive choice than a lanthanum glass.

Figure 17.4 was designed for use as a commercial 2-in *f*/1.6, 16-mm movie projection lens. Note the ordinary (SK4) glass in the front elements and the economical (i.e., thin) use of the lanthanum glass in the rear elements.

Figure 17.12 is unusual in that the negative elements are of potassium iodide, in order to take advantage of its very high dispersion and relatively low index. The aberration plot shows the excellent correction for both zonal spherical and spherochromatic which can result.

Two slower lenses are shown in Figs. 17.13 and 17.14, each covering about a 54° total field. Figure 17.13 is a relatively standard configuration at *f*/3.6; however, it is a good example of one of the common problems with the double-Gauss form, i.e., a strongly overcorrecting oblique spherical aberration. This problem can often be reduced by shaping the doublets to a more meniscus form so that the oblique rays pass more normally through them.

Figure 17.10 is one of the ultimate lenses which resulted from an extensive study of *f*/2, normal-focal-length 35-mm camera lenses on which Mandler reported at the 1980 International Lens Design Conference. This conference also featured a lens design contest, the original aim of which was to take a lens optimized at *f*/2 and, without human intervention, automatically redesign it to a narrow angle *f*/1, or to wider angle (53°) *f*/8. (Unfortunately, the starting design had to be constrained to be sure that the *f*/1 rays on one hand, and the 53° rays on the other, would get through the lens; this had the effect of limiting the quality of the starting lens. Nevertheless, some interesting lenses were produced.) A typical result for the wide-angle lens is shown in Fig. 17.14; many of the wide-angle entries took this rather uncommercial configuration. Figure 17.7 is an after-the-contest design; Brixner took the glasses and thicknesses of the contest starting lens and began his optimization with plano surfaces in order to demonstrate that (not too surprisingly) an almost identical lens would result.

Several varieties of six-element Biotar type lenses are shown in Figs. 17.15 through 17.20. In Fig. 17.15 the inner doublets are split and the front doublet is reversed; note also the reversal of the usual V-value arrangement in elements 2 and 3, with the negative element having the higher V value rather than the usual lower.

Figures 17.16, 17.17, and 17.20 have been designed to work at finite conjugates, with magnifications of $-0.25\times$ and $-0.11\times$. Figure 17.16 is designed for the near ultraviolet (0.36 to 0.44 nm) and uses BK7 and LF1 glasses, both of which transmit well in the near ultraviolet. But, for still shorter wavelengths, optical glasses do not transmit well enough; materials such as fused quartz (SiO_2), calcium fluoride, and lithium fluoride are appropriate. Unfortunately, they are even lower in index than the glasses in this example. In contrast, Fig. 17.20 is a lens designed for a narrow band in the very near infrared. Figures 17.18 and 17.19 have reversed doublets; note also the use of flint glasses in the positive rear singlets in these designs.

17.2 The Seven-Element Biotar–Split-Rear Singlet

Splitting the rear singlet of the Biotar is an excellent way to improve on the basic configuration. It is a commonly used technique to allow an increased speed, and is often seen in faster 35-mm camera lenses. Eight examples of the type are shown in Figs. 17.21 through 17.28, all of relatively high speed. Figure 17.22 is an older design using SK16 glass; compare it with Figs. 17.21, 17.23, etc., which utilize higher-index glasses. Most of these designs maintain the two rear elements at a minimal thickness and closely spaced. Figure 17.25 uses a larger airspace; this can be helpful in controlling the sagittal oblique spherical. Figure 17.27 is an interesting lens, in that it is a *simplification* of an older design (Fig. 17.43) by Tronnier which had a cemented doublet as the thick rear component. (In the original unmodified versions, the prescriptions and glasses were identical, with the exception of the last component.)

17.3 The Seven-Element Biotar–Broken Contact Front Doublet

The added freedom gained by breaking the cemented contact in the front doublet has been utilized recently in many camera lenses. Figures 17.29, 17.30, and 17.31 show three high-speed designs which are modifications of the split-rear-singlet form described in the preceding section. Figure 17.32 is a slower version, and Fig. 17.33 is a special-purpose, high-resolution version for use at finite conjugates. Figure 17.34 is based on

(Text continues on page 340.)

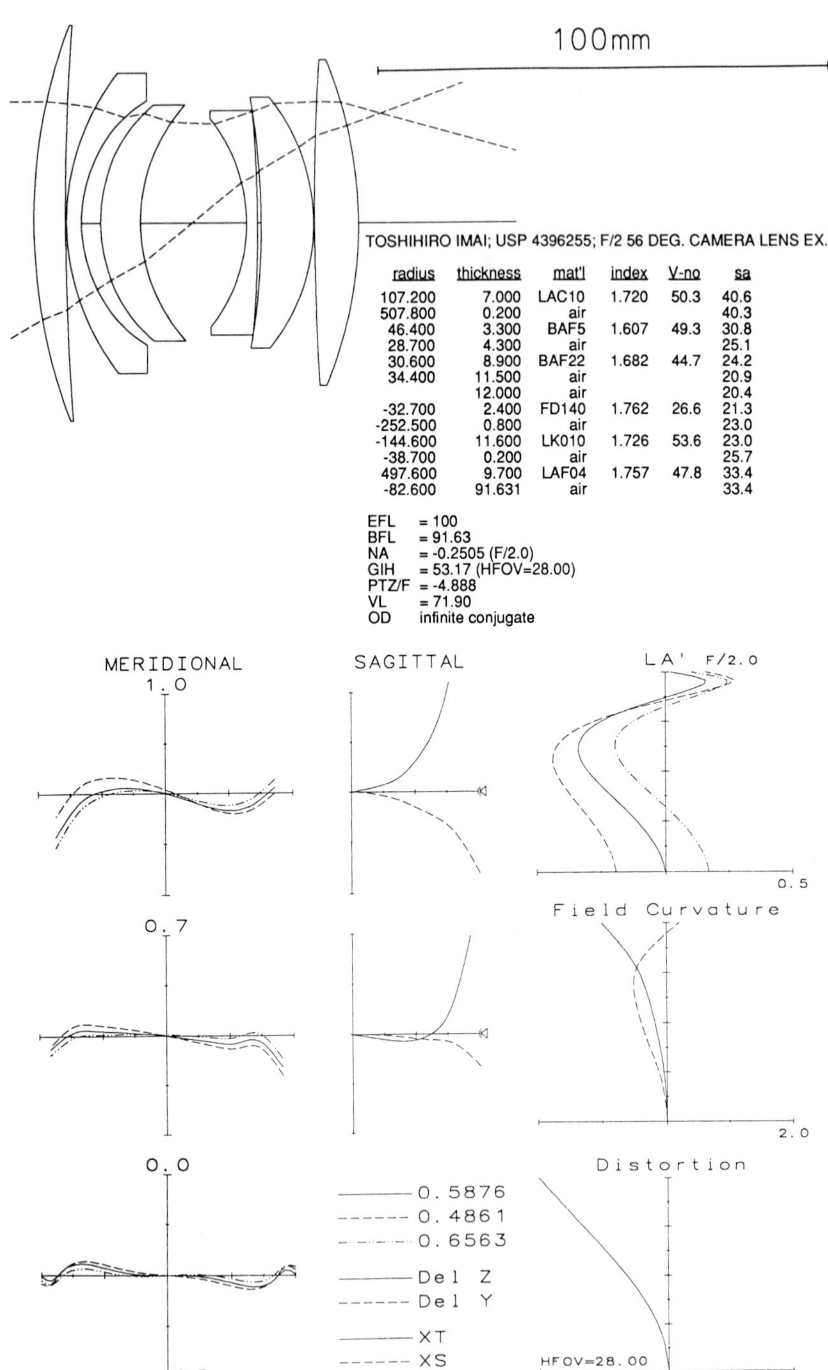

TOSHIHIRO IMAI; USP 4396255; F/2 56 DEG. CAMERA LENS EX. 1

radius	thickness	mat'l	index	V-no	sa
107.200	7.000	LAC10	1.720	50.3	40.6
507.800	0.200	air			40.3
46.400	3.300	BAF5	1.607	49.3	30.8
28.700	4.300	air			25.1
30.600	8.900	BAF22	1.682	44.7	24.2
34.400	11.500	air			20.9
	12.000	air			20.4
-32.700	2.400	FD140	1.762	26.6	21.3
-252.500	0.800	air			23.0
-144.600	11.600	LK010	1.726	53.6	23.0
-38.700	0.200	air			25.7
497.600	9.700	LAF04	1.757	47.8	33.4
-82.600	91.631	air			33.4

EFL = 100
BFL = 91.63
NA = -0.2505 (F/2.0)
GIH = 53.17 (HFOV=28.00)
PTZ/F = -4.888
VL = 71.90
OD infinite conjugate

Figure 17.15

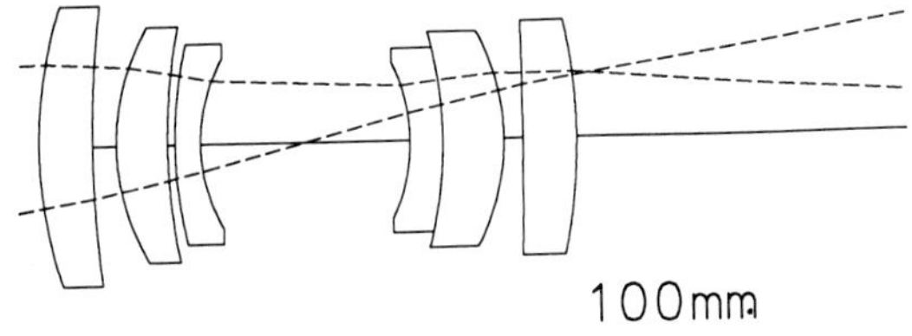

HERMAN LOWENTHAL; USP 3517979; F/5.6 18 DEG. NEAR UV LENS .25X

radius	thickness	mat'l	index	V-no	sa
40.920	5.700	BK7	1.531	48.6	13.9
102.980	2.530	air			13.0
21.730	5.460	BK7	1.531	48.6	11.6
54.400	0.820	air			10.4
39.510	2.720	LF1	1.598	28.5	9.9
16.010	11.230	air			8.4
	11.230	air			5.9
-16.840	3.490	LF1	1.598	28.5	7.8
-38.680	6.500	BK7	1.531	48.6	9.1
-21.380	2.010	air			10.7
-448.230	5.820	BK7	1.531	48.6	11.6
-51.080	89.441	air			12.2

EFL = 100.2
BFL = 89.44
NA = -0.0714 (F/7.0)
GIH = 19.68
PTZ/F = -6.61
VL = 57.51
OD = 458.71 (MAG = -0.249)

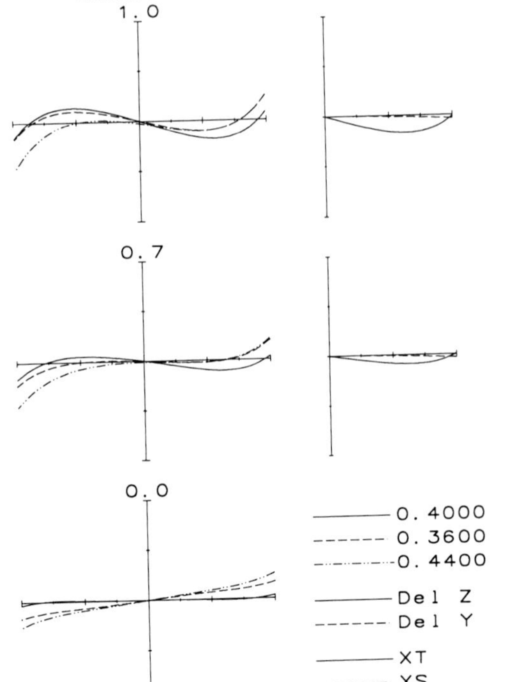

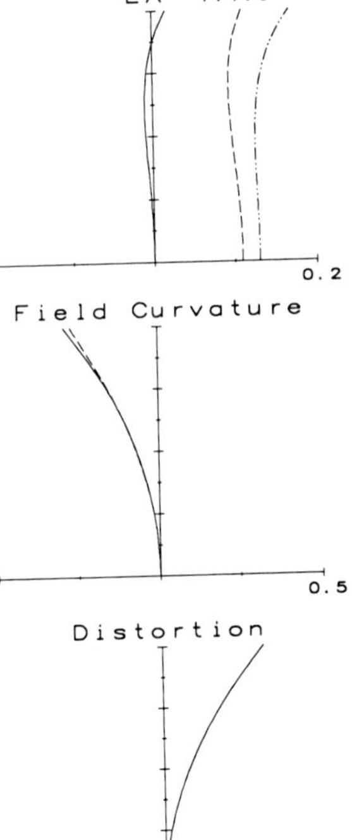

Figure 17.16

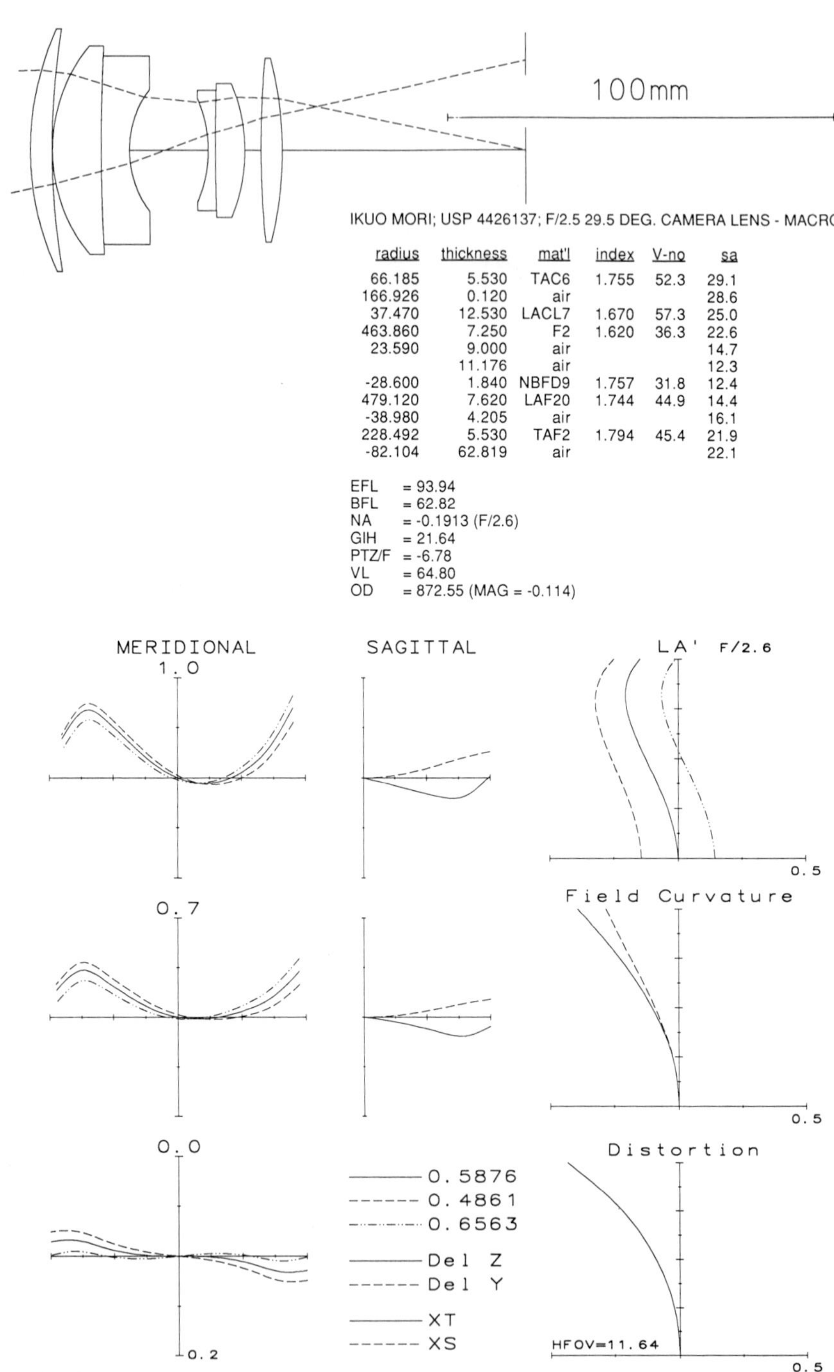

IKUO MORI; USP 4426137; F/2.5 29.5 DEG. CAMERA LENS - MACRO

radius	thickness	mat'l	index	V-no	sa
66.185	5.530	TAC6	1.755	52.3	29.1
166.926	0.120	air			28.6
37.470	12.530	LACL7	1.670	57.3	25.0
463.860	7.250	F2	1.620	36.3	22.6
23.590	9.000	air			14.7
	11.176	air			12.3
-28.600	1.840	NBFD9	1.757	31.8	12.4
479.120	7.620	LAF20	1.744	44.9	14.4
-38.980	4.205	air			16.1
228.492	5.530	TAF2	1.794	45.4	21.9
-82.104	62.819	air			22.1

EFL = 93.94
BFL = 62.82
NA = -0.1913 (F/2.6)
GIH = 21.64
PTZ/F = -6.78
VL = 64.80
OD = 872.55 (MAG = -0.114)

Figure 17.17

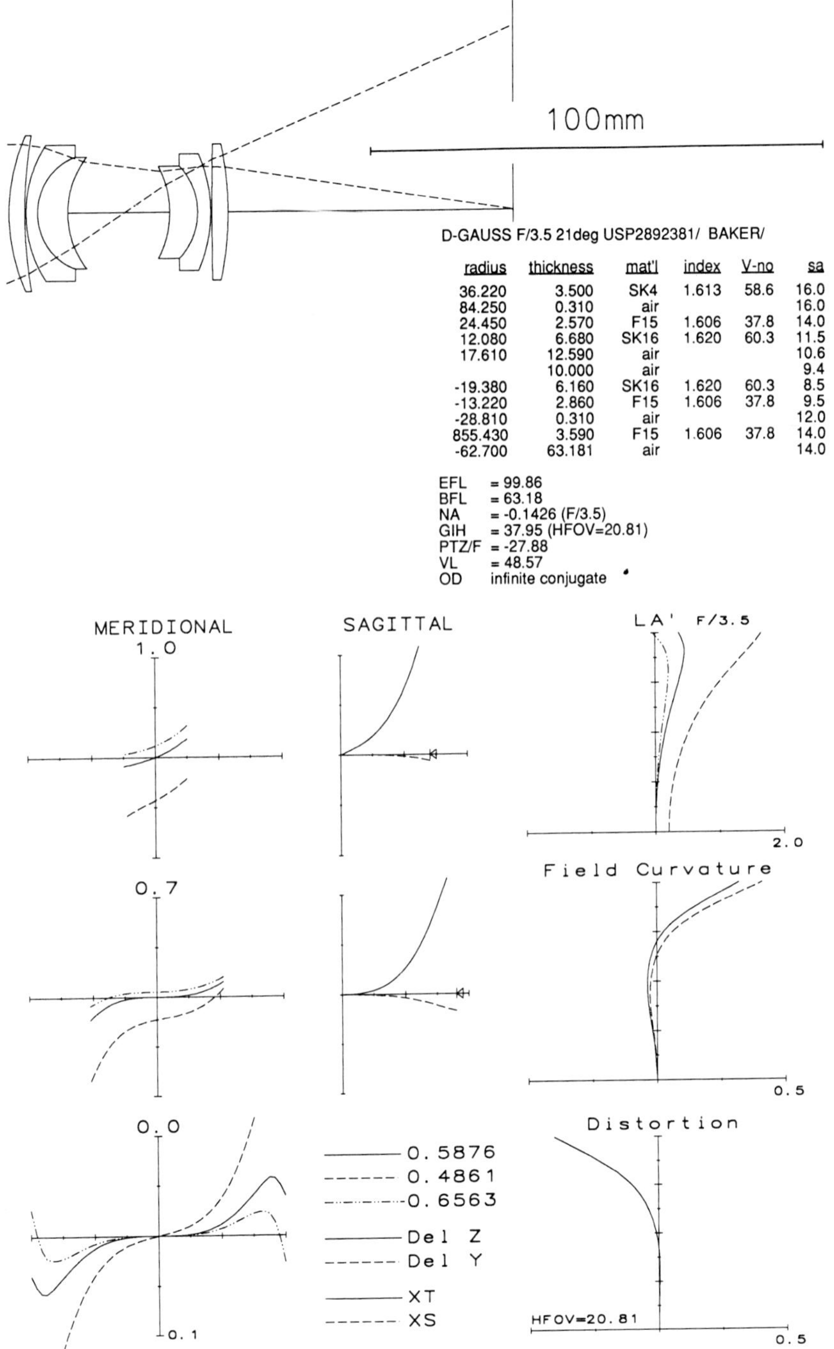

D-GAUSS F/3.5 21deg USP2892381/ BAKER/

radius	thickness	mat'l	index	V-no	sa
36.220	3.500	SK4	1.613	58.6	16.0
84.250	0.310	air			16.0
24.450	2.570	F15	1.606	37.8	14.0
12.080	6.680	SK16	1.620	60.3	11.5
17.610	12.590	air			10.6
	10.000	air			9.4
-19.380	6.160	SK16	1.620	60.3	8.5
-13.220	2.860	F15	1.606	37.8	9.5
-28.810	0.310	air			12.0
855.430	3.590	F15	1.606	37.8	14.0
-62.700	63.181	air			14.0

EFL = 99.86
BFL = 63.18
NA = -0.1426 (F/3.5)
GIH = 37.95 (HFOV=20.81)
PTZ/F = -27.88
VL = 48.57
OD infinite conjugate

Figure 17.18

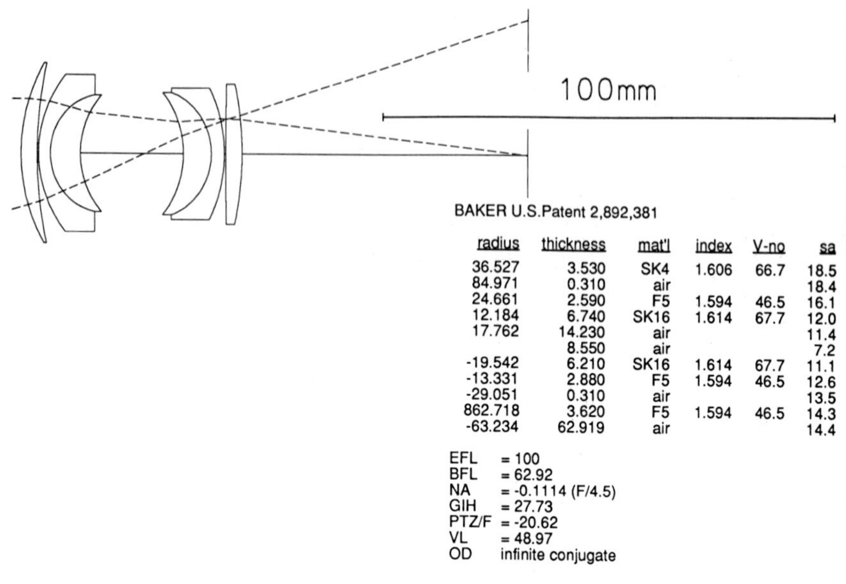

BAKER U.S.Patent 2,892,381

radius	thickness	mat'l	index	V-no	sa
36.527	3.530	SK4	1.606	66.7	18.5
84.971	0.310	air			18.4
24.661	2.590	F5	1.594	46.5	16.1
12.184	6.740	SK16	1.614	67.7	12.0
17.762	14.230	air			11.4
	8.550	air			7.2
-19.542	6.210	SK16	1.614	67.7	11.1
-13.331	2.880	F5	1.594	46.5	12.6
-29.051	0.310	air			13.5
862.718	3.620	F5	1.594	46.5	14.3
-63.234	62.919	air			14.4

EFL = 100
BFL = 62.92
NA = -0.1114 (F/4.5)
GIH = 27.73
PTZ/F = -20.62
VL = 48.97
OD infinite conjugate

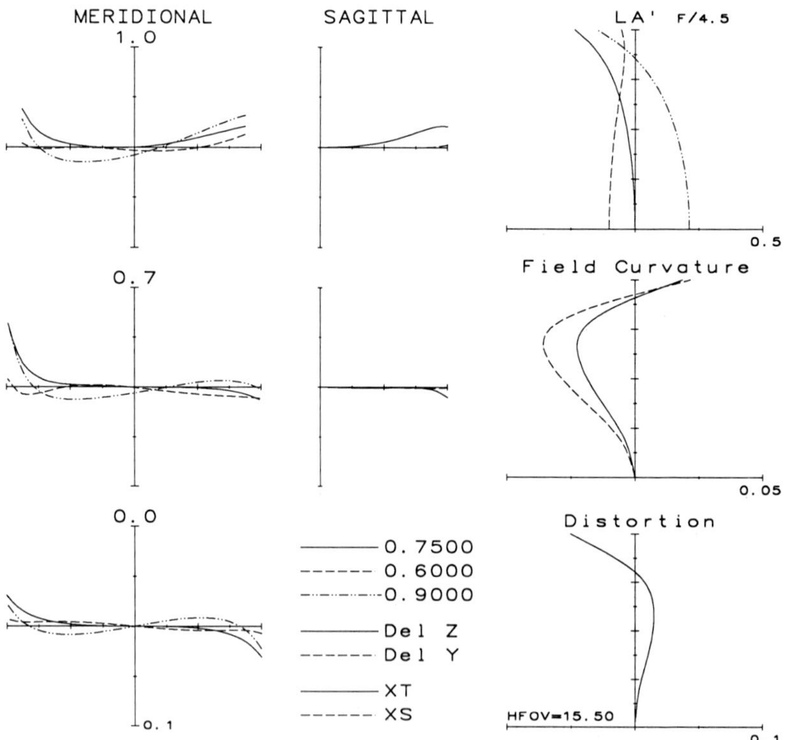

Figure 17.19

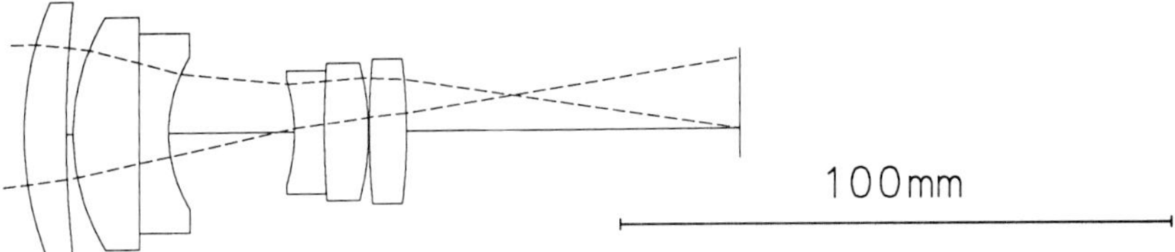

CCD LENS FROM MILLS COLLEGE LENS

radius	thickness	mat'l	index	V-no	sa
55.181	7.560	LAK9	1.682	910.4	22.3
191.267	1.247	air			21.4
35.668	12.150	LAK9	1.682	910.4	19.6
-11014.712	5.400	SF1	1.702	574.0	16.8
23.698	19.790	air			12.9
	3.213	air			8.6
-28.906	5.400	SF1	1.702	574.0	9.0
147.529	8.100	LAK9	1.682	910.4	10.4
-42.527	0.192	air			11.6
138.892	6.750	LAK9	1.682	910.4	12.0
-96.115	60.435	air			12.2
	0.000	air			13.5

EFL = 100.1
BFL = 60.44
NA = -0.1427 (F/3.5)
GIH = 11.90
PTZ/F = -6.457
VL = 69.80
OD = 939.24 (MAG = -0.113)

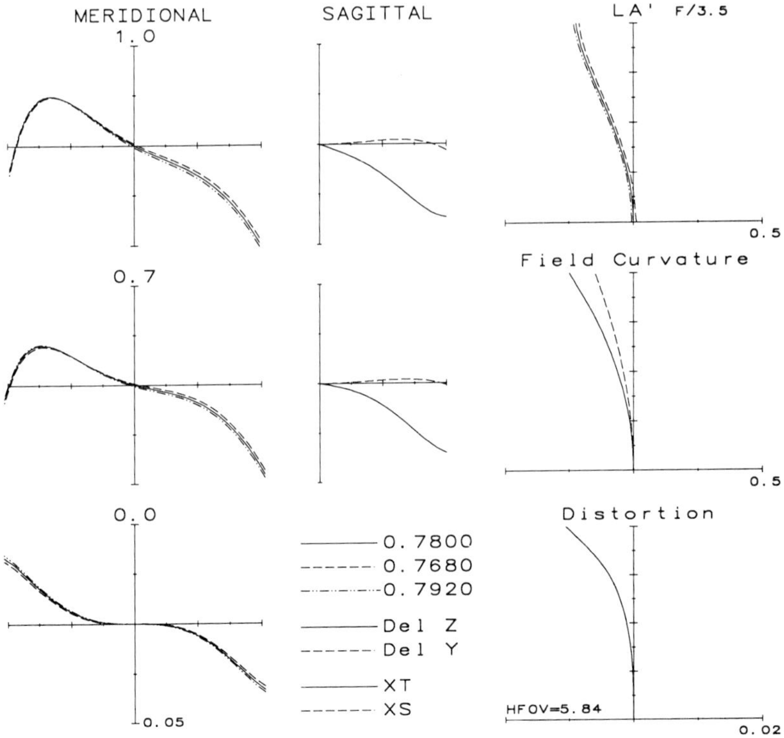

MERIDIONAL SAGITTAL LA' F/3.5
1.0
0.5
Field Curvature
0.7
0.5
0.0
Distortion
0.7800
0.7680
0.7920
Del Z
Del Y
XT
XS
HFOV=5.84
0.05
0.02

Figure 17.20

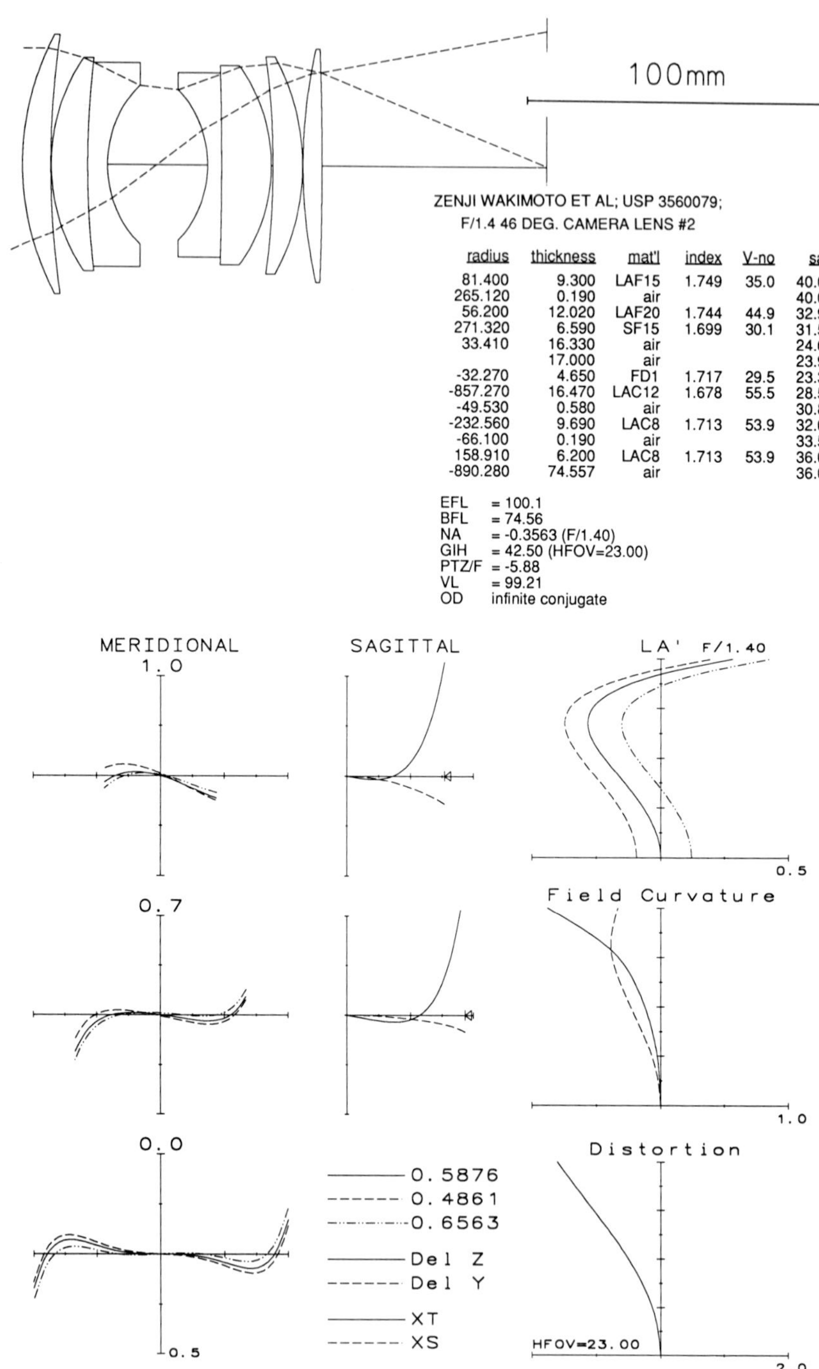

ZENJI WAKIMOTO ET AL; USP 3560079;
F/1.4 46 DEG. CAMERA LENS #2

radius	thickness	mat'l	index	V-no	sa
81.400	9.300	LAF15	1.749	35.0	40.0
265.120	0.190	air			40.0
56.200	12.020	LAF20	1.744	44.9	32.9
271.320	6.590	SF15	1.699	30.1	31.5
33.410	16.330	air			24.6
	17.000	air			23.9
-32.270	4.650	FD1	1.717	29.5	23.3
-857.270	16.470	LAC12	1.678	55.5	28.5
-49.530	0.580	air			30.8
-232.560	9.690	LAC8	1.713	53.9	32.0
-66.100	0.190	air			33.5
158.910	6.200	LAC8	1.713	53.9	36.0
-890.280	74.557	air			36.0

EFL = 100.1
BFL = 74.56
NA = -0.3563 (F/1.40)
GIH = 42.50 (HFOV=23.00)
PTZ/F = -5.88
VL = 99.21
OD infinite conjugate

Figure 17.21

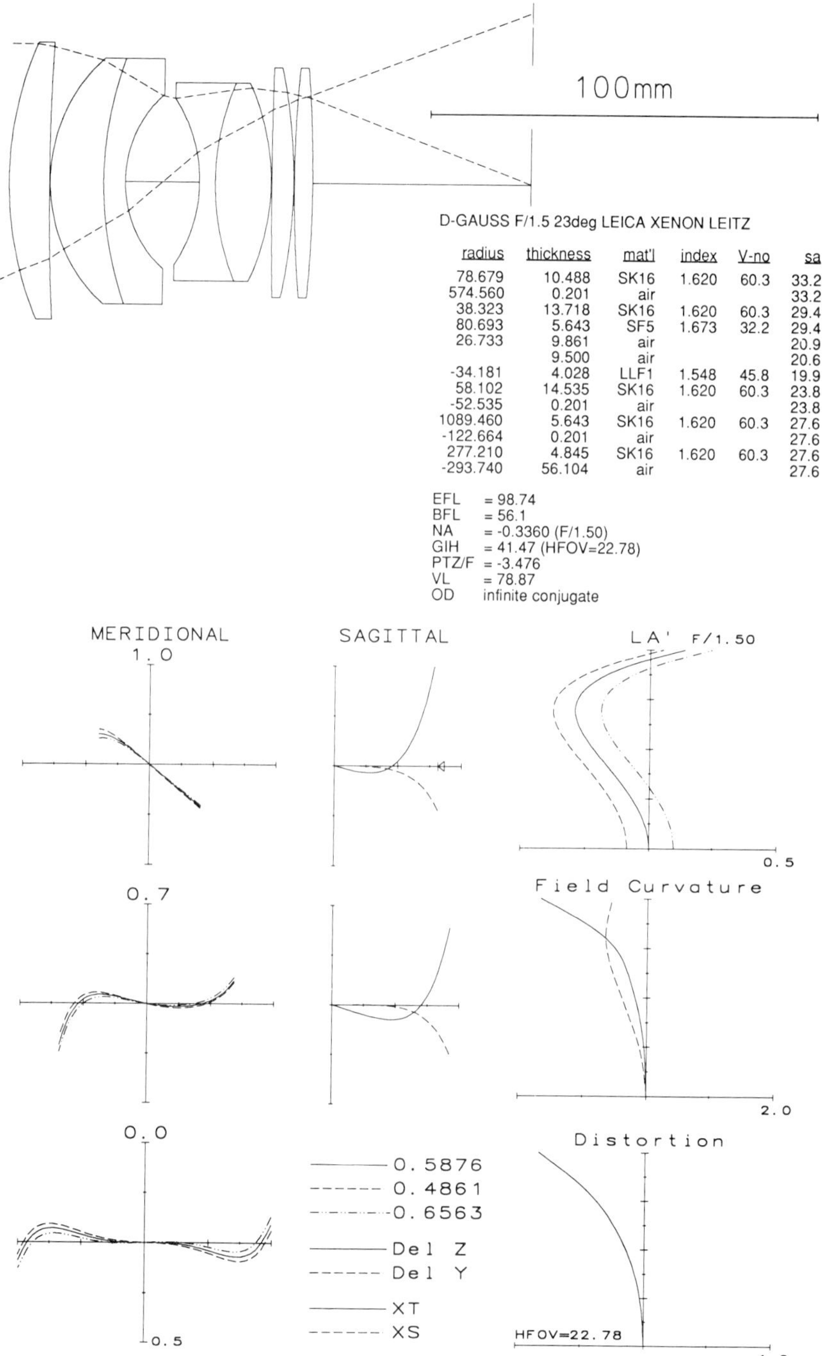

D-GAUSS F/1.5 23deg LEICA XENON LEITZ

radius	thickness	mat'l	index	V-no	sa
78.679	10.488	SK16	1.620	60.3	33.2
574.560	0.201	air			33.2
38.323	13.718	SK16	1.620	60.3	29.4
80.693	5.643	SF5	1.673	32.2	29.4
26.733	9.861	air			20.9
	9.500	air			20.6
-34.181	4.028	LLF1	1.548	45.8	19.9
58.102	14.535	SK16	1.620	60.3	23.8
-52.535	0.201	air			23.8
1089.460	5.643	SK16	1.620	60.3	27.6
-122.664	0.201	air			27.6
277.210	4.845	SK16	1.620	60.3	27.6
-293.740	56.104	air			27.6

EFL = 98.74
BFL = 56.1
NA = -0.3360 (F/1.50)
GIH = 41.47 (HFOV=22.78)
PTZ/F = -3.476
VL = 78.87
OD infinite conjugate

Figure 17.22

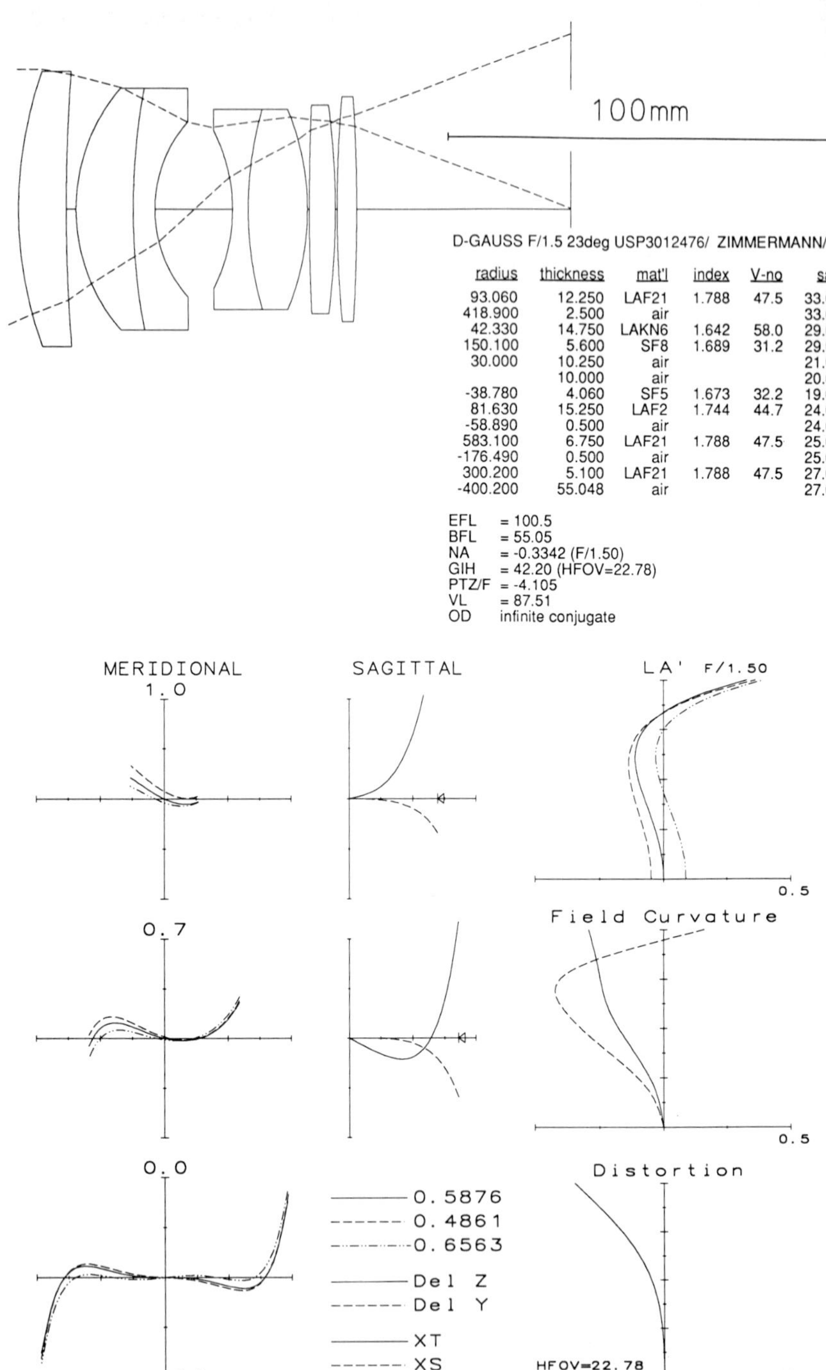

D-GAUSS F/1.5 23deg USP3012476/ ZIMMERMANN/

radius	thickness	mat'l	index	V-no	sa
93.060	12.250	LAF21	1.788	47.5	33.0
418.900	2.500	air			33.0
42.330	14.750	LAKN6	1.642	58.0	29.0
150.100	5.600	SF8	1.689	31.2	29.0
30.000	10.250	air			21.0
	10.000	air			20.6
-38.780	4.060	SF5	1.673	32.2	19.0
81.630	15.250	LAF2	1.744	44.7	24.0
-58.890	0.500	air			24.0
583.100	6.750	LAF21	1.788	47.5	25.0
-176.490	0.500	air			25.0
300.200	5.100	LAF21	1.788	47.5	27.0
-400.200	55.048	air			27.0

EFL = 100.5
BFL = 55.05
NA = -0.3342 (F/1.50)
GIH = 42.20 (HFOV=22.78)
PTZ/F = -4.105
VL = 87.51
OD infinite conjugate

Figure 17.23

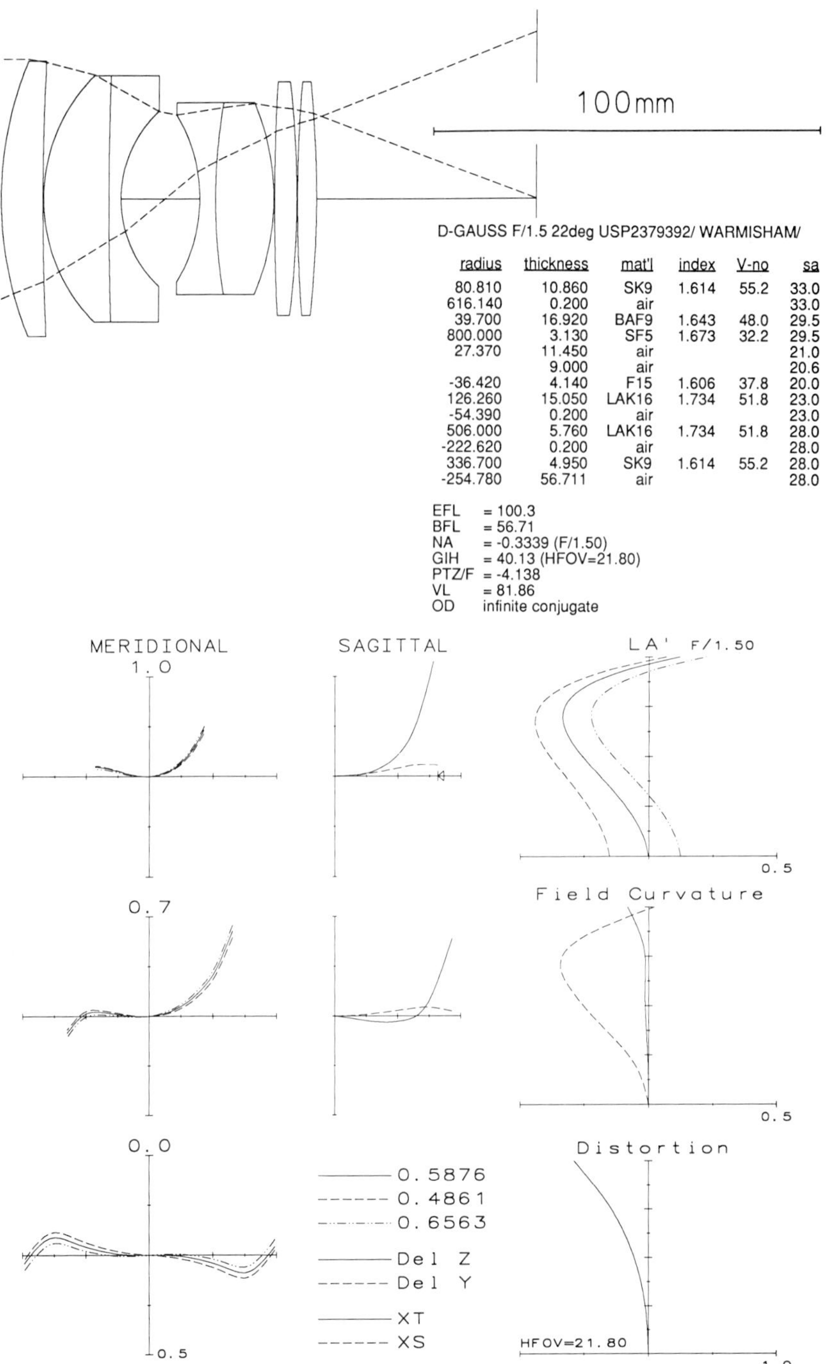

D-GAUSS F/1.5 22deg USP2379392/ WARMISHAM/

radius	thickness	mat'l	index	V-no	sa
80.810	10.860	SK9	1.614	55.2	33.0
616.140	0.200	air			33.0
39.700	16.920	BAF9	1.643	48.0	29.5
800.000	3.130	SF5	1.673	32.2	29.5
27.370	11.450	air			21.0
	9.000	air			20.6
-36.420	4.140	F15	1.606	37.8	20.0
126.260	15.050	LAK16	1.734	51.8	23.0
-54.390	0.200	air			23.0
506.000	5.760	LAK16	1.734	51.8	28.0
-222.620	0.200	air			28.0
336.700	4.950	SK9	1.614	55.2	28.0
-254.780	56.711	air			28.0

EFL = 100.3
BFL = 56.71
NA = -0.3339 (F/1.50)
GIH = 40.13 (HFOV=21.80)
PTZ/F = -4.138
VL = 81.86
OD infinite conjugate

Figure 17.24

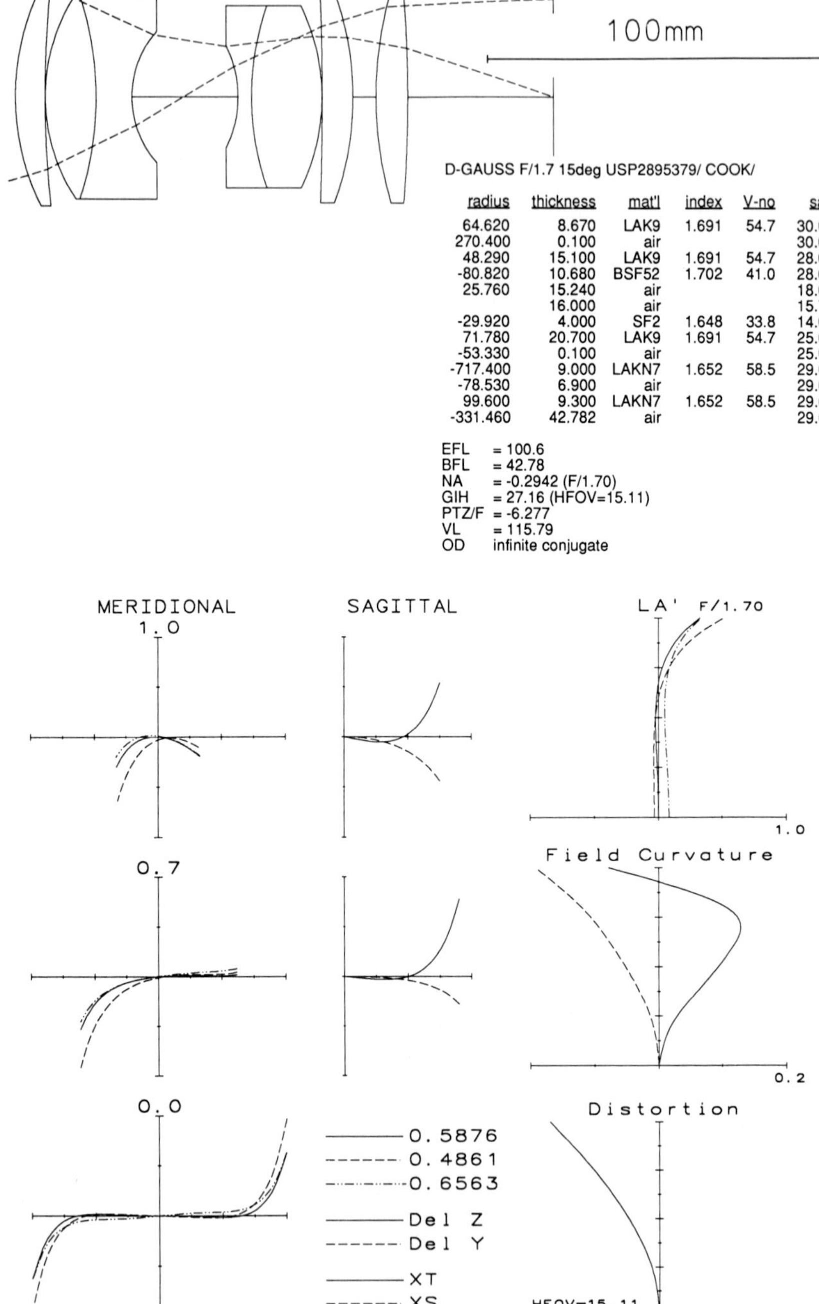

D-GAUSS F/1.7 15deg USP2895379/ COOK/

radius	thickness	mat'l	index	V-no	sa
64.620	8.670	LAK9	1.691	54.7	30.0
270.400	0.100	air			30.0
48.290	15.100	LAK9	1.691	54.7	28.0
-80.820	10.680	BSF52	1.702	41.0	28.0
25.760	15.240	air			18.0
	16.000	air			15.7
-29.920	4.000	SF2	1.648	33.8	14.0
71.780	20.700	LAK9	1.691	54.7	25.0
-53.330	0.100	air			25.0
-717.400	9.000	LAKN7	1.652	58.5	29.0
-78.530	6.900	air			29.0
99.600	9.300	LAKN7	1.652	58.5	29.0
-331.460	42.782	air			29.0

EFL = 100.6
BFL = 42.78
NA = -0.2942 (F/1.70)
GIH = 27.16 (HFOV=15.11)
PTZ/F = -6.277
VL = 115.79
OD infinite conjugate

Figure 17.25

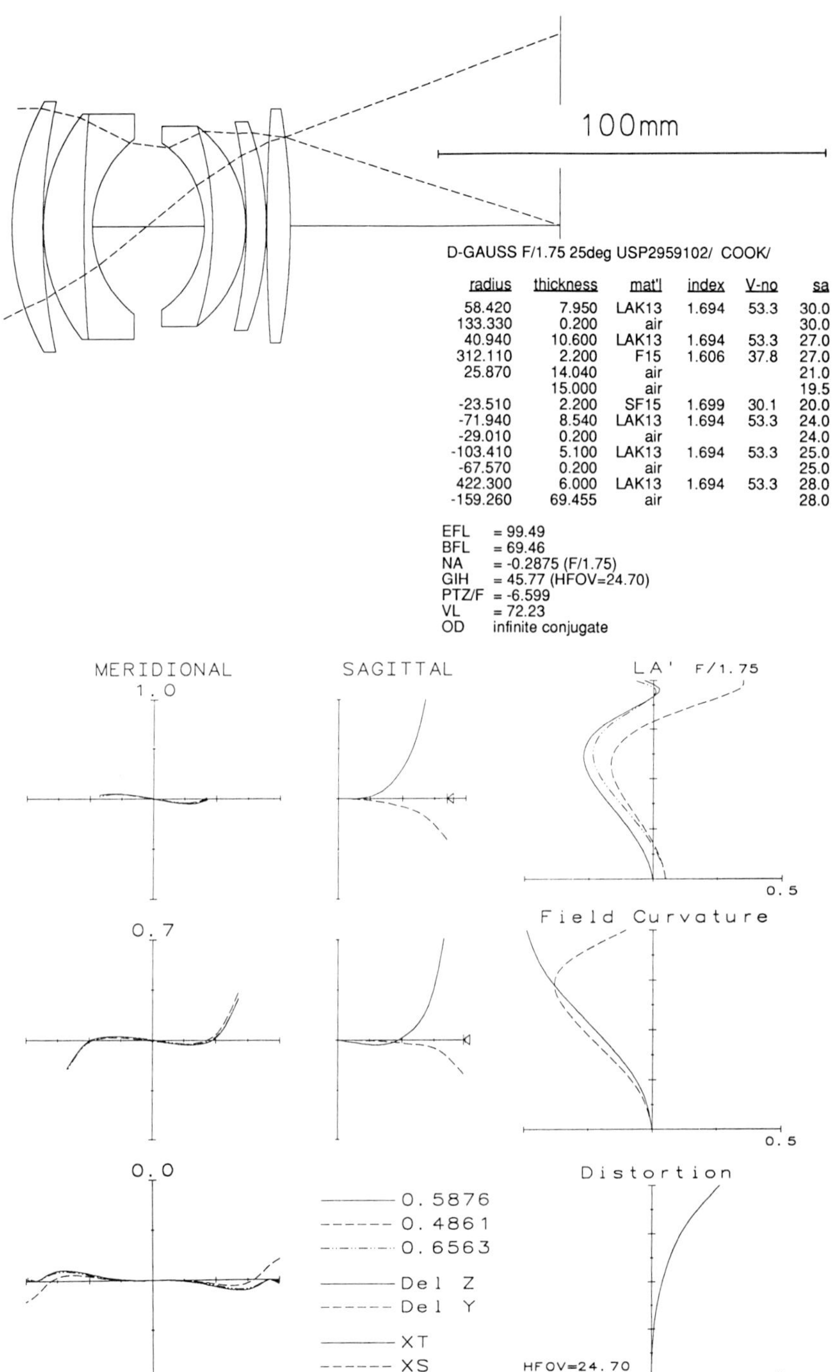

D-GAUSS F/1.75 25deg USP2959102/ COOK/

radius	thickness	mat'l	index	V-no	sa
58.420	7.950	LAK13	1.694	53.3	30.0
133.330	0.200	air			30.0
40.940	10.600	LAK13	1.694	53.3	27.0
312.110	2.200	F15	1.606	37.8	27.0
25.870	14.040	air			21.0
	15.000	air			19.5
-23.510	2.200	SF15	1.699	30.1	20.0
-71.940	8.540	LAK13	1.694	53.3	24.0
-29.010	0.200	air			24.0
-103.410	5.100	LAK13	1.694	53.3	25.0
-67.570	0.200	air			25.0
422.300	6.000	LAK13	1.694	53.3	28.0
-159.260	69.455	air			28.0

EFL = 99.49
BFL = 69.46
NA = -0.2875 (F/1.75)
GIH = 45.77 (HFOV=24.70)
PTZ/F = -6.599
VL = 72.23
OD infinite conjugate

Figure 17.26

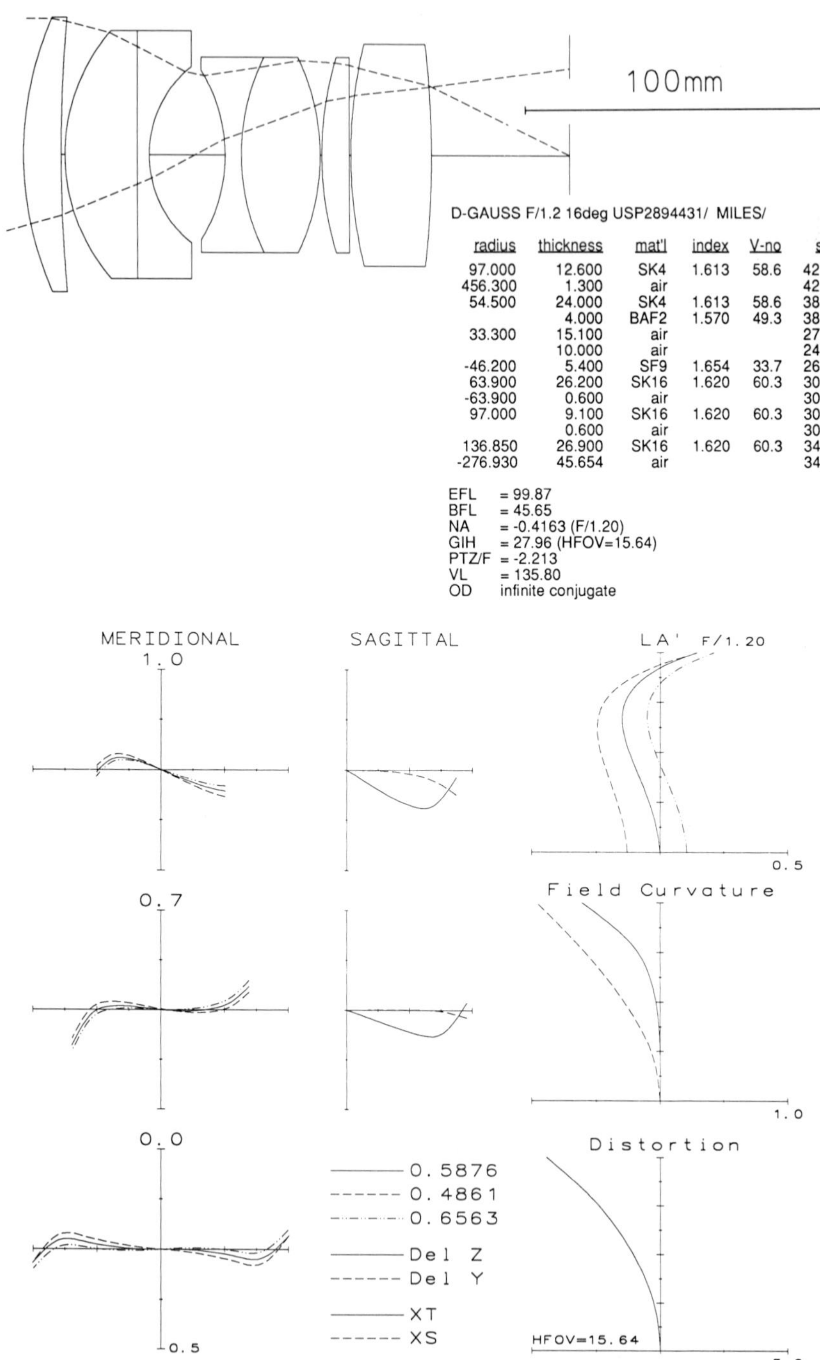

D-GAUSS F/1.2 16deg USP2894431/ MILES/

radius	thickness	mat'l	index	V-no	sa
97.000	12.600	SK4	1.613	58.6	42.0
456.300	1.300	air			42.0
54.500	24.000	SK4	1.613	58.6	38.0
	4.000	BAF2	1.570	49.3	38.0
33.300	15.100	air			27.0
	10.000	air			24.9
-46.200	5.400	SF9	1.654	33.7	26.0
63.900	26.200	SK16	1.620	60.3	30.0
-63.900	0.600	air			30.0
97.000	9.100	SK16	1.620	60.3	30.0
	0.600	air			30.0
136.850	26.900	SK16	1.620	60.3	34.0
-276.930	45.654	air			34.0

EFL = 99.87
BFL = 45.65
NA = -0.4163 (F/1.20)
GIH = 27.96 (HFOV=15.64)
PTZ/F = -2.213
VL = 135.80
OD infinite conjugate

Figure 17.27

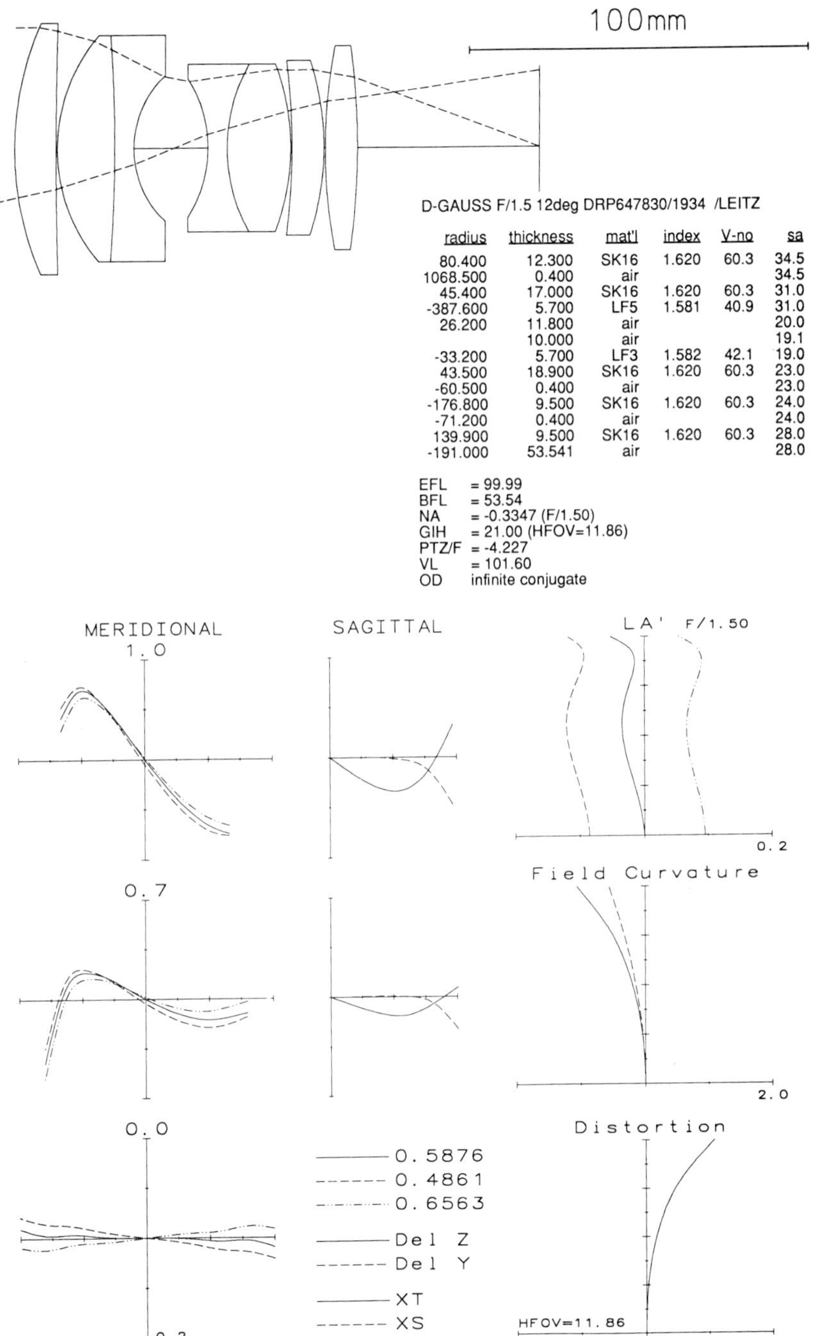

D-GAUSS F/1.5 12deg DRP647830/1934 /LEITZ

radius	thickness	mat'l	index	V-no	sa
80.400	12.300	SK16	1.620	60.3	34.5
1068.500	0.400	air			34.5
45.400	17.000	SK16	1.620	60.3	31.0
-387.600	5.700	LF5	1.581	40.9	31.0
26.200	11.800	air			20.0
	10.000	air			19.1
-33.200	5.700	LF3	1.582	42.1	19.0
43.500	18.900	SK16	1.620	60.3	23.0
-60.500	0.400	air			23.0
-176.800	9.500	SK16	1.620	60.3	24.0
-71.200	0.400	air			24.0
139.900	9.500	SK16	1.620	60.3	28.0
-191.000	53.541	air			28.0

EFL = 99.99
BFL = 53.54
NA = -0.3347 (F/1.50)
GIH = 21.00 (HFOV=11.86)
PTZ/F = -4.227
VL = 101.60
OD infinite conjugate

Figure 17.28

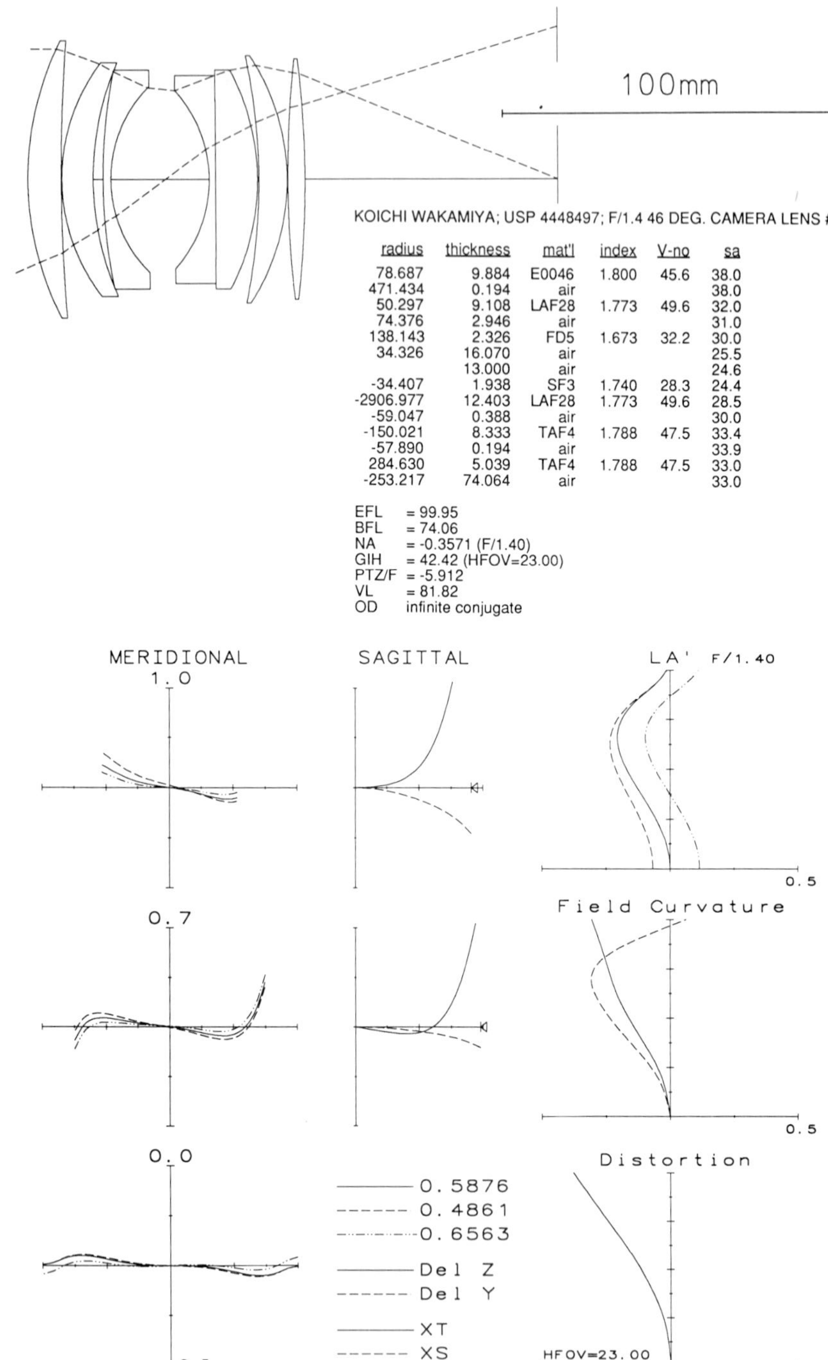

KOICHI WAKAMIYA; USP 4448497; F/1.4 46 DEG. CAMERA LENS #1

radius	thickness	mat'l	index	V-no	sa
78.687	9.884	E0046	1.800	45.6	38.0
471.434	0.194	air			38.0
50.297	9.108	LAF28	1.773	49.6	32.0
74.376	2.946	air			31.0
138.143	2.326	FD5	1.673	32.2	30.0
34.326	16.070	air			25.5
	13.000	air			24.6
-34.407	1.938	SF3	1.740	28.3	24.4
-2906.977	12.403	LAF28	1.773	49.6	28.5
-59.047	0.388	air			30.0
-150.021	8.333	TAF4	1.788	47.5	33.4
-57.890	0.194	air			33.9
284.630	5.039	TAF4	1.788	47.5	33.0
-253.217	74.064	air			33.0

EFL = 99.95
BFL = 74.06
NA = -0.3571 (F/1.40)
GIH = 42.42 (HFOV=23.00)
PTZ/F = -5.912
VL = 81.82
OD infinite conjugate

Figure 17.29

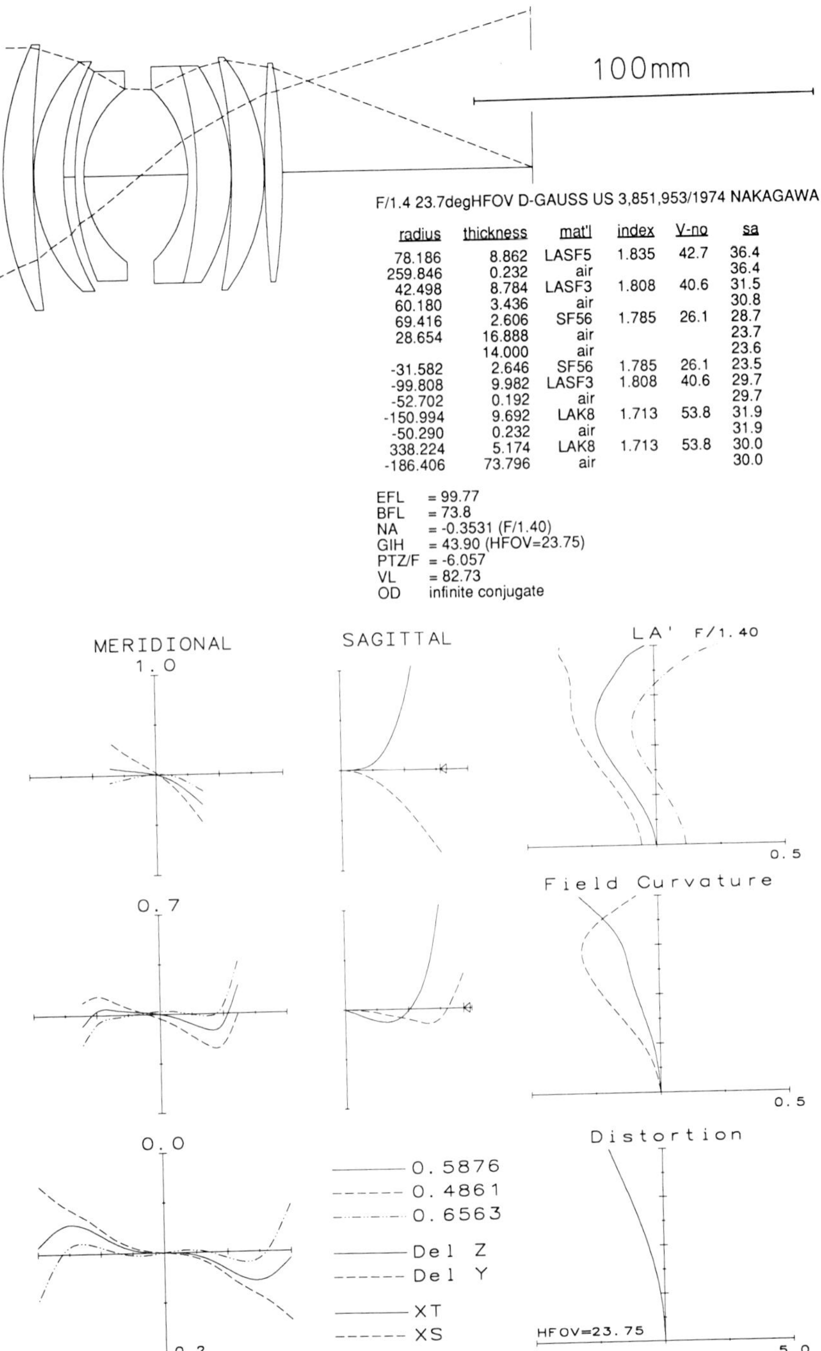

F/1.4 23.7degHFOV D-GAUSS US 3,851,953/1974 NAKAGAWA

radius	thickness	mat'l	index	V-no	sa
78.186	8.862	LASF5	1.835	42.7	36.4
259.846	0.232	air			36.4
42.498	8.784	LASF3	1.808	40.6	31.5
60.180	3.436	air			30.8
69.416	2.606	SF56	1.785	26.1	28.7
28.654	16.888	air			23.7
	14.000	air			23.6
-31.582	2.646	SF56	1.785	26.1	23.5
-99.808	9.982	LASF3	1.808	40.6	29.7
-52.702	0.192	air			29.7
-150.994	9.692	LAK8	1.713	53.8	31.9
-50.290	0.232	air			31.9
338.224	5.174	LAK8	1.713	53.8	30.0
-186.406	73.796	air			30.0

EFL = 99.77
BFL = 73.8
NA = -0.3531 (F/1.40)
GIH = 43.90 (HFOV=23.75)
PTZ/F = -6.057
VL = 82.73
OD infinite conjugate

Figure 17.30

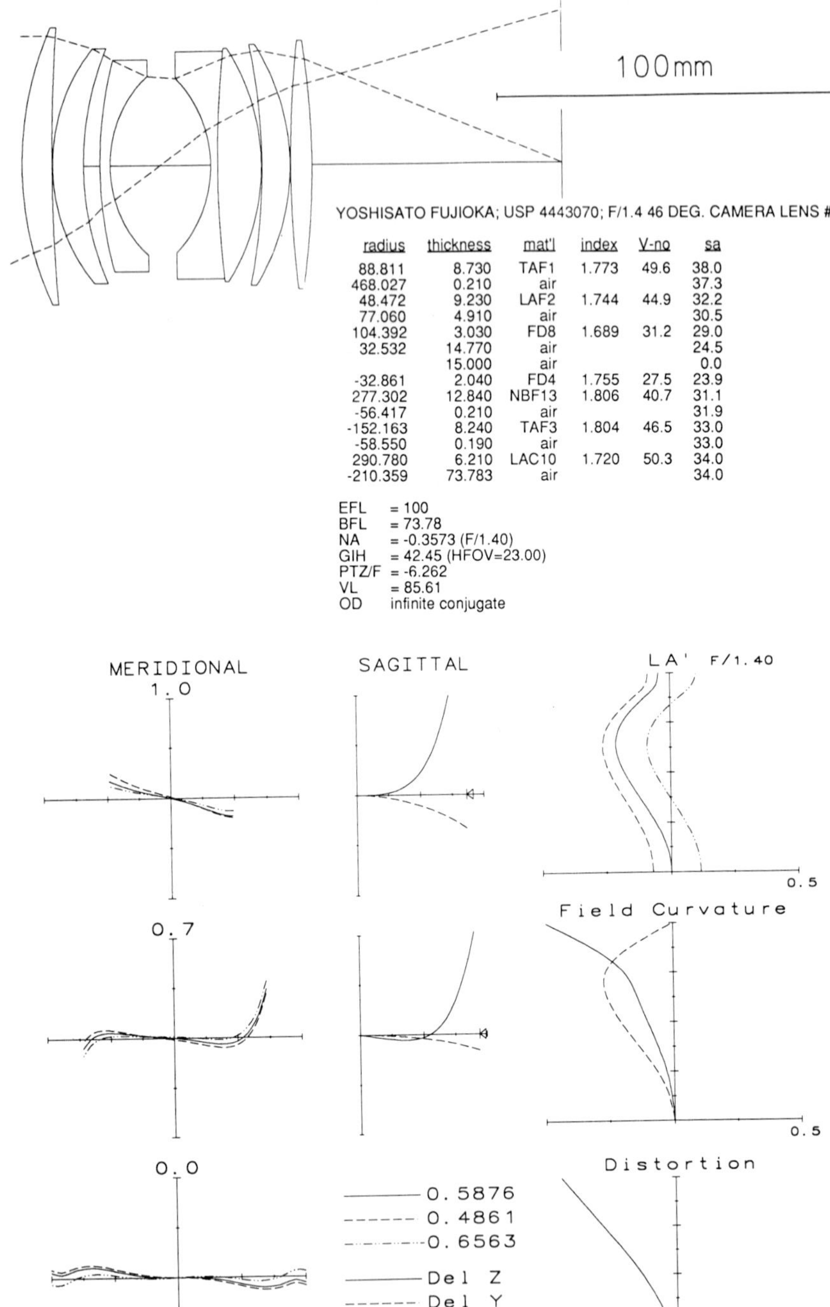

YOSHISATO FUJIOKA; USP 4443070; F/1.4 46 DEG. CAMERA LENS #1

radius	thickness	mat'l	index	V-no	sa
88.811	8.730	TAF1	1.773	49.6	38.0
468.027	0.210	air			37.3
48.472	9.230	LAF2	1.744	44.9	32.2
77.060	4.910	air			30.5
104.392	3.030	FD8	1.689	31.2	29.0
32.532	14.770	air			24.5
	15.000	air			0.0
-32.861	2.040	FD4	1.755	27.5	23.9
277.302	12.840	NBF13	1.806	40.7	31.1
-56.417	0.210	air			31.9
-152.163	8.240	TAF3	1.804	46.5	33.0
-58.550	0.190	air			33.0
290.780	6.210	LAC10	1.720	50.3	34.0
-210.359	73.783	air			34.0

EFL = 100
BFL = 73.78
NA = -0.3573 (F/1.40)
GIH = 42.45 (HFOV=23.00)
PTZ/F = -6.262
VL = 85.61
OD infinite conjugate

Figure 17.31

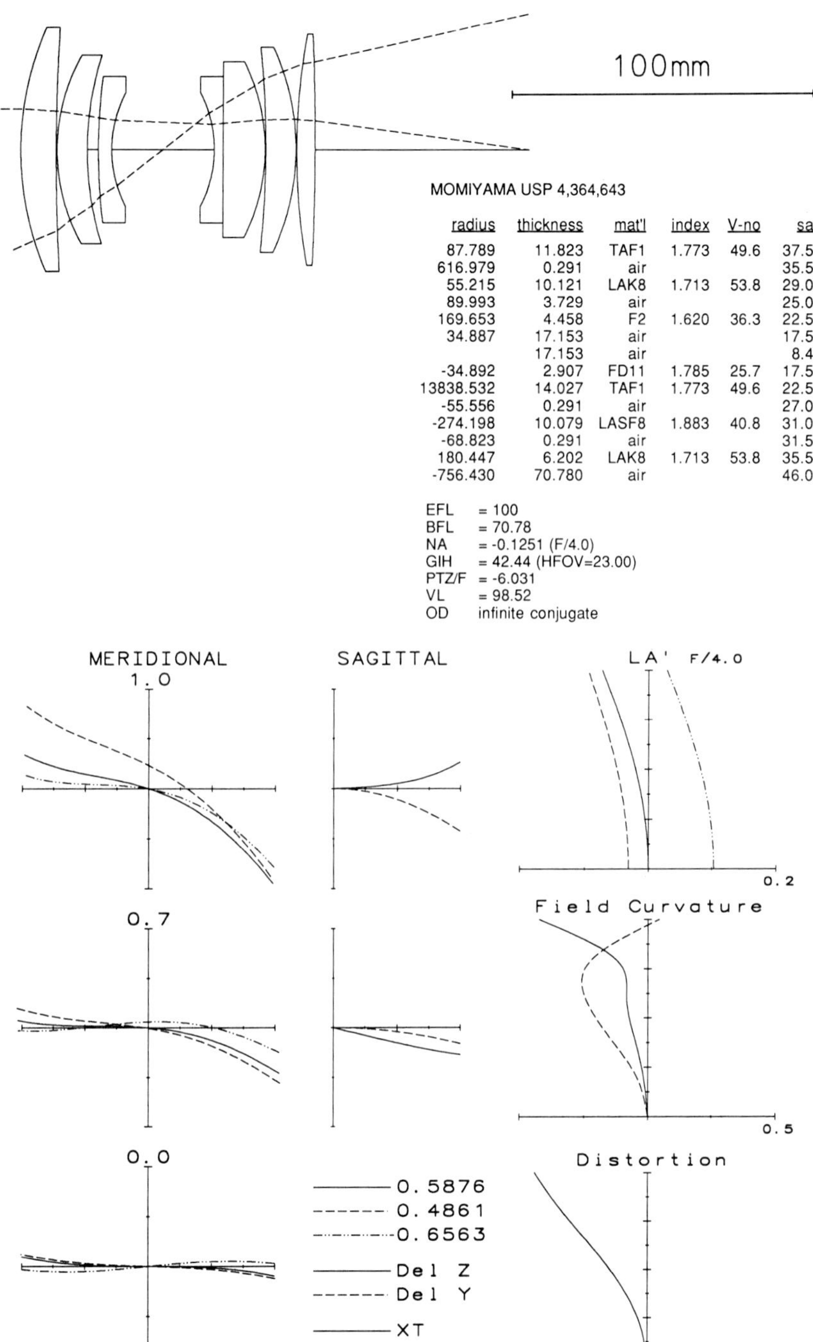

MOMIYAMA USP 4,364,643

radius	thickness	mat'l	index	V-no	sa
87.789	11.823	TAF1	1.773	49.6	37.5
616.979	0.291	air			35.5
55.215	10.121	LAK8	1.713	53.8	29.0
89.993	3.729	air			25.0
169.653	4.458	F2	1.620	36.3	22.5
34.887	17.153	air			17.5
	17.153	air			8.4
-34.892	2.907	FD11	1.785	25.7	17.5
13838.532	14.027	TAF1	1.773	49.6	22.5
-55.556	0.291	air			27.0
-274.198	10.079	LASF8	1.883	40.8	31.0
-68.823	0.291	air			31.5
180.447	6.202	LAK8	1.713	53.8	35.5
-756.430	70.780	air			46.0

EFL = 100
BFL = 70.78
NA = -0.1251 (F/4.0)
GIH = 42.44 (HFOV=23.00)
PTZ/F = -6.031
VL = 98.52
OD infinite conjugate

Figure 17.32

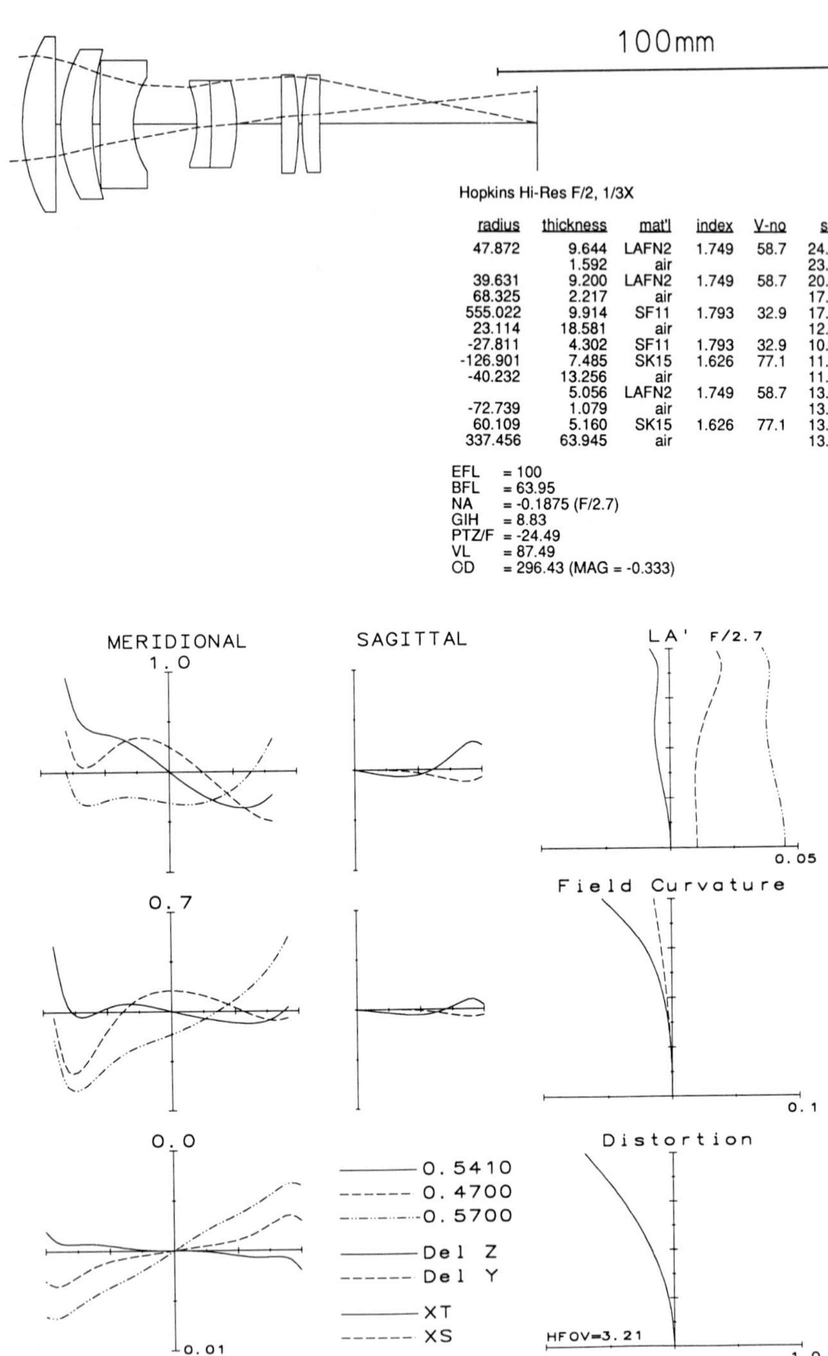

Hopkins Hi-Res F/2, 1/3X

radius	thickness	mat'l	index	V-no	sa
47.872	9.644	LAFN2	1.749	58.7	24.0
	1.592	air			23.4
39.631	9.200	LAFN2	1.749	58.7	20.5
68.325	2.217	air			17.7
555.022	9.914	SF11	1.793	32.9	17.5
23.114	18.581	air			12.9
-27.811	4.302	SF11	1.793	32.9	10.8
-126.901	7.485	SK15	1.626	77.1	11.9
-40.232	13.256	air			11.9
	5.056	LAFN2	1.749	58.7	13.2
-72.739	1.079	air			13.5
60.109	5.160	SK15	1.626	77.1	13.4
337.456	63.945	air			13.0

EFL = 100
BFL = 63.95
NA = -0.1875 (F/2.7)
GIH = 8.83
PTZ/F = -24.49
VL = 87.49
OD = 296.43 (MAG = -0.333)

Figure 17.33

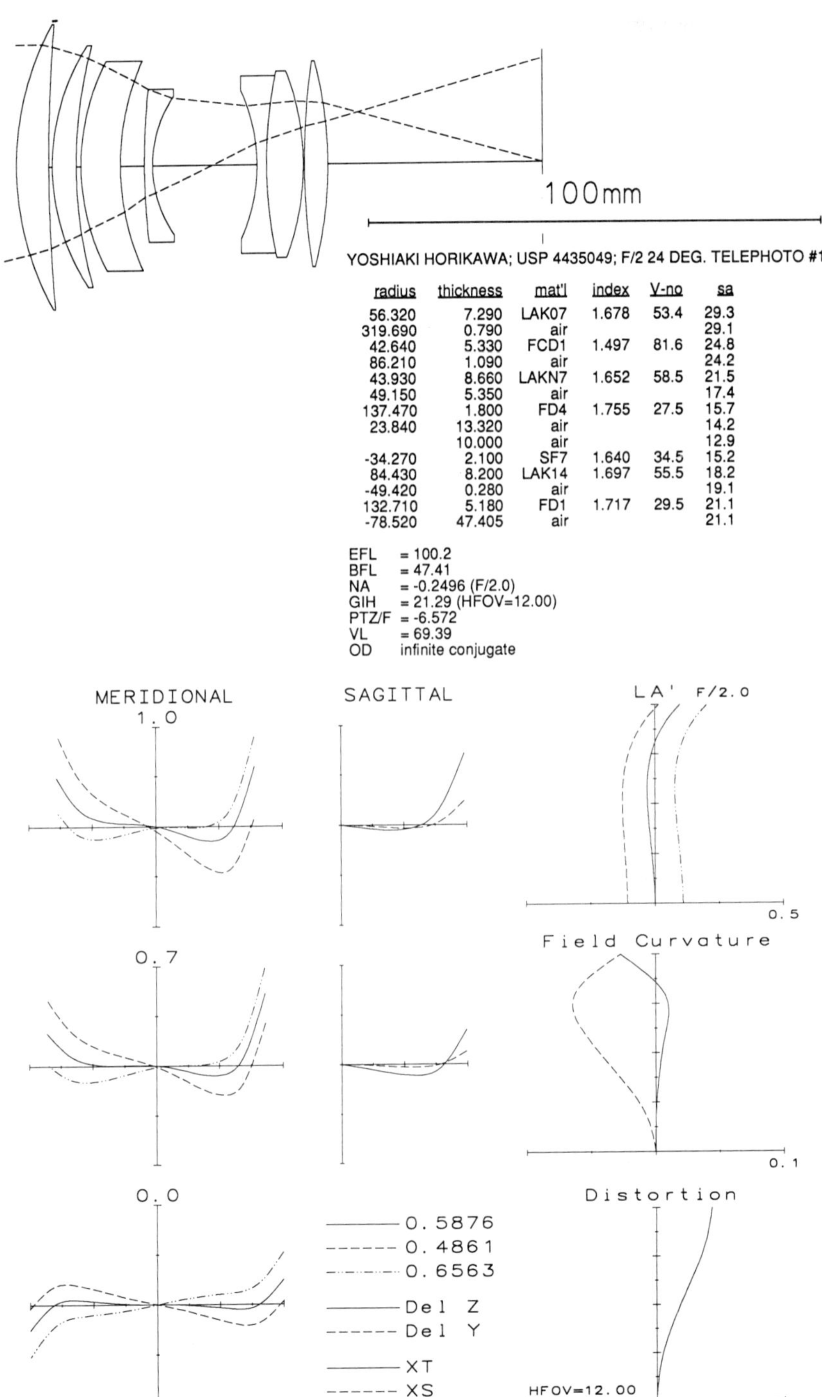

YOSHIAKI HORIKAWA; USP 4435049; F/2 24 DEG. TELEPHOTO #1

radius	thickness	mat'l	index	V-no	sa
56.320	7.290	LAK07	1.678	53.4	29.3
319.690	0.790	air			29.1
42.640	5.330	FCD1	1.497	81.6	24.8
86.210	1.090	air			24.2
43.930	8.660	LAKN7	1.652	58.5	21.5
49.150	5.350	air			17.4
137.470	1.800	FD4	1.755	27.5	15.7
23.840	13.320	air			14.2
	10.000	air			12.9
-34.270	2.100	SF7	1.640	34.5	15.2
84.430	8.200	LAK14	1.697	55.5	18.2
-49.420	0.280	air			19.1
132.710	5.180	FD1	1.717	29.5	21.1
-78.520	47.405	air			21.1

EFL = 100.2
BFL = 47.41
NA = -0.2496 (F/2.0)
GIH = 21.29 (HFOV=12.00)
PTZ/F = -6.572
VL = 69.39
OD infinite conjugate

Figure 17.34

the split-*front*-singlet form and incorporates an unusual partial dispersion glass to make possible an improved secondary spectrum (although, as is often the case with published designs, the sample design is not well-corrected, showing undercorrected axial chromatic).

17.4 The Seven-Element Biotar—One Compounded Outer Element

Figures 17.35 and 17.36 show *f*/1.4 lenses of modest angular coverage, one with a compounded front, the other with a compounded rear. Another pair at a speed of *f*/2 is shown in Figs. 17.37 and 17.38. Note that, in all cases (except Fig. 17.35), the doublet is a new achromat as described in Sec. 3.5. Although infrequently encountered, both outer singlets may be compounded. Figures 17.39 and 17.40 are compounded rear versions at an *f*/2.5 speed.

17.5 The Eight-Element Biotar

Four high-speed, eight-element designs are shown in Figs. 17.41 through 17.44. In Fig. 17.41, both outer elements are split and the capability for a reasonable state of correction is achieved using only medium-high-index glass. [Note that the undercorrected axial chromatic of this prescription can be corrected by using a glass such as BaSF6 (668:419) in the front element instead of SF5.] Two versions of compounding the rear singlet of a seven-element, split-rear single type are shown in Figs. 17.42 and 17.43. Note that Fig. 17.43 is the design which is presented in simplified form in Fig. 17.27. Figure 17.44, while a high-speed lens, covers only a narrow angular field; it uses unusual partial-dispersion glasses such as FK51 and PSK53 to reduce the secondary spectrum.

17.6 Miscellaneous Biotars

Figures 17.45 and 17.46 do not fall into a convenient classification. They are, however, high-speed (*f*/1.4) lenses of an interesting construction. Figure 17.47 is an example of a triplet inner component, a form rarely encountered outside the patent literature.

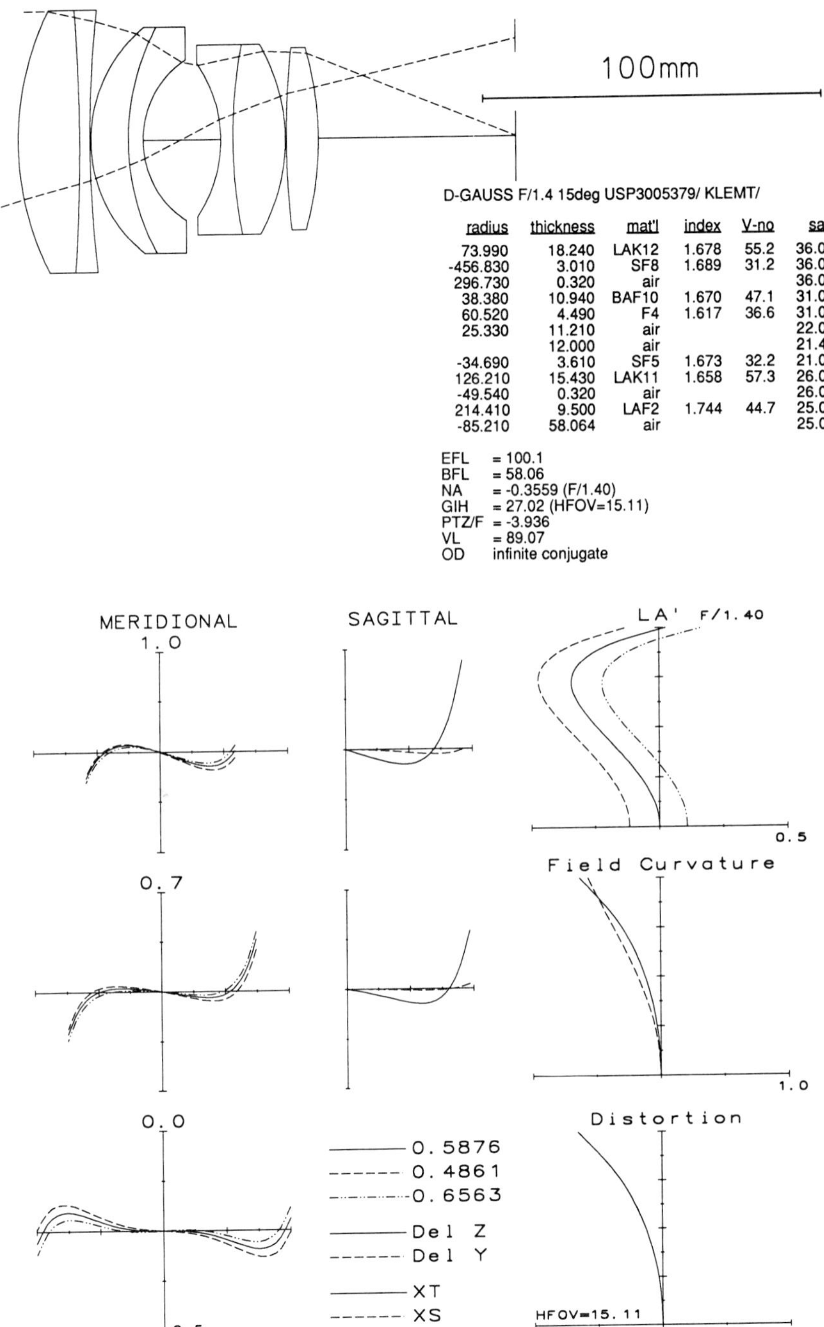

D-GAUSS F/1.4 15deg USP3005379/ KLEMT/

radius	thickness	mat'l	index	V-no	sa
73.990	18.240	LAK12	1.678	55.2	36.0
-456.830	3.010	SF8	1.689	31.2	36.0
296.730	0.320	air			36.0
38.380	10.940	BAF10	1.670	47.1	31.0
60.520	4.490	F4	1.617	36.6	31.0
25.330	11.210	air			22.0
	12.000	air			21.4
-34.690	3.610	SF5	1.673	32.2	21.0
126.210	15.430	LAK11	1.658	57.3	26.0
-49.540	0.320	air			26.0
214.410	9.500	LAF2	1.744	44.7	25.0
-85.210	58.064	air			25.0

EFL = 100.1
BFL = 58.06
NA = -0.3559 (F/1.40)
GIH = 27.02 (HFOV=15.11)
PTZ/F = -3.936
VL = 89.07
OD infinite conjugate

Figure 17.35

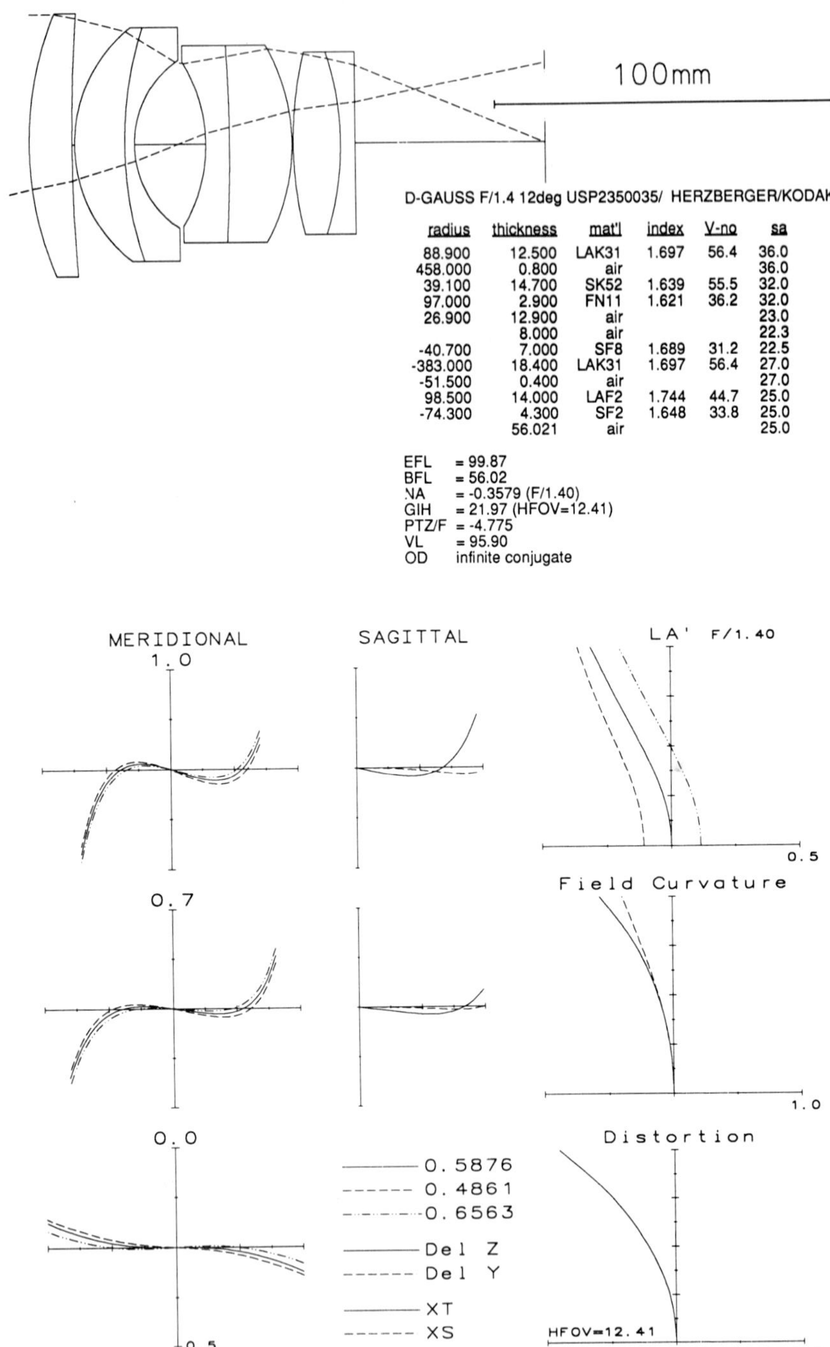

D-GAUSS F/1.4 12deg USP2350035/ HERZBERGER/KODAK

radius	thickness	mat'l	index	V-no	sa
88.900	12.500	LAK31	1.697	56.4	36.0
458.000	0.800	air			36.0
39.100	14.700	SK52	1.639	55.5	32.0
97.000	2.900	FN11	1.621	36.2	32.0
26.900	12.900	air			23.0
	8.000	air			22.3
-40.700	7.000	SF8	1.689	31.2	22.5
-383.000	18.400	LAK31	1.697	56.4	27.0
-51.500	0.400	air			27.0
98.500	14.000	LAF2	1.744	44.7	25.0
-74.300	4.300	SF2	1.648	33.8	25.0
	56.021	air			25.0

EFL = 99.87
BFL = 56.02
NA = -0.3579 (F/1.40)
GIH = 21.97 (HFOV=12.41)
PTZ/F = -4.775
VL = 95.90
OD infinite conjugate

Figure 17.36

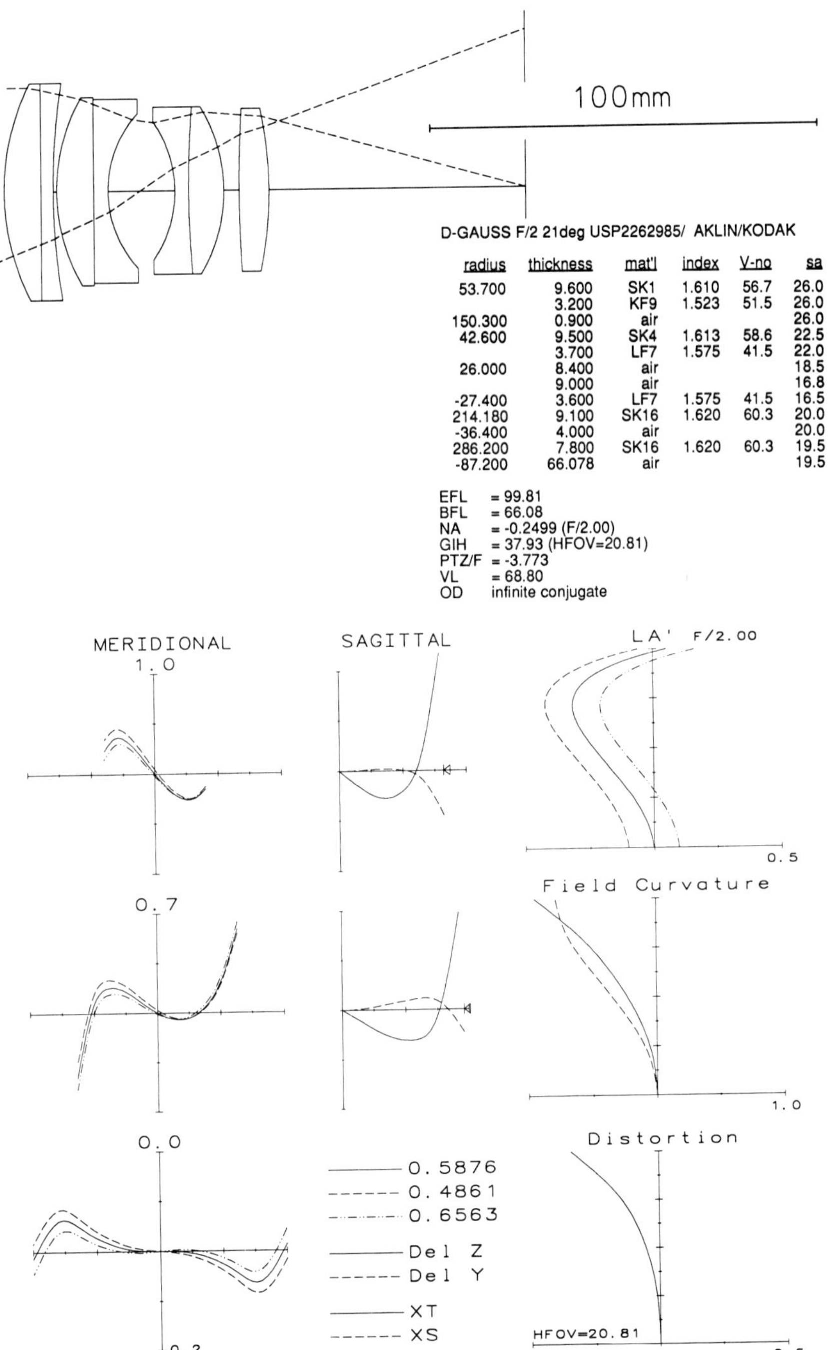

D-GAUSS F/2 21deg USP2262985/ AKLIN/KODAK

radius	thickness	mat'l	index	V-no	sa
53.700	9.600	SK1	1.610	56.7	26.0
	3.200	KF9	1.523	51.5	26.0
150.300	0.900	air			26.0
42.600	9.500	SK4	1.613	58.6	22.5
	3.700	LF7	1.575	41.5	22.0
26.000	8.400	air			18.5
	9.000	air			16.8
-27.400	3.600	LF7	1.575	41.5	16.5
214.180	9.100	SK16	1.620	60.3	20.0
-36.400	4.000	air			20.0
286.200	7.800	SK16	1.620	60.3	19.5
-87.200	66.078	air			19.5

EFL = 99.81
BFL = 66.08
NA = -0.2499 (F/2.00)
GIH = 37.93 (HFOV=20.81)
PTZ/F = -3.773
VL = 68.80
OD infinite conjugate

Figure 17.37

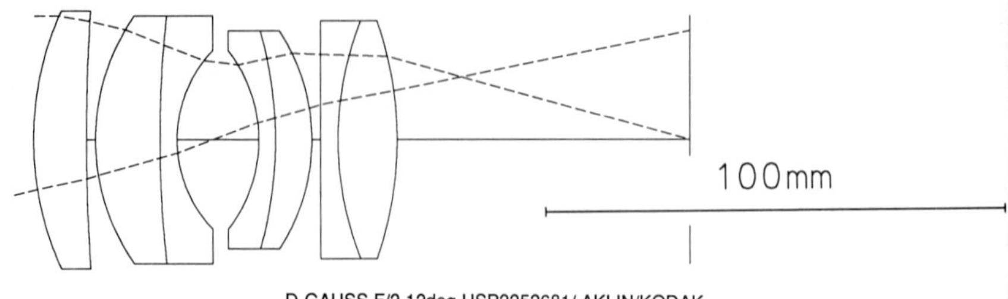

D-GAUSS F/2 12deg USP2252681/ AKLIN/KODAK

radius	thickness	mat'l	index	V-no	sa
60.200	11.300	BAK1	1.572	57.5	26.0
379.220	1.900	air			26.0
42.000	13.900	SK1	1.610	56.7	25.0
214.000	3.700	F5	1.603	38.0	25.0
24.900	7.800	air			18.0
	10.000	air			15.8
-27.400	3.700	SF8	1.689	31.2	18.0
-73.100	7.900	LAF2	1.744	44.7	22.0
-37.800	1.900	air			22.0
	3.800	SF8	1.689	31.2	24.0
60.800	12.800	LAF2	1.744	44.7	24.0
-68.300	63.175	air			24.0

EFL = 99.95
BFL = 63.17
NA = -0.2496 (F/2.0)
GIH = 21.99 (HFOV=12.41)
PTZ/F = -5.932
VL = 78.70
OD infinite conjugate

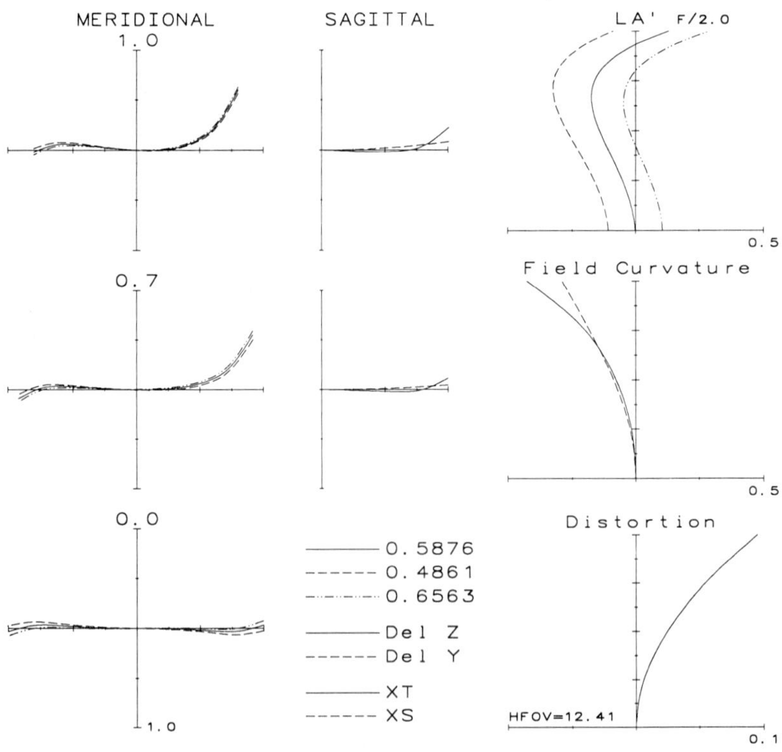

Figure 17.38

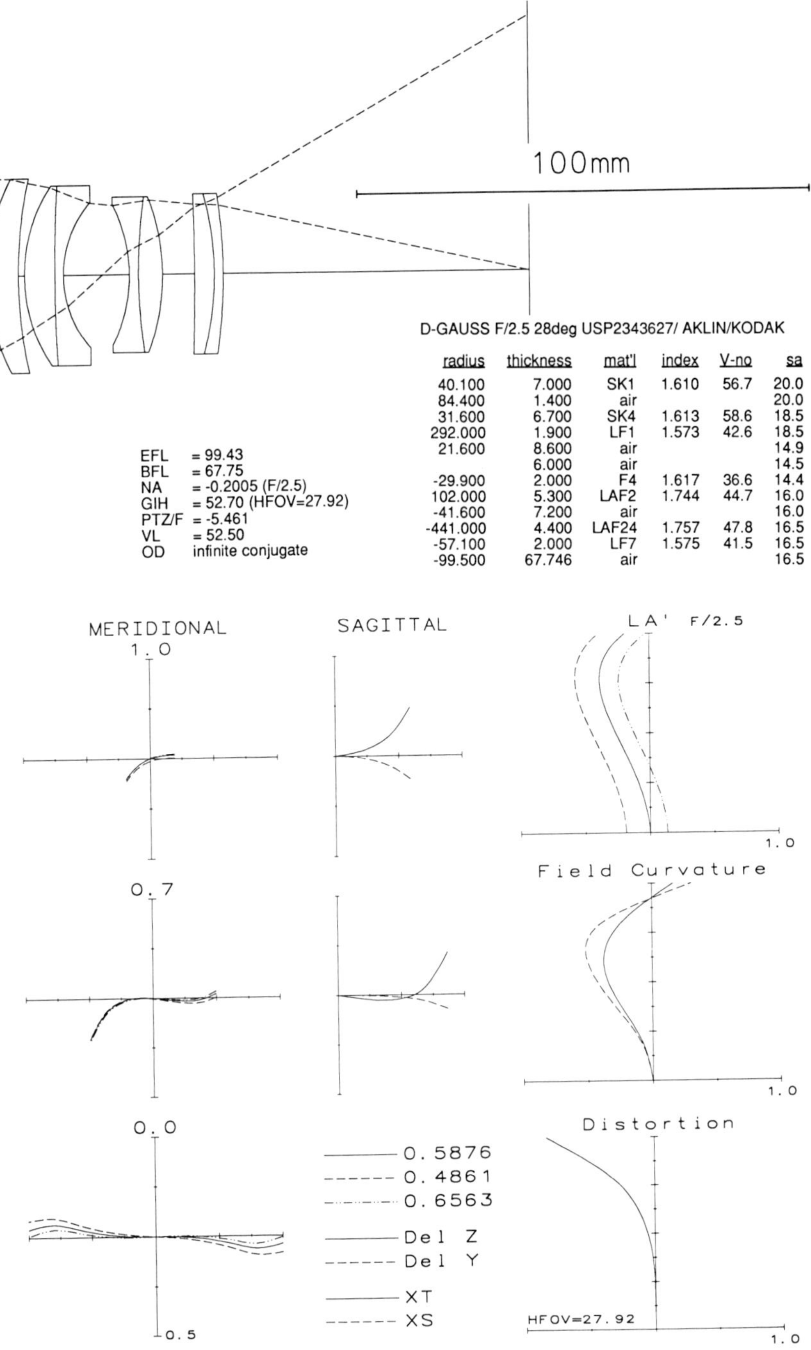

D-GAUSS F/2.5 28deg USP2343627/ AKLIN/KODAK

radius	thickness	mat'l	index	V-no	sa
40.100	7.000	SK1	1.610	56.7	20.0
84.400	1.400	air			20.0
31.600	6.700	SK4	1.613	58.6	18.5
292.000	1.900	LF1	1.573	42.6	18.5
21.600	8.600	air			14.9
	6.000	air			14.5
-29.900	2.000	F4	1.617	36.6	14.4
102.000	5.300	LAF2	1.744	44.7	16.0
-41.600	7.200	air			16.0
-441.000	4.400	LAF24	1.757	47.8	16.5
-57.100	2.000	LF7	1.575	41.5	16.5
-99.500	67.746	air			16.5

Figure 17.39

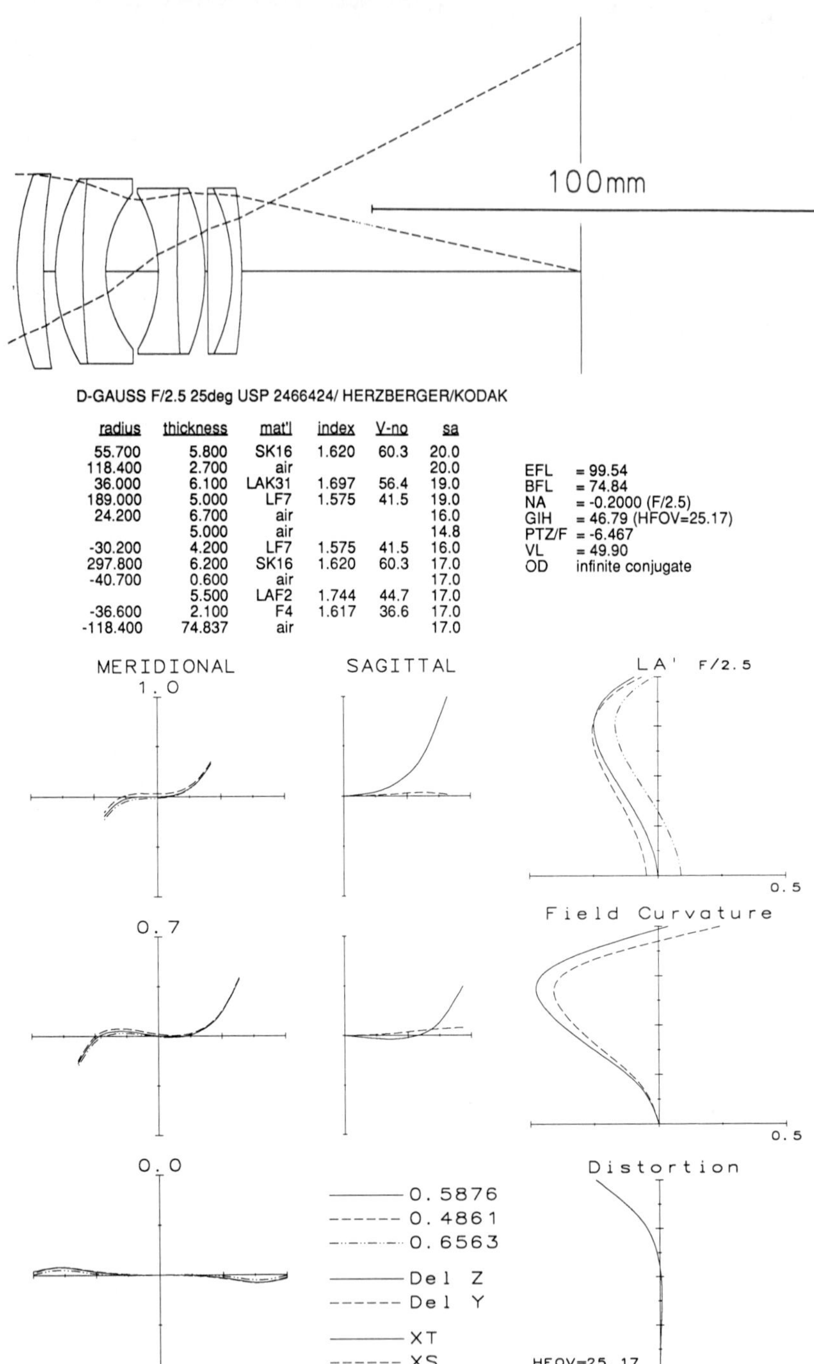

D-GAUSS F/2.5 25deg USP 2466424/ HERZBERGER/KODAK

radius	thickness	mat'l	index	V-no	sa
55.700	5.800	SK16	1.620	60.3	20.0
118.400	2.700	air			20.0
36.000	6.100	LAK31	1.697	56.4	19.0
189.000	5.000	LF7	1.575	41.5	19.0
24.200	6.700	air			16.0
	5.000	air			14.8
-30.200	4.200	LF7	1.575	41.5	16.0
297.800	6.200	SK16	1.620	60.3	17.0
-40.700	0.600	air			17.0
	5.500	LAF2	1.744	44.7	17.0
-36.600	2.100	F4	1.617	36.6	17.0
-118.400	74.837	air			17.0

EFL = 99.54
BFL = 74.84
NA = -0.2000 (F/2.5)
GIH = 46.79 (HFOV=25.17)
PTZ/F = -6.467
VL = 49.90
OD infinite conjugate

Figure 17.40

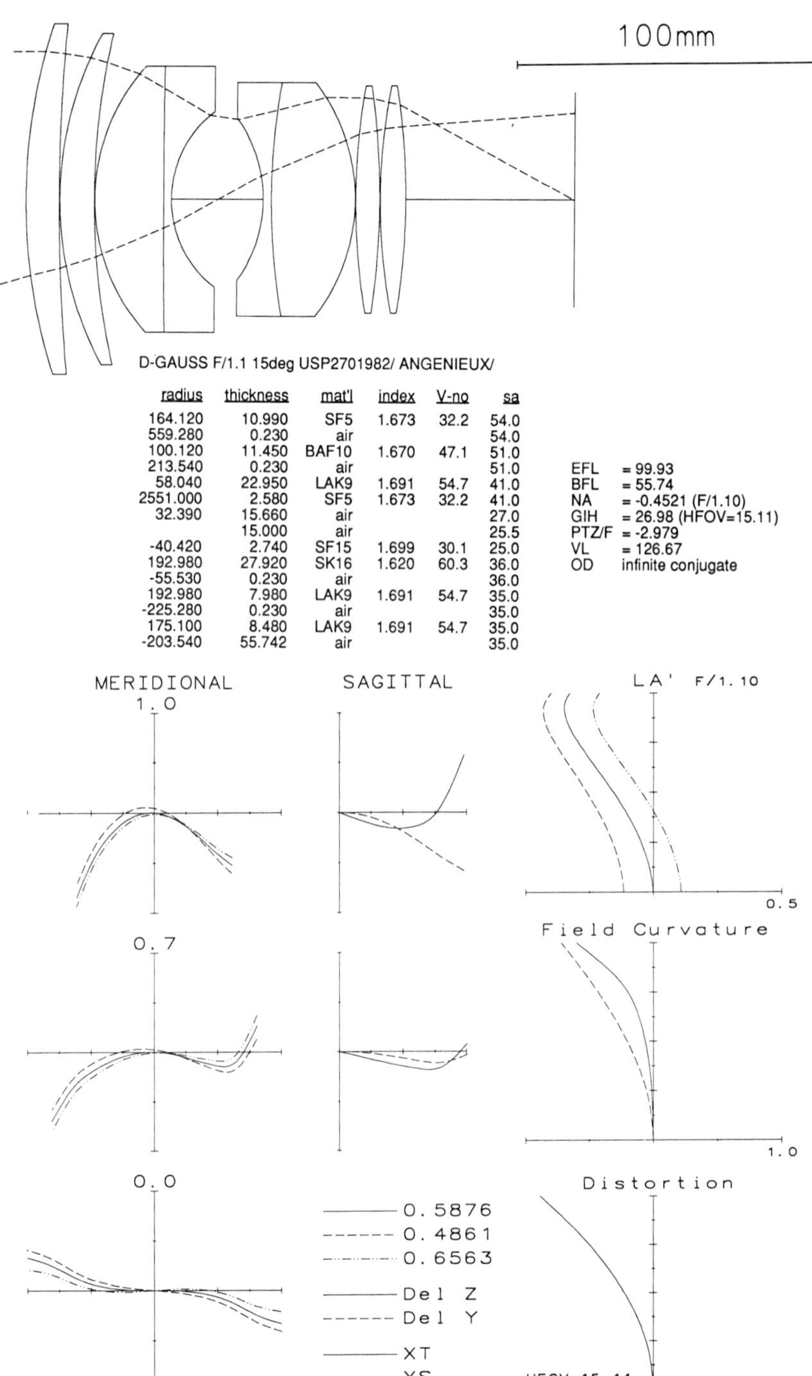

D-GAUSS F/1.1 15deg USP2701982/ ANGENIEUX/

radius	thickness	mat'l	index	V-no	sa
164.120	10.990	SF5	1.673	32.2	54.0
559.280	0.230	air			54.0
100.120	11.450	BAF10	1.670	47.1	51.0
213.540	0.230	air			51.0
58.040	22.950	LAK9	1.691	54.7	41.0
2551.000	2.580	SF5	1.673	32.2	41.0
32.390	15.660	air			27.0
	15.000	air			25.5
-40.420	2.740	SF15	1.699	30.1	25.0
192.980	27.920	SK16	1.620	60.3	36.0
-55.530	0.230	air			36.0
192.980	7.980	LAK9	1.691	54.7	35.0
-225.280	0.230	air			35.0
175.100	8.480	LAK9	1.691	54.7	35.0
-203.540	55.742	air			35.0

EFL = 99.93
BFL = 55.74
NA = -0.4521 (F/1.10)
GIH = 26.98 (HFOV=15.11)
PTZ/F = -2.979
VL = 126.67
OD infinite conjugate

Figure 17.41

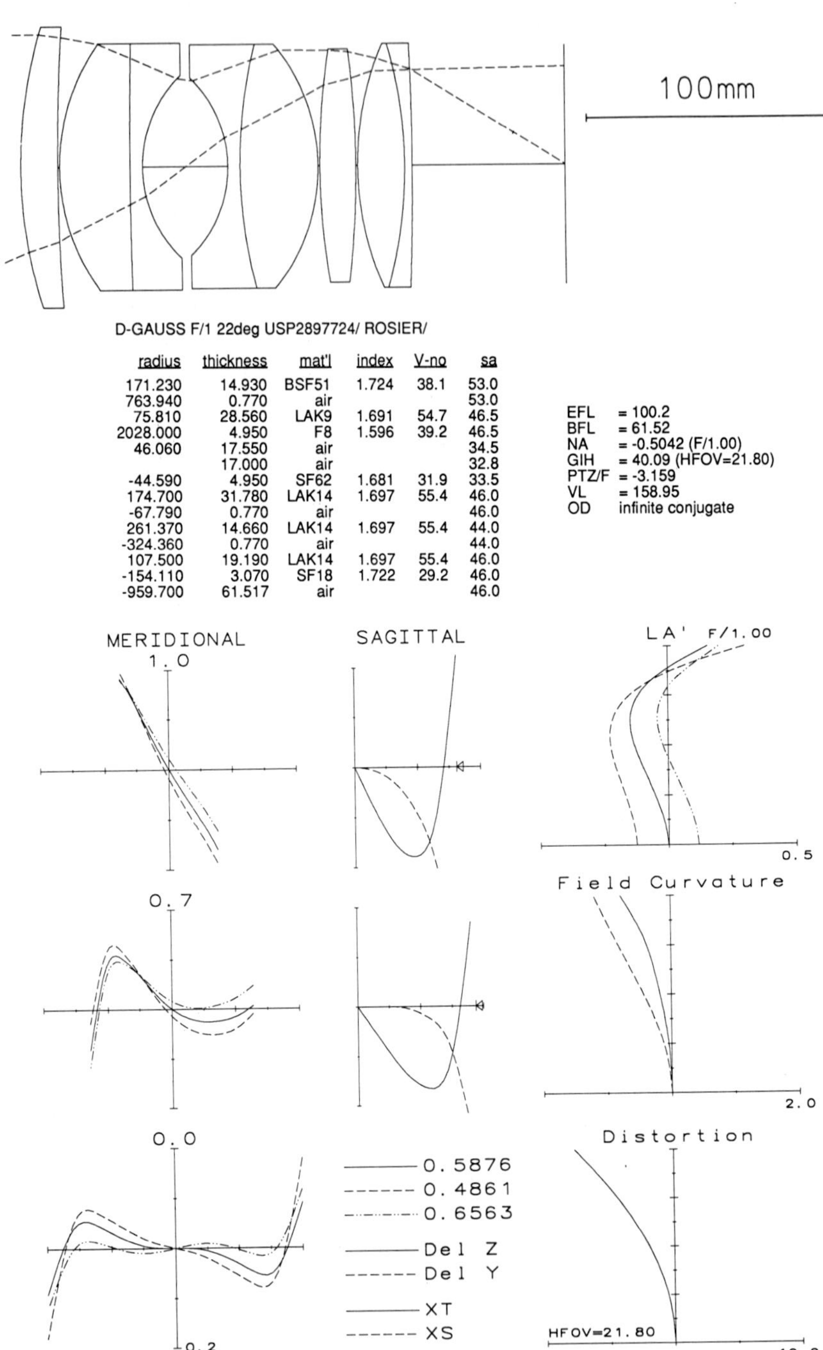

D-GAUSS F/1 22deg USP2897724/ ROSIER/

radius	thickness	mat'l	index	V-no	sa
171.230	14.930	BSF51	1.724	38.1	53.0
763.940	0.770	air			53.0
75.810	28.560	LAK9	1.691	54.7	46.5
2028.000	4.950	F8	1.596	39.2	46.5
46.060	17.550	air			34.5
	17.000	air			32.8
-44.590	4.950	SF62	1.681	31.9	33.5
174.700	31.780	LAK14	1.697	55.4	46.0
-67.790	0.770	air			46.0
261.370	14.660	LAK14	1.697	55.4	44.0
-324.360	0.770	air			44.0
107.500	19.190	LAK14	1.697	55.4	46.0
-154.110	3.070	SF18	1.722	29.2	46.0
-959.700	61.517	air			46.0

EFL = 100.2
BFL = 61.52
NA = -0.5042 (F/1.00)
GIH = 40.09 (HFOV=21.80)
PTZ/F = -3.159
VL = 158.95
OD infinite conjugate

Figure 17.42

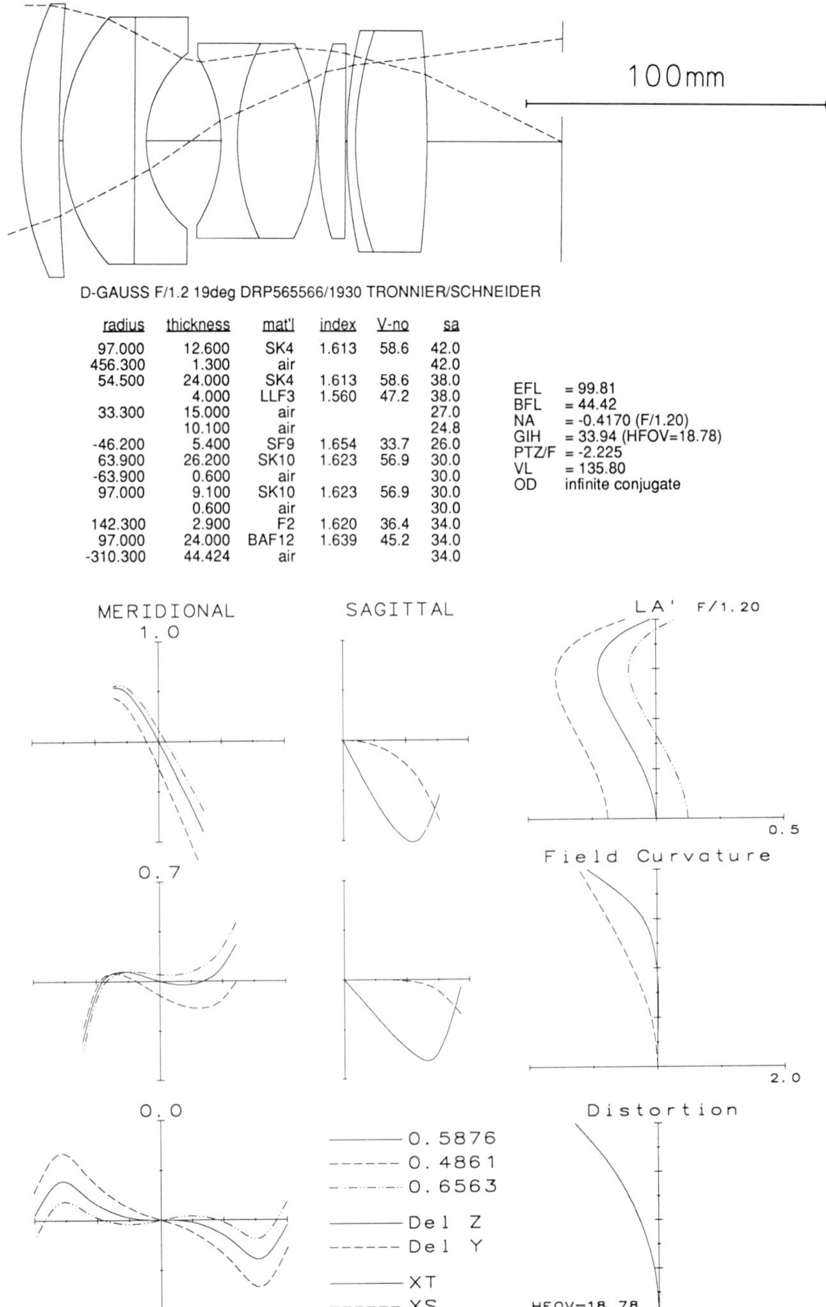

D-GAUSS F/1.2 19deg DRP565566/1930 TRONNIER/SCHNEIDER

radius	thickness	mat'l	index	V-no	sa
97.000	12.600	SK4	1.613	58.6	42.0
456.300	1.300	air			42.0
54.500	24.000	SK4	1.613	58.6	38.0
	4.000	LLF3	1.560	47.2	38.0
33.300	15.000	air			27.0
	10.100	air			24.8
-46.200	5.400	SF9	1.654	33.7	26.0
63.900	26.200	SK10	1.623	56.9	30.0
-63.900	0.600	air			30.0
97.000	9.100	SK10	1.623	56.9	30.0
	0.600	air			30.0
142.300	2.900	F2	1.620	36.4	34.0
97.000	24.000	BAF12	1.639	45.2	34.0
-310.300	44.424	air			34.0

EFL = 99.81
BFL = 44.42
NA = -0.4170 (F/1.20)
GIH = 33.94 (HFOV=18.78)
PTZ/F = -2.225
VL = 135.80
OD infinite conjugate

Figure 17.43

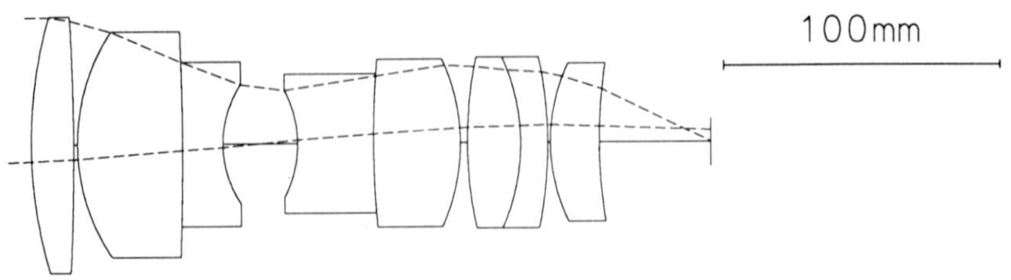

F/1.2 2.3degHFOV D-GAUSS US 3,925,910/1975 MATSUBARA

radius	thickness	mat'l	index	V-no	sa
142.569	15.300	LAK31	1.697	56.4	42.7
-715.194	1.440	air			42.7
64.476	37.800	FK51	1.487	84.5	37.6
-855.297	14.580	SF2	1.648	33.8	27.4
33.363	13.500	air			19.9
	13.680	air			19.0
-36.198	27.270	F1	1.626	35.7	17.8
281.889	30.960	PSK53	1.620	63.5	23.2
-63.513	2.520	air			28.0
126.126	19.260	FK51	1.487	84.5	28.4
-64.278	9.810	SF7	1.640	34.6	28.4
-131.175	0.990	air			28.4
57.033	17.370	LAF21	1.788	47.5	26.4
156.870	39.751	air			22.6

EFL = 102.1
BFL = 39.75
NA = -0.4151 (F/1.21)
GIH = 4.08 (HFOV=2.29)
PTZ/F = -6.273
VL = 204.48
OD infinite conjugate

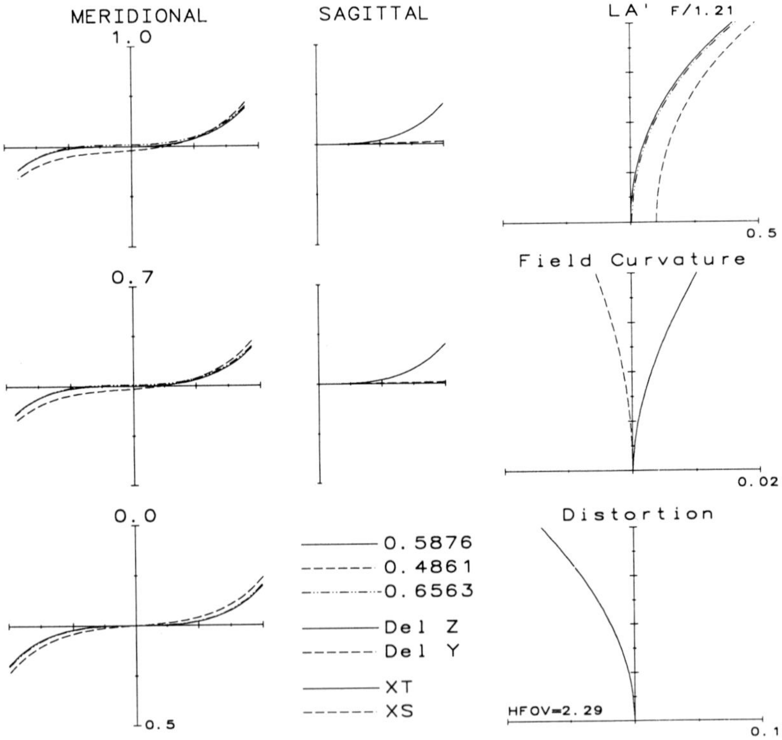

MERIDIONAL SAGITTAL LA' F/1.21
1.0
0.5
Field Curvature
0.7
0.02
0.0 Distortion
0.5876
0.4861
0.6563
Del Z
Del Y
XT
XS
0.5 HFOV=2.29 0.1

Figure 17.44

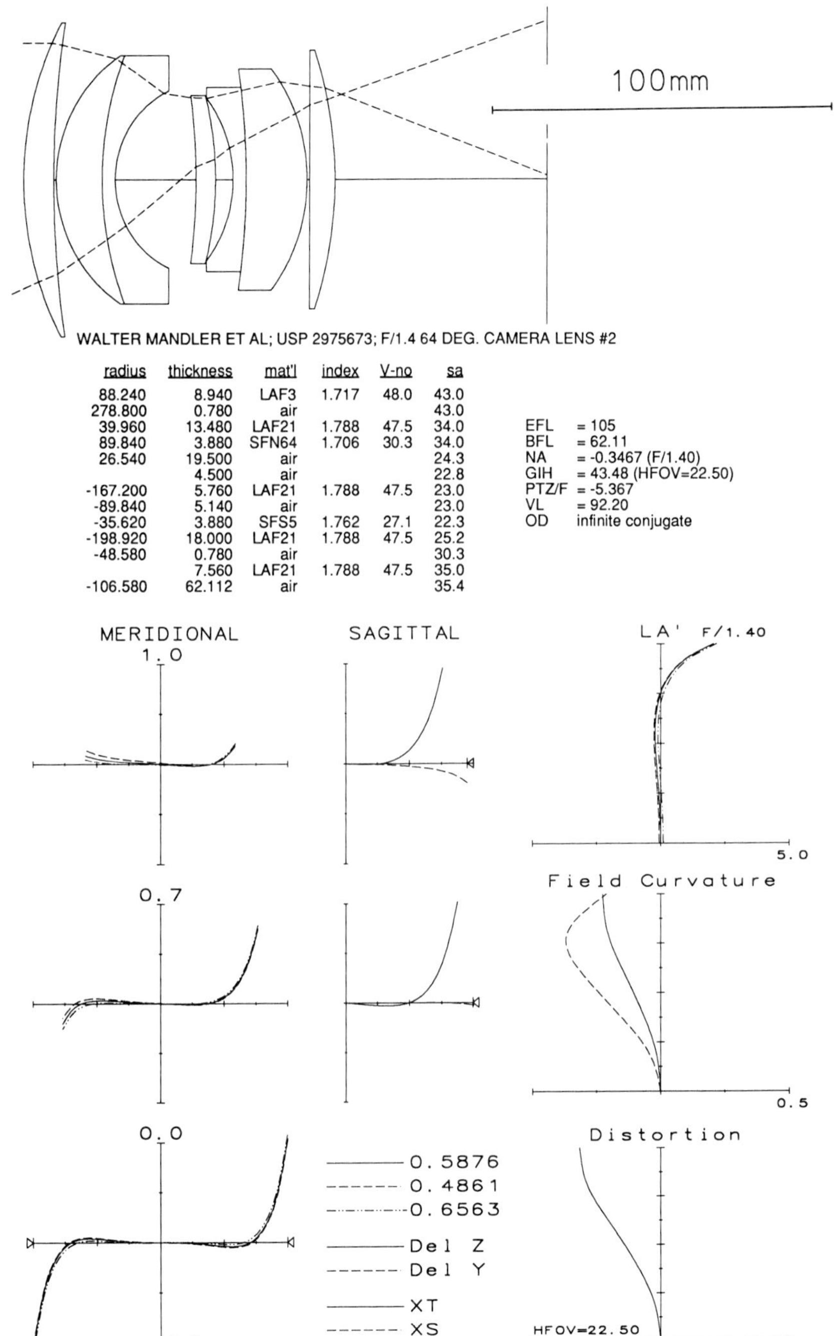

WALTER MANDLER ET AL; USP 2975673; F/1.4 64 DEG. CAMERA LENS #2

radius	thickness	mat'l	index	V-no	sa
88.240	8.940	LAF3	1.717	48.0	43.0
278.800	0.780	air			43.0
39.960	13.480	LAF21	1.788	47.5	34.0
89.840	3.880	SFN64	1.706	30.3	34.0
26.540	19.500	air			24.3
	4.500	air			22.8
-167.200	5.760	LAF21	1.788	47.5	23.0
-89.840	5.140	air			23.0
-35.620	3.880	SFS5	1.762	27.1	22.3
-198.920	18.000	LAF21	1.788	47.5	25.2
-48.580	0.780	air			30.3
	7.560	LAF21	1.788	47.5	35.0
-106.580	62.112	air			35.4

EFL = 105
BFL = 62.11
NA = -0.3467 (F/1.40)
GIH = 43.48 (HFOV=22.50)
PTZ/F = -5.367
VL = 92.20
OD infinite conjugate

Figure 17.45

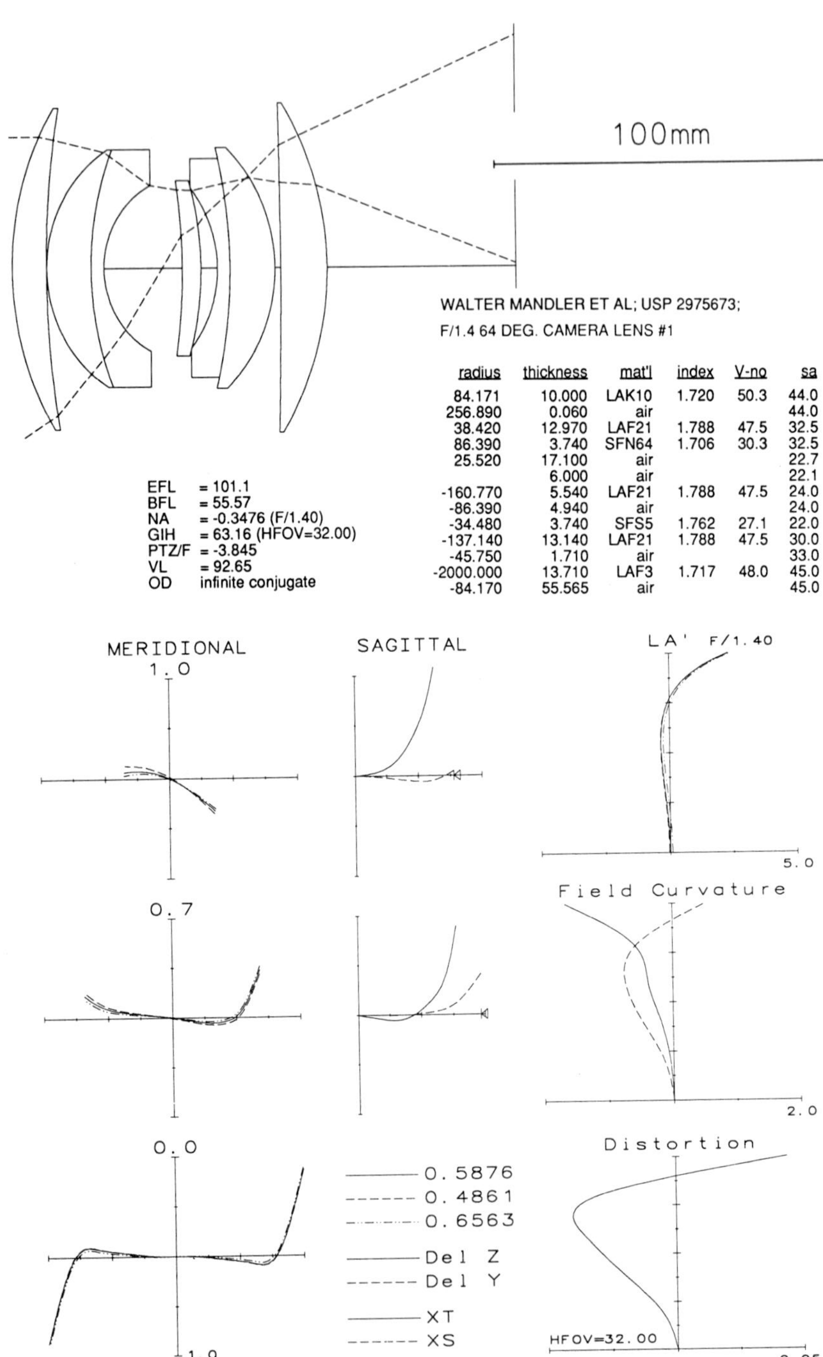

WALTER MANDLER ET AL; USP 2975673;
F/1.4 64 DEG. CAMERA LENS #1

radius	thickness	mat'l	index	V-no	sa
84.171	10.000	LAK10	1.720	50.3	44.0
256.890	0.060	air			44.0
38.420	12.970	LAF21	1.788	47.5	32.5
86.390	3.740	SFN64	1.706	30.3	32.5
25.520	17.100	air			22.7
	6.000	air			22.1
-160.770	5.540	LAF21	1.788	47.5	24.0
-86.390	4.940	air			24.0
-34.480	3.740	SFS5	1.762	27.1	22.0
-137.140	13.140	LAF21	1.788	47.5	30.0
-45.750	1.710	air			33.0
-2000.000	13.710	LAF3	1.717	48.0	45.0
-84.170	55.565	air			45.0

EFL = 101.1
BFL = 55.57
NA = -0.3476 (F/1.40)
GIH = 63.16 (HFOV=32.00)
PTZ/F = -3.845
VL = 92.65
OD infinite conjugate

Figure 17.46

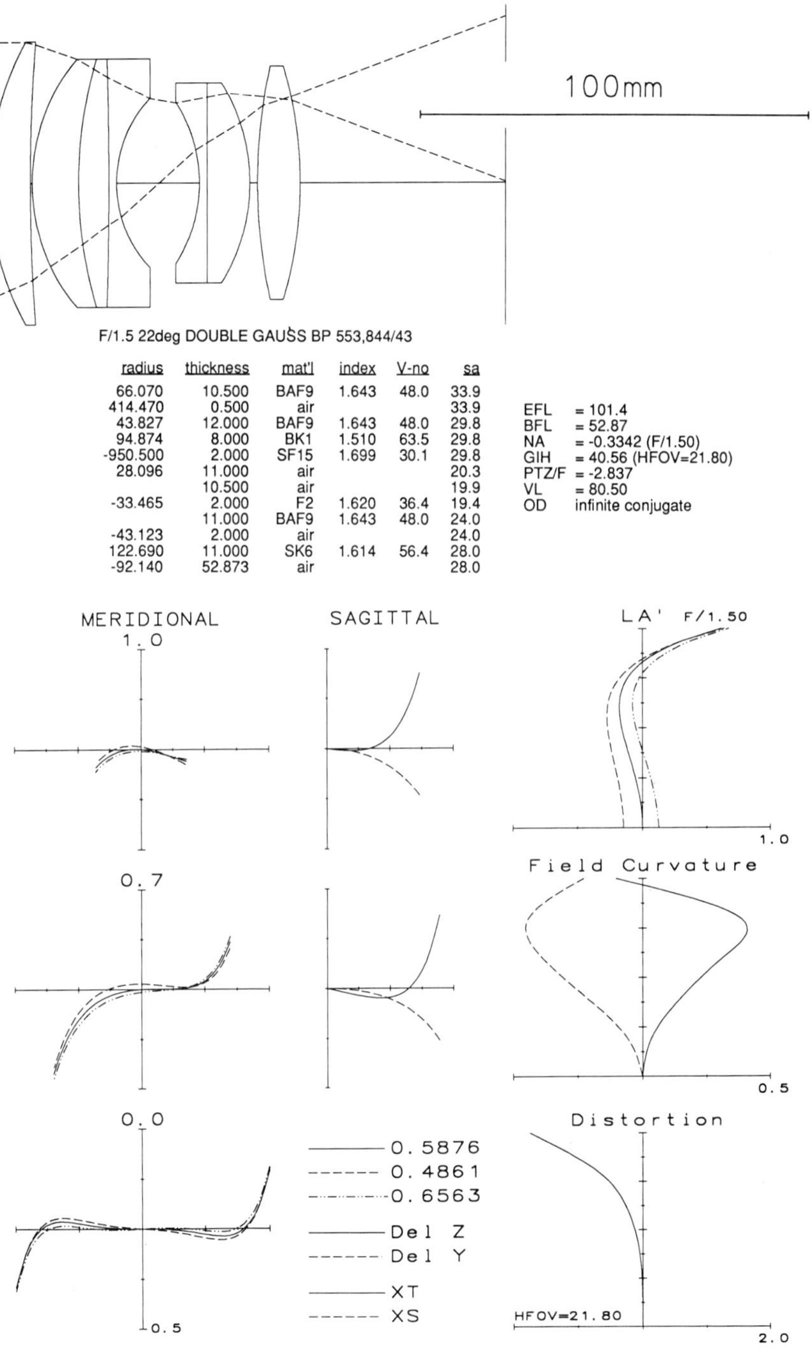

F/1.5 22deg DOUBLE GAUSS BP 553,844/43

radius	thickness	mat'l	index	V-no	sa
66.070	10.500	BAF9	1.643	48.0	33.9
414.470	0.500	air			33.9
43.827	12.000	BAF9	1.643	48.0	29.8
94.874	8.000	BK1	1.510	63.5	29.8
-950.500	2.000	SF15	1.699	30.1	29.8
28.096	11.000	air			20.3
	10.500	air			19.9
-33.465	2.000	F2	1.620	36.4	19.4
	11.000	BAF9	1.643	48.0	24.0
-43.123	2.000	air			24.0
122.690	11.000	SK6	1.614	56.4	28.0
-92.140	52.873	air			28.0

EFL = 101.4
BFL = 52.87
NA = -0.3342 (F/1.50)
GIH = 40.56 (HFOV=21.80)
PTZ/F = -2.837
VL = 80.50
OD infinite conjugate

Figure 17.47

Chapter

18

Wide-Angle Lenses with Negative Outer Elements

This chapter presents several lenses (Figs. 18.1 to 18.8) which derive their ability to cover a wide angular field from a construction which incorporates negative meniscus elements as the outer members. A glance at the path of the principal ray in any of the lens drawings will indicate the purpose of this construction. The outer negative lenses take a wide angular field in object and/or image space and convert it to a much smaller angular field inside the lens, a smaller field which is within the ability of the inner elements to handle. In addition, the wide spacing between the positive center member and the negative outer members provides the spacing between positive and negative power which is used to flatten the Petzval curvature.

To some extent these negative components can be viewed in the same light as the negative front member of the retrofocus type discussed in Chap. 9. However, the reverse telephoto lens has a long back focal length and absolutely no symmetry. In the lenses of this chapter the possibility of at least a roughly symmetrical construction is apparent, since (with one exception) they all have meniscus negative outer elements on both sides of the lens, and the inner members are of a roughly symmetrical construction as well. The back focal length of a retrofocus lens is typically greater than its focal length; in these lenses the back focal length ranges from 23 to 65 percent of the focal length.

The unusual design of Fig. 18.7 can be regarded as (1) a member of this family with only one outer negative element, (2) a telephoto lens, or (3) a triplet anastigmat with an aspheric field corrector.

Note that there are also lenses of wide angular coverage included in Chaps. 9 and 11.

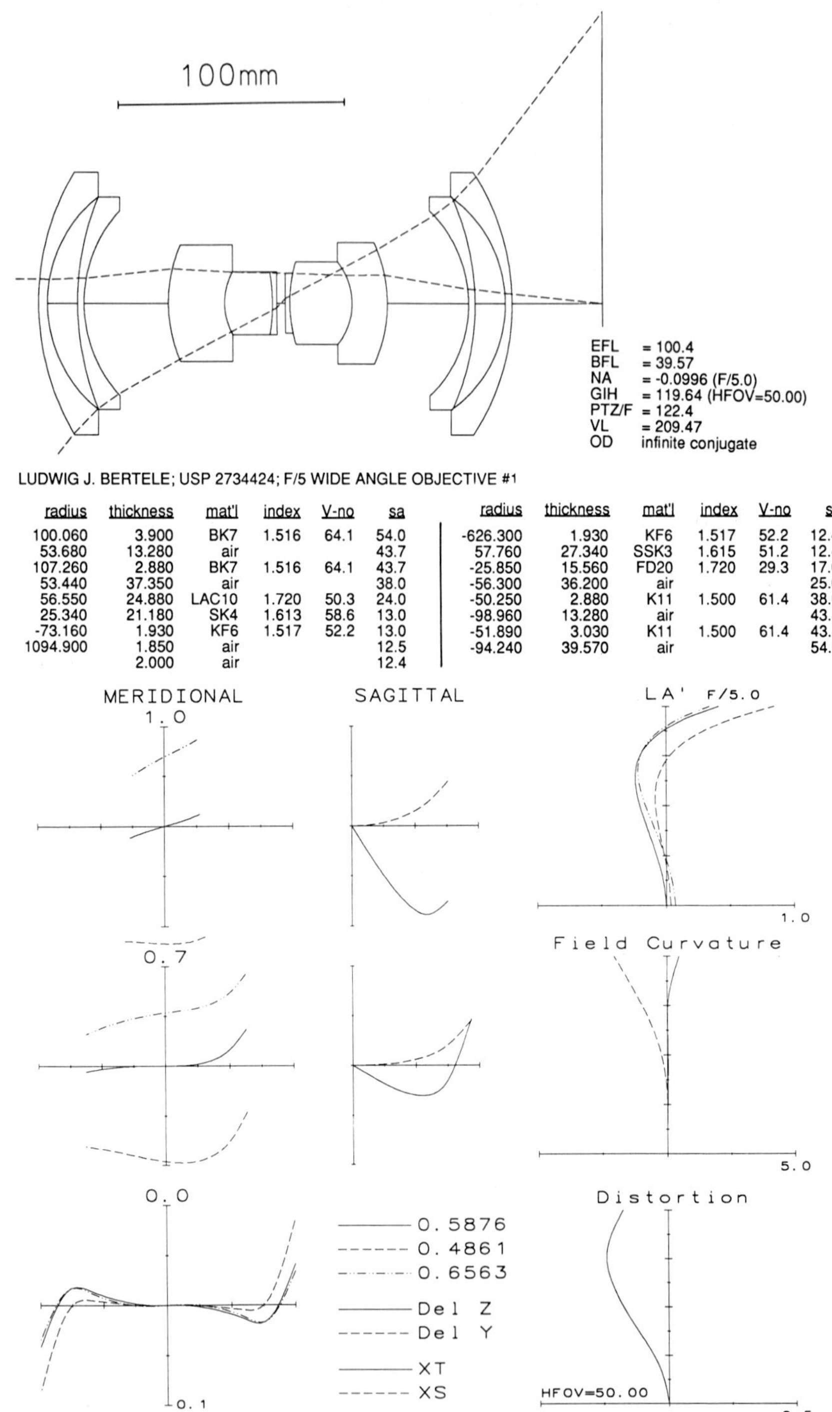

LUDWIG J. BERTELE; USP 2734424; F/5 WIDE ANGLE OBJECTIVE #1

radius	thickness	mat'l	index	V-no	sa
100.060	3.900	BK7	1.516	64.1	54.0
53.680	13.280	air			43.7
107.260	2.880	BK7	1.516	64.1	43.7
53.440	37.350	air			38.0
56.550	24.880	LAC10	1.720	50.3	24.0
25.340	21.180	SK4	1.613	58.6	13.0
-73.160	1.930	KF6	1.517	52.2	13.0
1094.900	1.850	air			12.5
	2.000	air			12.4

radius	thickness	mat'l	index	V-no	sa
-626.300	1.930	KF6	1.517	52.2	12.4
57.760	27.340	SSK3	1.615	51.2	12.4
-25.850	15.560	FD20	1.720	29.3	17.0
-56.300	36.200	air			25.0
-50.250	2.880	K11	1.500	61.4	38.0
-98.960	13.280	air			43.8
-51.890	3.030	K11	1.500	61.4	43.8
-94.240	39.570	air			54.0

Figure 18.1

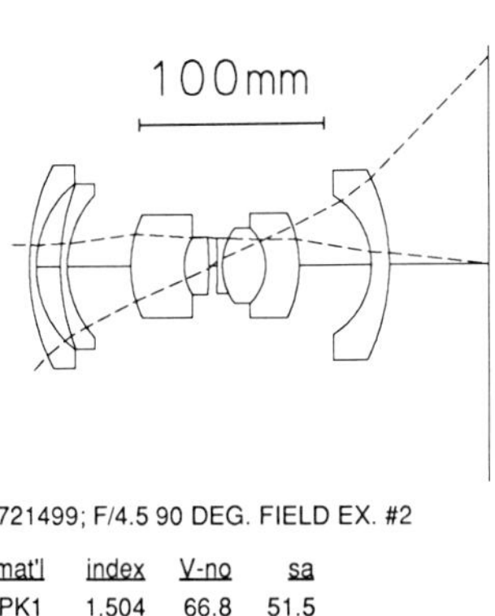

LUDWIG BERTELE; USP 2721499; F/4.5 90 DEG. FIELD EX. #2

radius	thickness	mat'l	index	V-no	sa
109.140	3.700	PK1	1.504	66.8	51.5
52.630	13.200	air			42.0
110.250	3.700	FK5	1.487	70.2	41.6
50.720	35.000	air			36.0
56.240	29.300	LAC10	1.720	50.3	26.0
25.370	13.300	BACD7	1.607	59.5	14.4
-194.920	1.700	air			14.3
	3.000	air			14.1
-252.700	2.800	BAK1	1.572	57.5	14.0
30.590	23.700	SSK2	1.622	53.2	13.8
-25.510	18.200	FD20	1.720	29.3	19.0
-57.380	39.000	air			26.5
-40.150	9.800	SBC2	1.642	58.1	35.0
-102.640	53.863	air			48.0

EFL = 104.6
BFL = 53.86
NA = -0.1052 (F/4.7)
GIH = 104.59 (HFOV=45.00)
PTZ/F = -52.14
VL = 196.40
OD infinite conjugate

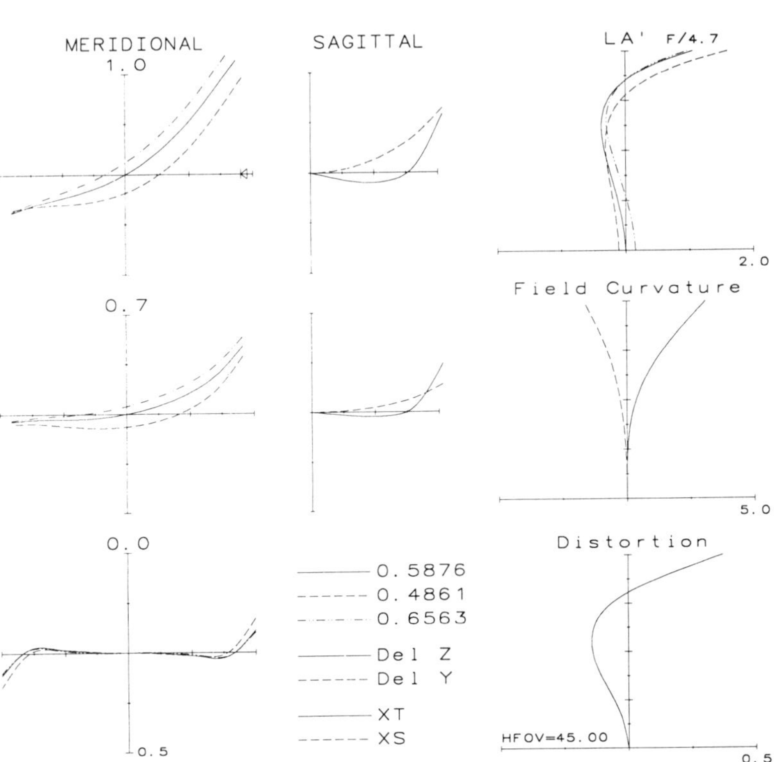

Figure 18.2

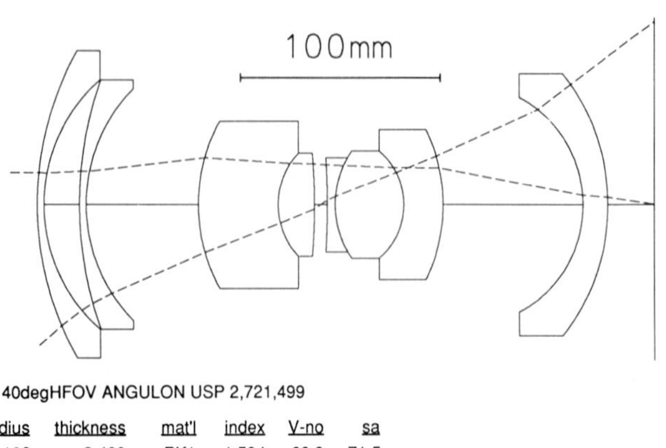

F/3.3 40degHFOV ANGULON USP 2,721,499

radius	thickness	mat'l	index	V-no	sa
138.198	3.400	PK1	1.504	66.9	71.5
73.801	17.800	air			58.0
162.285	3.400	FK5	1.487	70.4	58.0
73.540	56.200	air			54.0
76.383	40.100	LAK10	1.720	50.4	39.0
33.620	18.200	BAF5	1.607	49.4	24.0
-267.237	3.200	air			24.0
	3.200	air			18.8
-346.380	3.900	BAK4	1.569	56.1	22.0
41.930	34.300	SK15	1.623	58.1	22.0
-31.480	19.600	SF18	1.722	29.2	25.0
-76.359	69.600	air			35.0
-55.279	12.500	LAKN6	1.642	58.0	50.0
-93.624	23.093	air			61.0

EFL = 101.7
BFL = 23.09
NA = -0.1466 (F/3.4)
GIH = 85.43 (HFOV=40.03)
PTZ/F = -4.717
VL = 285.40
OD infinite conjugate

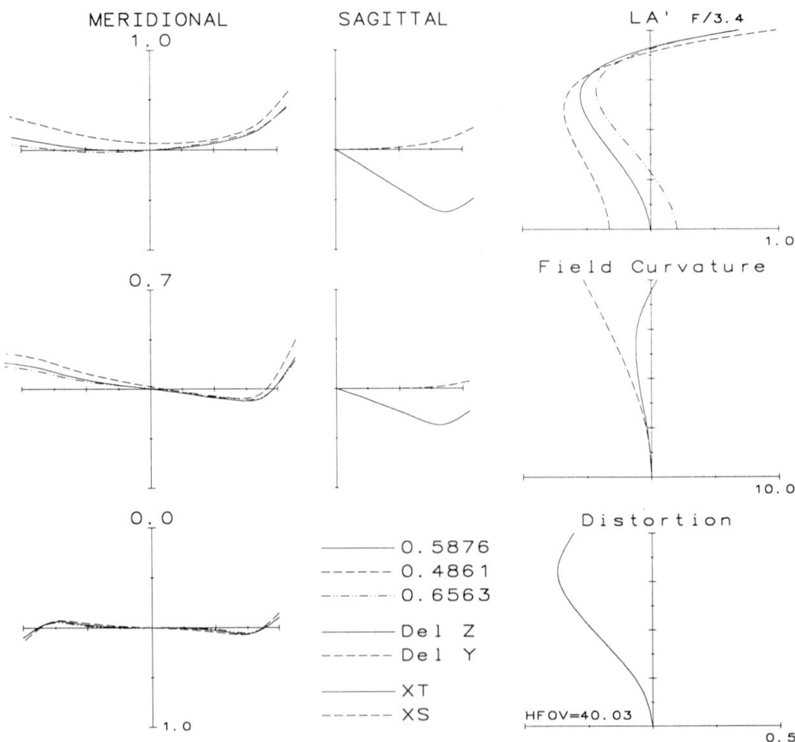

Figure 18.3

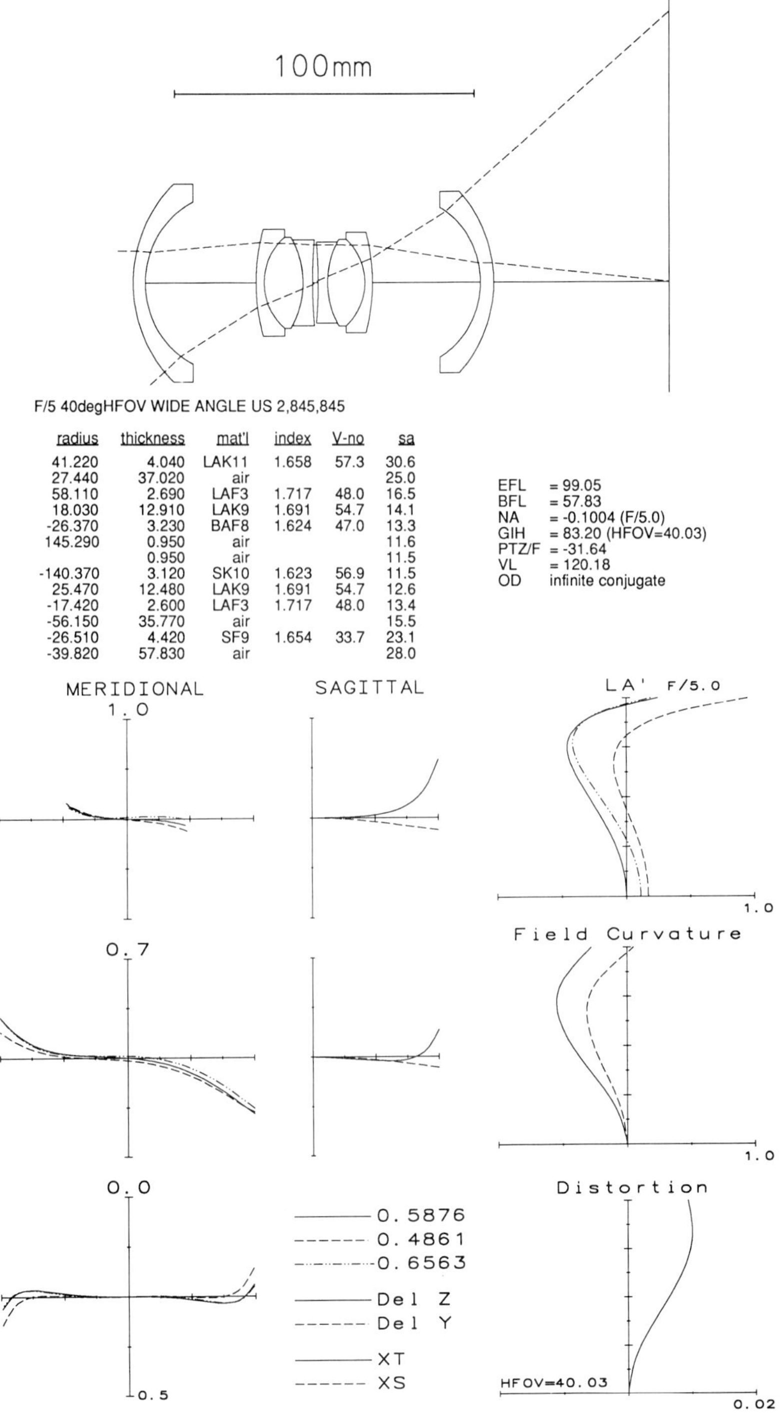

F/5 40degHFOV WIDE ANGLE US 2,845,845

radius	thickness	mat'l	index	V-no	sa
41.220	4.040	LAK11	1.658	57.3	30.6
27.440	37.020	air			25.0
58.110	2.690	LAF3	1.717	48.0	16.5
18.030	12.910	LAK9	1.691	54.7	14.1
-26.370	3.230	BAF8	1.624	47.0	13.3
145.290	0.950	air			11.6
	0.950	air			11.5
-140.370	3.120	SK10	1.623	56.9	11.5
25.470	12.480	LAK9	1.691	54.7	12.6
-17.420	2.600	LAF3	1.717	48.0	13.4
-56.150	35.770	air			15.5
-26.510	4.420	SF9	1.654	33.7	23.1
-39.820	57.830	air			28.0

EFL	= 99.05
BFL	= 57.83
NA	= -0.1004 (F/5.0)
GIH	= 83.20 (HFOV=40.03)
PTZ/F	= -31.64
VL	= 120.18
OD	infinite conjugate

Figure 18.4

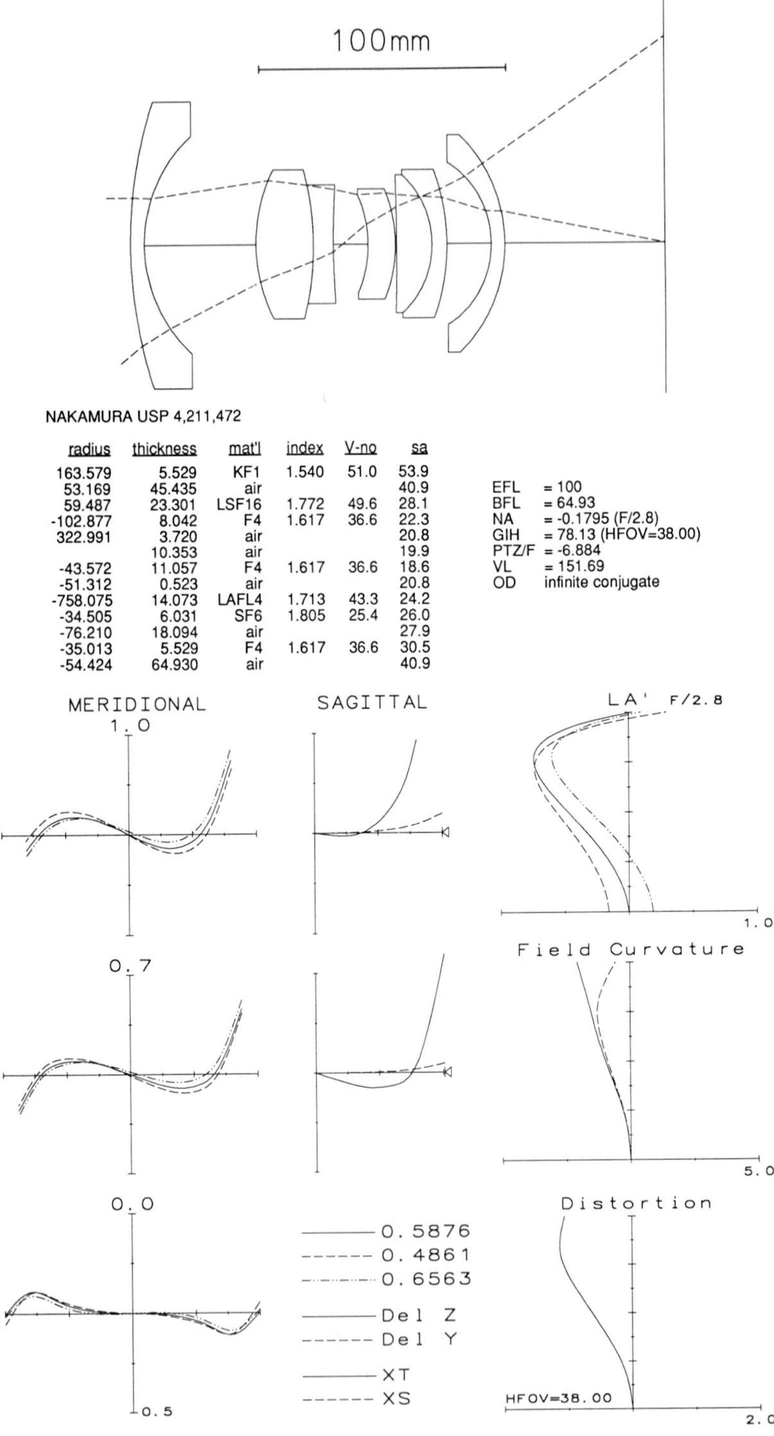

NAKAMURA USP 4,211,472

radius	thickness	mat'l	index	V-no	sa
163.579	5.529	KF1	1.540	51.0	53.9
53.169	45.435	air			40.9
59.487	23.301	LSF16	1.772	49.6	28.1
-102.877	8.042	F4	1.617	36.6	22.3
322.991	3.720	air			20.8
	10.353	air			19.9
-43.572	11.057	F4	1.617	36.6	18.6
-51.312	0.523	air			20.8
-758.075	14.073	LAFL4	1.713	43.3	24.2
-34.505	6.031	SF6	1.805	25.4	26.0
-76.210	18.094	air			27.9
-35.013	5.529	F4	1.617	36.6	30.5
-54.424	64.930	air			40.9

EFL = 100
BFL = 64.93
NA = -0.1795 (F/2.8)
GIH = 78.13 (HFOV=38.00)
PTZ/F = -6.884
VL = 151.69
OD infinite conjugate

Figure 18.5

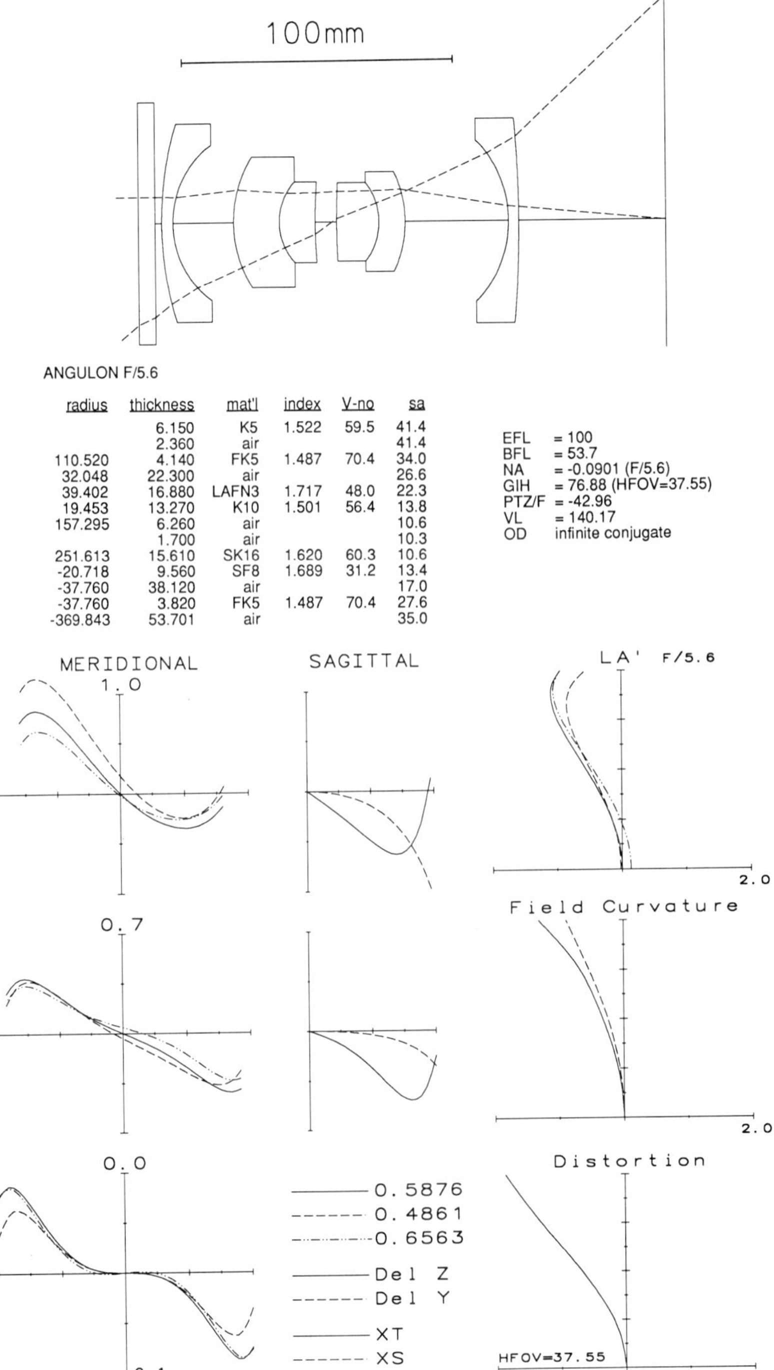

ANGULON F/5.6

radius	thickness	mat'l	index	V-no	sa
	6.150	K5	1.522	59.5	41.4
	2.360	air			41.4
110.520	4.140	FK5	1.487	70.4	34.0
32.048	22.300	air			26.6
39.402	16.880	LAFN3	1.717	48.0	22.3
19.453	13.270	K10	1.501	56.4	13.8
157.295	6.260	air			10.6
	1.700	air			10.3
251.613	15.610	SK16	1.620	60.3	10.6
-20.718	9.560	SF8	1.689	31.2	13.4
-37.760	38.120	air			17.0
-37.760	3.820	FK5	1.487	70.4	27.6
-369.843	53.701	air			35.0

EFL = 100
BFL = 53.7
NA = -0.0901 (F/5.6)
GIH = 76.88 (HFOV=37.55)
PTZ/F = -42.96
VL = 140.17
OD infinite conjugate

Figure 18.6

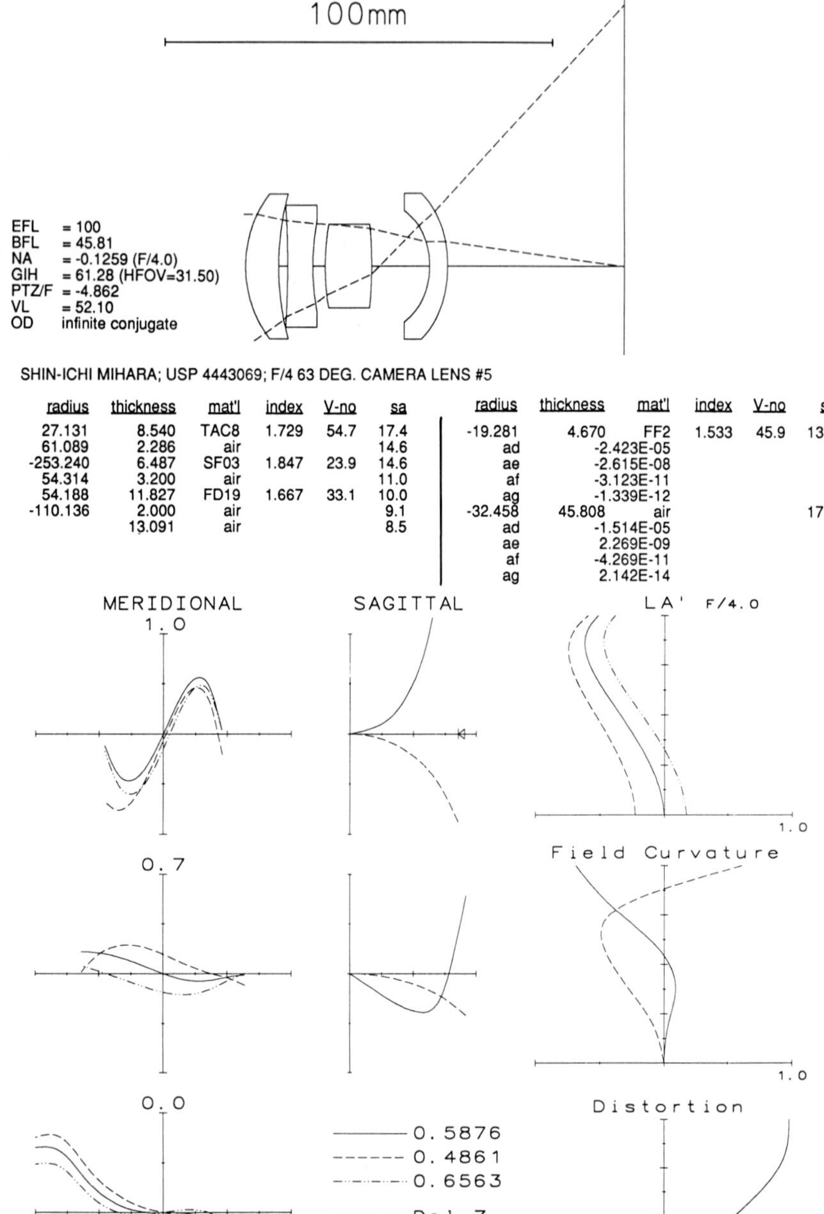

SHIN-ICHI MIHARA; USP 4443069; F/4 63 DEG. CAMERA LENS #5

radius	thickness	mat'l	index	V-no	sa
27.131	8.540	TAC8	1.729	54.7	17.4
61.089	2.286	air			14.6
-253.240	6.487	SF03	1.847	23.9	14.6
54.314	3.200	air			11.0
54.188	11.827	FD19	1.667	33.1	10.0
-110.136	2.000	air			9.1
	13.091	air			8.5

radius	thickness	mat'l	index	V-no	sa
-19.281	4.670	FF2	1.533	45.9	13.3
ad	-2.423E-05				
ae	-2.615E-08				
af	-3.123E-11				
ag	-1.339E-12				
-32.458	45.808	air			17.4
ad	-1.514E-05				
ae	2.269E-09				
af	-4.269E-11				
ag	2.142E-14				

Figure 18.7

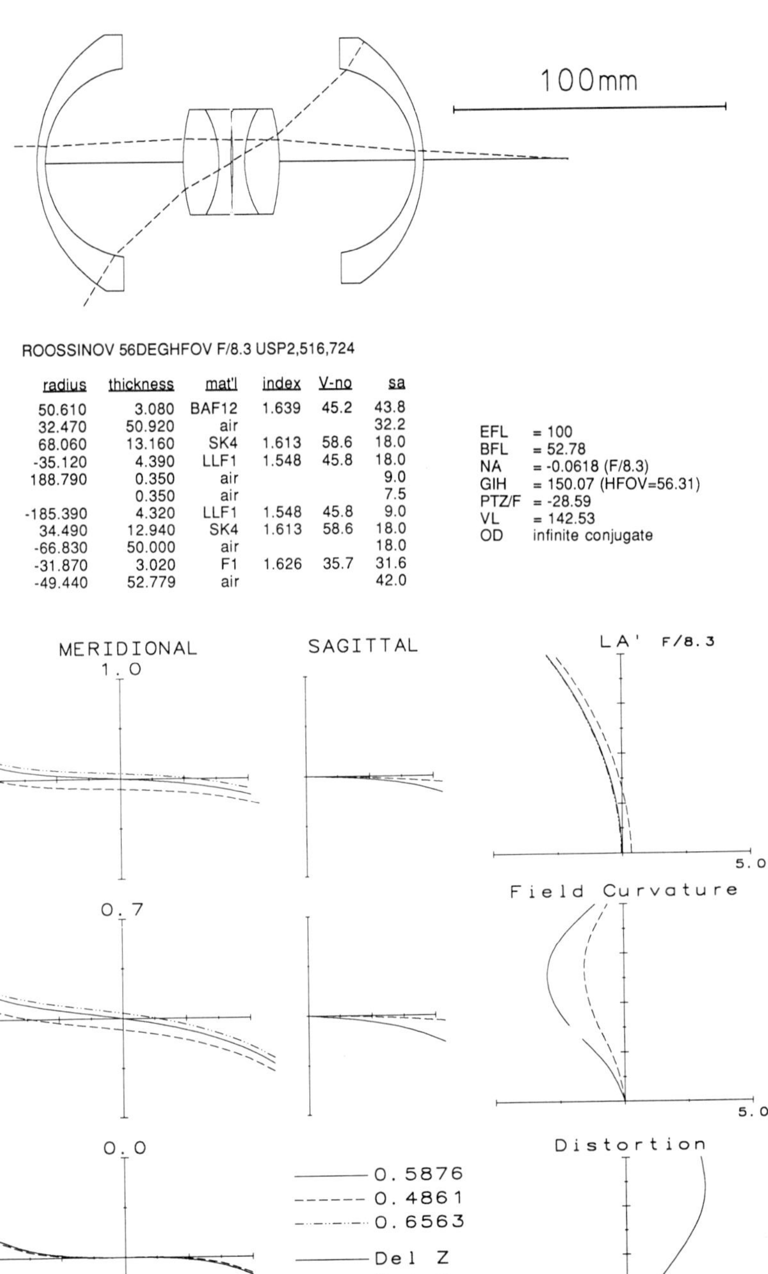

ROOSSINOV 56DEGHFOV F/8.3 USP2,516,724

radius	thickness	mat'l	index	V-no	sa
50.610	3.080	BAF12	1.639	45.2	43.8
32.470	50.920	air			32.2
68.060	13.160	SK4	1.613	58.6	18.0
-35.120	4.390	LLF1	1.548	45.8	18.0
188.790	0.350	air			9.0
	0.350	air			7.5
-185.390	4.320	LLF1	1.548	45.8	9.0
34.490	12.940	SK4	1.613	58.6	18.0
-66.830	50.000	air			18.0
-31.870	3.020	F1	1.626	35.7	31.6
-49.440	52.779	air			42.0

EFL = 100
BFL = 52.78
NA = -0.0618 (F/8.3)
GIH = 150.07 (HFOV=56.31)
PTZ/F = -28.59
VL = 142.53
OD infinite conjugate

Figure 18.8

Chapter

19

Projection TV Lenses and Macro Lenses

19.1 Projection TV Lenses

The lenses used to project large-screen television must have a high speed (usually *f*/1 or faster) in order to project a bright enough image, and must also cover a relatively wide field in order to produce a large picture in a short throw. However, they do not need to be fully chromatically corrected, since the three color images are independently projected and each lens covers only a limited spectral band. The resolution requirements are relatively modest, in keeping with the low resolution of television, but, of course, a high MTF is needed to maintain the image quality at this low resolution. Plastic elements are used for low cost and light weight, and (primarily) because they allow for the economical inclusion of aspheric surfaces, which are absolutely essential to the design. The resultant high tooling and start-up costs are readily amortized over the large production runs.

The basic concept behind most of these lenses begins with a positive element and a field flattener, plus some aspheric surfaces. The two sample designs shown here are fairly typical. Figure 19.1 is a simple three-element version with a weak, aspheric-surfaced leading element and an aspheric field flattener, both made of acrylic (which is the crown of the plastic materials). The primary function of the leading element is to act as a corrector for the rest of the system. The strong positive element is made of glass and has spherical surfaces, since an aspheric glass element would be prohibitively expensive. A problem encountered with plastic optics is that they shift focus with temperature because of their high thermal expansion coefficient and their large negative change of index with temperature (dN/dT) both of which work to increase the focal length as the temperature rises.

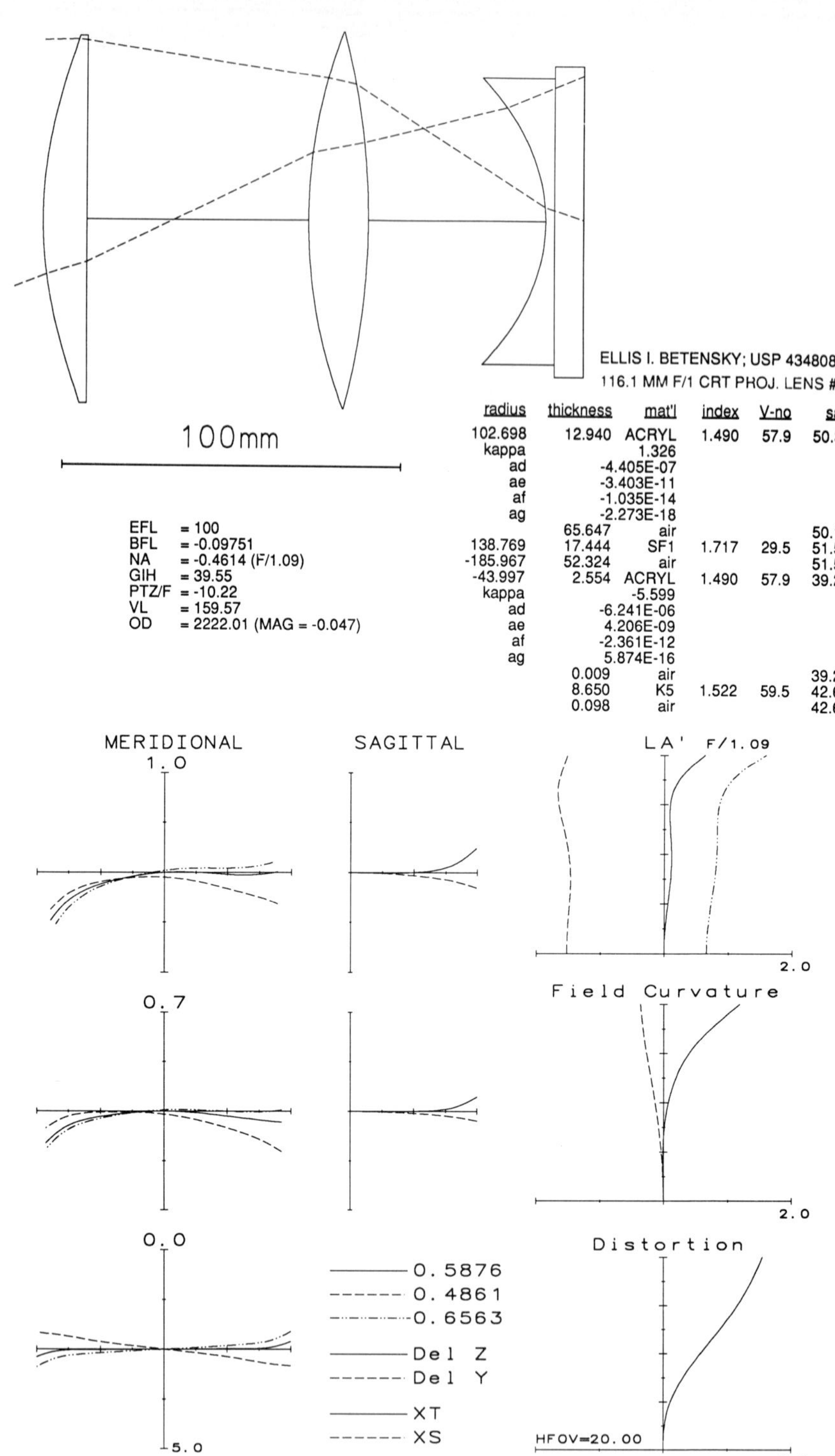

radius	thickness	mat'l	index	V-no	sa
102.698	12.940	ACRYL	1.490	57.9	50.3
kappa	1.326				
ad	-4.405E-07				
ae	-3.403E-11				
af	-1.035E-14				
ag	-2.273E-18				
	65.647	air			50.1
138.769	17.444	SF1	1.717	29.5	51.5
-185.967	52.324	air			51.5
-43.997	2.554	ACRYL	1.490	57.9	39.2
kappa	-5.599				
ad	-6.241E-06				
ae	4.206E-09				
af	-2.361E-12				
ag	5.874E-16				
	0.009	air			39.2
	8.650	K5	1.522	59.5	42.6
	0.098	air			42.6

Figure 19.1

Thus, putting most of the lens power into a much more stable glass element, as in this design, tends to correct this problem. Figure 19.2 uses four acrylic elements, each with one aspheric surface.

19.2 Macro Lenses

The aberration correction of a lens changes with object distance. Some design forms are more sensitive in this regard than others. High-speed lenses are usually more subject to this effect, and, for a given object distance, the effect is obviously more pronounced for longer-focal-length lenses. The two sample lenses in this section are the double-Gauss or Biotar form, which is a very high quality lens, but one which is relatively sensitive to changes in conjugate distances. The designs have been stabilized as discussed in Sec. 10.2 by changing airspaces as the lens is focused on nearby objects.

Figures 19.3 and 19.4 show the same lens with object distances of infinity and 40 focal lengths, respectively; the last airspace is changed to maintain correction. In effect, this lens is focused simply by moving all but the last element, which remains fixed in place. In Figs. 19.5 and 19.6 a similar lens is shown, but with a more complex motion and a shorter object distance of less than nine focal lengths.

Figure 20.5 in Chap. 20 shows a zoom lens with macro capability.

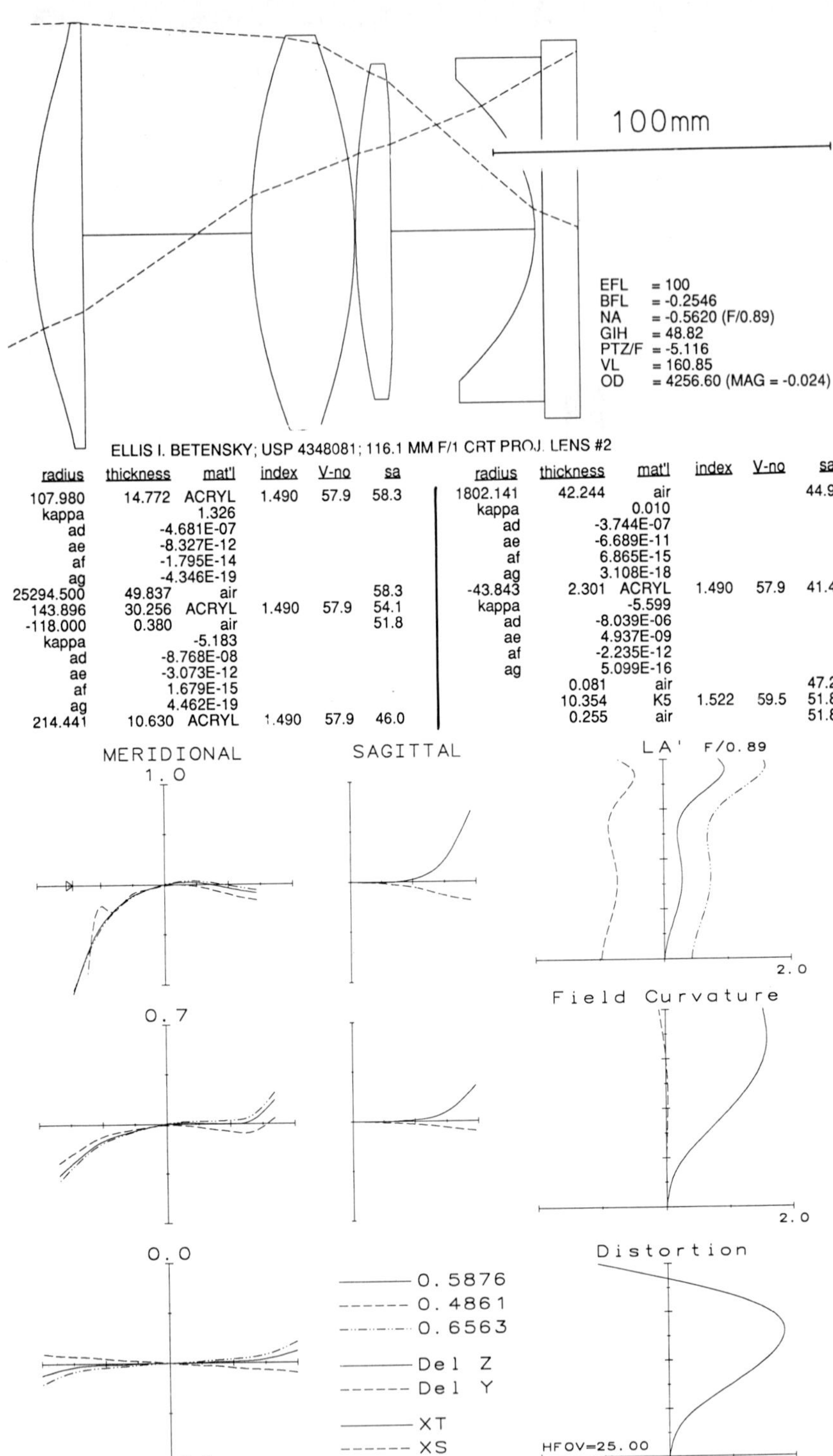

ELLIS I. BETENSKY; USP 4348081; 116.1 MM F/1 CRT PROJ. LENS #2

radius	thickness	mat'l	index	V-no	sa
107.980	14.772	ACRYL	1.490	57.9	58.3
kappa		1.326			
ad		-4.681E-07			
ae		-8.327E-12			
af		-1.795E-14			
ag		-4.346E-19			
25294.500	49.837	air			58.3
143.896	30.256	ACRYL	1.490	57.9	54.1
-118.000	0.380	air			51.8
kappa		-5.183			
ad		-8.768E-08			
ae		-3.073E-12			
af		1.679E-15			
ag		4.462E-19			
214.441	10.630	ACRYL	1.490	57.9	46.0

radius	thickness	mat'l	index	V-no	sa
1802.141	42.244	air			44.9
kappa		0.010			
ad		-3.744E-07			
ae		-6.689E-11			
af		6.865E-15			
ag		3.108E-18			
-43.843	2.301	ACRYL	1.490	57.9	41.4
kappa		-5.599			
ad		-8.039E-06			
ae		4.937E-09			
af		-2.235E-12			
ag		5.099E-16			
	0.081	air			47.2
	10.354	K5	1.522	59.5	51.8
	0.255	air			51.8

Figure 19.2

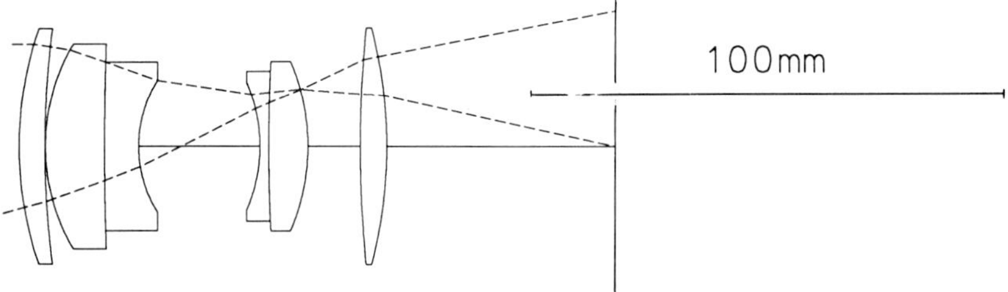

IKUO MORI; USP 4390252; F/2.5 30 DEG. TELEPHOTO LENS #1 INF.

radius	thickness	mat'l	index	V-no	sa
65.731	5.500	TAC6	1.755	52.3	23.1
166.522	0.130	air			22.2
37.250	12.500	LACL7	1.670	57.3	20.1
575.000	7.250	F3	1.613	37.0	16.6
23.130	9.000	air			12.9
	16.750	air			0.0
-28.630	1.880	LAF11	1.757	31.7	12.5
312.500	8.380	LAF20	1.744	44.9	14.7
-38.880	11.000	air			16.6
218.151	5.630	TAF2	1.794	45.4	23.0
-84.847	48.274	air			23.2

EFL = 99.99
BFL = 48.27
NA = -0.2003 (F/2.5)
GIH = 26.79 (HFOV=15.00)
PTZ/F = -7.04
VL = 78.02
OD infinite conjugate

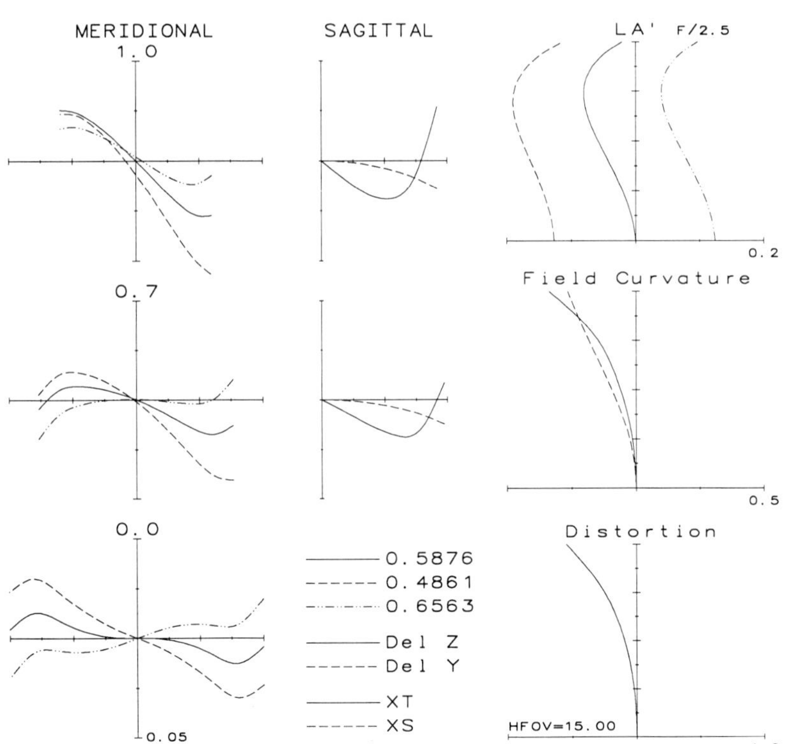

Figure 19.3

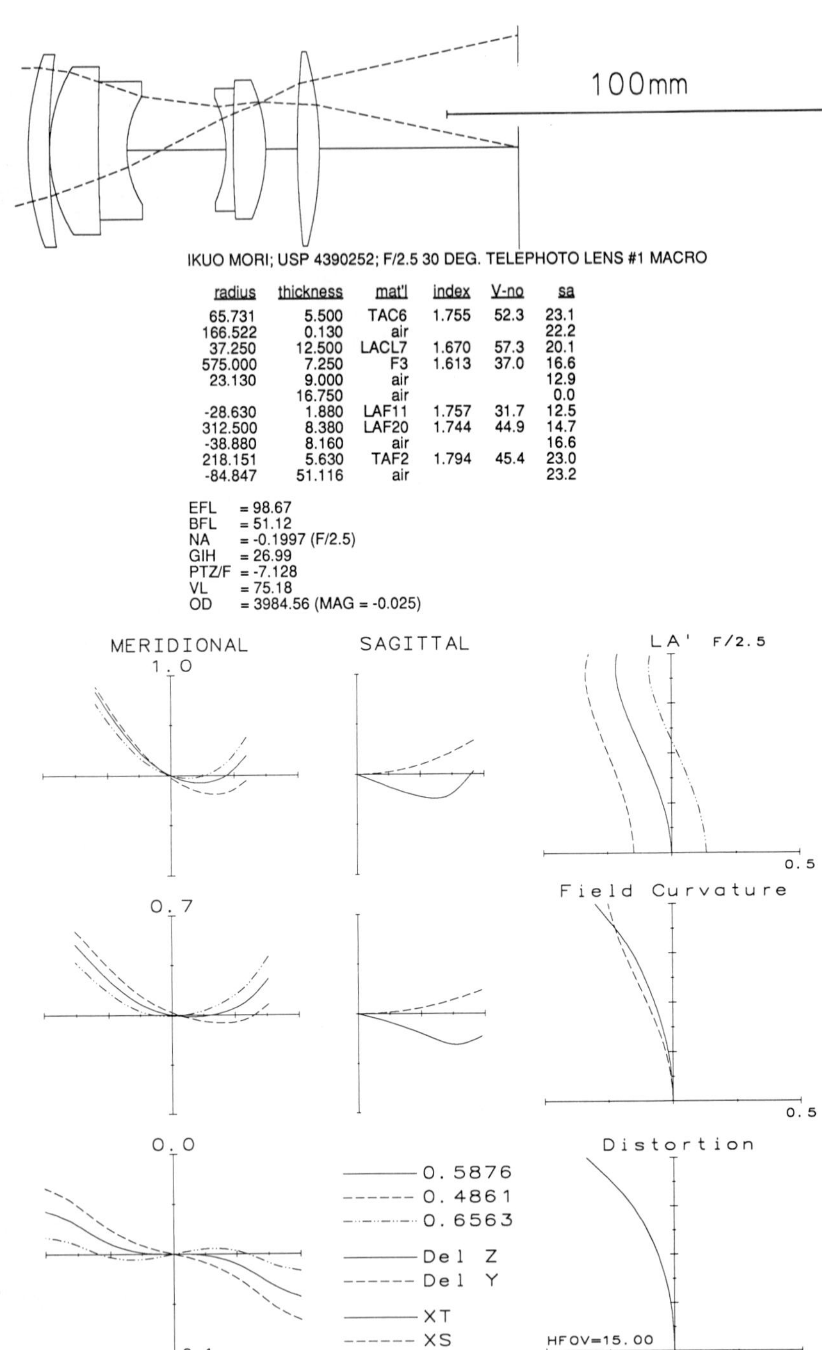

IKUO MORI; USP 4390252; F/2.5 30 DEG. TELEPHOTO LENS #1 MACRO

radius	thickness	mat'l	index	V-no	sa
65.731	5.500	TAC6	1.755	52.3	23.1
166.522	0.130	air			22.2
37.250	12.500	LACL7	1.670	57.3	20.1
575.000	7.250	F3	1.613	37.0	16.6
23.130	9.000	air			12.9
	16.750	air			0.0
-28.630	1.880	LAF11	1.757	31.7	12.5
312.500	8.380	LAF20	1.744	44.9	14.7
-38.880	8.160	air			16.6
218.151	5.630	TAF2	1.794	45.4	23.0
-84.847	51.116	air			23.2

EFL = 98.67
BFL = 51.12
NA = -0.1997 (F/2.5)
GIH = 26.99
PTZ/F = -7.128
VL = 75.18
OD = 3984.56 (MAG = -0.025)

Figure 19.4

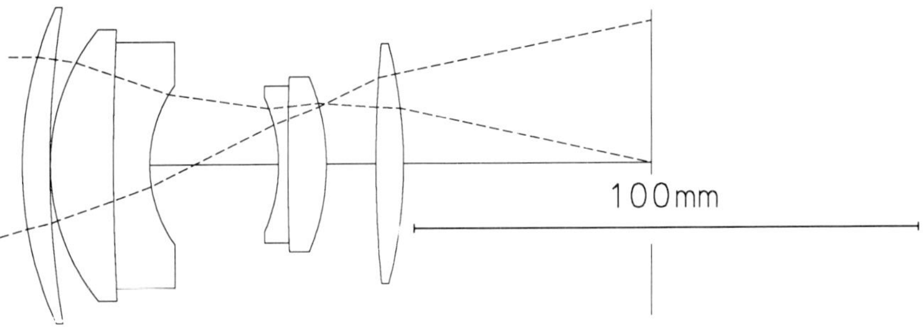

IKUO MORI; USP 4426137; F/2.5 29.5 DEG. CAMERA LENS - INFINITE

radius	thickness	mat'l	index	V-no	sa
66.185	5.530	TAC6	1.755	52.3	29.1
166.926	0.120	air			28.6
37.470	12.530	LACL7	1.670	57.3	25.0
463.860	7.250	F2	1.620	36.3	22.6
23.590	9.000	air			14.7
	16.799	air			12.3
-28.600	1.840	NBFD9	1.757	31.8	12.4
479.120	7.620	LAF20	1.744	44.9	14.4
-38.980	9.828	air			16.1
228.492	5.530	TAF2	1.794	45.4	21.9
-82.104	49.216	air			22.1

EFL = 99.99
BFL = 49.22
NA = -0.2002 (F/2.5)
GIH = 26.33 (HFOV=14.75)
PTZ/F = -6.389
VL = 76.05
OD infinite conjugate

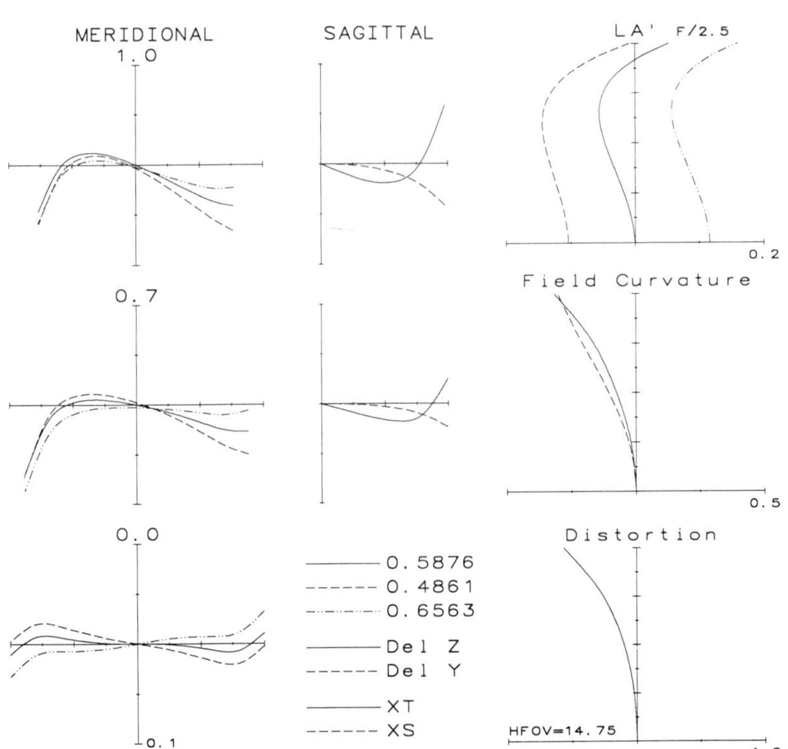

Figure 19.5

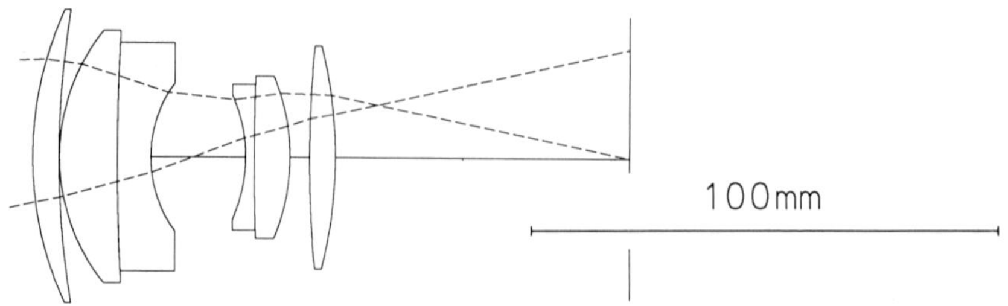

IKUO MORI; USP 4426137; F/2.5 29.5 DEG. CAMERA LENS - MACRO

radius	thickness	mat'l	index	V-no	sa
66.185	5.530	TAC6	1.755	52.3	29.1
166.926	0.120	air			28.6
37.470	12.530	LACL7	1.670	57.3	25.0
463.860	7.250	F2	1.620	36.3	22.6
23.590	9.000	air			14.7
	11.176	air			12.3
-28.600	1.840	NBFD9	1.757	31.8	12.4
479.120	7.620	LAF20	1.744	44.9	14.4
-38.980	4.205	air			16.1
228.492	5.530	TAF2	1.794	45.4	21.9
-82.104	62.819	air			22.1

EFL = 93.94
BFL = 62.82
NA = -0.1913 (F/2.6)
GIH = 21.64
PTZ/F = -6.78
VL = 64.80
OD = 872.55 (MAG = -0.114)

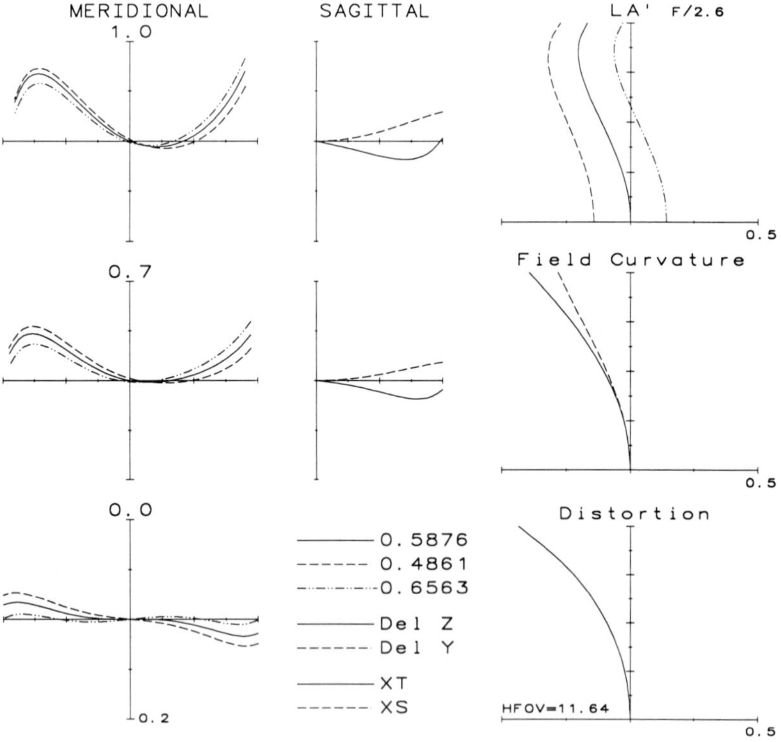

Figure 19.6

Chapter

20

Zoom Lenses

Other than a single component working near unit magnification, the simplest zoom lens consists of two components with a variable airspace between them to change the focal length, plus, of course, a provision to shift the entire system to keep it in focus. In order to get the maximum change of focal length, one component is usually made positive and the other negative. Thus the two possible arrangements are similar to either a telephoto or a retrofocus, depending on whether the positive or negative component is on the object side of the lens. The telephoto arrangement is obviously more suited to longer focal lengths, and the retrofocus arrangement, to shorter focal lengths and wider angular coverage. Figures 20.1 and 20.2 are systems of the latter type, with the negative component facing the object. Figure 20.1 has a modest zoom ratio (long efl/short efl) of only 1.46 but covers a wide field angle of 83° at its short-focal-length setting. It maintains a speed of *f*/3.5 throughout the zoom. There are nine elements, and the eighth surface is a general aspheric.

Figure 20.2 has a larger zoom ratio of 1.9 and covers a somewhat smaller angle of 61° at the short focus. Its *f* number changes from *f*/3.5 to *f*/4.5 as the lens is zoomed from short to long focal length. In a modern camera with automatic exposure control this change in lens speed is not a serious problem. But without automatic exposure control, the iris must be located in a fixed position relative to the image plane so that the *f* number and the image illumination remain constant throughout the zoom. Note that, in this lens, the designer inserted a fixed, weak meniscus lens behind the two moving components. Its weak power (only 9 percent of the midrange power) leads one to wonder if it could be removed from the design.

(*Text continues on page 379.*)

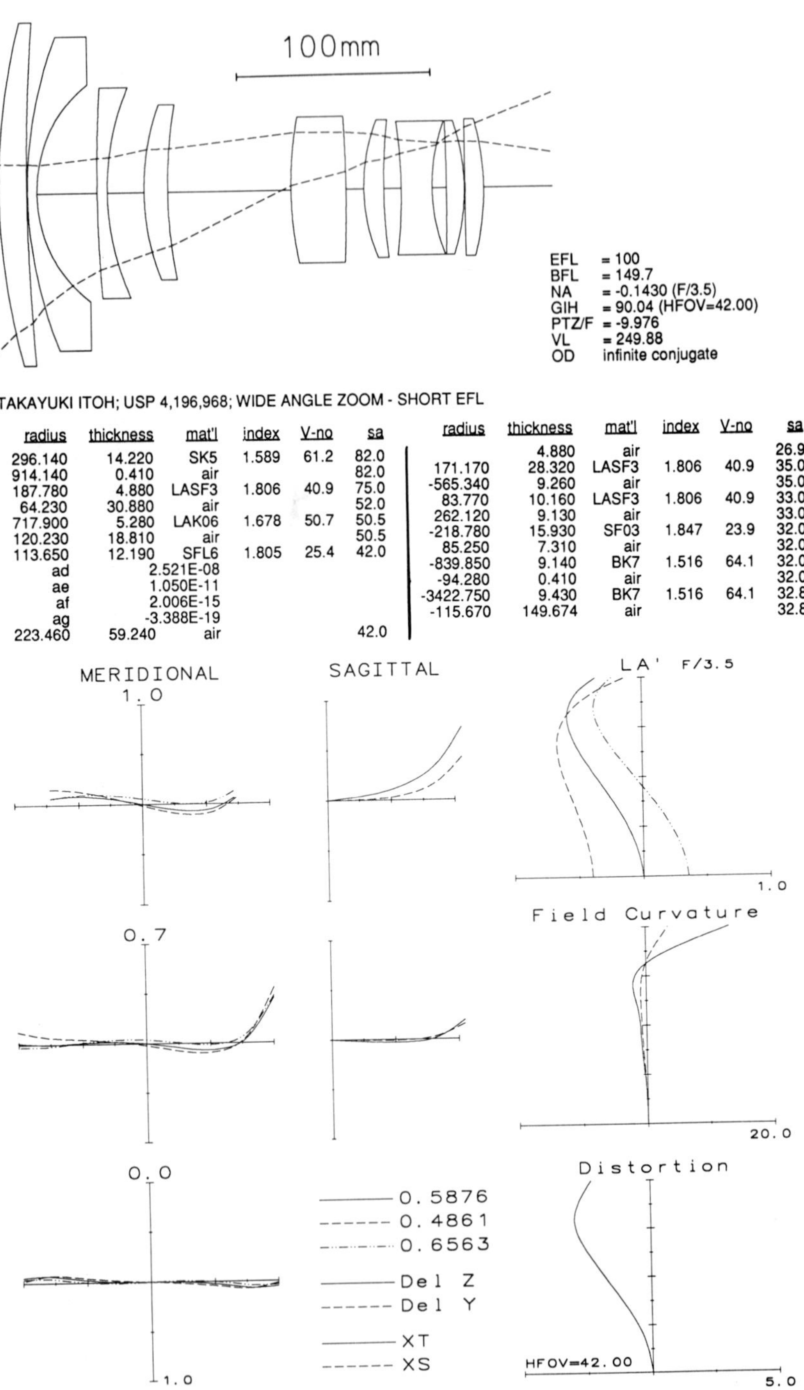

TAKAYUKI ITOH; USP 4,196,968; WIDE ANGLE ZOOM - SHORT EFL

radius	thickness	mat'l	index	V-no	sa
296.140	14.220	SK5	1.589	61.2	82.0
914.140	0.410	air			82.0
187.780	4.880	LASF3	1.806	40.9	75.0
64.230	30.880	air			52.0
717.900	5.280	LAK06	1.678	50.7	50.5
120.230	18.810	air			50.5
113.650	12.190	SFL6	1.805	25.4	42.0
ad	2.521E-08				
ae	1.050E-11				
af	2.006E-15				
ag	-3.388E-19				
223.460	59.240	air			42.0

radius	thickness	mat'l	index	V-no	sa
	4.880	air			26.9
171.170	28.320	LASF3	1.806	40.9	35.0
-565.340	9.260	air			35.0
83.770	10.160	LASF3	1.806	40.9	33.0
262.120	9.130	air			33.0
-218.780	15.930	SF03	1.847	23.9	32.0
85.250	7.310	air			32.0
-839.850	9.140	BK7	1.516	64.1	32.0
-94.280	0.410	air			32.0
-3422.750	9.430	BK7	1.516	64.1	32.8
-115.670	149.674	air			32.8

Figure 20.1

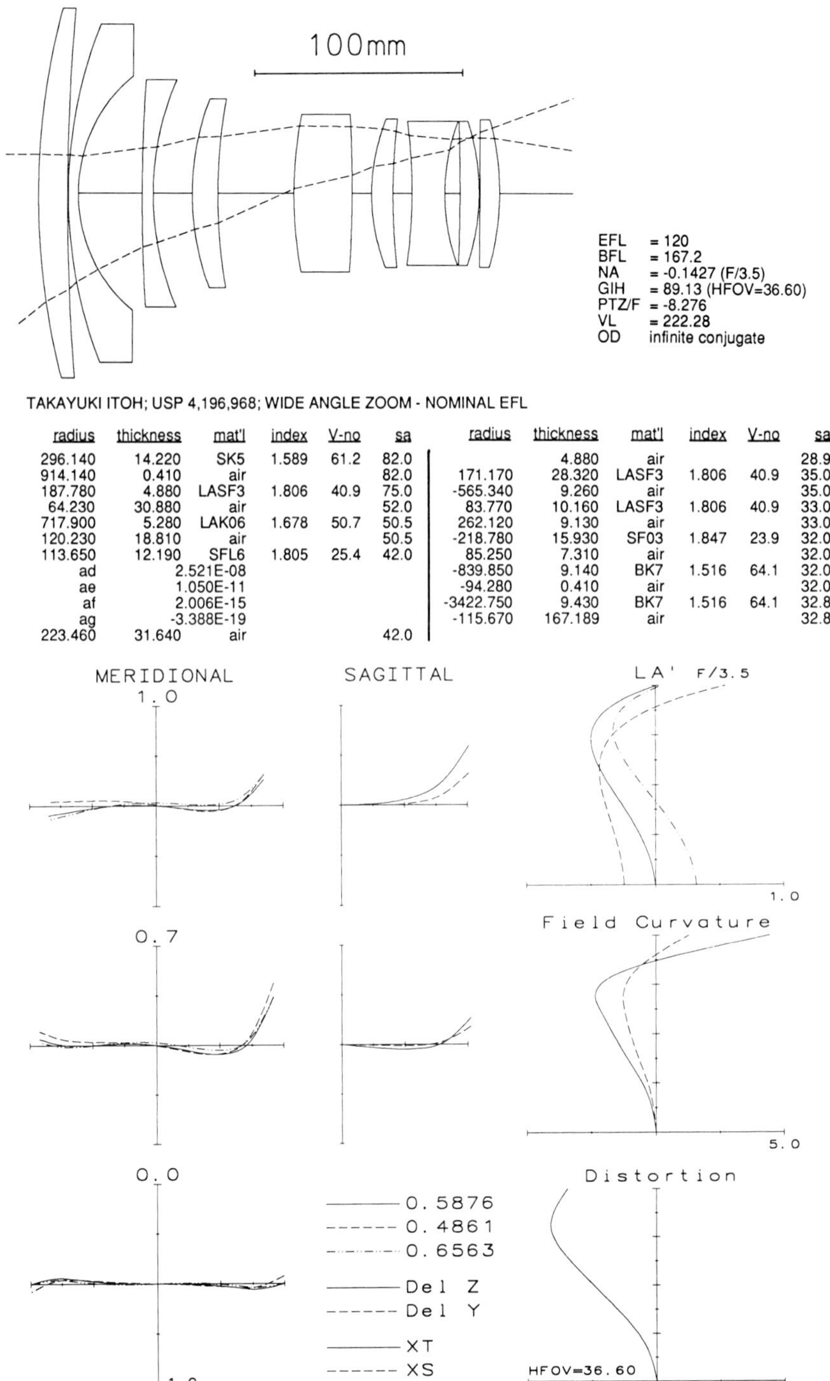

TAKAYUKI ITOH; USP 4,196,968; WIDE ANGLE ZOOM - NOMINAL EFL

radius	thickness	mat'l	index	V-no	sa
296.140	14.220	SK5	1.589	61.2	82.0
914.140	0.410	air			82.0
187.780	4.880	LASF3	1.806	40.9	75.0
64.230	30.880	air			52.0
717.900	5.280	LAK06	1.678	50.7	50.5
120.230	18.810	air			50.5
113.650	12.190	SFL6	1.805	25.4	42.0
ad	2.521E-08				
ae	1.050E-11				
af	2.006E-15				
ag	-3.388E-19				
223.460	31.640	air			42.0
	4.880	air			28.9
171.170	28.320	LASF3	1.806	40.9	35.0
-565.340	9.260	air			35.0
83.770	10.160	LASF3	1.806	40.9	33.0
262.120	9.130	air			33.0
-218.780	15.930	SF03	1.847	23.9	32.0
85.250	7.310	air			32.0
-839.850	9.140	BK7	1.516	64.1	32.0
-94.280	0.410	air			32.0
-3422.750	9.430	BK7	1.516	64.1	32.8
-115.670	167.189	air			32.8

Figure 20.1 *(Continued)*

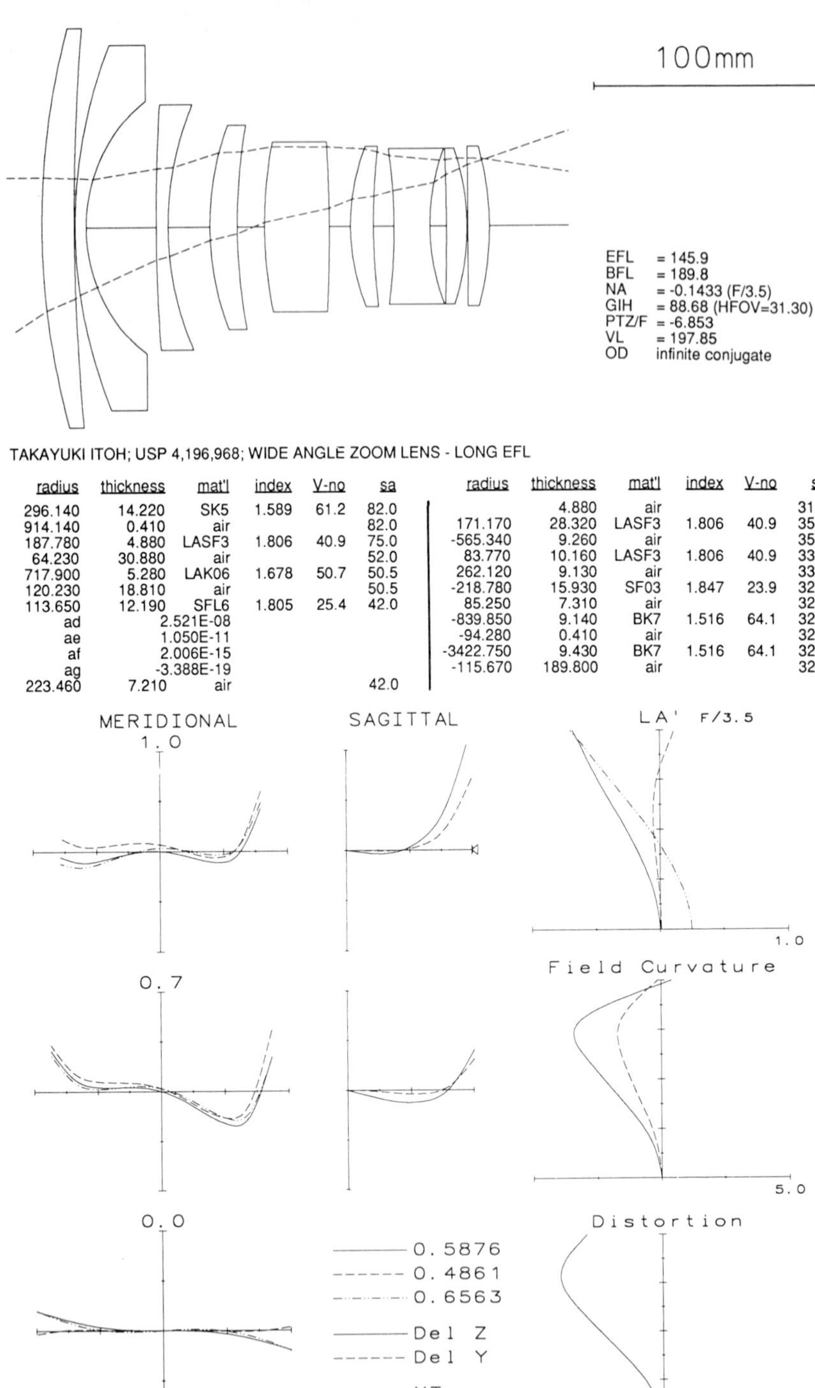

TAKAYUKI ITOH; USP 4,196,968; WIDE ANGLE ZOOM LENS - LONG EFL

radius	thickness	mat'l	index	V-no	sa
296.140	14.220	SK5	1.589	61.2	82.0
914.140	0.410	air			82.0
187.780	4.880	LASF3	1.806	40.9	75.0
64.230	30.880	air			52.0
717.900	5.280	LAK06	1.678	50.7	50.5
120.230	18.810	air			50.5
113.650	12.190	SFL6	1.805	25.4	42.0
ad	2.521E-08				
ae	1.050E-11				
af	2.006E-15				
ag	-3.388E-19				
223.460	7.210	air			42.0
	4.880	air			31.8
171.170	28.320	LASF3	1.806	40.9	35.0
-565.340	9.260	air			35.0
83.770	10.160	LASF3	1.806	40.9	33.0
262.120	9.130	air			33.0
-218.780	15.930	SF03	1.847	23.9	32.0
85.250	7.310	air			32.0
-839.850	9.140	BK7	1.516	64.1	32.0
-94.280	0.410	air			32.0
-3422.750	9.430	BK7	1.516	64.1	32.8
-115.670	189.800	air			32.8

Figure 20.1 (*Continued*)

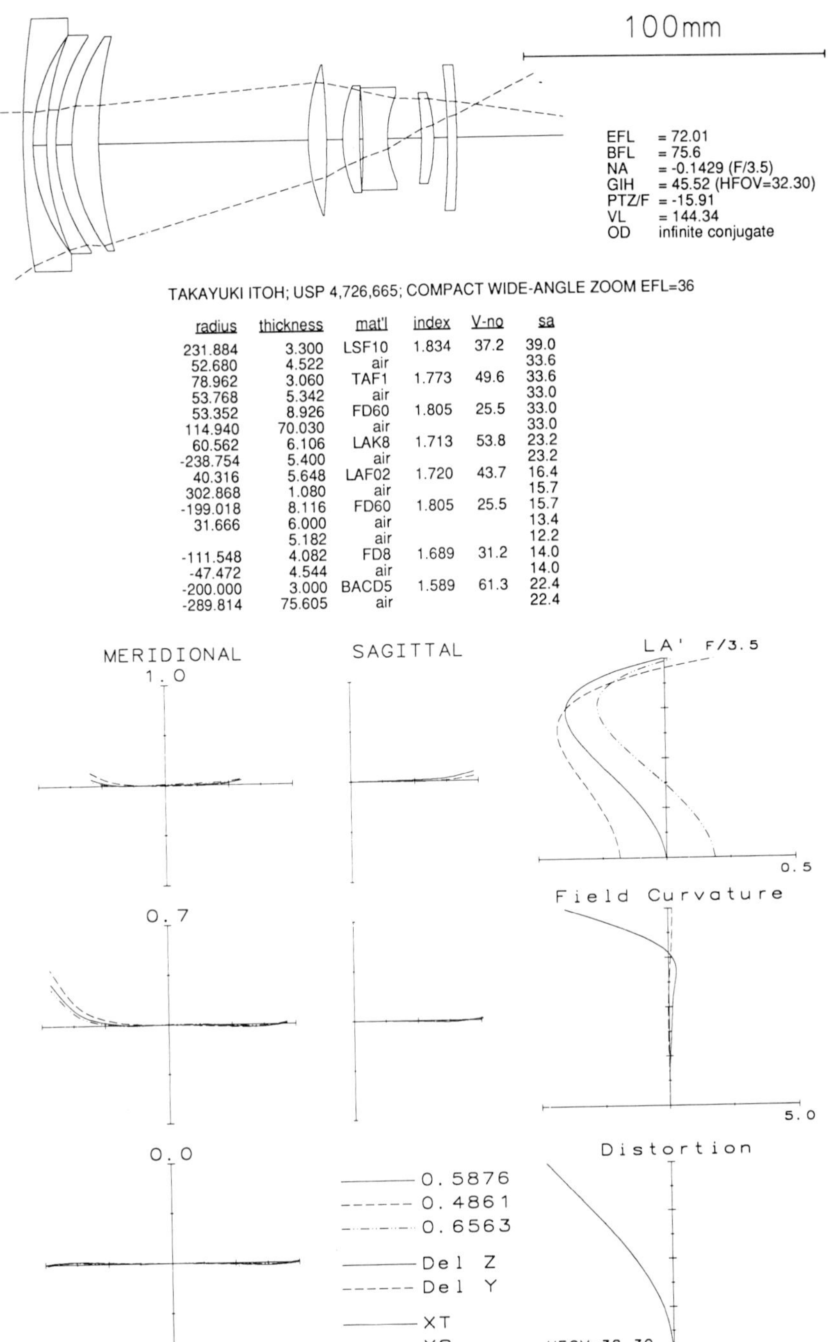

TAKAYUKI ITOH; USP 4,726,665; COMPACT WIDE-ANGLE ZOOM EFL=36

radius	thickness	mat'l	index	V-no	sa
231.884	3.300	LSF10	1.834	37.2	39.0
52.680	4.522	air			33.6
78.962	3.060	TAF1	1.773	49.6	33.6
53.768	5.342	air			33.0
53.352	8.926	FD60	1.805	25.5	33.0
114.940	70.030	air			33.0
60.562	6.106	LAK8	1.713	53.8	23.2
-238.754	5.400	air			23.2
40.316	5.648	LAF02	1.720	43.7	16.4
302.868	1.080	air			15.7
-199.018	8.116	FD60	1.805	25.5	15.7
31.666	6.000	air			13.4
	5.182	air			12.2
-111.548	4.082	FD8	1.689	31.2	14.0
-47.472	4.544	air			14.0
-200.000	3.000	BACD5	1.589	61.3	22.4
-289.814	75.605	air			22.4

Figure 20.2

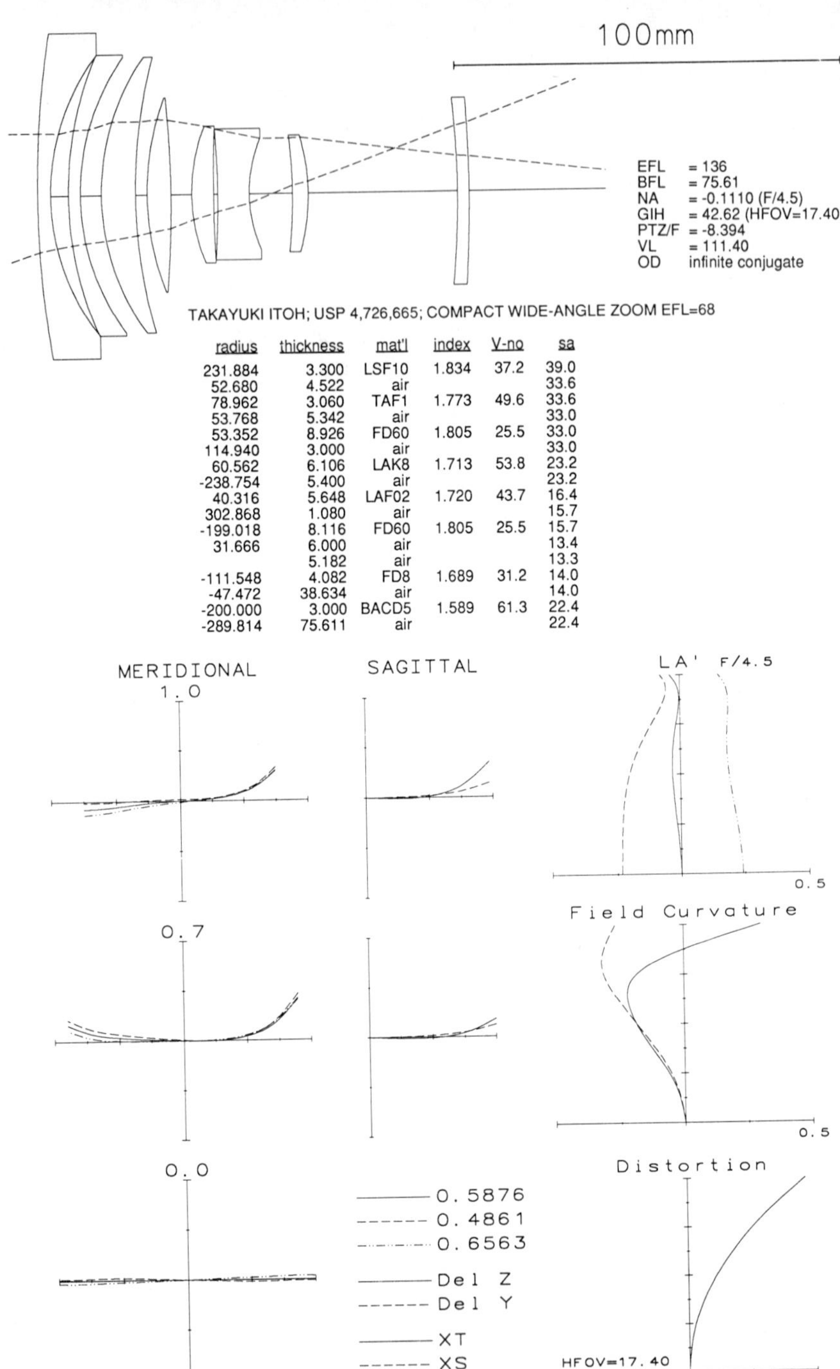

TAKAYUKI ITOH; USP 4,726,665; COMPACT WIDE-ANGLE ZOOM EFL=68

radius	thickness	mat'l	index	V-no	sa
231.884	3.300	LSF10	1.834	37.2	39.0
52.680	4.522	air			33.6
78.962	3.060	TAF1	1.773	49.6	33.6
53.768	5.342	air			33.0
53.352	8.926	FD60	1.805	25.5	33.0
114.940	3.000	air			33.0
60.562	6.106	LAK8	1.713	53.8	23.2
-238.754	5.400	air			23.2
40.316	5.648	LAF02	1.720	43.7	16.4
302.868	1.080	air			15.7
-199.018	8.116	FD60	1.805	25.5	15.7
31.666	6.000	air			13.4
	5.182	air			13.3
-111.548	4.082	FD8	1.689	31.2	14.0
-47.472	38.634	air			14.0
-200.000	3.000	BACD5	1.589	61.3	22.4
-289.814	75.611	air			22.4

Figure 20.2 *(Continued)*

Figure 20.3 is a 12-element, 2.5-zoom-ratio system which works at *f*/2.8 and covers about 30° at its short focal length. It is an almost classic example of a positive-negative-positive (+ − +) afocal zoom unit followed by a four-element focusing lens. The three-element inner negative component moves linearly along the axis to zoom the focal length, and the following positive doublet component moves in a nonlinear path to compensate for the image shift which the zooming component produces. Many lenses with large zoom ratios utilize this + − + configuration to produce the zoom and to compensate for the focus shift.

Figure 20.4 covers a wider field (72°) and has a somewhat larger zoom ratio (2.9), but a significantly slower speed (*f*/4.0 to *f*/5.5) than Fig. 20.3. It also has a considerably more complex set of component motions. Each of the four components has a different motion. Note that the component powers are arranged − + − +; in Fig. 20.3 the zoom part of the system was + − +.

Figure 20.5 is more complex still, in that it incorporates a macro capability. The system has a zoom ratio of 3.4 and covers a field of 72° at a speed of *f*/4 to *f*/5.3, like Fig. 20.4. Its zoom components are also arranged − + − +, followed by a fixed rear doublet. Note that the small third component is also fixed. The macro version is shown for the short focal length in Fig. 20.5 on page 387.

Figure 20.6 shows a zoom scanner lens. It uses a fixed input scan angle of ±14° and a constant HeNe laser beam diameter; as a result, its speed and the scan line length vary throughout the zoom. The external stop is located at the scan mirror. Here again we see an almost classic arrangement of + − +, with the negative component producing the zoom and the third (positive) component moving to maintain the focus. The first and last elements are fixed and the two inner components move. Note that in this example the positive focus compensator element is not quite in the right location to maintain the focus distance constant for the midrange efl setting (Fig. 20.6 on page 391).

In a zoom camera lens which must be focused for various object distances, the focusing cannot be conveniently accomplished by shifting the entire lens, because the amount of focus shift needed will vary with the focal length. For this reason, zoom camera lenses are focused by a spacing change between the outer front elements, so that the object position for the zoom kernel of the lens is a constant one regardless of the actual object position.

Figures 22.15 and 22.16 in Chap. 22 show a zooming lens for use as a laser collimator or focusing lens.

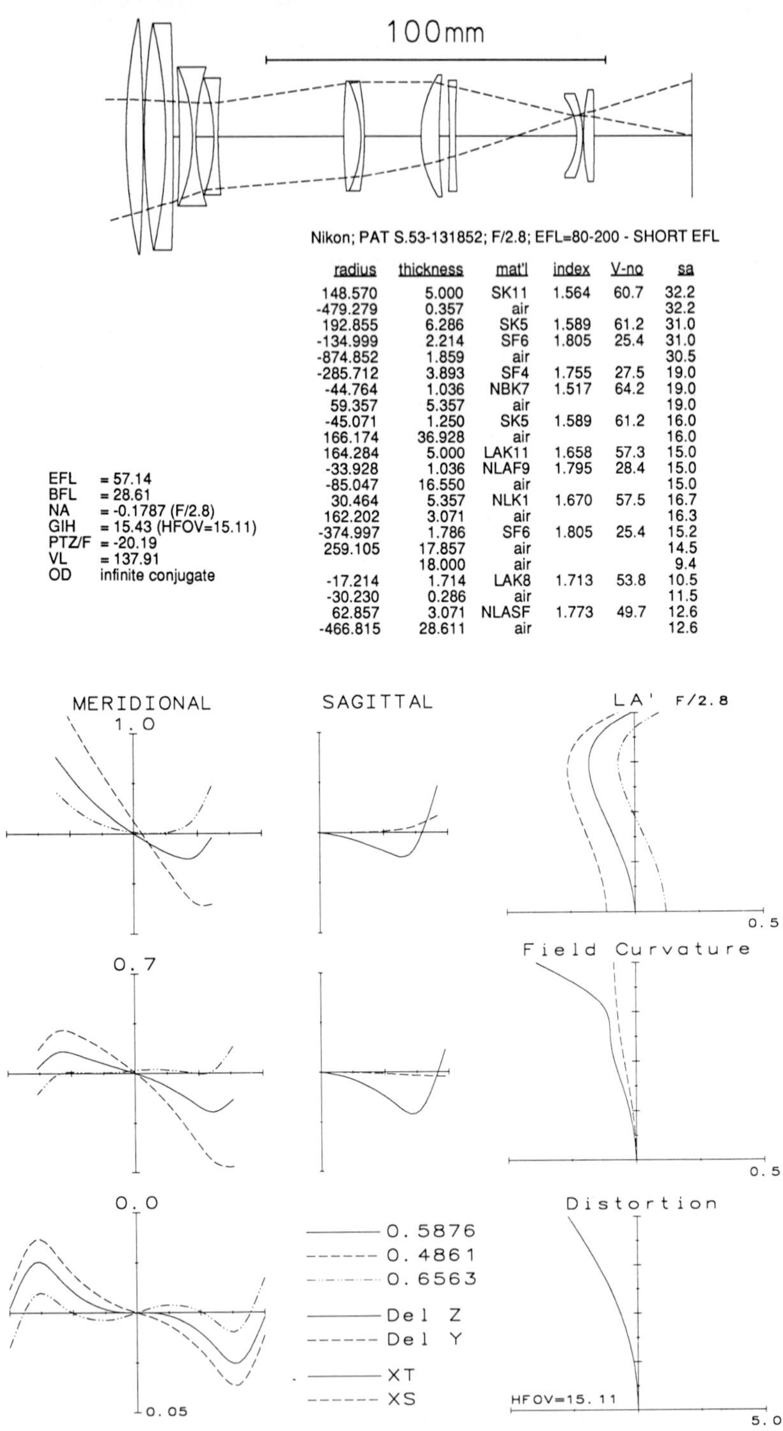

Nikon; PAT S.53-131852; F/2.8; EFL=80-200 - SHORT EFL

radius	thickness	mat'l	index	V-no	sa
148.570	5.000	SK11	1.564	60.7	32.2
-479.279	0.357	air			32.2
192.855	6.286	SK5	1.589	61.2	31.0
-134.999	2.214	SF6	1.805	25.4	31.0
-874.852	1.859	air			30.5
-285.712	3.893	SF4	1.755	27.5	19.0
-44.764	1.036	NBK7	1.517	64.2	19.0
59.357	5.357	air			19.0
-45.071	1.250	SK5	1.589	61.2	16.0
166.174	36.928	air			16.0
164.284	5.000	LAK11	1.658	57.3	15.0
-33.928	1.036	NLAF9	1.795	28.4	15.0
-85.047	16.550	air			15.0
30.464	5.357	NLK1	1.670	57.5	16.7
162.202	3.071	air			16.3
-374.997	1.786	SF6	1.805	25.4	15.2
259.105	17.857	air			14.5
	18.000	air			9.4
-17.214	1.714	LAK8	1.713	53.8	10.5
-30.230	0.286	air			11.5
62.857	3.071	NLASF	1.773	49.7	12.6
-466.815	28.611	air			12.6

EFL	= 57.14
BFL	= 28.61
NA	= -0.1787 (F/2.8)
GIH	= 15.43 (HFOV=15.11)
PTZ/F	= -20.19
VL	= 137.91
OD	infinite conjugate

Figure 20.3

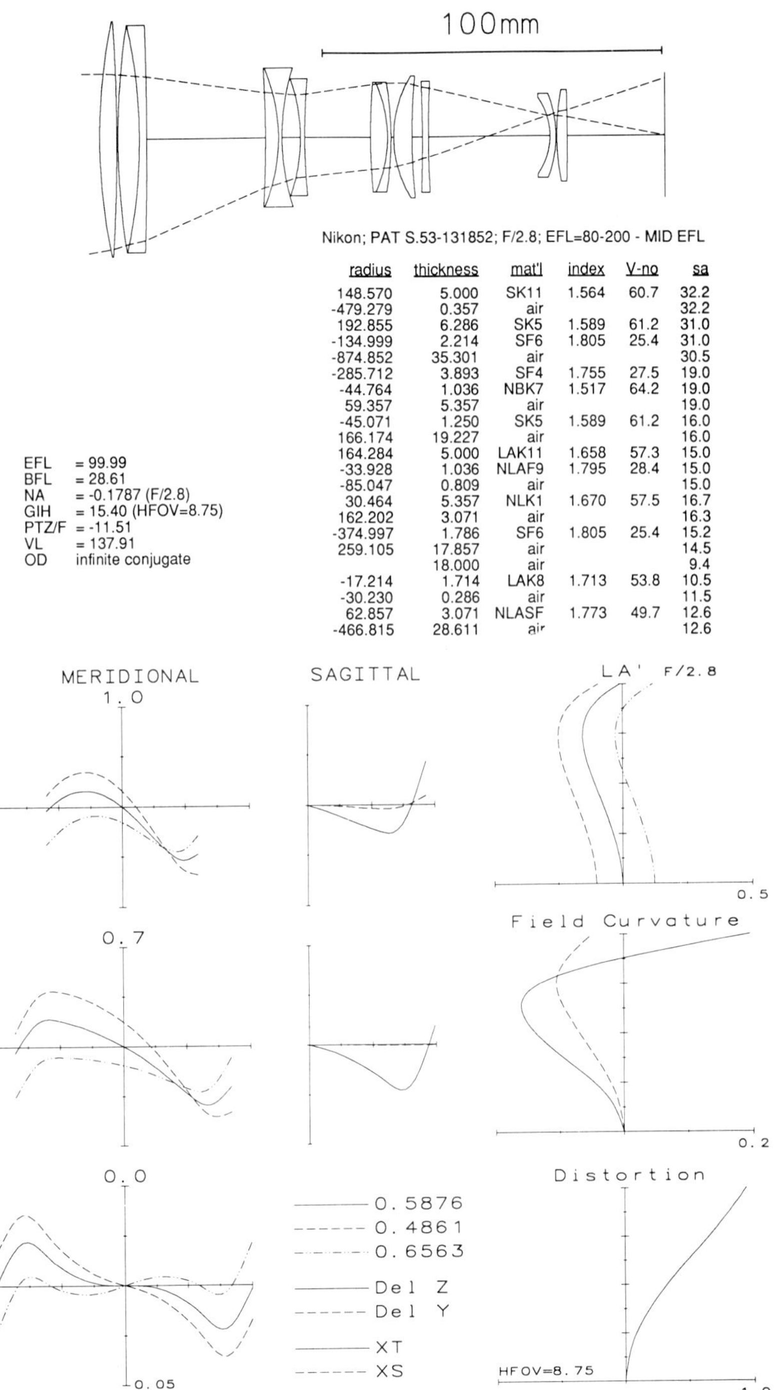

Nikon; PAT S.53-131852; F/2.8; EFL=80-200 - MID EFL

radius	thickness	mat'l	index	V-no	sa
148.570	5.000	SK11	1.564	60.7	32.2
-479.279	0.357	air			32.2
192.855	6.286	SK5	1.589	61.2	31.0
-134.999	2.214	SF6	1.805	25.4	31.0
-874.852	35.301	air			30.5
-285.712	3.893	SF4	1.755	27.5	19.0
-44.764	1.036	NBK7	1.517	64.2	19.0
59.357	5.357	air			19.0
-45.071	1.250	SK5	1.589	61.2	16.0
166.174	19.227	air			16.0
164.284	5.000	LAK11	1.658	57.3	15.0
-33.928	1.036	NLAF9	1.795	28.4	15.0
-85.047	0.809	air			15.0
30.464	5.357	NLK1	1.670	57.5	16.7
162.202	3.071	air			16.3
-374.997	1.786	SF6	1.805	25.4	15.2
259.105	17.857	air			14.5
	18.000	air			9.4
-17.214	1.714	LAK8	1.713	53.8	10.5
-30.230	0.286	air			11.5
62.857	3.071	NLASF	1.773	49.7	12.6
-466.815	28.611	air			12.6

EFL = 99.99
BFL = 28.61
NA = -0.1787 (F/2.8)
GIH = 15.40 (HFOV=8.75)
PTZ/F = -11.51
VL = 137.91
OD infinite conjugate

Figure 20.3 (*Continued*)

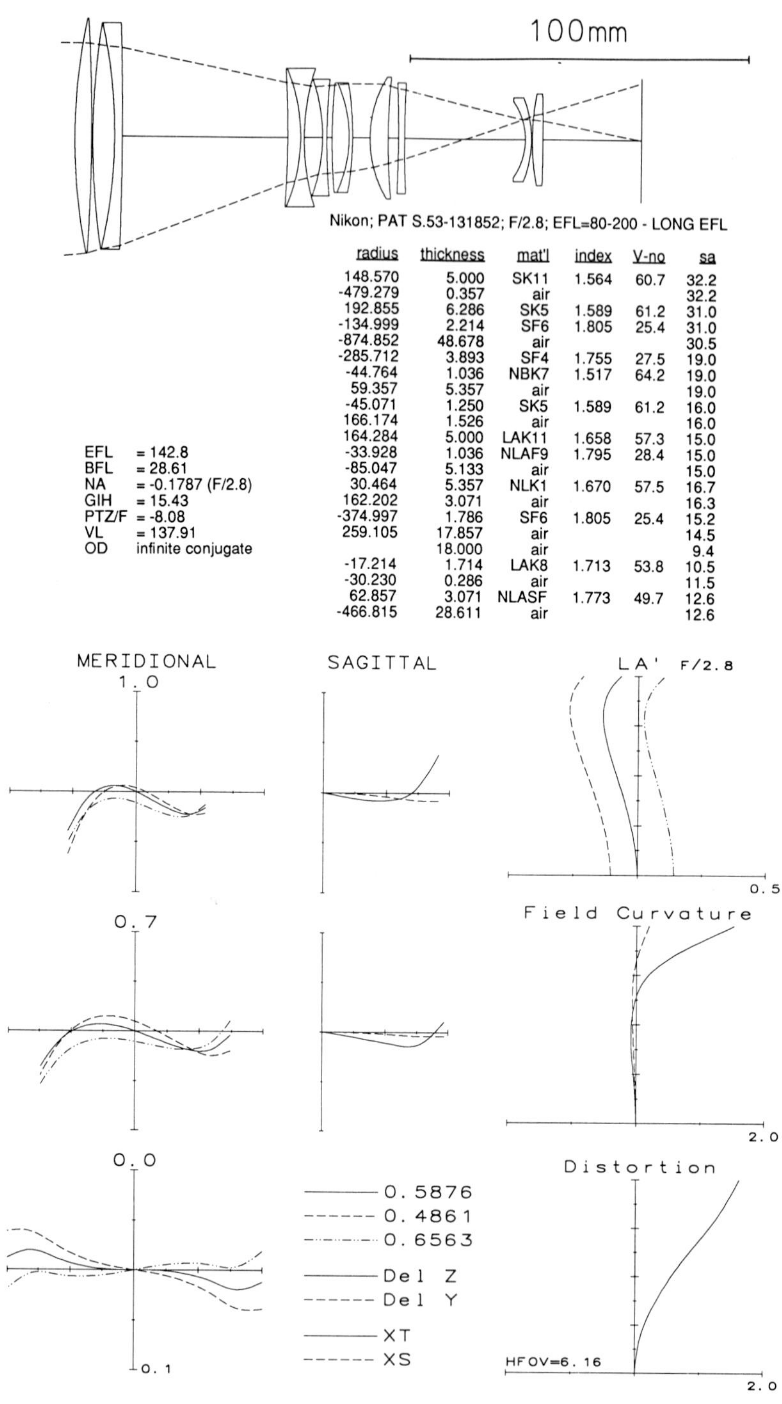

Nikon; PAT S.53-131852; F/2.8; EFL=80-200 - LONG EFL

radius	thickness	mat'l	index	V-no	sa
148.570	5.000	SK11	1.564	60.7	32.2
-479.279	0.357	air			32.2
192.855	6.286	SK5	1.589	61.2	31.0
-134.999	2.214	SF6	1.805	25.4	31.0
-874.852	48.678	air			30.5
-285.712	3.893	SF4	1.755	27.5	19.0
-44.764	1.036	NBK7	1.517	64.2	19.0
59.357	5.357	air			19.0
-45.071	1.250	SK5	1.589	61.2	16.0
166.174	1.526	air			16.0
164.284	5.000	LAK11	1.658	57.3	15.0
-33.928	1.036	NLAF9	1.795	28.4	15.0
-85.047	5.133	air			15.0
30.464	5.357	NLK1	1.670	57.5	16.7
162.202	3.071	air			16.3
-374.997	1.786	SF6	1.805	25.4	15.2
259.105	17.857	air			14.5
	18.000	air			9.4
-17.214	1.714	LAK8	1.713	53.8	10.5
-30.230	0.286	air			11.5
62.857	3.071	NLASF	1.773	49.7	12.6
-466.815	28.611	air			12.6

EFL = 142.8
BFL = 28.61
NA = -0.1787 (F/2.8)
GIH = 15.43
PTZ/F = -8.08
VL = 137.91
OD infinite conjugate

Figure 20.3 *(Continued)*

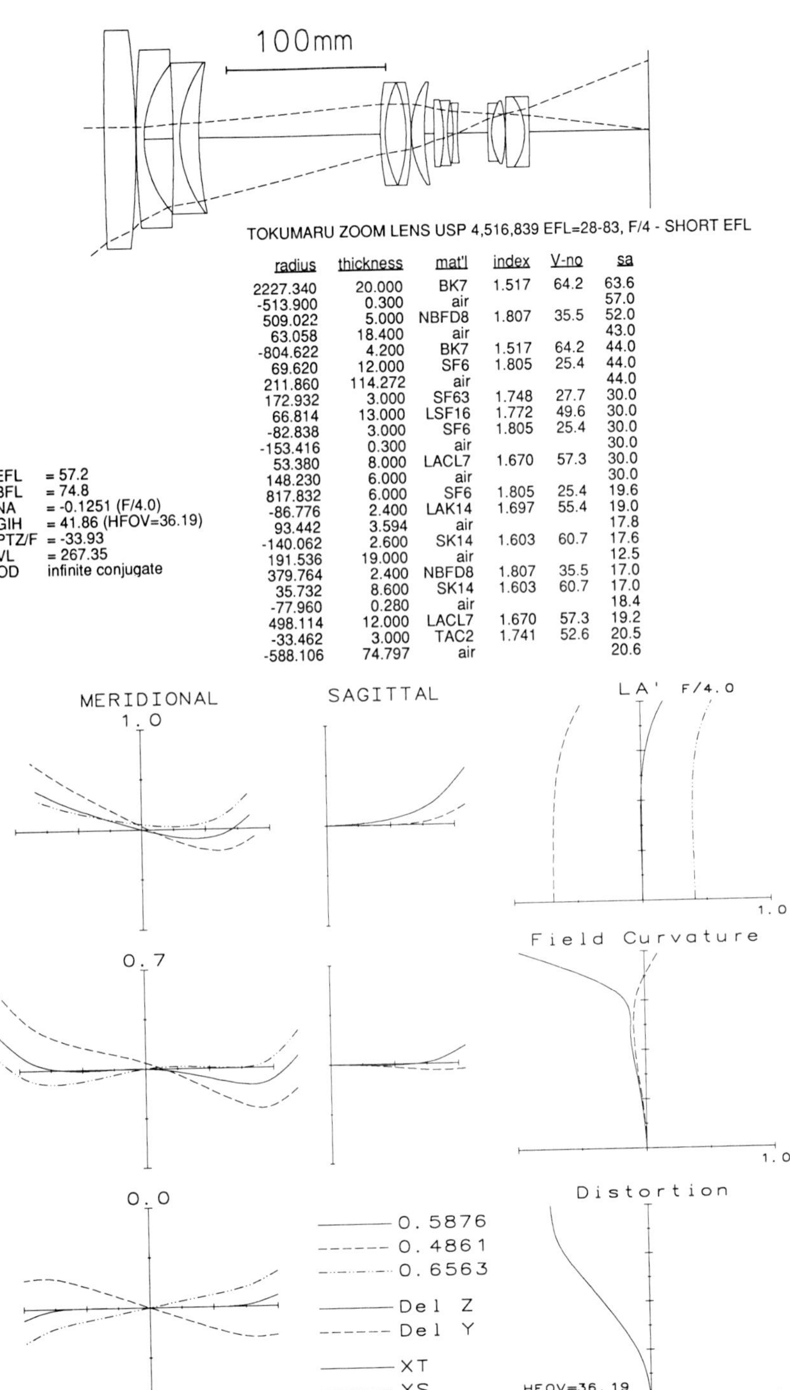

TOKUMARU ZOOM LENS USP 4,516,839 EFL=28-83, F/4 - SHORT EFL

radius	thickness	mat'l	index	V-no	sa
2227.340	20.000	BK7	1.517	64.2	63.6
-513.900	0.300	air			57.0
509.022	5.000	NBFD8	1.807	35.5	52.0
63.058	18.400	air			43.0
-804.622	4.200	BK7	1.517	64.2	44.0
69.620	12.000	SF6	1.805	25.4	44.0
211.860	114.272	air			44.0
172.932	3.000	SF63	1.748	27.7	30.0
66.814	13.000	LSF16	1.772	49.6	30.0
-82.838	3.000	SF6	1.805	25.4	30.0
-153.416	0.300	air			30.0
53.380	8.000	LACL7	1.670	57.3	30.0
148.230	6.000	air			30.0
817.832	6.000	SF6	1.805	25.4	19.6
-86.776	2.400	LAK14	1.697	55.4	19.0
93.442	3.594	air			17.8
-140.062	2.600	SK14	1.603	60.7	17.6
191.536	19.000	air			12.5
379.764	2.400	NBFD8	1.807	35.5	17.0
35.732	8.600	SK14	1.603	60.7	17.0
-77.960	0.280	air			18.4
498.114	12.000	LACL7	1.670	57.3	19.2
-33.462	3.000	TAC2	1.741	52.6	20.5
-588.106	74.797	air			20.6

EFL = 57.2
BFL = 74.8
NA = -0.1251 (F/4.0)
GIH = 41.86 (HFOV=36.19)
PTZ/F = -33.93
VL = 267.35
OD infinite conjugate

Figure 20.4

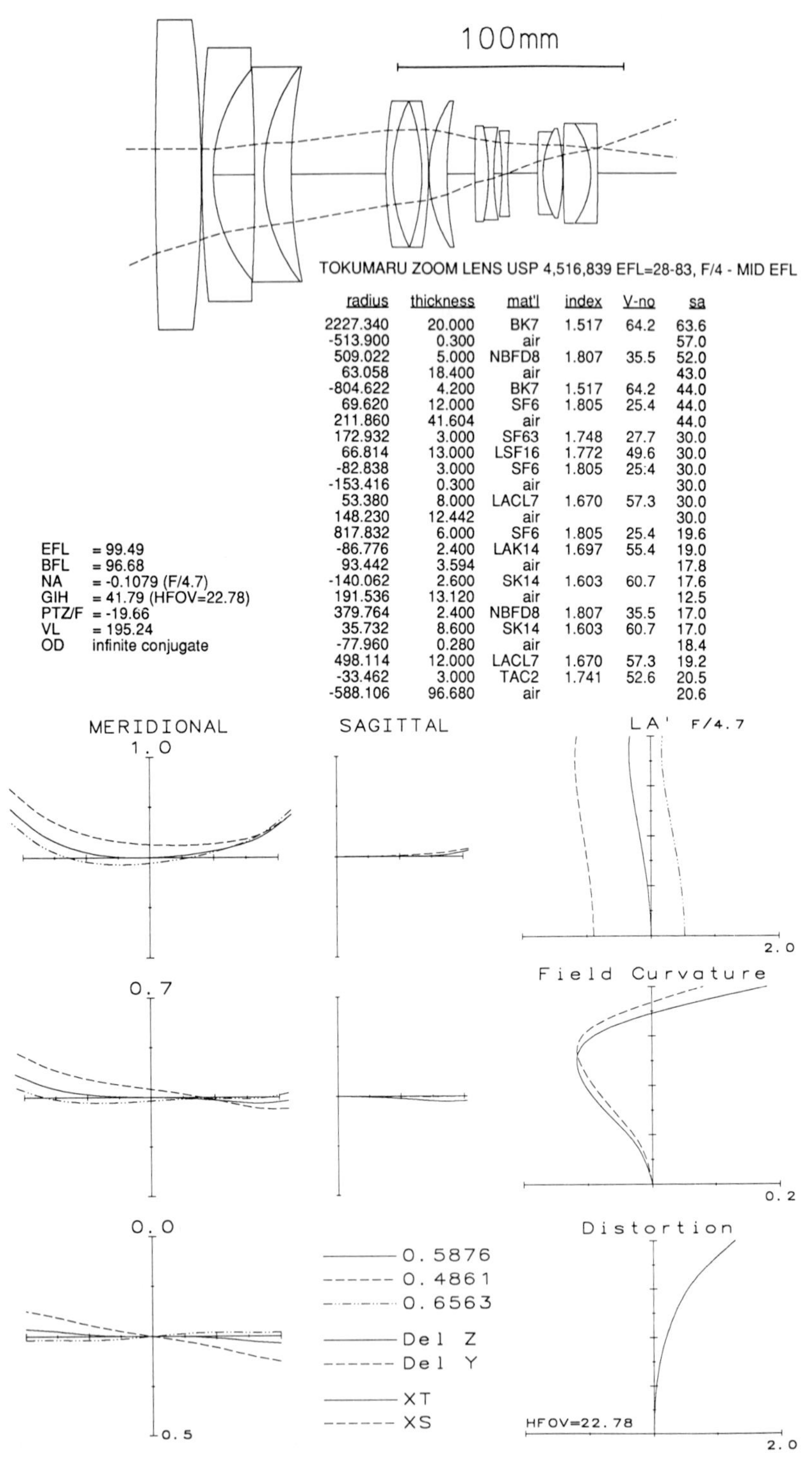

TOKUMARU ZOOM LENS USP 4,516,839 EFL=28-83, F/4 - MID EFL

radius	thickness	mat'l	index	V-no	sa
2227.340	20.000	BK7	1.517	64.2	63.6
-513.900	0.300	air			57.0
509.022	5.000	NBFD8	1.807	35.5	52.0
63.058	18.400	air			43.0
-804.622	4.200	BK7	1.517	64.2	44.0
69.620	12.000	SF6	1.805	25.4	44.0
211.860	41.604	air			44.0
172.932	3.000	SF63	1.748	27.7	30.0
66.814	13.000	LSF16	1.772	49.6	30.0
-82.838	3.000	SF6	1.805	25.4	30.0
-153.416	0.300	air			30.0
53.380	8.000	LACL7	1.670	57.3	30.0
148.230	12.442	air			30.0
817.832	6.000	SF6	1.805	25.4	19.6
-86.776	2.400	LAK14	1.697	55.4	19.0
93.442	3.594	air			17.8
-140.062	2.600	SK14	1.603	60.7	17.6
191.536	13.120	air			12.5
379.764	2.400	NBFD8	1.807	35.5	17.0
35.732	8.600	SK14	1.603	60.7	17.0
-77.960	0.280	air			18.4
498.114	12.000	LACL7	1.670	57.3	19.2
-33.462	3.000	TAC2	1.741	52.6	20.5
-588.106	96.680	air			20.6

EFL = 99.49
BFL = 96.68
NA = -0.1079 (F/4.7)
GIH = 41.79 (HFOV=22.78)
PTZ/F = -19.66
VL = 195.24
OD infinite conjugate

Figure 20.4 *(Continued)*

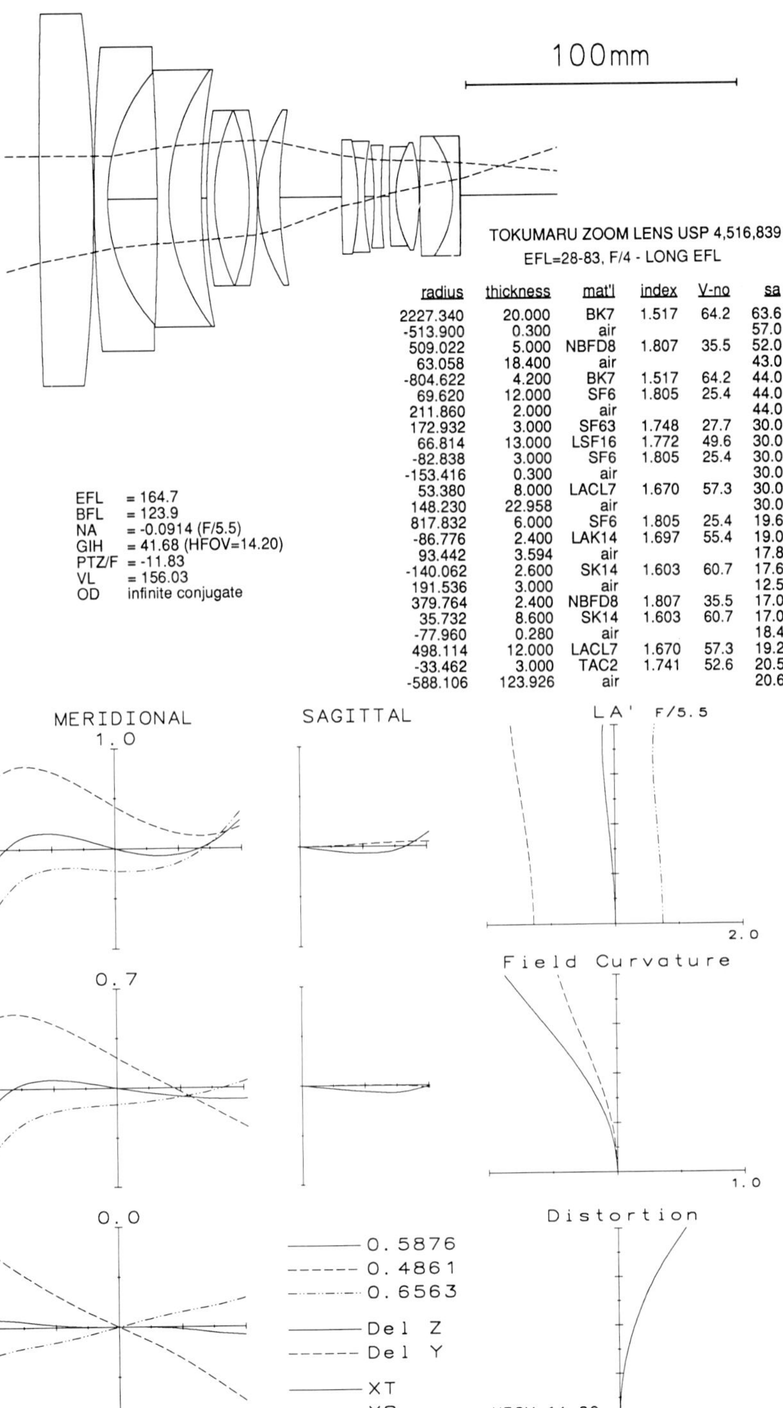

TOKUMARU ZOOM LENS USP 4,516,839
EFL=28-83, F/4 - LONG EFL

radius	thickness	mat'l	index	V-no	sa
2227.340	20.000	BK7	1.517	64.2	63.6
-513.900	0.300	air			57.0
509.022	5.000	NBFD8	1.807	35.5	52.0
63.058	18.400	air			43.0
-804.622	4.200	BK7	1.517	64.2	44.0
69.620	12.000	SF6	1.805	25.4	44.0
211.860	2.000	air			44.0
172.932	3.000	SF63	1.748	27.7	30.0
66.814	13.000	LSF16	1.772	49.6	30.0
-82.838	3.000	SF6	1.805	25.4	30.0
-153.416	0.300	air			30.0
53.380	8.000	LACL7	1.670	57.3	30.0
148.230	22.958	air			30.0
817.832	6.000	SF6	1.805	25.4	19.6
-86.776	2.400	LAK14	1.697	55.4	19.0
93.442	3.594	air			17.8
-140.062	2.600	SK14	1.603	60.7	17.6
191.536	3.000	air			12.5
379.764	2.400	NBFD8	1.807	35.5	17.0
35.732	8.600	SK14	1.603	60.7	17.0
-77.960	0.280	air			18.4
498.114	12.000	LACL7	1.670	57.3	19.2
-33.462	3.000	TAC2	1.741	52.6	20.5
-588.106	123.926	air			20.6

EFL = 164.7
BFL = 123.9
NA = -0.0914 (F/5.5)
GIH = 41.68 (HFOV=14.20)
PTZ/F = -11.83
VL = 156.03
OD infinite conjugate

Figure 20.4 *(Continued)*

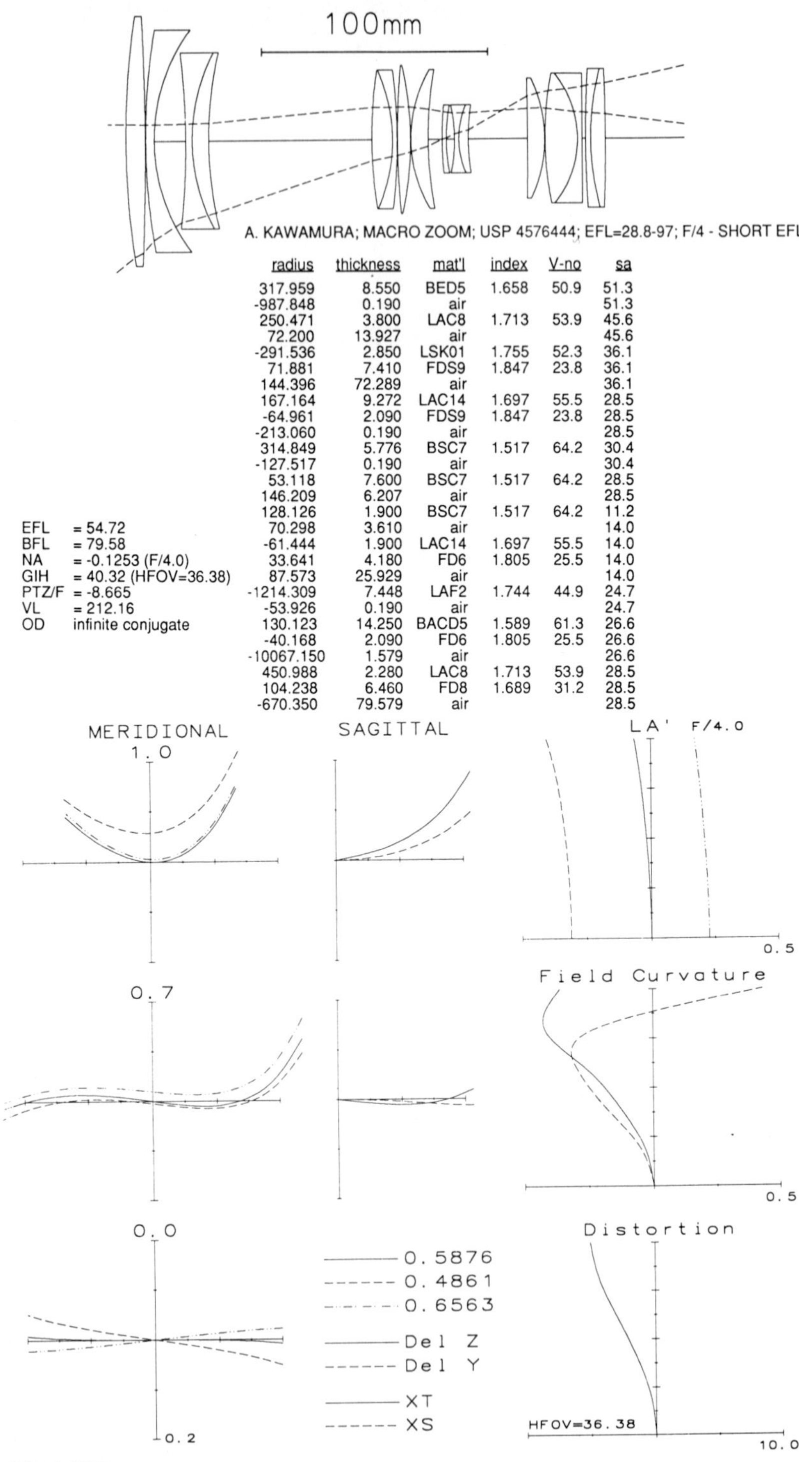

A. KAWAMURA; MACRO ZOOM; USP 4576444; EFL=28.8-97; F/4 - SHORT EFL

radius	thickness	mat'l	index	V-no	sa
317.959	8.550	BED5	1.658	50.9	51.3
-987.848	0.190	air			51.3
250.471	3.800	LAC8	1.713	53.9	45.6
72.200	13.927	air			45.6
-291.536	2.850	LSK01	1.755	52.3	36.1
71.881	7.410	FDS9	1.847	23.8	36.1
144.396	72.289	air			36.1
167.164	9.272	LAC14	1.697	55.5	28.5
-64.961	2.090	FDS9	1.847	23.8	28.5
-213.060	0.190	air			28.5
314.849	5.776	BSC7	1.517	64.2	30.4
-127.517	0.190	air			30.4
53.118	7.600	BSC7	1.517	64.2	28.5
146.209	6.207	air			28.5
128.126	1.900	BSC7	1.517	64.2	11.2
70.298	3.610	air			14.0
-61.444	1.900	LAC14	1.697	55.5	14.0
33.641	4.180	FD6	1.805	25.5	14.0
87.573	25.929	air			14.0
-1214.309	7.448	LAF2	1.744	44.9	24.7
-53.926	0.190	air			24.7
130.123	14.250	BACD5	1.589	61.3	26.6
-40.168	2.090	FD6	1.805	25.5	26.6
-10067.150	1.579	air			26.6
450.988	2.280	LAC8	1.713	53.9	28.5
104.238	6.460	FD8	1.689	31.2	28.5
-670.350	79.579	air			28.5

EFL = 54.72
BFL = 79.58
NA = -0.1253 (F/4.0)
GIH = 40.32 (HFOV=36.38)
PTZ/F = -8.665
VL = 212.16
OD infinite conjugate

Figure 20.5

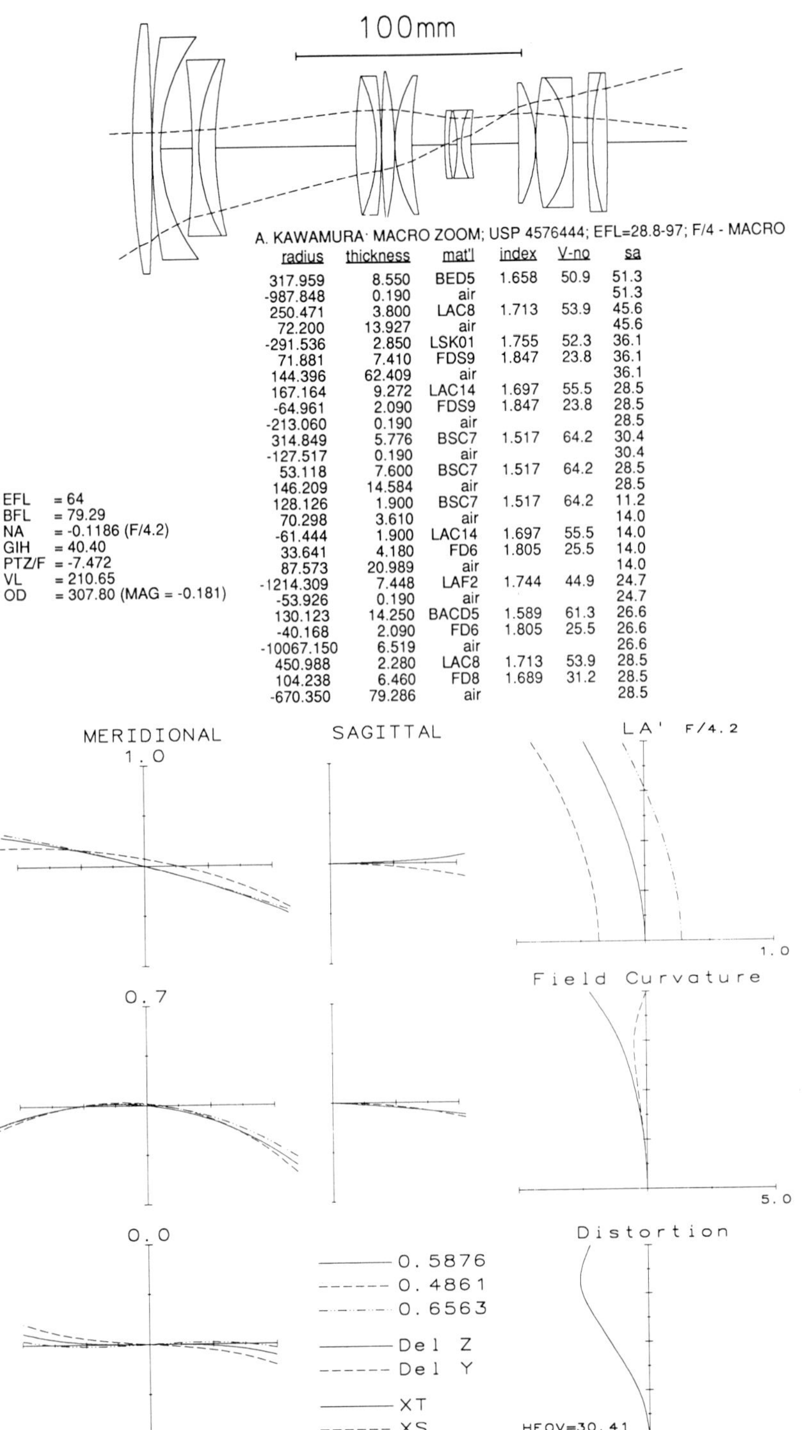

A. KAWAMURA· MACRO ZOOM; USP 4576444; EFL=28.8-97; F/4 - MACRO

radius	thickness	mat'l	index	V-no	sa
317.959	8.550	BED5	1.658	50.9	51.3
-987.848	0.190	air			51.3
250.471	3.800	LAC8	1.713	53.9	45.6
72.200	13.927	air			45.6
-291.536	2.850	LSK01	1.755	52.3	36.1
71.881	7.410	FDS9	1.847	23.8	36.1
144.396	62.409	air			36.1
167.164	9.272	LAC14	1.697	55.5	28.5
-64.961	2.090	FDS9	1.847	23.8	28.5
-213.060	0.190	air			28.5
314.849	5.776	BSC7	1.517	64.2	30.4
-127.517	0.190	air			30.4
53.118	7.600	BSC7	1.517	64.2	28.5
146.209	14.584	air			28.5
128.126	1.900	BSC7	1.517	64.2	11.2
70.298	3.610	air			14.0
-61.444	1.900	LAC14	1.697	55.5	14.0
33.641	4.180	FD6	1.805	25.5	14.0
87.573	20.989	air			14.0
-1214.309	7.448	LAF2	1.744	44.9	24.7
-53.926	0.190	air			24.7
130.123	14.250	BACD5	1.589	61.3	26.6
-40.168	2.090	FD6	1.805	25.5	26.6
-10067.150	6.519	air			26.6
450.988	2.280	LAC8	1.713	53.9	28.5
104.238	6.460	FD8	1.689	31.2	28.5
-670.350	79.286	air			28.5

EFL = 64
BFL = 79.29
NA = -0.1186 (F/4.2)
GIH = 40.40
PTZ/F = -7.472
VL = 210.65
OD = 307.80 (MAG = -0.181)

Figure 20.5 *(Continued)*

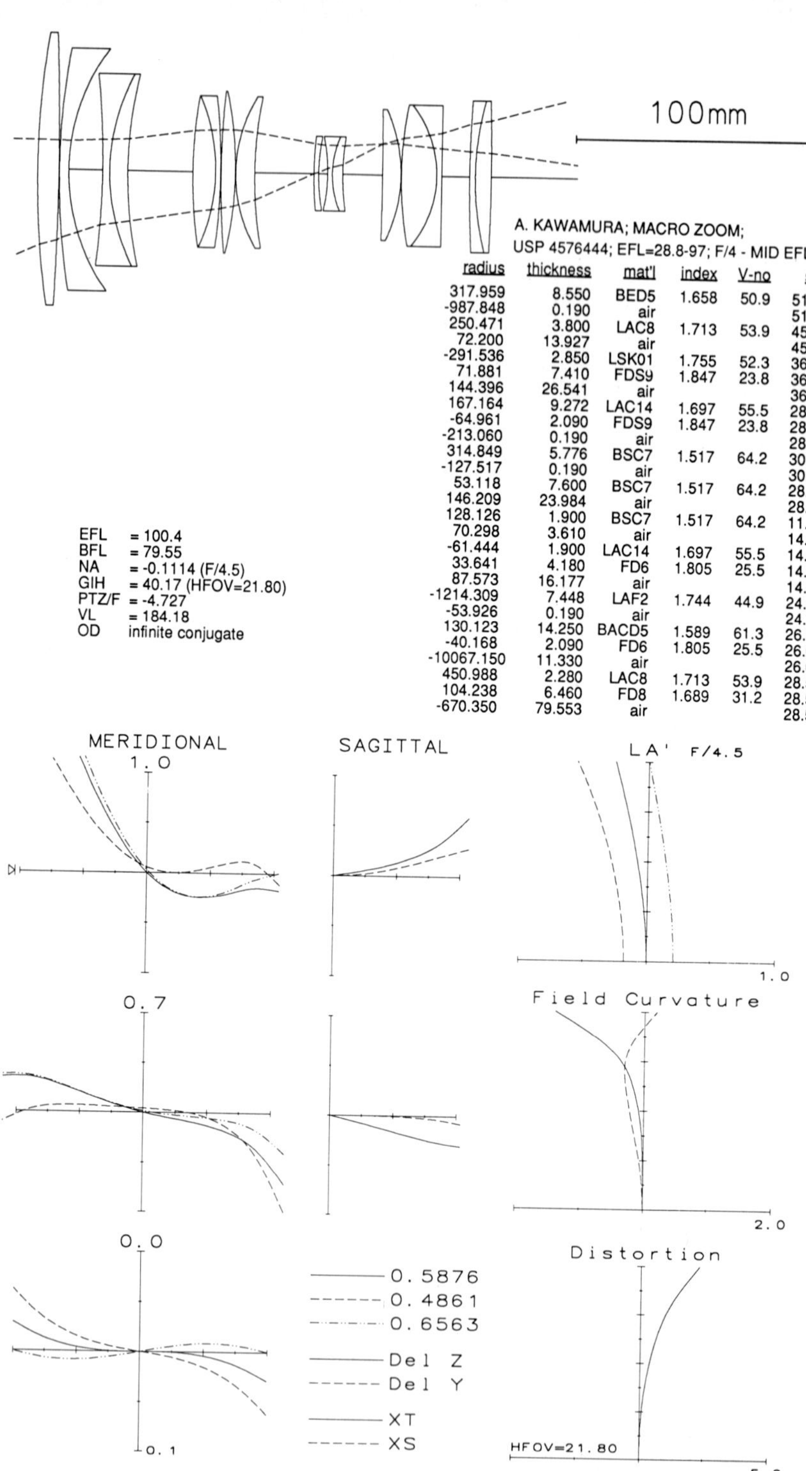

radius	thickness	mat'l	index	V-no	sa
317.959	8.550	BED5	1.658	50.9	51.3
-987.848	0.190	air			51.3
250.471	3.800	LAC8	1.713	53.9	45.6
72.200	13.927	air			45.6
-291.536	2.850	LSK01	1.755	52.3	36.1
71.881	7.410	FDS9	1.847	23.8	36.1
144.396	26.541	air			36.1
167.164	9.272	LAC14	1.697	55.5	28.5
-64.961	2.090	FDS9	1.847	23.8	28.5
-213.060	0.190	air			28.5
314.849	5.776	BSC7	1.517	64.2	30.4
-127.517	0.190	air			30.4
53.118	7.600	BSC7	1.517	64.2	28.5
146.209	23.984	air			28.5
128.126	1.900	BSC7	1.517	64.2	11.2
70.298	3.610	air			14.0
-61.444	1.900	LAC14	1.697	55.5	14.0
33.641	4.180	FD6	1.805	25.5	14.0
87.573	16.177	air			14.0
-1214.309	7.448	LAF2	1.744	44.9	24.7
-53.926	0.190	air			24.7
130.123	14.250	BACD5	1.589	61.3	26.6
-40.168	2.090	FD6	1.805	25.5	26.6
-10067.150	11.330	air			26.6
450.988	2.280	LAC8	1.713	53.9	28.5
104.238	6.460	FD8	1.689	31.2	28.5
-670.350	79.553	air			28.5

Figure 20.5 (*Continued*)

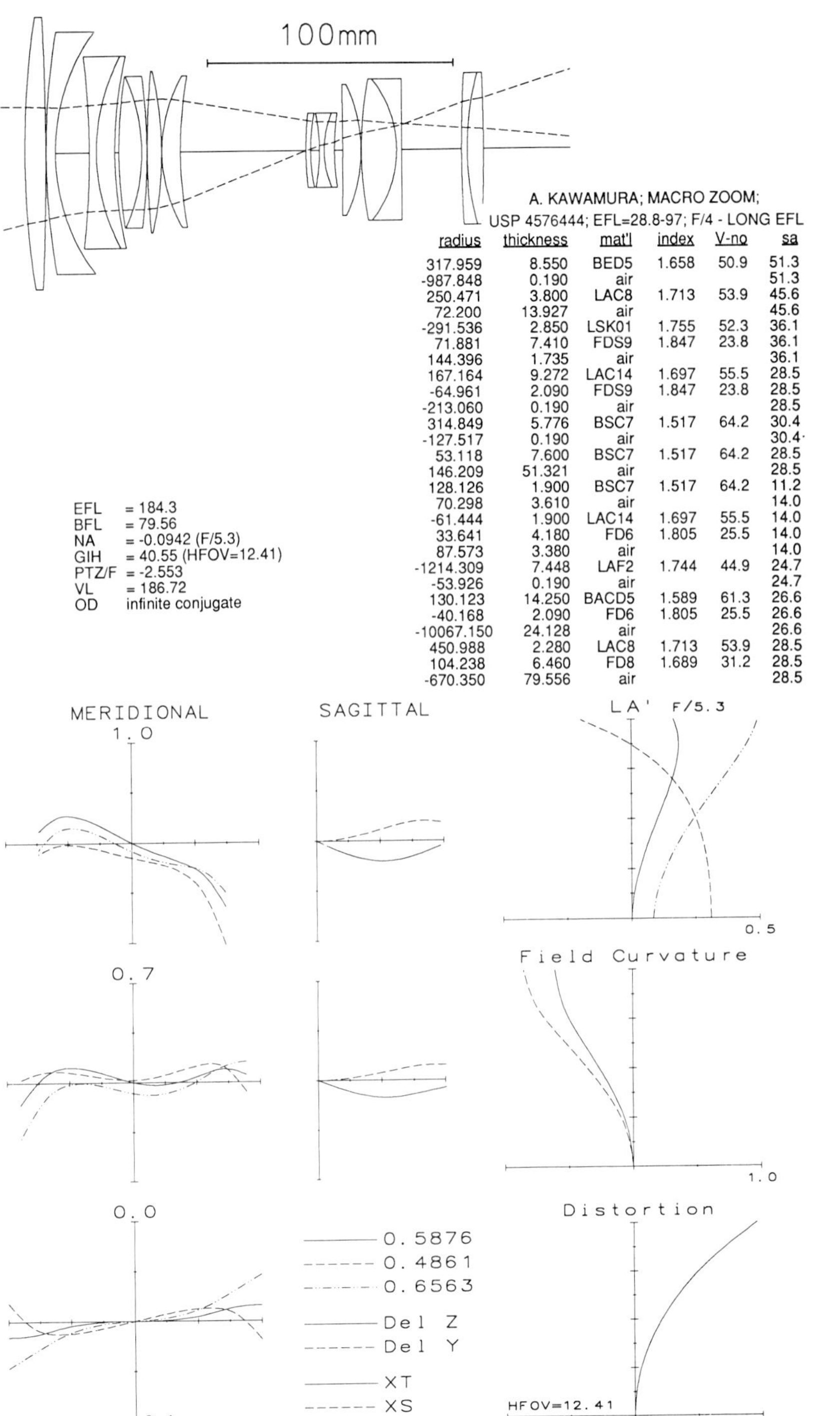

A. KAWAMURA; MACRO ZOOM;
USP 4576444; EFL=28.8-97; F/4 - LONG EFL

radius	thickness	mat'l	index	V-no	sa
317.959	8.550	BED5	1.658	50.9	51.3
-987.848	0.190	air			51.3
250.471	3.800	LAC8	1.713	53.9	45.6
72.200	13.927	air			45.6
-291.536	2.850	LSK01	1.755	52.3	36.1
71.881	7.410	FDS9	1.847	23.8	36.1
144.396	1.735	air			36.1
167.164	9.272	LAC14	1.697	55.5	28.5
-64.961	2.090	FDS9	1.847	23.8	28.5
-213.060	0.190	air			28.5
314.849	5.776	BSC7	1.517	64.2	30.4
-127.517	0.190	air			30.4
53.118	7.600	BSC7	1.517	64.2	28.5
146.209	51.321	air			28.5
128.126	1.900	BSC7	1.517	64.2	11.2
70.298	3.610	air			14.0
-61.444	1.900	LAC14	1.697	55.5	14.0
33.641	4.180	FD6	1.805	25.5	14.0
87.573	3.380	air			14.0
-1214.309	7.448	LAF2	1.744	44.9	24.7
-53.926	0.190	air			24.7
130.123	14.250	BACD5	1.589	61.3	26.6
-40.168	2.090	FD6	1.805	25.5	26.6
-10067.150	24.128	air			26.6
450.988	2.280	LAC8	1.713	53.9	28.5
104.238	6.460	FD8	1.689	31.2	28.5
-670.350	79.556	air			28.5

EFL = 184.3
BFL = 79.56
NA = -0.0942 (F/5.3)
GIH = 40.55 (HFOV=12.41)
PTZ/F = -2.553
VL = 186.72
OD infinite conjugate

Figure 20.5 *(Continued)*

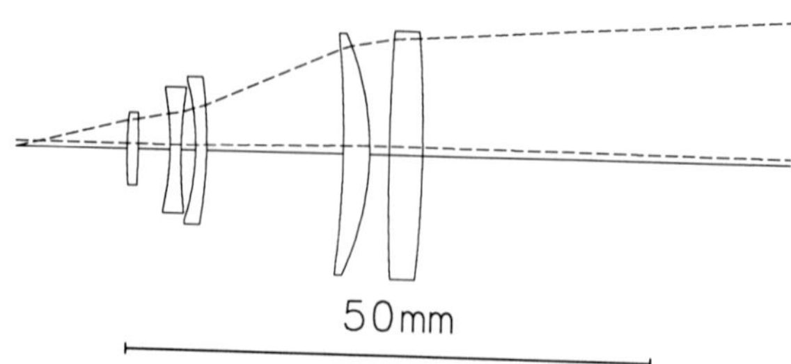

SCAN ZOOM LENS - SHORT EFL

radius	thickness	mat'l	index	V-no	sa
	10.573	air			0.0
40.064	1.057	SF4	1.750		3.2
-118.428	3.058	air			3.2
-26.621	1.057	SF4	1.750		5.5
42.000	1.374	air			5.6
-21.698	1.057	SF4	1.750		6.2
-43.028	13.049	air			6.5
-89.711	2.501	SF4	1.750		10.1
-24.680	17.973	air			10.7
	3.058	air			0.0
	13.049	air			0.0
215.914	3.172	SF4	1.750		11.0
-118.803	101.953	air			11.2

EFL = 70.58
BFL = 102
NA = -0.0075 (F/66.8)
GIH = 17.17
PTZ/F = 3.94
VL = 38.76
OD infinite conjugate

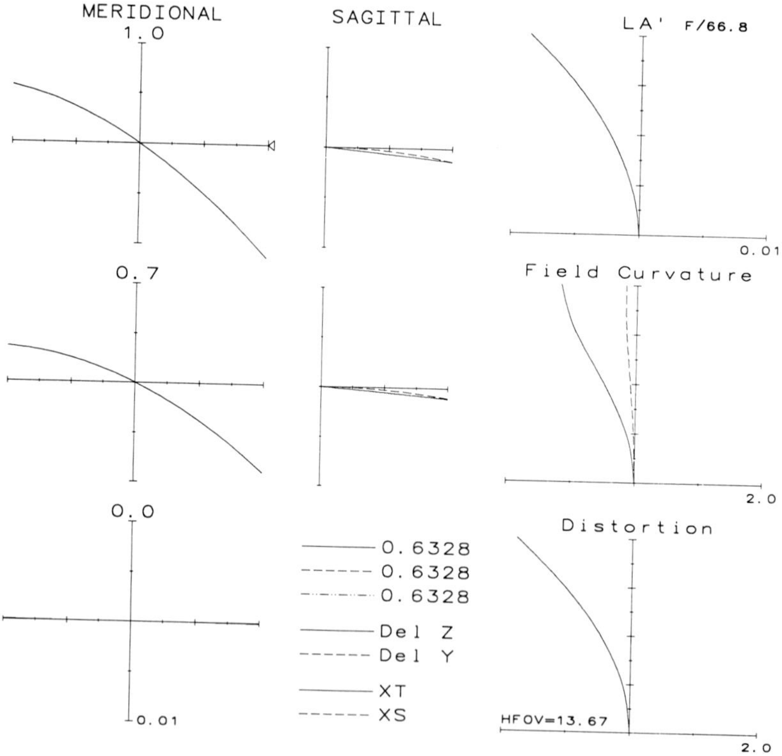

MERIDIONAL 1.0
SAGITTAL
LA' F/66.8
0.01
0.7
Field Curvature
2.0
0.0
Distortion
0.6328
0.6328
0.6328
Del Z
Del Y
XT
XS
0.01
HFOV=13.67
2.0

Figure 20.6

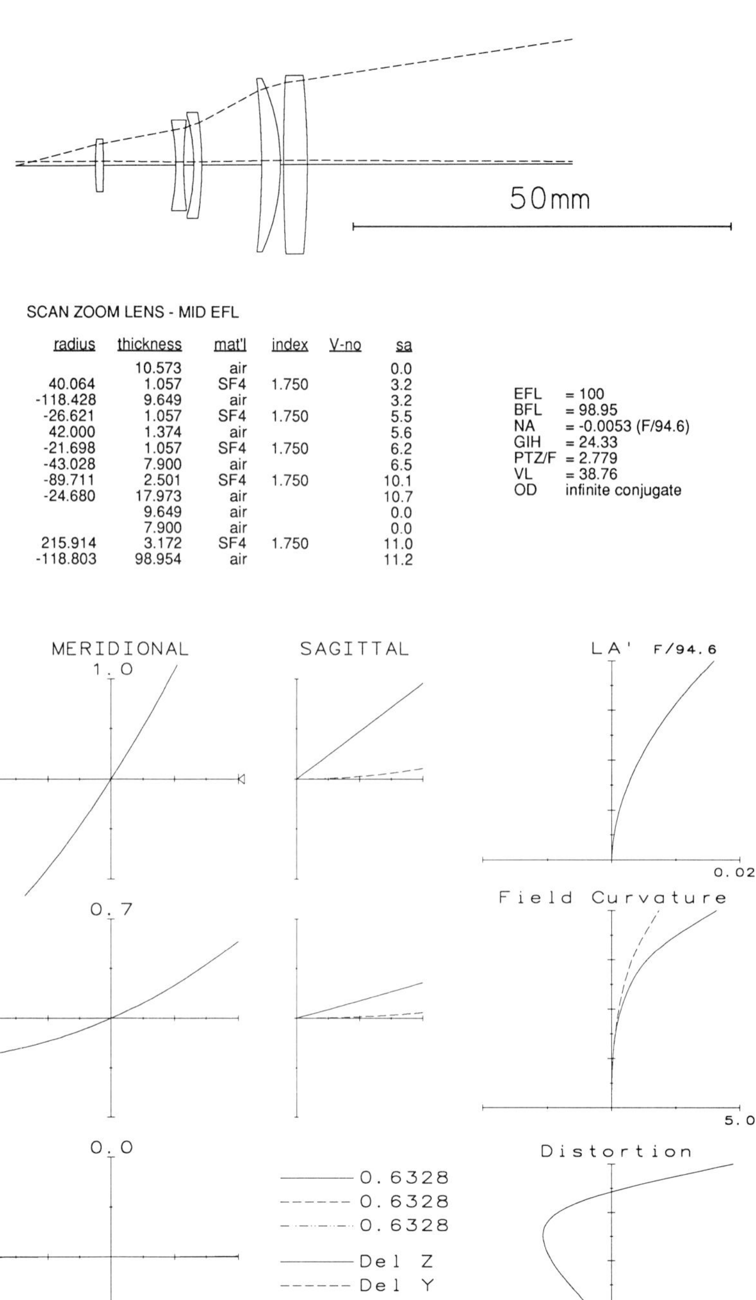

SCAN ZOOM LENS - MID EFL

radius	thickness	mat'l	index	V-no	sa
	10.573	air			0.0
40.064	1.057	SF4	1.750		3.2
-118.428	9.649	air			3.2
-26.621	1.057	SF4	1.750		5.5
42.000	1.374	air			5.6
-21.698	1.057	SF4	1.750		6.2
-43.028	7.900	air			6.5
-89.711	2.501	SF4	1.750		10.1
-24.680	17.973	air			10.7
	9.649	air			0.0
	7.900	air			0.0
215.914	3.172	SF4	1.750		11.0
-118.803	98.954	air			11.2

EFL = 100
BFL = 98.95
NA = -0.0053 (F/94.6)
GIH = 24.33
PTZ/F = 2.779
VL = 38.76
OD infinite conjugate

Figure 20.6 (*Continued*)

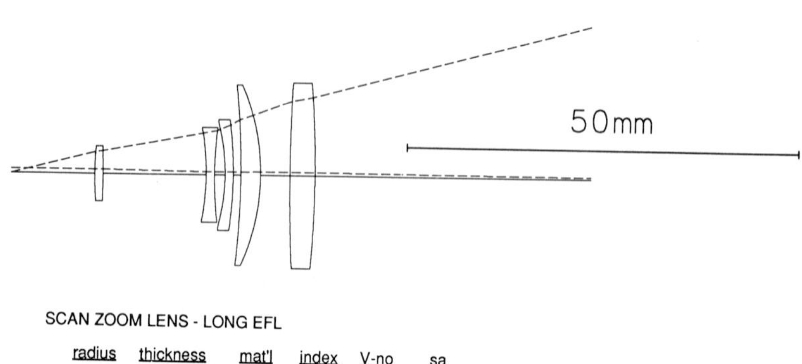

SCAN ZOOM LENS - LONG EFL

radius	thickness	mat'l	index	V-no	sa
	10.573	air			0.0
40.064	1.057	SF4	1.750		3.2
-118.428	13.278	air			3.2
-26.621	1.057	SF4	1.750		5.5
42.000	1.374	air			5.6
-21.698	1.057	SF4	1.750		6.2
-43.028	0.972	air			6.5
-89.711	2.501	SF4	1.750		10.1
-24.680	17.973	air			10.7
	13.278	air			0.0
	0.972	air			0.0
215.914	3.172	SF4	1.750		11.0
-118.803	101.910	air			11.2

EFL = 141.6
BFL = 101.9
NA = -0.0037 (F/133.9)
GIH = 34.44
PTZ/F = 1.963
VL = 38.76
OD infinite conjugate

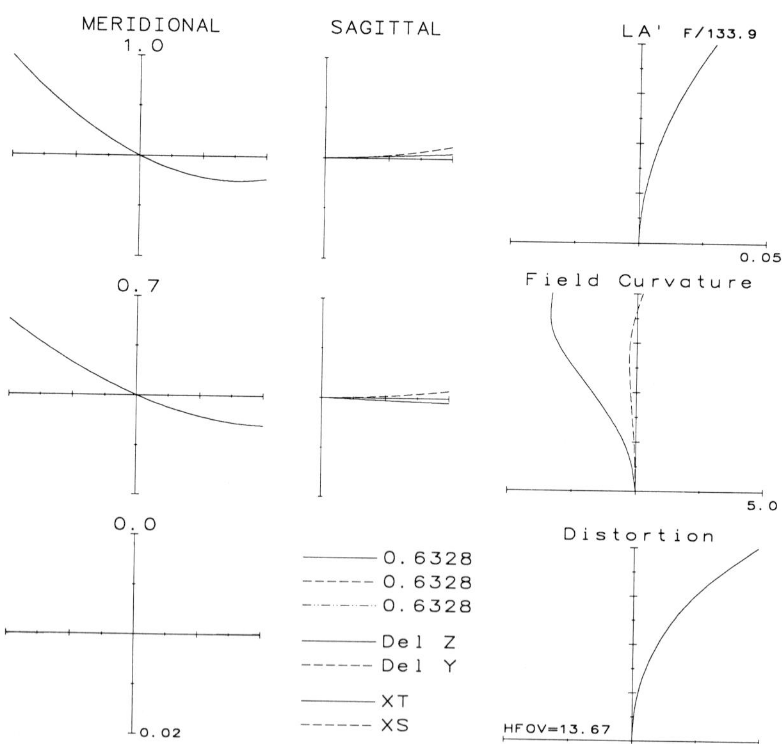

Figure 20.6 *(Continued)*

Chapter

21

Infrared Systems

21.1 Infrared Optics

Infrared optical systems differ from ordinary (visual) systems principally in the materials which are used for the optics. Out to a wavelength of about 2 μm, ordinary optical glass has a satisfactory transmission. However, crown and flint glasses reverse their roles at about 1.5 μm; at the longer wavelengths the crown glasses have a greater relative dispersion than do the flints. Thus the wavelength region between 1.0 and 1.5 μm is a transition zone wherein crowns and flints have the same relative dispersion (V value), and achromatism is not possible with optical glasses. An achromat *is* possible in this spectral region if (for example) calcium fluoride is used for the positive element.

Many of the other halogen salts (chlorides, iodides, bromides, and fluorides) are excellent transmitters in the infrared (IR) and have attractive dispersion characteristics, but their physical properties leave a lot to be desired. For this reason their use is almost entirely limited to laboratory-type environments.

The workhorse materials for the IR are silicon, germanium, zinc sulfide, zinc selenide, AMTIR (Ge/As/Se), chalcogenides, etc. These materials have significantly higher indices than optical glass and the crystals mentioned above. Of course, a higher index means smaller aberrations (see Fig. 3.1, for example), and the low relative dispersions of silicon and germanium allow reasonably good image quality to be obtained from a singlet element at a relatively high speed. In many IR systems, no attempt is made at achromatism because of this.

Another factor affecting infrared system design is that the longer wavelength both limits the resolution of an infrared system and simultaneously reduces the impact of the aberrations. As can be seen from the equations in Sec. F.12, the longer the wavelength, the larger the aberration can be before it produces a given fraction of a wavelength deformation of the wavefront. This in itself allows IR systems to be of comparatively simple construction.

21.2 IR Objective Lenses

Three achromatic doublets for the 8 to 12 μm window are shown in Figs. 21.1, 21.2, and 21.3. These are essentially telescope objectives and, despite the fact that high-index materials yield a relatively flat Petzval field, the negative astigmatism which is present in every thin, stop-in-contact optical system severely limits the off-axis image quality and the useful field of view of these designs. However, the speed of these simple lenses is *f*/1.5; this is made possible by the high indices of refraction. Note that the sign of the secondary spectrum is reversed from that of the usual (glass) lens. A doublet of silicon and germanium makes an excellent achromat, but the transmission of silicon usually limits its usefulness to the shorter wavelengths (e.g., the 3- to 5-μm atmospheric window).

Two *f*/0.75 infrared triplets are shown in Figs. 21.4 and 21.5. These can be regarded as the IR equivalent of the Cooke triplet. The increased length (compared to Figs. 21.2, 21.2, and 21.3) allows the introduction of some overcorrected astigmatism (by the elements which are located away from the stop) to offset the negative astigmatism of the components of the stop. Many IR triplets take the form of a closely spaced (+ −) doublet followed by a positive singlet which is spaced well away from the stop; this particular arrangement can be regarded as a Petzval lens with a rear singlet instead of the usual doublet. The high index and low dispersion of silicon and germanium produce many configurations which are quite different from the usual optical glass systems to which they correspond.

An assortment of four-element designs is shown in Figs. 21.6 through 21.11. Figure 21.6 is a highly unusual design. It may be regarded as a reverse telephoto lens with a field flattener, but a more likely view is to realize that the germanium element (#3) has almost all the effective positive power and the first two Irtran (ZnS) elements operate as a corrector for its aberrations without introducing much power—a sort of spherical Schmidt corrector. Figure 21.7 is really a four-element telescope objective. The silicon-germanium combination works well in the 3- to 5-μm spectral region, but this compact (short) configuration does not allow for correction of the negative astigmatism.

Figures 21.8 and 21.9 are effectively IR Petzval lenses, consisting as they do of two widely spaced achromatic doublets. At *f*/1.5 and over a 6° field, four elements allow a high level of correction. The lenses of Figs. 21.10 and 21.11 are quite sophisticated constructions, and, while the level of correction does not match that of the two preceding de-

(Text continues on page 406.)

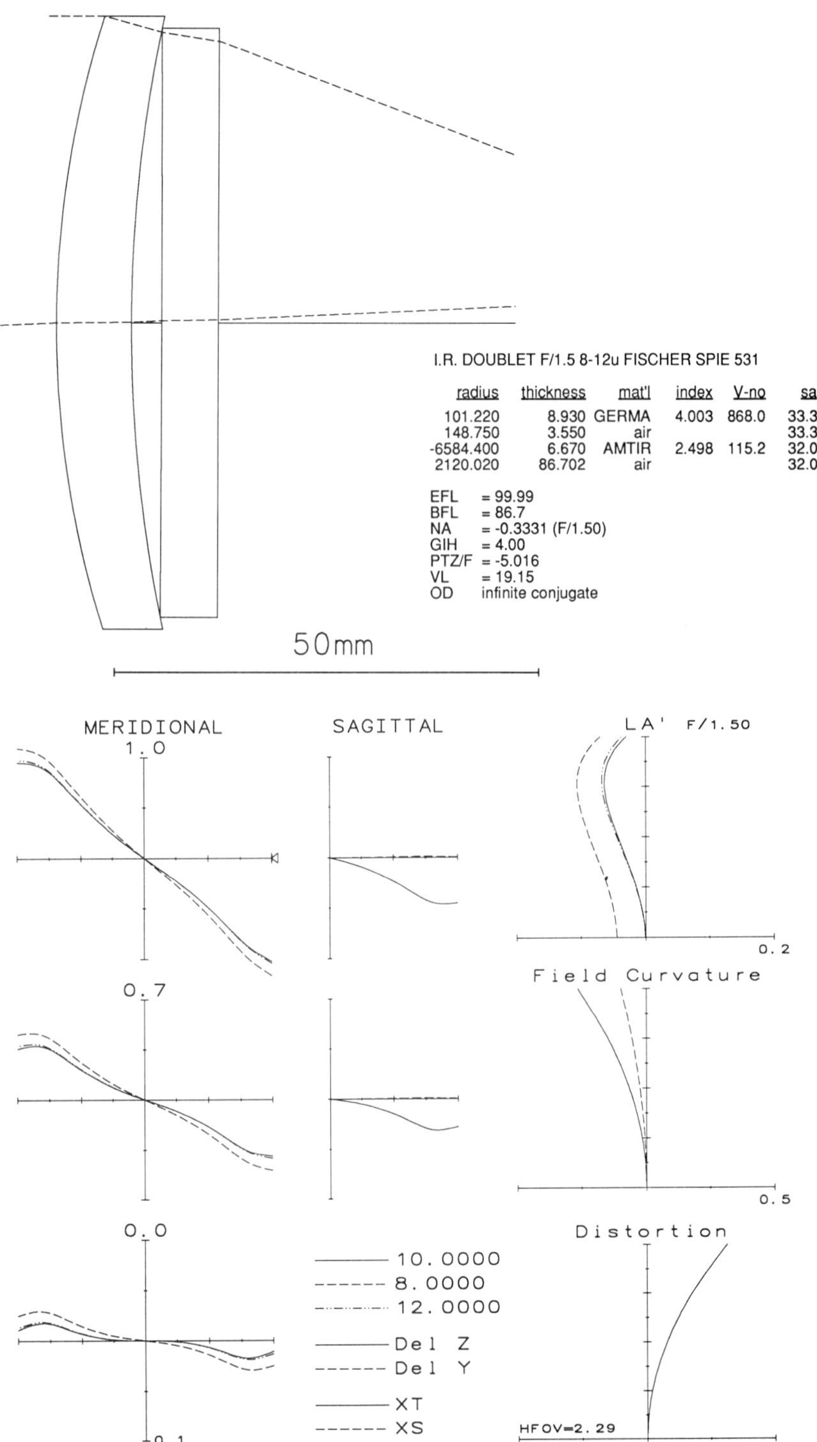

I.R. DOUBLET F/1.5 8-12u FISCHER SPIE 531

radius	thickness	mat'l	index	V-no	sa
101.220	8.930	GERMA	4.003	868.0	33.3
148.750	3.550	air			33.3
-6584.400	6.670	AMTIR	2.498	115.2	32.0
2120.020	86.702	air			32.0

EFL = 99.99
BFL = 86.7
NA = -0.3331 (F/1.50)
GIH = 4.00
PTZ/F = -5.016
VL = 19.15
OD infinite conjugate

Figure 21.1

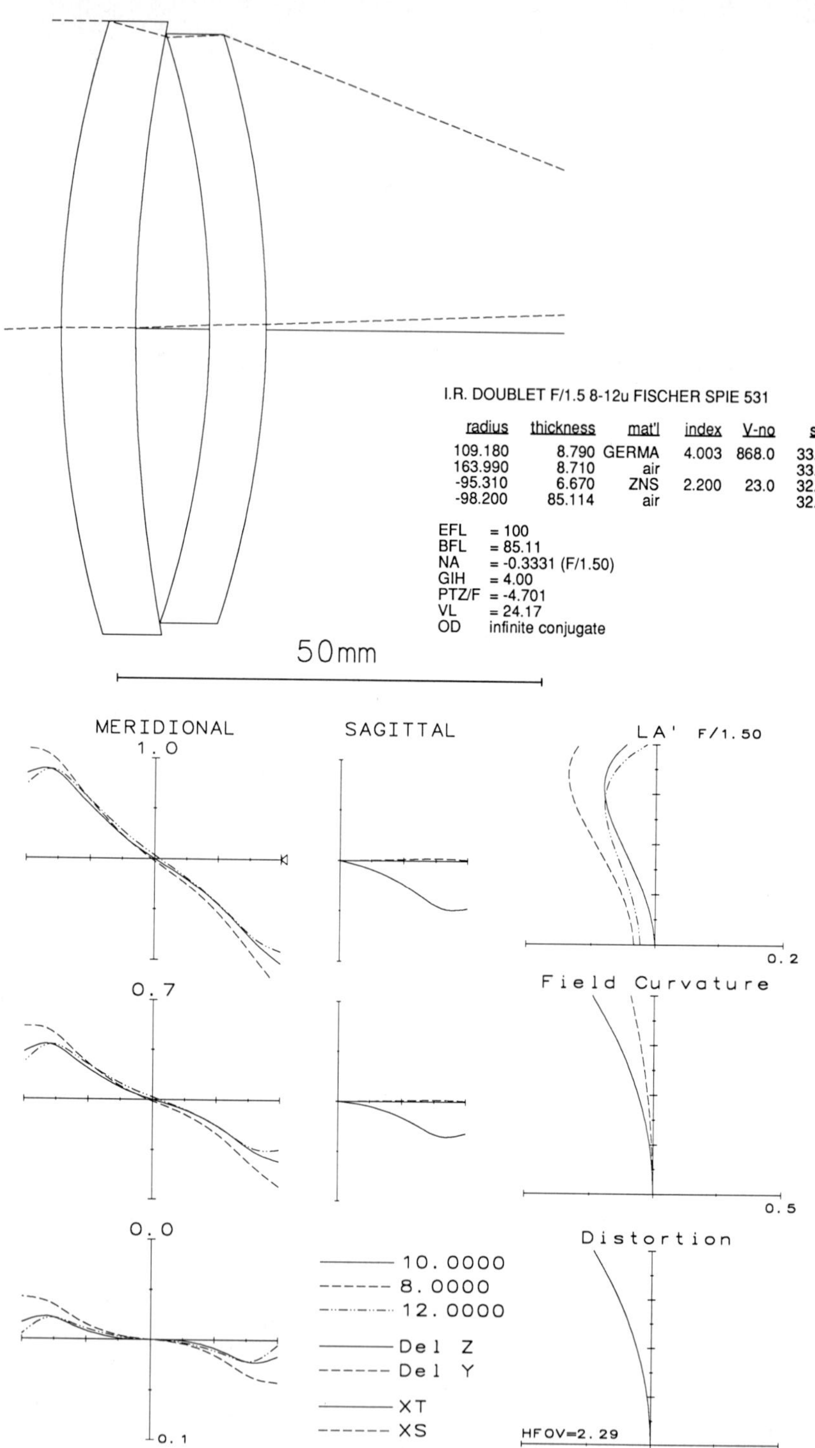
I.R. DOUBLET F/1.5 8-12u FISCHER SPIE 531
radius thickness mat'l index V-no sa
109.180 8.790 GERMA 4.003 868.0 33.3
163.990 8.710 air 33.3
-95.310 6.670 ZNS 2.200 23.0 32.0
-98.200 85.114 air 32.0
EFL = 100
BFL = 85.11
NA = -0.3331 (F/1.50)
GIH = 4.00
PTZ/F = -4.701
VL = 24.17
OD infinite conjugate
50mm
MERIDIONAL
SAGITTAL
1.0
0.7
0.0
0.1
LA' F/1.50
0.2
Field Curvature
0.5
Distortion
HFOV=2.29
0.005
10.0000
8.0000
12.0000
Del Z
Del Y
XT
XS

Figure 21.2

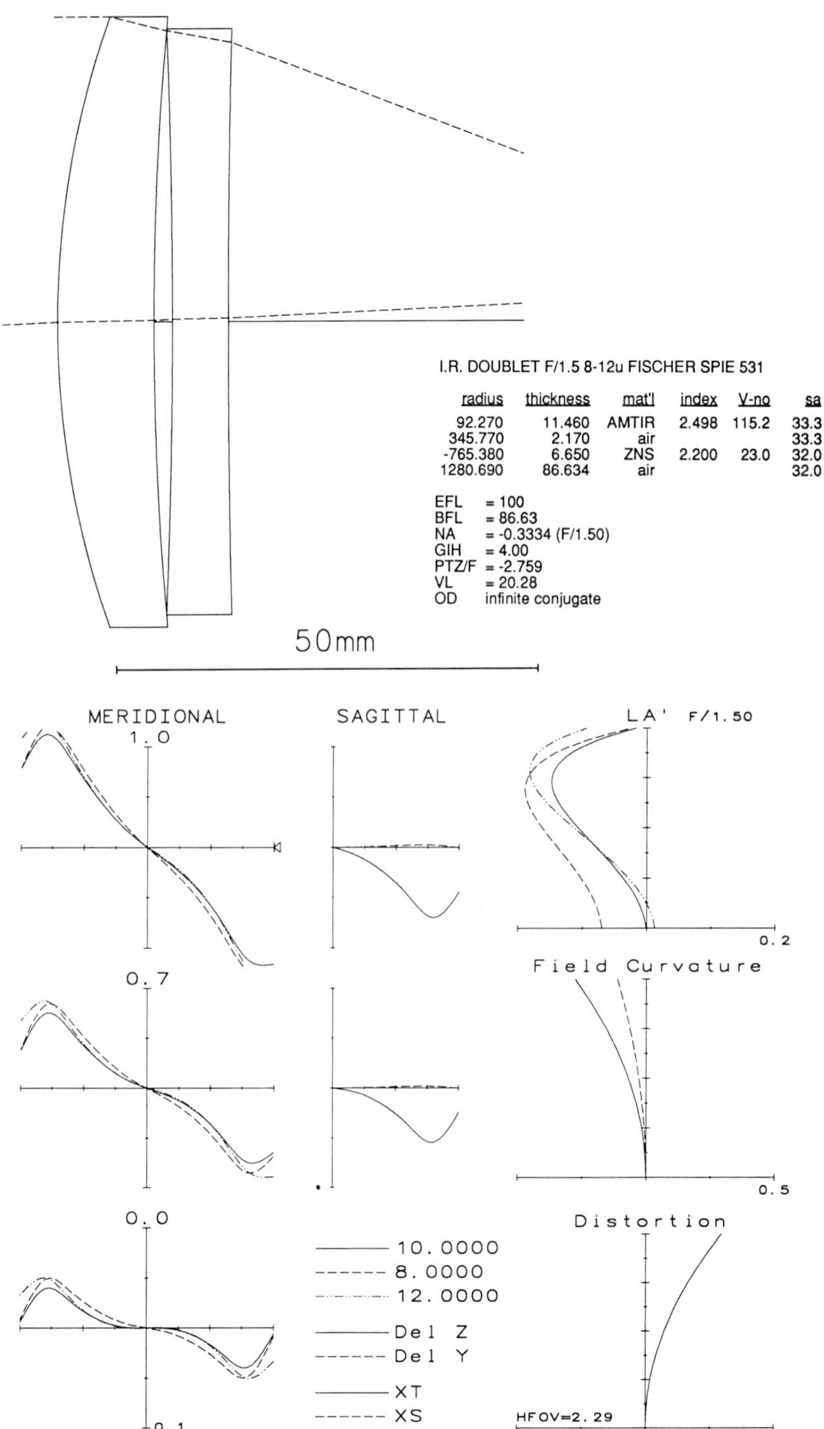

I.R. DOUBLET F/1.5 8-12u FISCHER SPIE 531

radius	thickness	mat'l	index	V-no	sa
92.270	11.460	AMTIR	2.498	115.2	33.3
345.770	2.170	air			33.3
-765.380	6.650	ZNS	2.200	23.0	32.0
1280.690	86.634	air			32.0

EFL = 100
BFL = 86.63
NA = -0.3334 (F/1.50)
GIH = 4.00
PTZ/F = -2.759
VL = 20.28
OD infinite conjugate

Figure 21.3

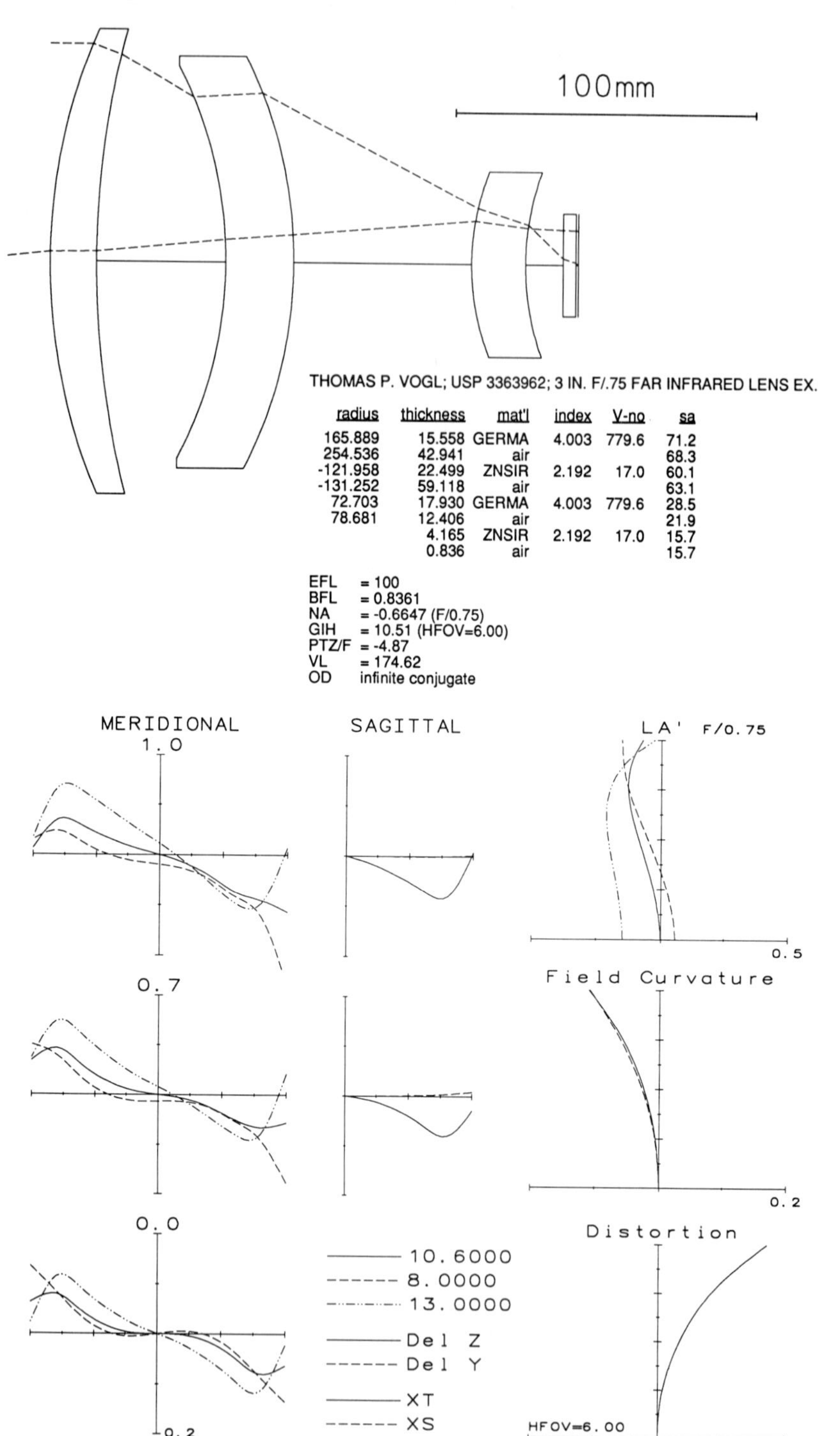

THOMAS P. VOGL; USP 3363962; 3 IN. F/.75 FAR INFRARED LENS EX. 1

radius	thickness	mat'l	index	V-no	sa
165.889	15.558	GERMA	4.003	779.6	71.2
254.536	42.941	air			68.3
-121.958	22.499	ZNSIR	2.192	17.0	60.1
-131.252	59.118	air			63.1
72.703	17.930	GERMA	4.003	779.6	28.5
78.681	12.406	air			21.9
	4.165	ZNSIR	2.192	17.0	15.7
	0.836	air			15.7

EFL = 100
BFL = 0.8361
NA = -0.6647 (F/0.75)
GIH = 10.51 (HFOV=6.00)
PTZ/F = -4.87
VL = 174.62
OD infinite conjugate

Figure 21.4

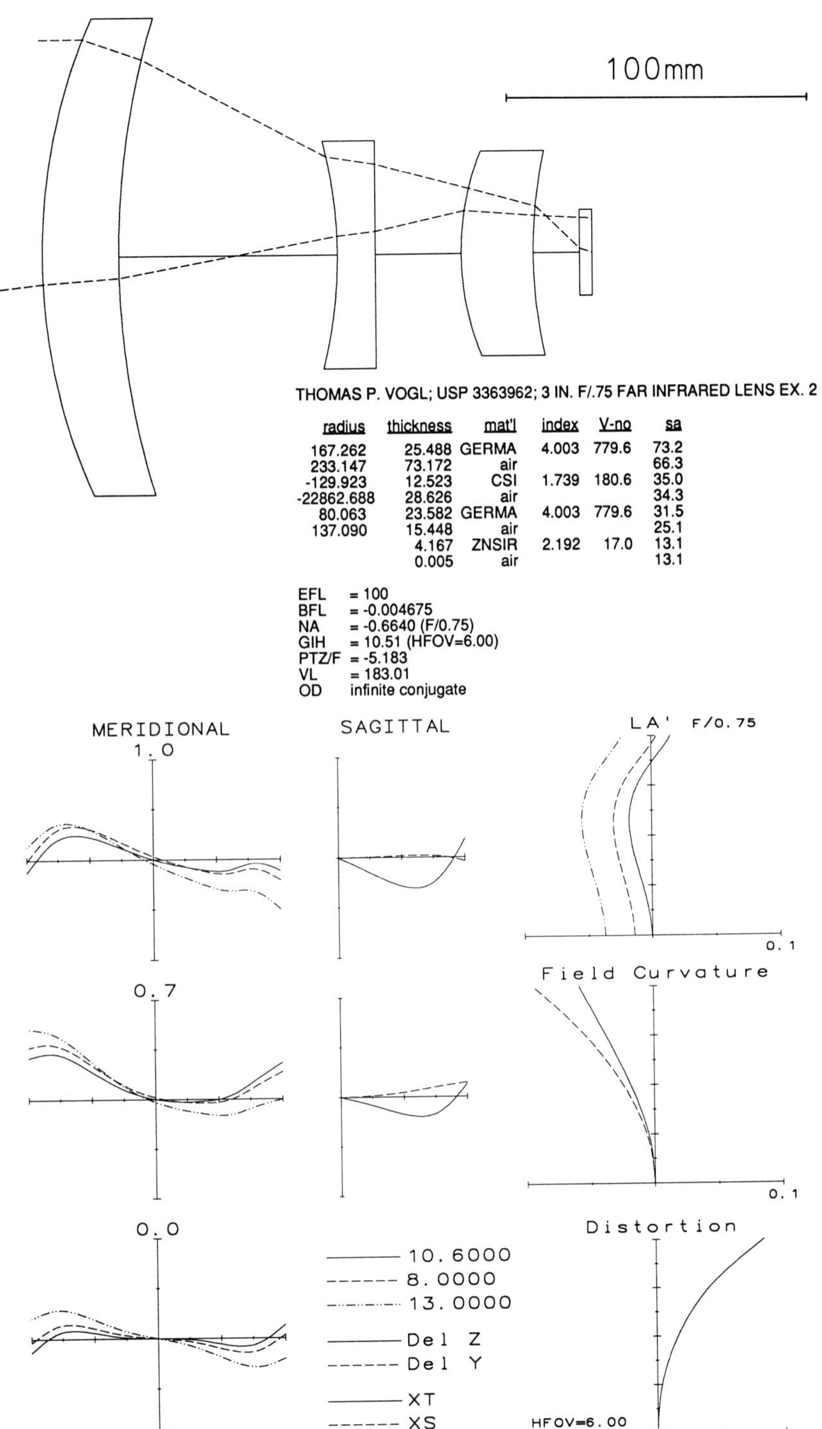

THOMAS P. VOGL; USP 3363962; 3 IN. F/.75 FAR INFRARED LENS EX. 2

radius	thickness	mat'l	index	V-no	sa
167.262	25.488	GERMA	4.003	779.6	73.2
233.147	73.172	air			66.3
-129.923	12.523	CSI	1.739	180.6	35.0
-22862.688	28.626	air			34.3
80.063	23.582	GERMA	4.003	779.6	31.5
137.090	15.448	air			25.1
	4.167	ZNSIR	2.192	17.0	13.1
	0.005	air			13.1

EFL = 100
BFL = -0.004675
NA = -0.6640 (F/0.75)
GIH = 10.51 (HFOV=6.00)
PTZ/F = -5.183
VL = 183.01
OD infinite conjugate

Figure 21.5

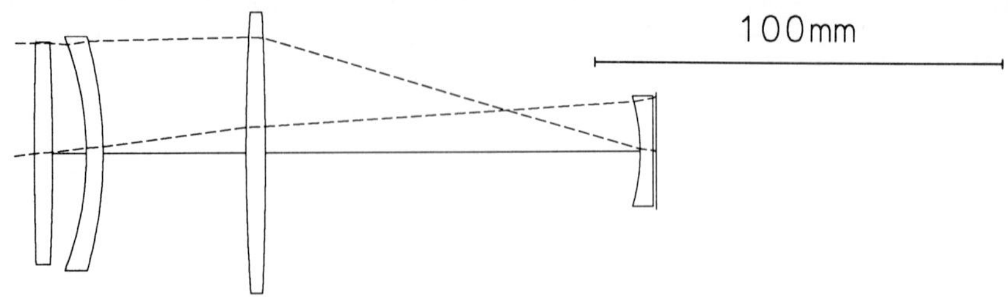

ARTIE D. KIRKPATRICK; USP 3439969; 3 IN. F/2 FAR INFRARED LENS

radius	thickness	mat'l	index	V-no	sa
934.850	4.438	IRTRN	2.185	16.8	25.2
-574.627	8.241	air			25.2
-69.352	4.121	IRTRN	2.185	16.8	26.6
-90.951	34.868	air			26.6
484.959	4.755	GERMA	4.003	779.5	31.9
-636.622	91.612	air			31.9
-42.252	3.169	IRTRN	2.185	16.8	12.6
	0.842	air			12.6

EFL = 100
BFL = 0.8421
NA = -0.2512 (F/2.0)
GIH = 12.28 (HFOV=7.00)
PTZ/F = 0.967
VL = 151.20
OD infinite conjugate

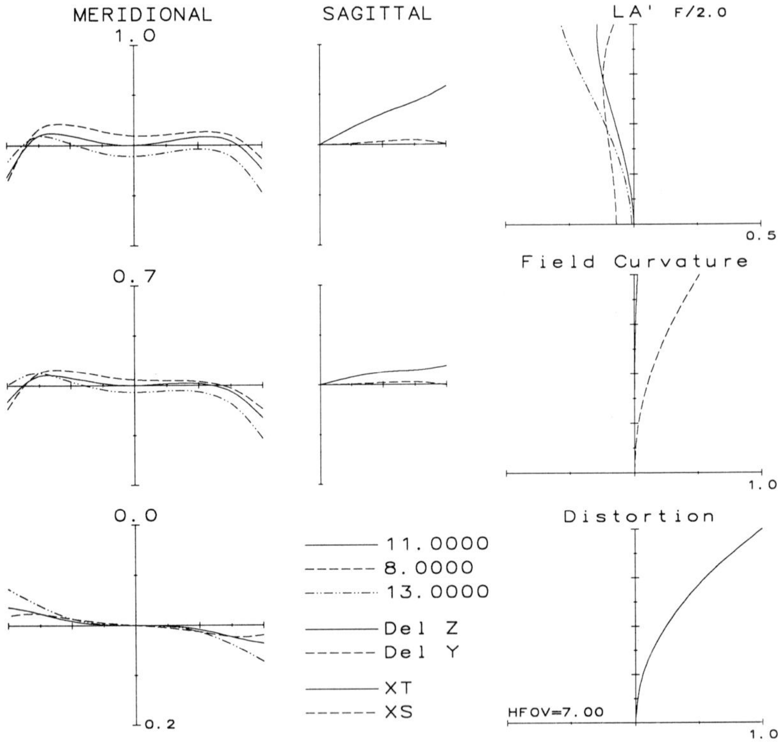

Figure 21.6

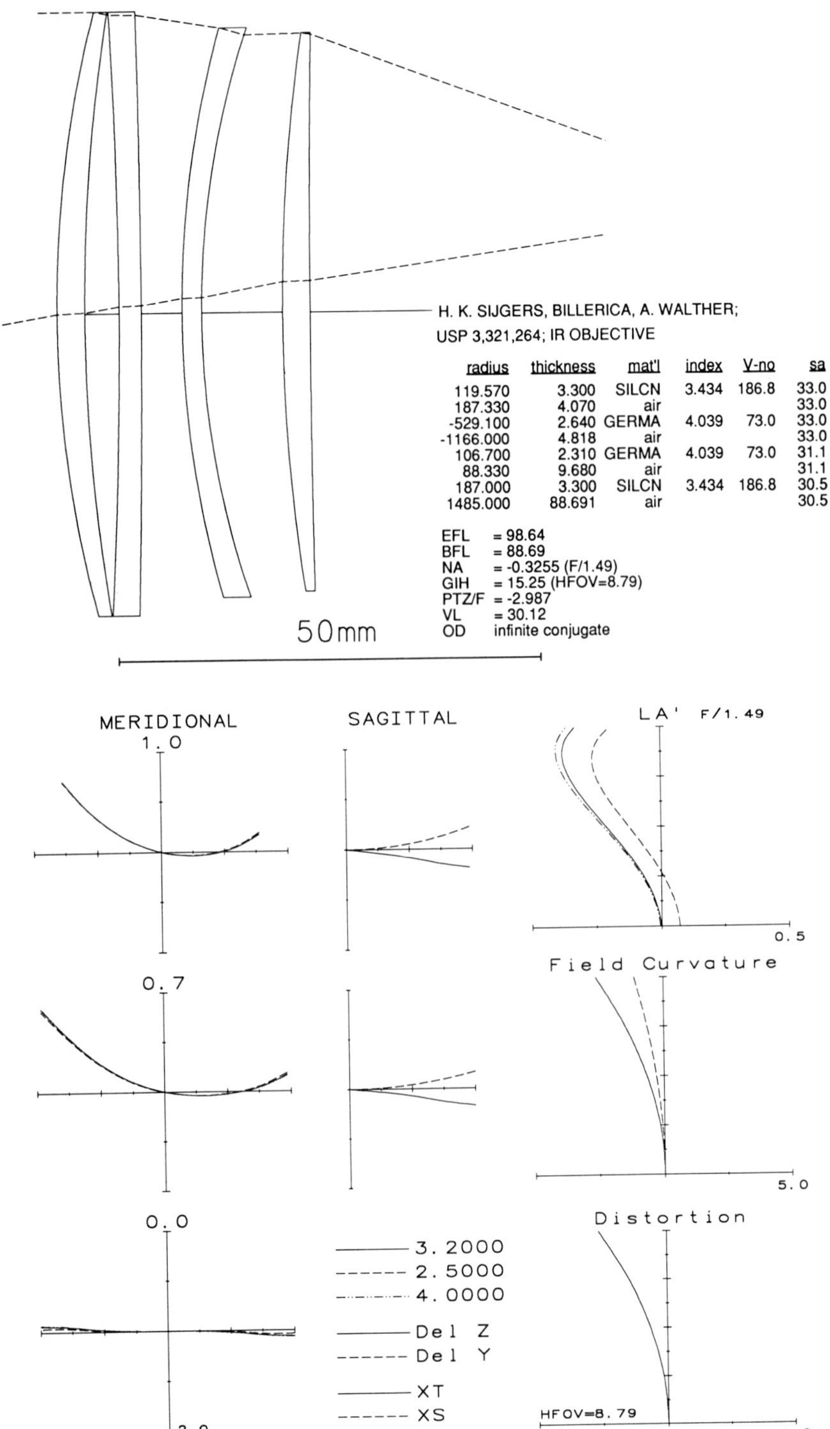

H. K. SIJGERS, BILLERICA, A. WALTHER;
USP 3,321,264; IR OBJECTIVE

radius	thickness	mat'l	index	V-no	sa
119.570	3.300	SILCN	3.434	186.8	33.0
187.330	4.070	air			33.0
-529.100	2.640	GERMA	4.039	73.0	33.0
-1166.000	4.818	air			33.0
106.700	2.310	GERMA	4.039	73.0	31.1
88.330	9.680	air			31.1
187.000	3.300	SILCN	3.434	186.8	30.5
1485.000	88.691	air			30.5

EFL = 98.64
BFL = 88.69
NA = -0.3255 (F/1.49)
GIH = 15.25 (HFOV=8.79)
PTZ/F = -2.987
VL = 30.12
OD infinite conjugate

Figure 21.7

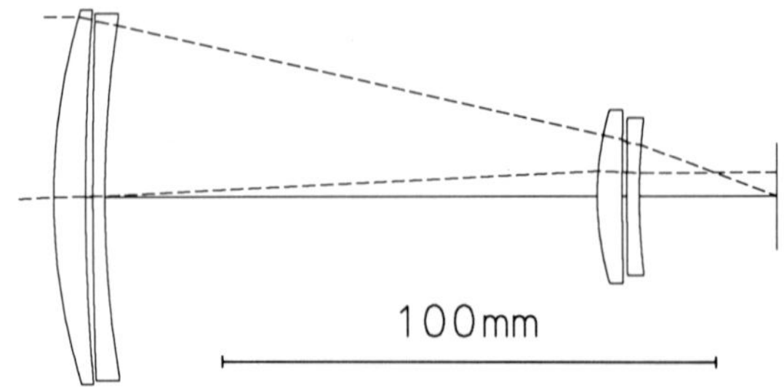

IAIN A. NEIL; USP 4505535; 51 MM F/1.5 IR OBJECTIVE LENS

radius	thickness	mat'l	index	V-no	sa
114.480	6.500	AS2SE	2.779		35.2
419.780	1.500	air			34.8
1178.759	2.500	ZNSE	2.407		34.6
199.940	100.000	air			33.7
54.920	5.000	AS2SE	2.779		16.4
684.341	1.000	air			15.7
1062.304	2.500	GE	4.003		14.9
118.460	28.034	air			14.3

EFL = 101.9
BFL = 28.03
NA = -0.3330 (F/1.50)
GIH = 4.45
PTZ/F = -1.455
VL = 119.00
OD infinite conjugate

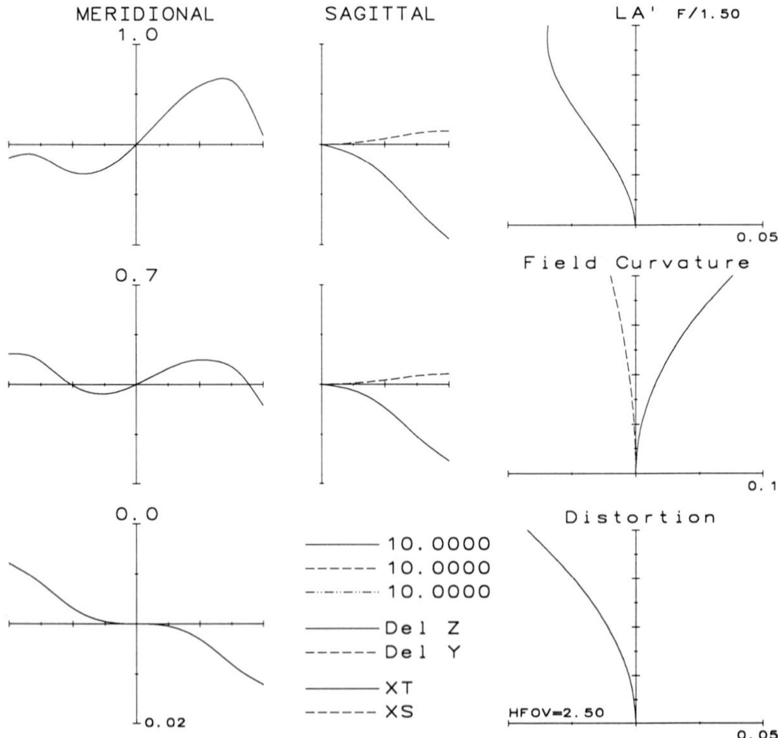

Figure 21.8

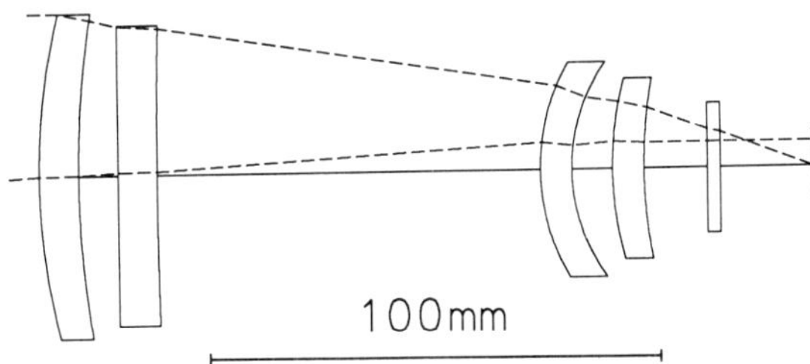

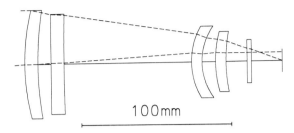

THOMAS P. FJELDSTED; USP 4380363; FOUR ELEMENT F/1.5 IR OBJECTIVE

radius	thickness	mat'l	index	V-no	sa
128.471	8.737	SILCN	3.431	488.9	33.3
186.905	8.737	air			32.1
8457.375	8.737	GERMA	4.029	207.1	30.8
1040.215	85.070	air			30.8
41.030	6.990	GERMA	4.029	207.1	22.0
35.228	9.017	air			22.0
63.397	6.990	SILCN	3.431	488.9	18.5
90.098	13.979	air			18.5
	2.796	SAPIR	1.688	18.9	13.3
	20.477	air			13.3

EFL = 100
BFL = 20.48
NA = -0.3334 (F/1.50)
GIH = 5.24
PTZ/F = -7.231
VL = 151.05
OD infinite conjugate

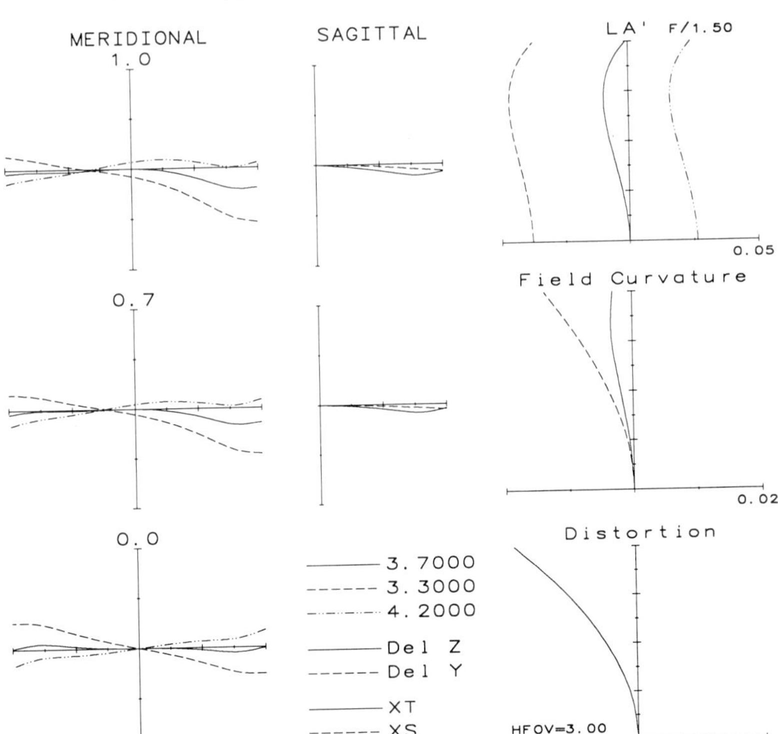

Figure 21.9

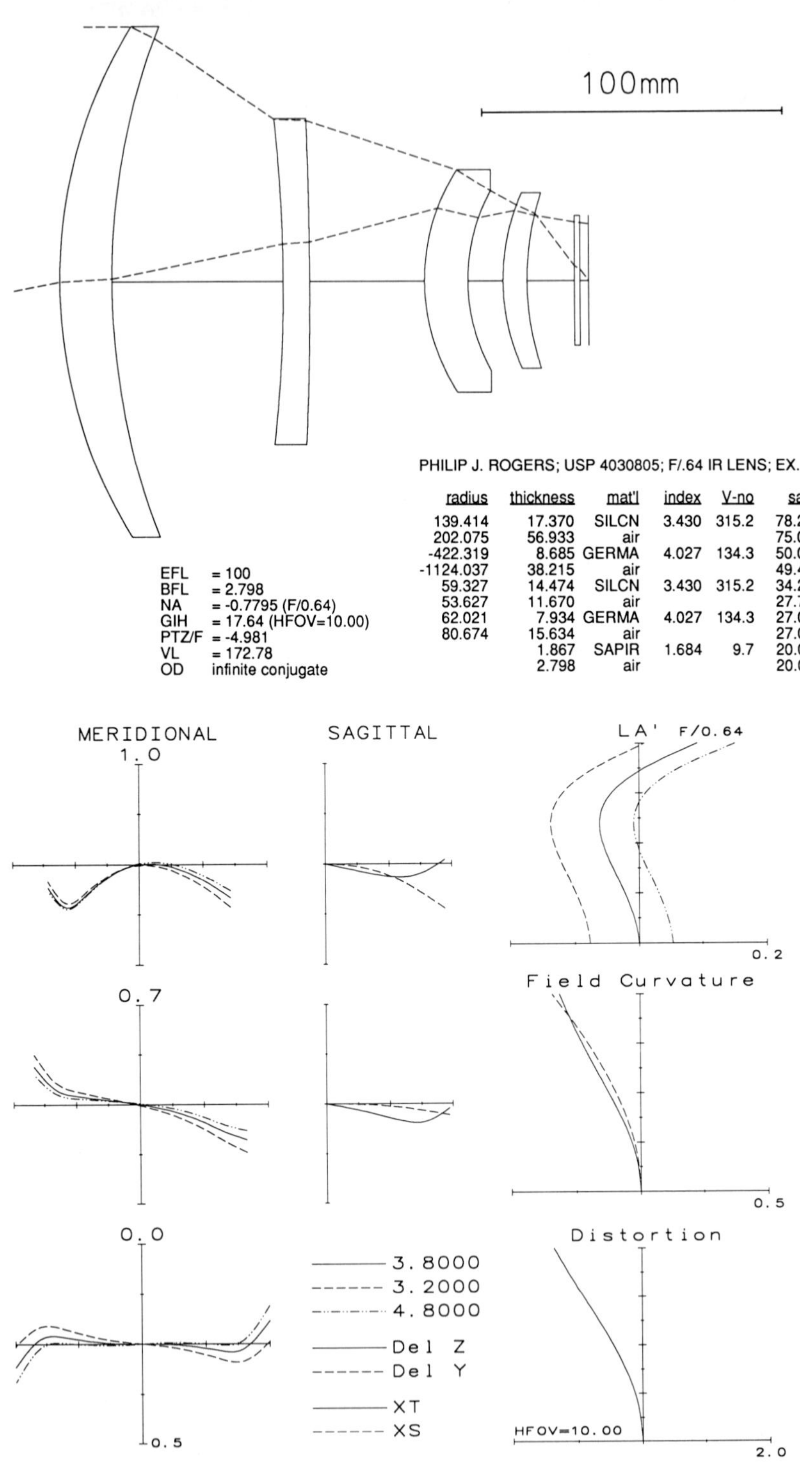

PHILIP J. ROGERS; USP 4030805; F/.64 IR LENS; EX. 5

radius	thickness	mat'l	index	V-no	sa
139.414	17.370	SILCN	3.430	315.2	78.2
202.075	56.933	air			75.0
-422.319	8.685	GERMA	4.027	134.3	50.0
-1124.037	38.215	air			49.4
59.327	14.474	SILCN	3.430	315.2	34.2
53.627	11.670	air			27.7
62.021	7.934	GERMA	4.027	134.3	27.0
80.674	15.634	air			27.0
	1.867	SAPIR	1.684	9.7	20.0
	2.798	air			20.0

EFL = 100
BFL = 2.798
NA = -0.7795 (F/0.64)
GIH = 17.64 (HFOV=10.00)
PTZ/F = -4.981
VL = 172.78
OD infinite conjugate

Figure 21.10

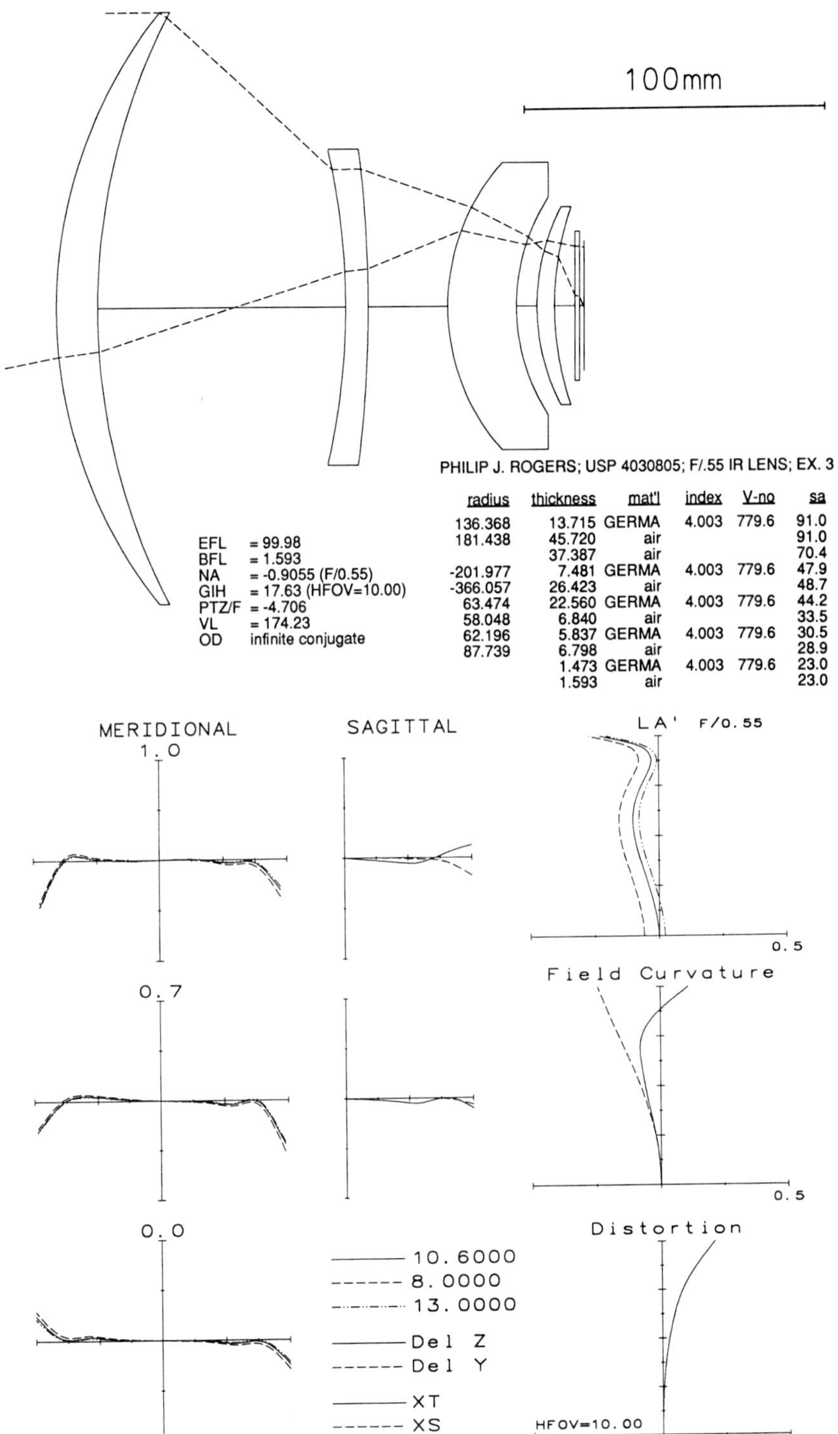

PHILIP J. ROGERS; USP 4030805; F/.55 IR LENS; EX. 3

radius	thickness	mat'l	index	V-no	sa
136.368	13.715	GERMA	4.003	779.6	91.0
181.438	45.720	air			91.0
	37.387	air			70.4
-201.977	7.481	GERMA	4.003	779.6	47.9
-366.057	26.423	air			48.7
63.474	22.560	GERMA	4.003	779.6	44.2
58.048	6.840	air			33.5
62.196	5.837	GERMA	4.003	779.6	30.5
87.739	6.798	air			28.9
	1.473	GERMA	4.003	779.6	23.0
	1.593	air			23.0

Figure 21.11

signs, the coverage of a 20° field at speeds of f/0.6 and f/0.55 is no mean feat in itself. (It is often difficult just to find a system at wide field and high speed which will allow all the rays to get through without missing a surface or encountering TIR.)

21.3 IR Telescopes

The infrared telescope usually serves as a collector for a following smaller-diameter system, or as a front for a FLIR (forward looking infrared) scanning system. As a collector, a galilean telescope or set of confocal paraboloids is appropriate. But the purpose of the system as a front for a scanner is to reduce the size of the beam in order to reduce the size and inertia of the scanning mirror. The mirror is placed at the exit pupil of the telescope; the need for an accessible exit pupil obviously requires a positive eyepiece and rules out the galilean configuration for this purpose. Note that all of the eyepieces in the following designs are very simple, consisting of just two meniscus germanium elements with their convex surfaces facing each other. The telescope aberrations are shown as if there were a perfect 1000-mm focal-length lens following the afocal telescope. Thus the transverse aberrations are shown in units of milliradians.

The 12× telescope of Fig. 21.12 has a catadioptric objective with a concentric meniscus corrector dome (which allows the balance of the system to be pivoted about the dome's center of curvature if required) and a Mangin-mirror-type secondary. Both reflecting surfaces are aspheric, as is the third surface of the eyepiece. Figure 21.13 can be regarded as having an objective similar to Fig. 21.6, plus the standard two-meniscus eyepiece. The negative field lens both flattens the Petzval field and increases the distance between the optics and the pupil/scanner mirror.

The simple telescope of Fig. 21.14 manages a magnification of 16× with the standard eyepiece and a sort of telephoto-type construction for the objective. Figure 21.15 is a similar, but more complex, construction, at a 5× magnification.

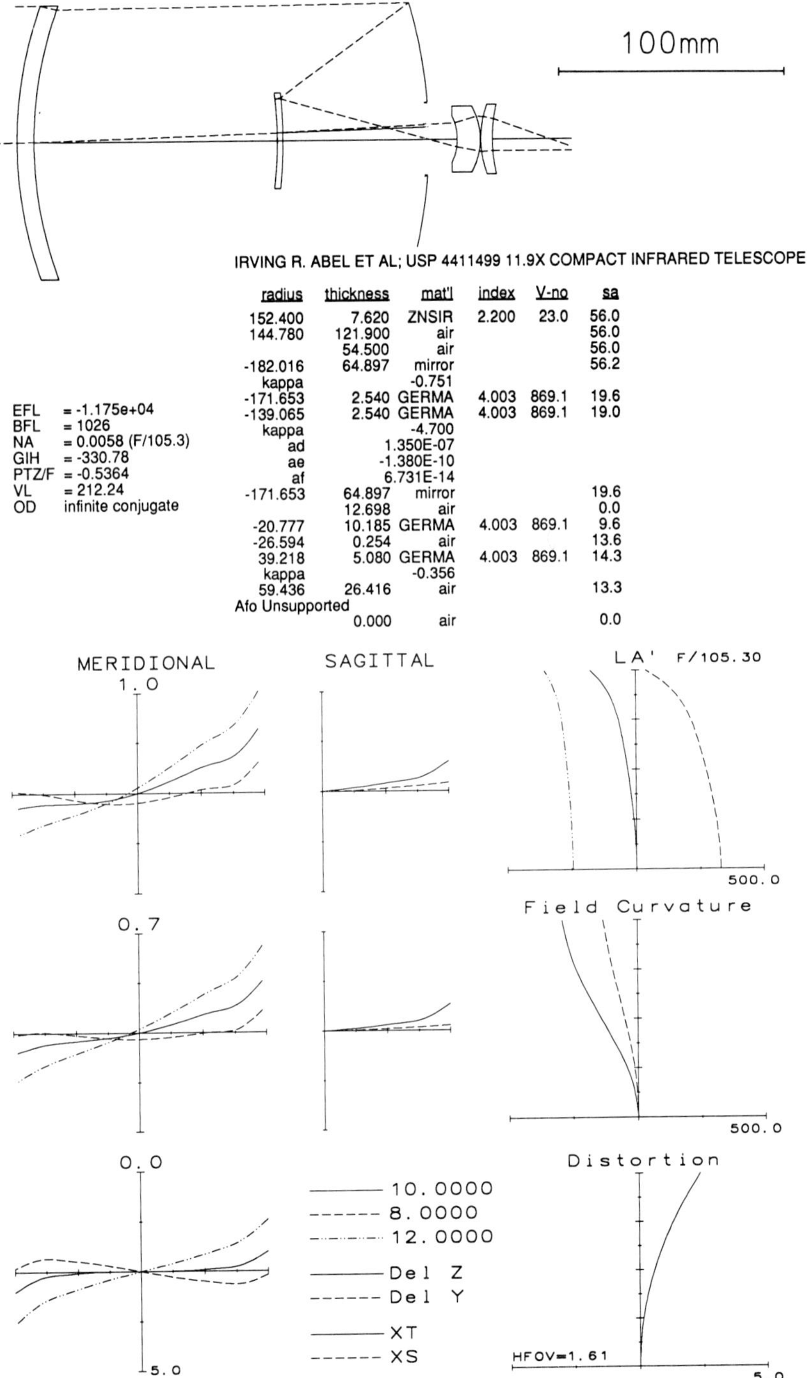

IRVING R. ABEL ET AL; USP 4411499 11.9X COMPACT INFRARED TELESCOPE

radius	thickness	mat'l	index	V-no	sa
152.400	7.620	ZNSIR	2.200	23.0	56.0
144.780	121.900	air			56.0
	54.500	air			56.0
-182.016	64.897	mirror			56.2
kappa		-0.751			
-171.653	2.540	GERMA	4.003	869.1	19.6
-139.065	2.540	GERMA	4.003	869.1	19.0
kappa		-4.700			
ad		1.350E-07			
ae		-1.380E-10			
af		6.731E-14			
-171.653	64.897	mirror			19.6
	12.698	air			0.0
-20.777	10.185	GERMA	4.003	869.1	9.6
-26.594	0.254	air			13.6
39.218	5.080	GERMA	4.003	869.1	14.3
kappa		-0.356			
59.436	26.416	air			13.3
Afo Unsupported					
	0.000	air			0.0

EFL = -1.175e+04
BFL = 1026
NA = 0.0058 (F/105.3)
GIH = -330.78
PTZ/F = -0.5364
VL = 212.24
OD infinite conjugate

Figure 21.12

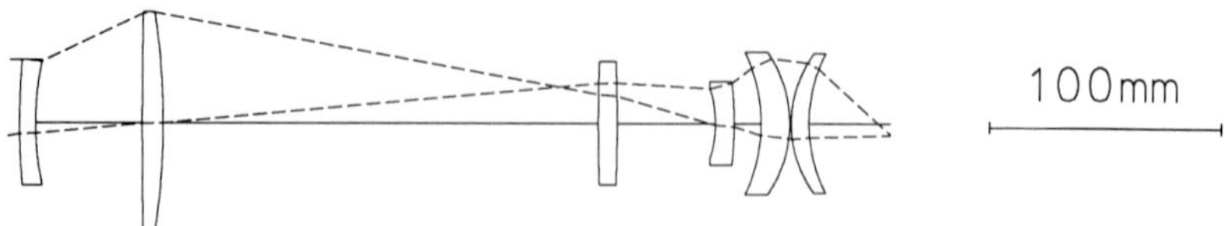

PHILIP J. ROGERS; USP 4383727 5X INFRARED TELESCOPE EXAMPLE 3

radius	thickness	mat'l	index	V-no	sa
269.337	6.858	GERMA	4.003	779.7	25.0
113.406	45.720	air			24.5
2202.662	9.144	GERMA	4.003	779.7	44.0
-263.157	186.385	air			44.0
699.526	8.001	GERMA	4.003	779.7	24.5
-678.667	41.910	air			24.5
-43.336	8.001	GERMA	4.003	779.7	14.0
-146.465	11.430	air			16.6
-63.154	12.276	GERMA	4.003	779.7	24.2
-46.862	0.640	air			28.3
44.034	7.620	GERMA	4.003	779.7	27.7
56.889	32.390	air			25.4
Afo	1000				

EFL = -4997
BFL = 1032
NA = 0.0246 (F/100.0)
VL = 337.98
OD infinite conjugate
MAG = 5

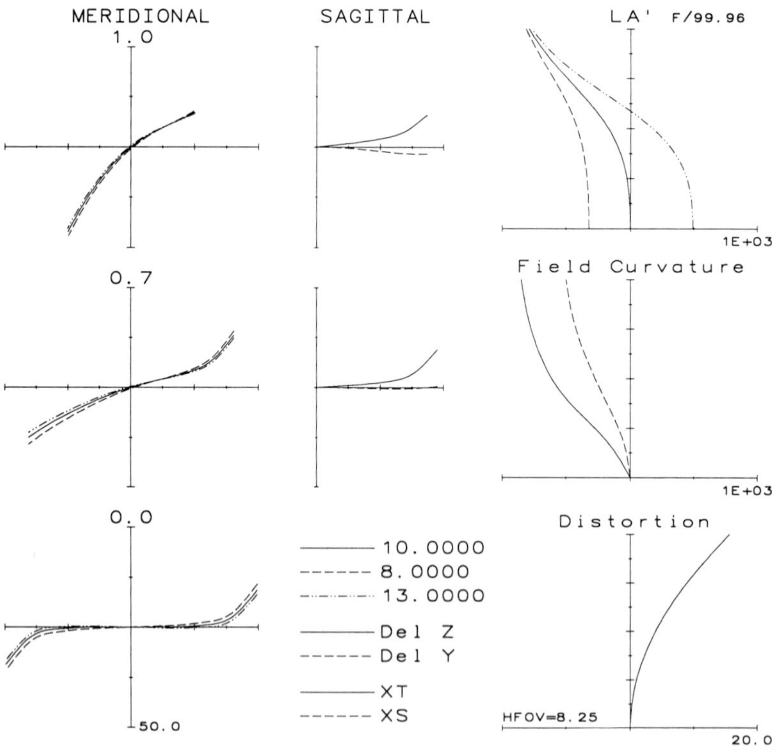

Figure 21.13

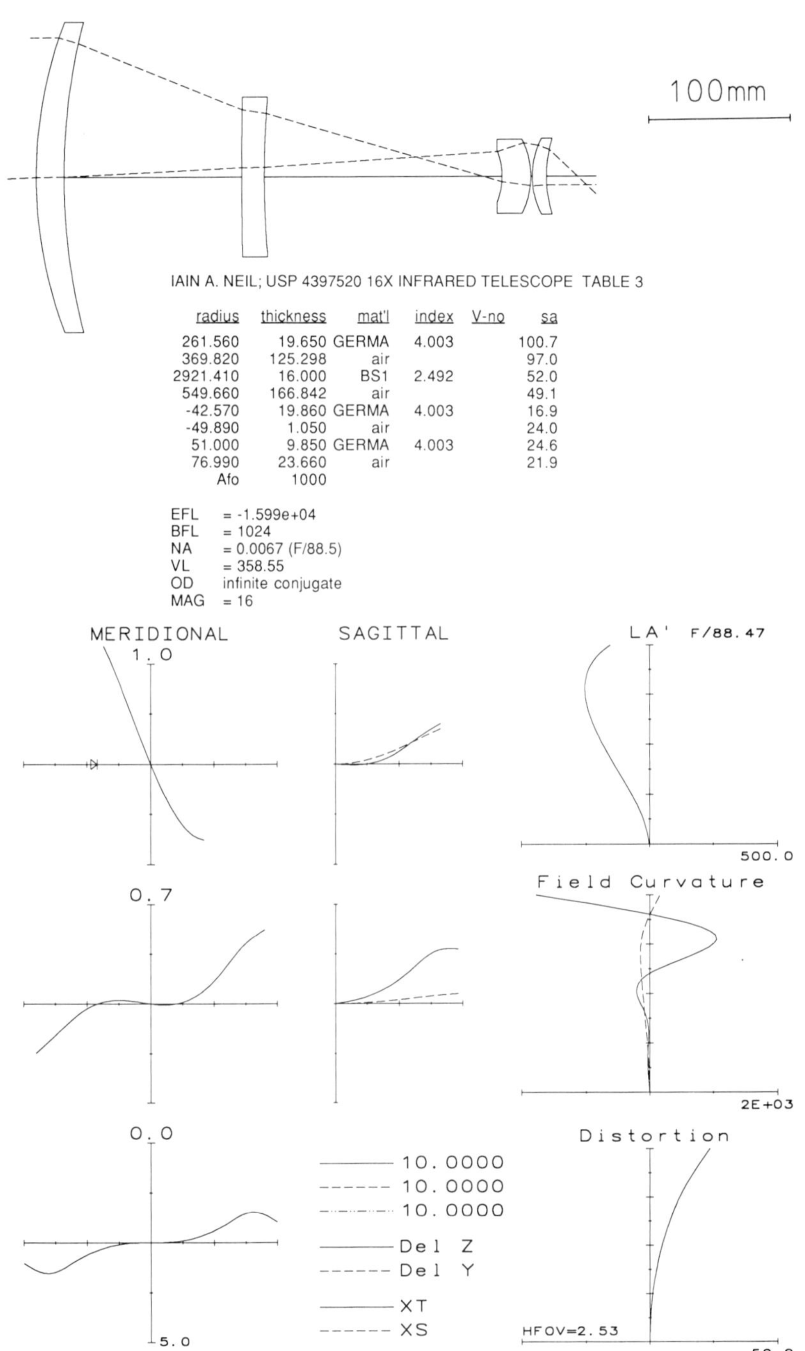

IAIN A. NEIL; USP 4397520 16X INFRARED TELESCOPE TABLE 3

radius	thickness	mat'l	index	V-no	sa
261.560	19.650	GERMA	4.003		100.7
369.820	125.298	air			97.0
2921.410	16.000	BS1	2.492		52.0
549.660	166.842	air			49.1
-42.570	19.860	GERMA	4.003		16.9
-49.890	1.050	air			24.0
51.000	9.850	GERMA	4.003		24.6
76.990	23.660	air			21.9
Afo	1000				

EFL = -1.599e+04
BFL = 1024
NA = 0.0067 (F/88.5)
VL = 358.55
OD infinite conjugate
MAG = 16

Figure 21.14

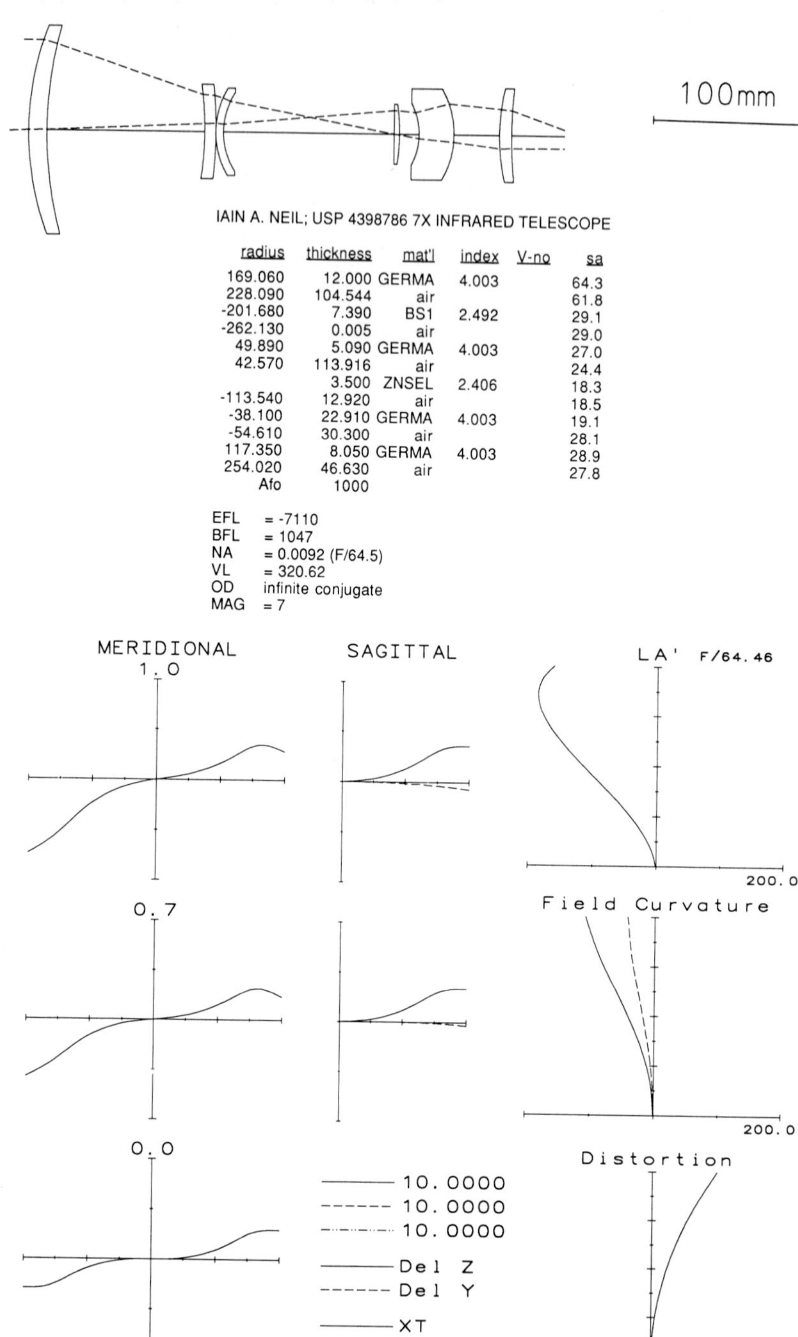

IAIN A. NEIL; USP 4398786 7X INFRARED TELESCOPE

radius	thickness	mat'l	index	V-no	sa
169.060	12.000	GERMA	4.003		64.3
228.090	104.544	air			61.8
-201.680	7.390	BS1	2.492		29.1
-262.130	0.005	air			29.0
49.890	5.090	GERMA	4.003		27.0
42.570	113.916	air			24.4
	3.500	ZNSEL	2.406		18.3
-113.540	12.920	air			18.5
-38.100	22.910	GERMA	4.003		19.1
-54.610	30.300	air			28.1
117.350	8.050	GERMA	4.003		28.9
254.020	46.630	air			27.8
Afo	1000				

EFL = -7110
BFL = 1047
NA = 0.0092 (F/64.5)
VL = 320.62
OD infinite conjugate
MAG = 7

Figure 21.15

Chapter

22

Scanner/f-θ and Laser Disk/Collimator Lenses

22.1 Monochromatic Systems

In a system which is truly monochromatic, the designer is no longer constrained by the need to achromatize the lens system. Thus high-index flint glasses can be used in positive elements and low-index crown glasses can be used in negative elements. This is obviously beneficial as regards the Petzval curvature, and obviates any need to use the expensive lanthanum glasses for the high-index elements. The resulting lens is, of course, a hyperchromat and is suitable only for use with very monochromatic light sources.

22.2 Scanner Lenses

The scanner lens operates with an oscillating mirror which scans the image across the field. To minimize the size of the scanning mirror, the pupil of the system is located at the mirror. An ordinary distortion-free lens has an image height (distance from the axis) which follows the rule $h = f \tan \theta$. When the image is scanned across the field by a mirror with a constant angular velocity, its linear velocity changes; the exposure produced will vary with the velocity. In order to achieve a uniform exposure across the field, distortion is deliberately introduced so that the image position relationship becomes $h = f\theta$. Note that all of the designs in this section have a negative distortion of this type.

The simplest scanner lens is a single meniscus lens, similar to the meniscus landscape lens. A two-element lens with the negative element facing the scan mirror is the next step of complexity. Figure 22.1 is a simple three-element lens, in a − + + configuration, with the negative element made of low-index BK7 glass (517-642) and the positive elements made of SF11, an inexpensive, stable, high-index flint. This basic configuration (− + +) is not only nearly ubiquitous, but quite versatile; it has even been used for long-wavelength (10.6-μm) scanners (with suitable materials). A slower, wider-angle scanner lens is shown in Fig. 22.2; here all three elements have the same index, and the order is + − +. Figure 22.3 achieves a higher speed and wider angle by splitting the last element (and shifting some power from the front element). Figure 22.4 is a similar configuration, except that the negative element is low-index BK7. The last of these examples is Fig. 22.5, with a fifth element added on the image side.

The last two scanner lenses. Figs. 22.6 and 22.7, are examples of telecentric systems in which the exit pupil is located at infinity, so that the principal ray of the imaging cone is always normal to the focal plane as the image is scanned across the field. As can be seen, telecentricity not only tends to require a complex design, but also requires that the lens aperture be larger than the image field.

Note that Fig. 20.6 shows a scanner lens with zooming capability.

22.3 Laser Disk, Focusing, and Collimator Lenses

Figure 22.8 shows a typical molded glass laser disk lens. Both surfaces are conics with general aspheric deformations. The lens thickness is important to the design in that it allows for some correction of the astigmatism. At the speed of *f*/0.9 of this example, it is of course vital that the plastic cladding on the disk be included in the design. The actual focal length of this type of lens is to the order of 5 mm, at which focal length the design wavefront aberration is a tiny fraction of a wavelength. This type of lens is often molded in plastic as well as glass.

Figure 22.9 is an airspaced doublet, whose correction is based on the same principles as outlined in Chap. 6 except that, as a monochromatic system, both elements can be made from a high-index flint glass (SF6 805-254). Note that if the configuration is chosen so that the spherochromatic aberration is well-corrected, then the lens can be used for several different wavelengths, although it will require refocusing for each wavelength.

Figures 22.10, 22.11, and 22.12 are examples of spherical-surfaced laser disk lenses. Note that, in each case, the final positive element is spaced well away from the aperture stop in order to allow for a modest correction of the astigmatism. In Fig. 22.12 the designer has com-

(*Text continues on page 425.*)

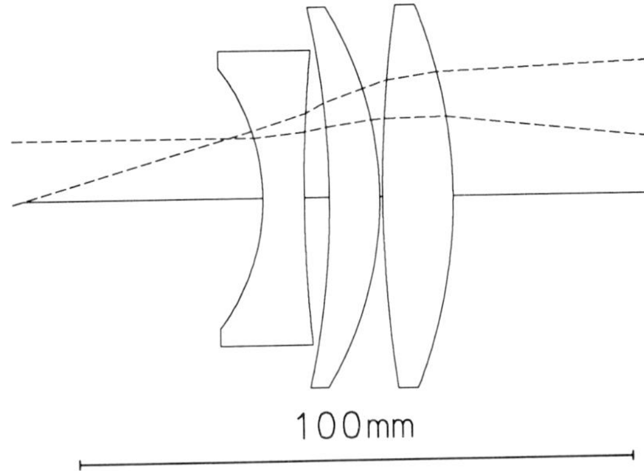

HOPKINS LASER DIODE SCAN LENS F/5, EFL=55, H'=14.31, FOV=30

radius	thickness	mat'l	index	V-no	sa
	43.550	air			10.0
-33.679	7.349	BK7	1.511		21.7
227.078	4.536	air			24.6
-137.219	9.073	SF11	1.765		27.6
-57.486	0.544	air			31.7
207.716	12.702	SF11	1.765		31.9
-80.622	129.740	air			33.8

EFL = 100
BFL = 129.7
NA = -0.1002 (F/5.0)
GIH = 26.79 (HFOV=15.00)
PTZ/F = -31.85
VL = 77.76
OD infinite conjugate

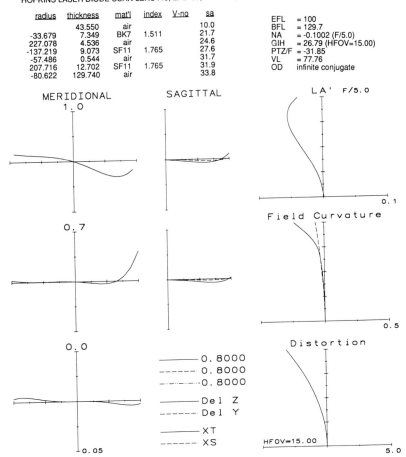

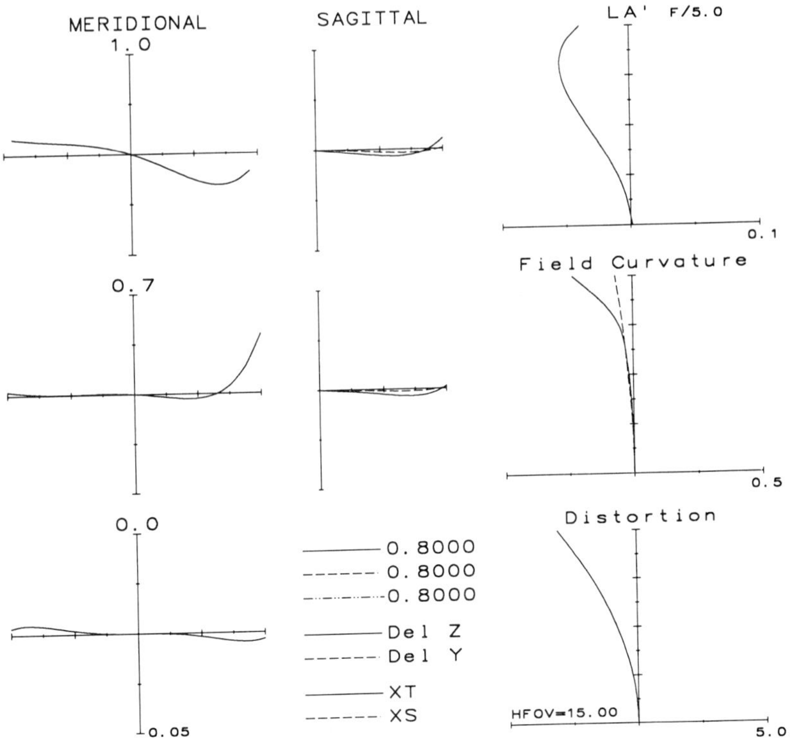

MERIDIONAL SAGITTAL LA' F/5.0

1.0 0.1

Field Curvature

0.7 0.5

Distortion

0.0

0.8000
0.8000
0.8000
Del Z
Del Y
XT
XS

0.05 HFOV=15.00 5.0

Figure 22.1

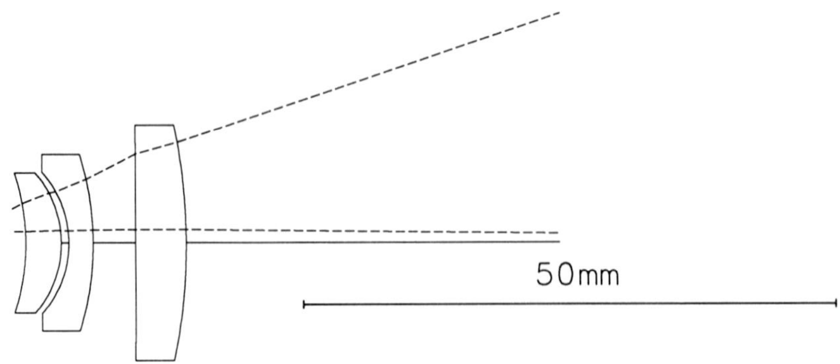

HARUO MAEDA; USP 4401362; F/50 51 DEG. F-THETA LENS EX. 2

radius	thickness	mat'l	index	V-no	sa
-17.210	3.330	FD10	1.723		5.4
-8.780	0.720	air			6.1
-8.680	2.280	FD10	1.723		6.1
-23.940	4.040	air			7.7
-1220.040	4.690	FD10	1.723		10.2
-47.120	115.045	air			11.1

EFL = 100.1
BFL = 115
NA = -0.0100 (F/50.1)
GIH = 47.73 (HFOV=25.50)
PTZ/F = -8.712
VL = 15.06
OD infinite conjugate

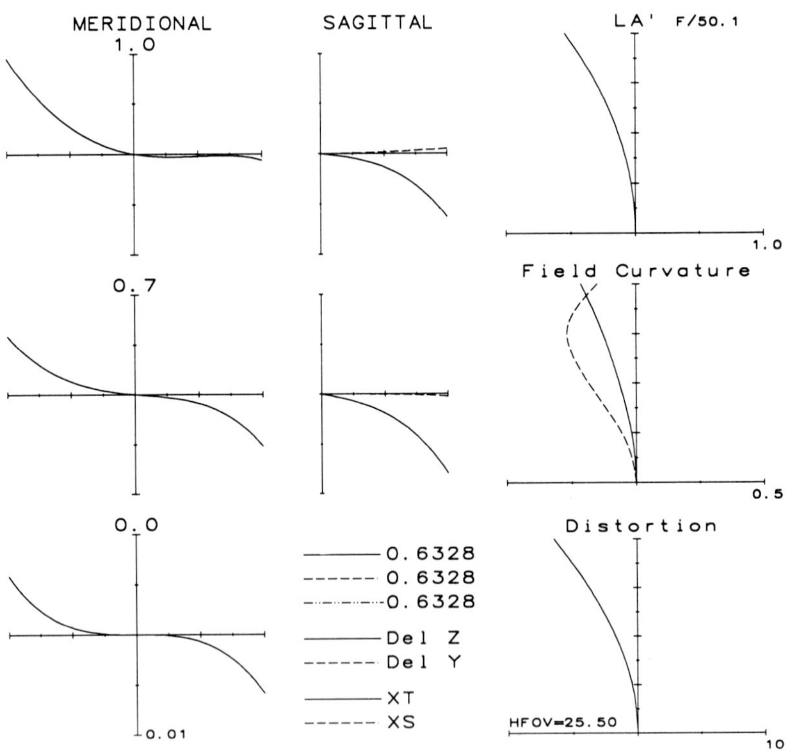

Figure 22.2

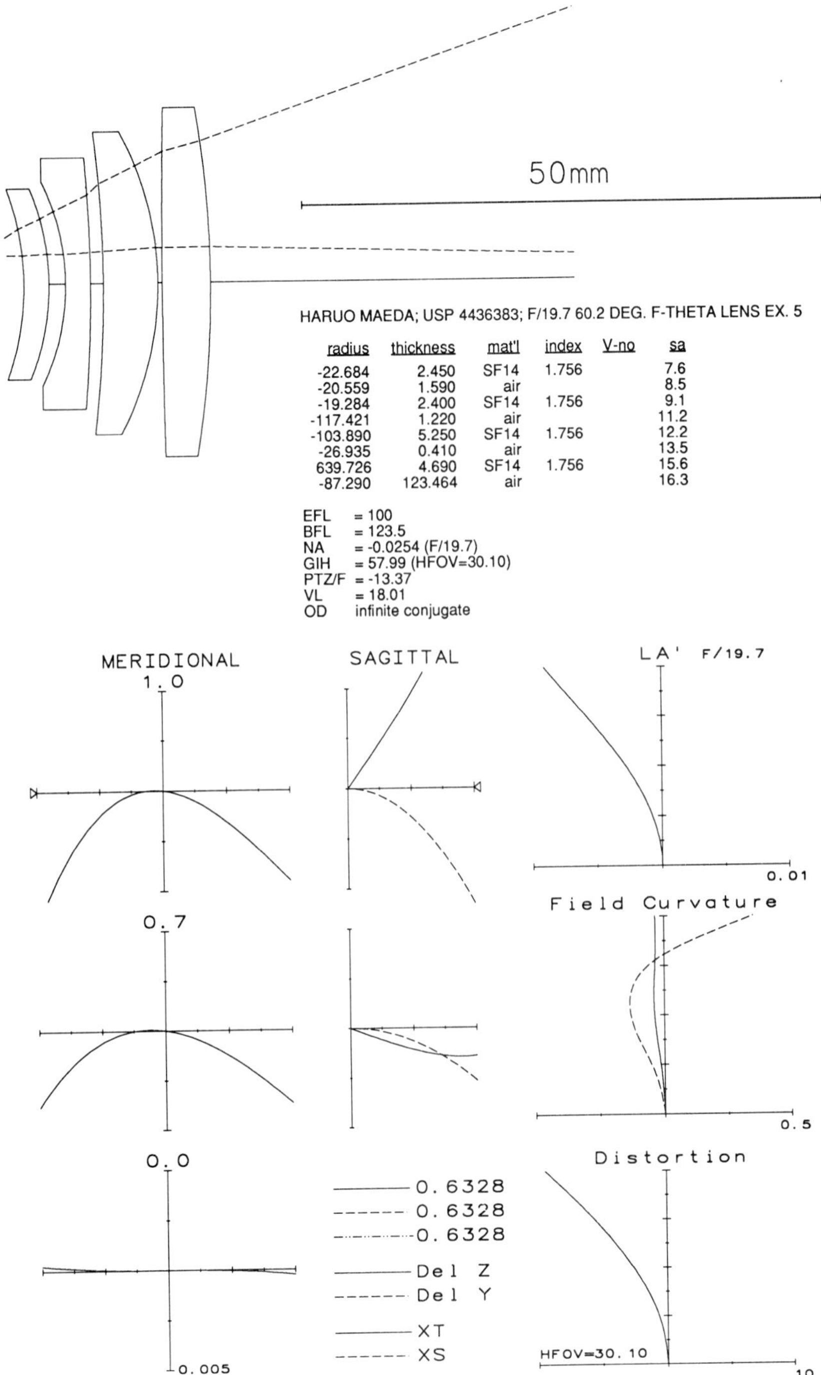

HARUO MAEDA; USP 4436383; F/19.7 60.2 DEG. F-THETA LENS EX. 5

radius	thickness	mat'l	index	V-no	sa
-22.684	2.450	SF14	1.756		7.6
-20.559	1.590	air			8.5
-19.284	2.400	SF14	1.756		9.1
-117.421	1.220	air			11.2
-103.890	5.250	SF14	1.756		12.2
-26.935	0.410	air			13.5
639.726	4.690	SF14	1.756		15.6
-87.290	123.464	air			16.3

EFL = 100
BFL = 123.5
NA = -0.0254 (F/19.7)
GIH = 57.99 (HFOV=30.10)
PTZ/F = -13.37
VL = 18.01
OD infinite conjugate

Figure 22.3

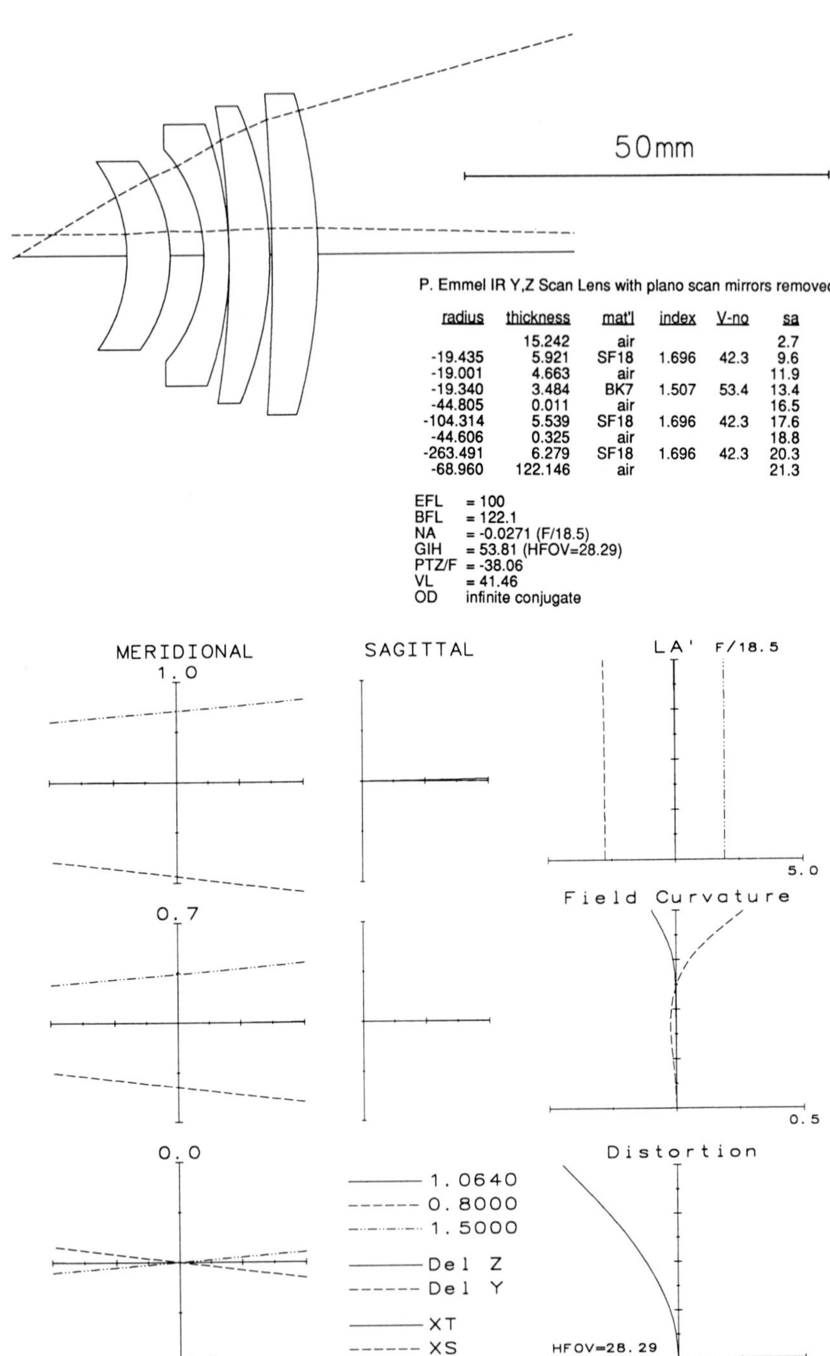

P. Emmel IR Y,Z Scan Lens with plano scan mirrors removed

radius	thickness	mat'l	index	V-no	sa
	15.242	air			2.7
-19.435	5.921	SF18	1.696	42.3	9.6
-19.001	4.663	air			11.9
-19.340	3.484	BK7	1.507	53.4	13.4
-44.805	0.011	air			16.5
-104.314	5.539	SF18	1.696	42.3	17.6
-44.606	0.325	air			18.8
-263.491	6.279	SF18	1.696	42.3	20.3
-68.960	122.146	air			21.3

EFL = 100
BFL = 122.1
NA = -0.0271 (F/18.5)
GIH = 53.81 (HFOV=28.29)
PTZ/F = -38.06
VL = 41.46
OD infinite conjugate

Figure 22.4

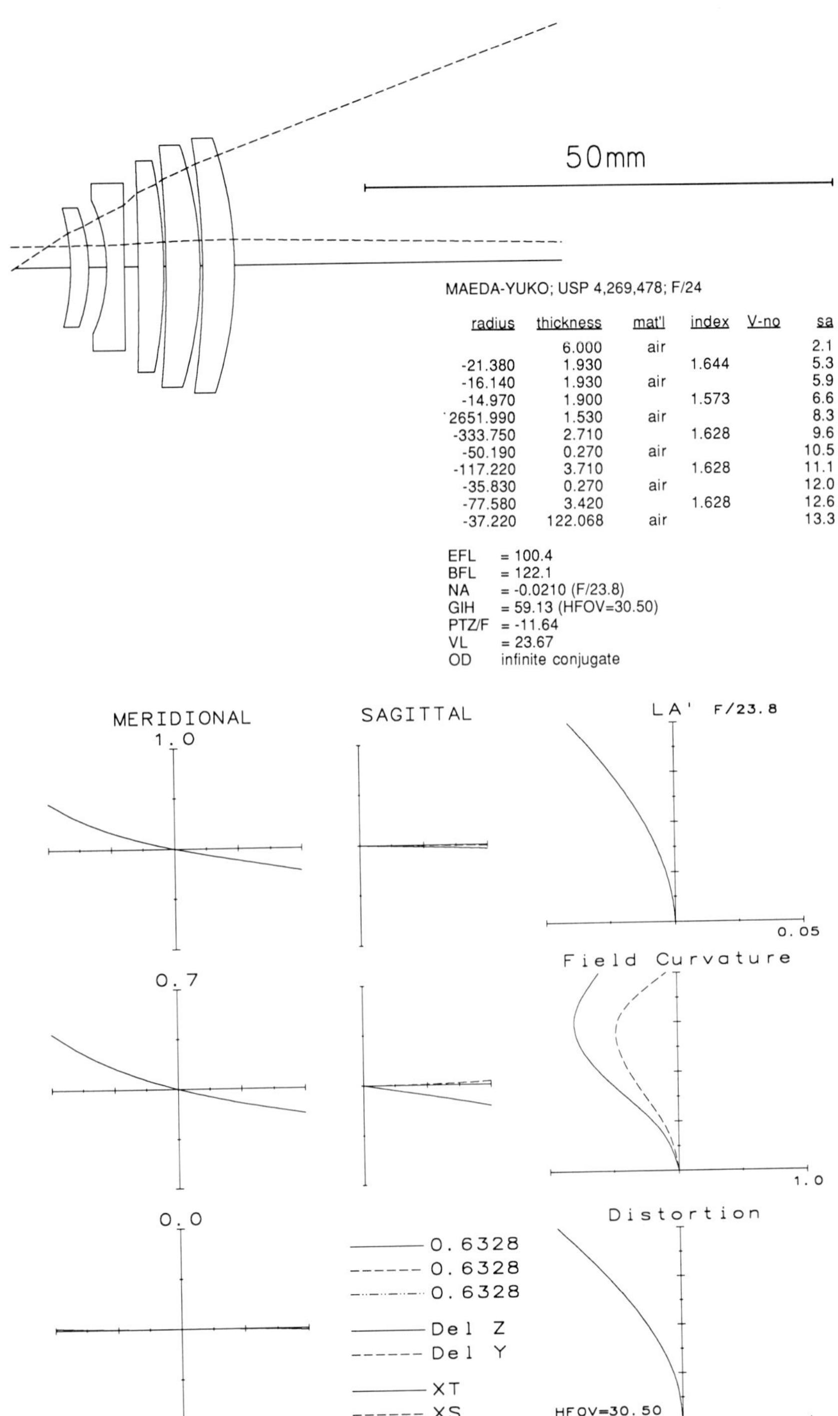

MAEDA-YUKO; USP 4,269,478; F/24

radius	thickness	mat'l	index	V-no	sa
	6.000	air			2.1
-21.380	1.930		1.644		5.3
-16.140	1.930	air			5.9
-14.970	1.900		1.573		6.6
2651.990	1.530	air			8.3
-333.750	2.710		1.628		9.6
-50.190	0.270	air			10.5
-117.220	3.710		1.628		11.1
-35.830	0.270	air			12.0
-77.580	3.420		1.628		12.6
-37.220	122.068	air			13.3

EFL = 100.4
BFL = 122.1
NA = -0.0210 (F/23.8)
GIH = 59.13 (HFOV=30.50)
PTZ/F = -11.64
VL = 23.67
OD infinite conjugate

Figure 22.5

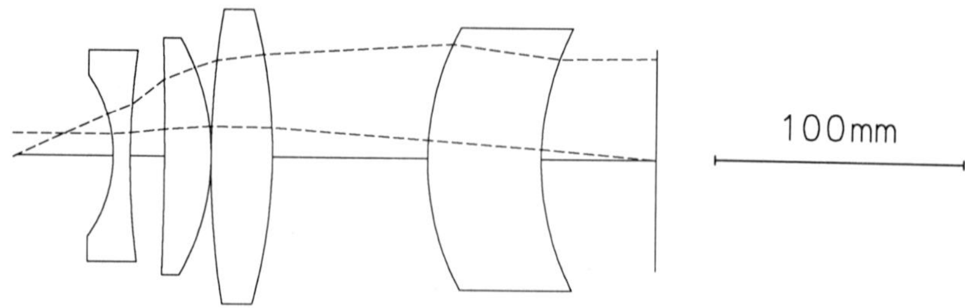

HOPKINS DIODE; F/6.1, Y'=12.5, FOV=45.4, TELECENTRIC IMAGE SPACE

radius	thickness	mat'l	index	V-no	sa
	39.301	air			8.2
-48.237	6.729	BK7	1.511		29.9
283.340	13.710	air			39.1
-1155.949	18.411	SF6	1.784		40.5
-84.616	0.003	air			43.9
322.746	24.553	SF6	1.784		54.5
-192.547	61.960	air			54.5
96.625	45.631	BK7	1.511		46.0
100.698	45.660	air			48.6

EFL = 100
BFL = 45.66
NA = -0.0818 (F/6.1)
GIH = 41.89 (HFOV=22.73)
PTZ/F = -25.06
VL = 210.30
OD infinite conjugate

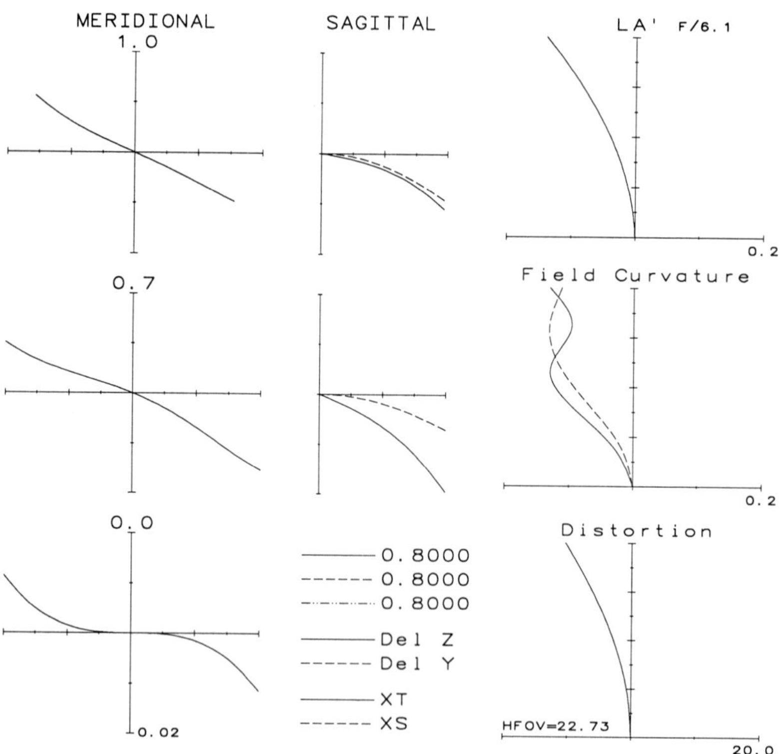

Figure 22.6

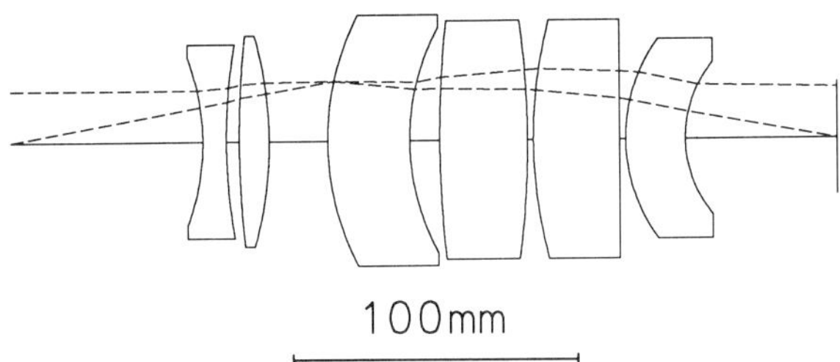

HOPKINS; HE-CD LASER SCAN LENS; F/3.0, EFL=48, FORMAT 16 FOV=19

radius	thickness	mat'l	index	V-no	sa
	67.830	air			0.0
-70.067	8.334	BK7	1.526		27.2
169.615	4.564	air			31.5
293.611	10.438	SF10	1.760		33.5
-119.709	20.835	air			34.3
83.487	28.762	SF10	1.760		40.7
72.819	10.418	air			36.7
326.930	30.839	SF10	1.760		37.2
-264.395	2.084	air			38.8
138.471	30.322	SF10	1.760		38.7
	2.131	air			35.5
52.286	20.835	BK7	1.526		32.5
37.045	52.897	air			24.8

EFL = 100
BFL = 52.9
NA = -0.1667 (F/3.0)
GIH = 16.82 (HFOV=9.55)
PTZ/F = -13.77
VL = 237.39
OD infinite conjugate

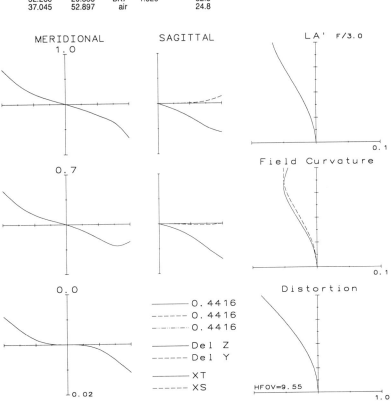

MERIDIONAL SAGITTAL LA' F/3.0
1.0
0.1
Field Curvature
0.7
0.1
0.0 Distortion
——— 0.4416
- - - - 0.4416
-·-·- 0.4416
——— Del Z
- - - - Del Y
——— XT
- - - - XS
0.02 HFOV=9.55 1.0

Figure 22.7

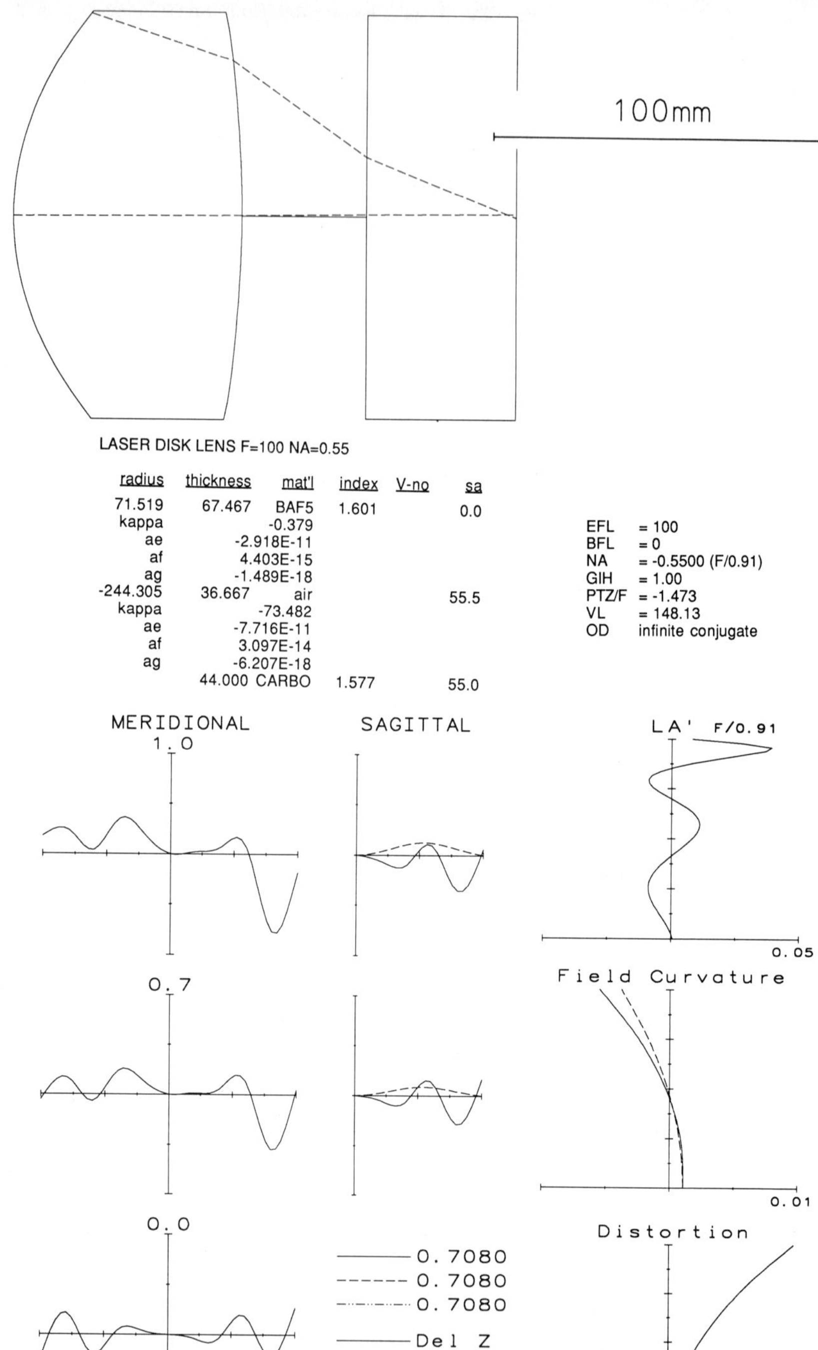

radius	thickness	mat'l	index	V-no	sa
71.519	67.467	BAF5	1.601		0.0
kappa		-0.379			
ae		-2.918E-11			
af		4.403E-15			
ag		-1.489E-18			
-244.305	36.667	air			55.5
kappa		-73.482			
ae		-7.716E-11			
af		3.097E-14			
ag		-6.207E-18			
	44.000	CARBO	1.577		55.0

Figure 22.8

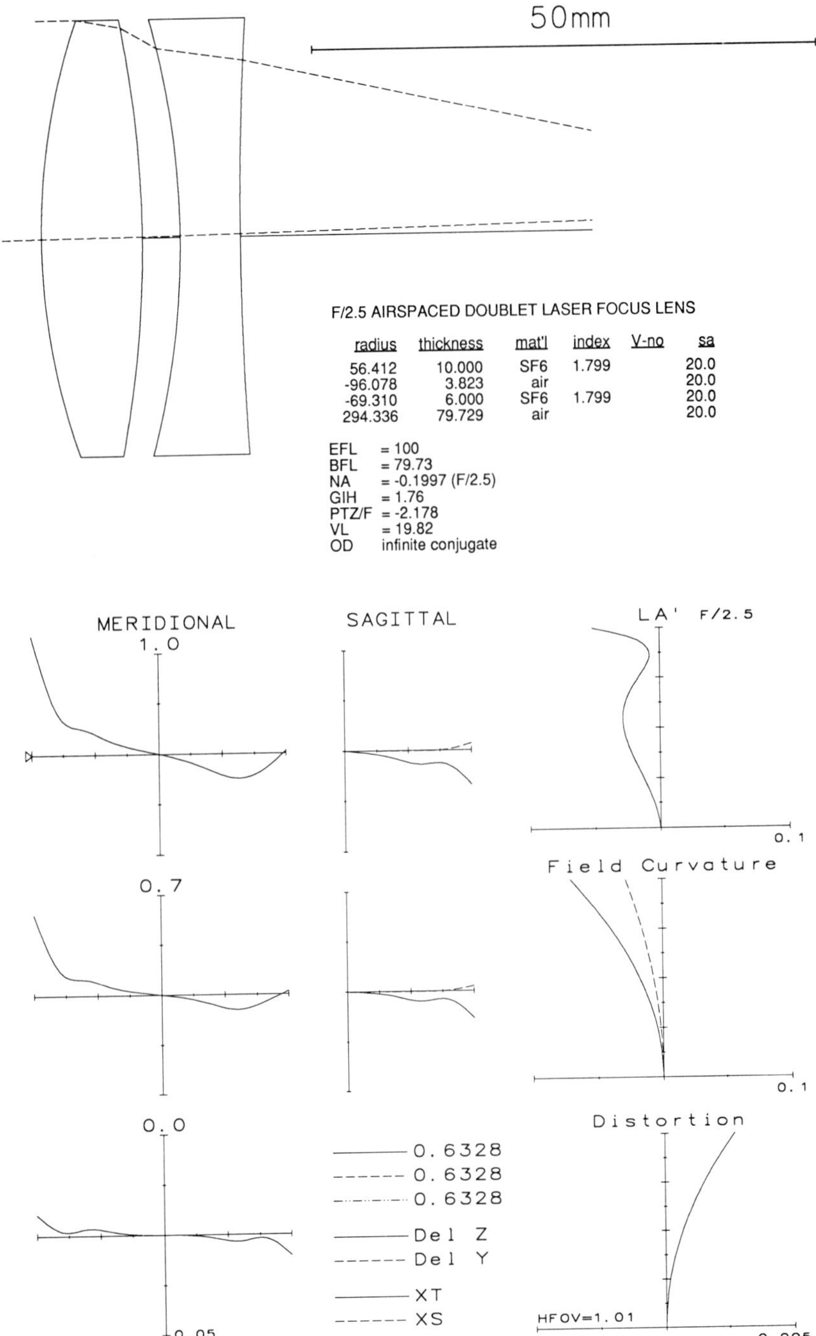

F/2.5 AIRSPACED DOUBLET LASER FOCUS LENS

radius	thickness	mat'l	index	V-no	sa
56.412	10.000	SF6	1.799		20.0
-96.078	3.823	air			20.0
-69.310	6.000	SF6	1.799		20.0
294.336	79.729	air			20.0

EFL = 100
BFL = 79.73
NA = -0.1997 (F/2.5)
GIH = 1.76
PTZ/F = -2.178
VL = 19.82
OD infinite conjugate

Figure 22.9

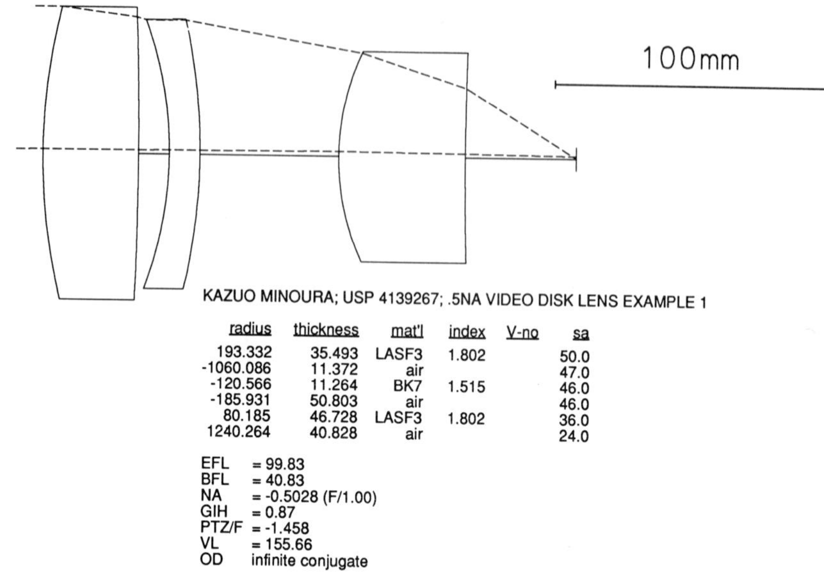

KAZUO MINOURA; USP 4139267; .5NA VIDEO DISK LENS EXAMPLE 1

radius	thickness	mat'l	index	V-no	sa
193.332	35.493	LASF3	1.802		50.0
-1060.086	11.372	air			47.0
-120.566	11.264	BK7	1.515		46.0
-185.931	50.803	air			46.0
80.185	46.728	LASF3	1.802		36.0
1240.264	40.828	air			24.0

EFL = 99.83
BFL = 40.83
NA = -0.5028 (F/1.00)
GIH = 0.87
PTZ/F = -1.458
VL = 155.66
OD infinite conjugate

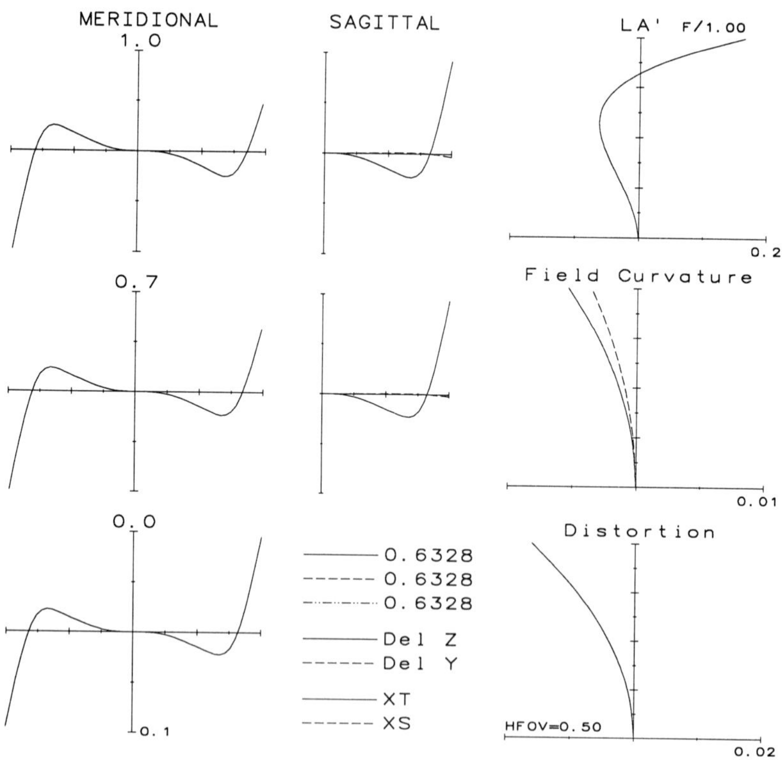

Figure 22.10

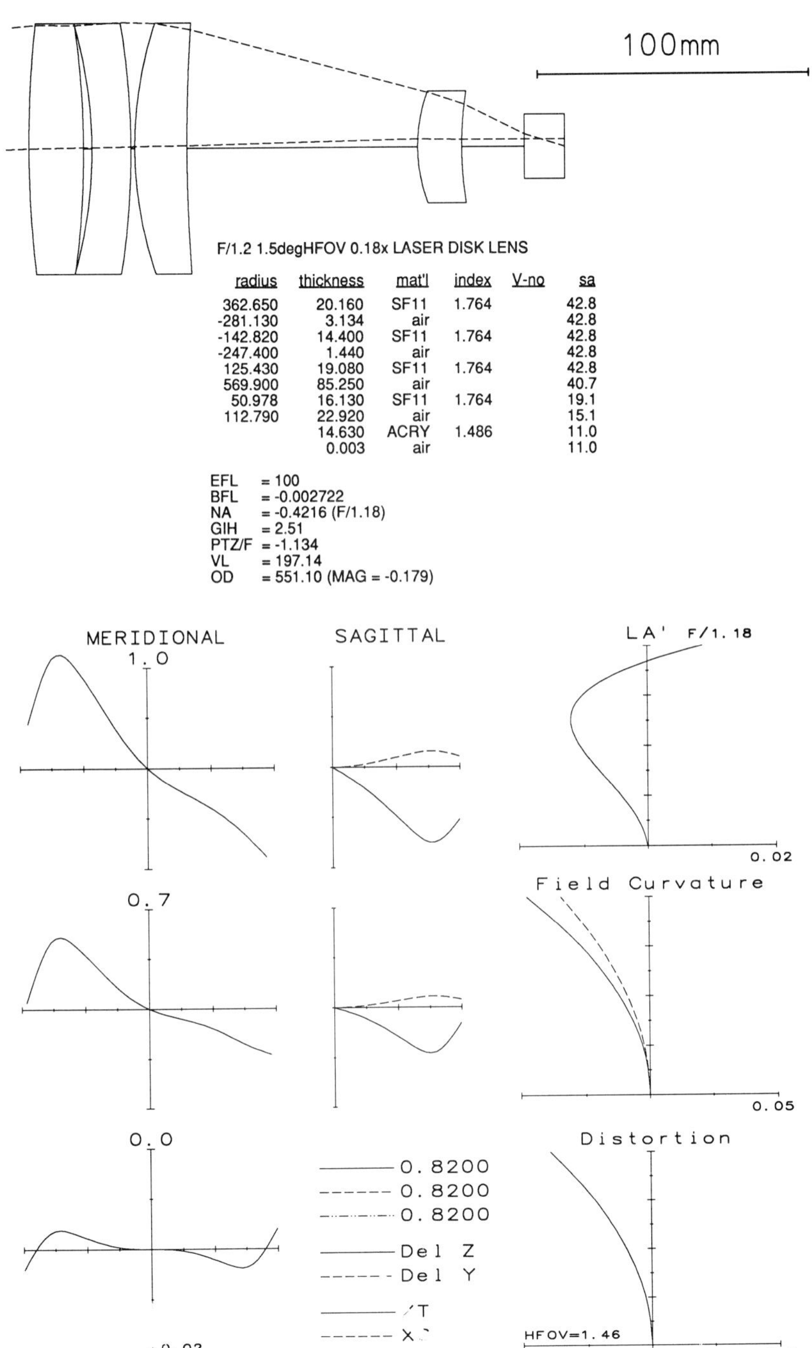

F/1.2 1.5degHFOV 0.18x LASER DISK LENS

radius	thickness	mat'l	index	V-no	sa
362.650	20.160	SF11	1.764		42.8
-281.130	3.134	air			42.8
-142.820	14.400	SF11	1.764		42.8
-247.400	1.440	air			42.8
125.430	19.080	SF11	1.764		42.8
569.900	85.250	air			40.7
50.978	16.130	SF11	1.764		19.1
112.790	22.920	air			15.1
	14.630	ACRY	1.486		11.0
	0.003	air			11.0

EFL = 100
BFL = -0.002722
NA = -0.4216 (F/1.18)
GIH = 2.51
PTZ/F = -1.134
VL = 197.14
OD = 551.10 (MAG = -0.179)

Figure 22.11

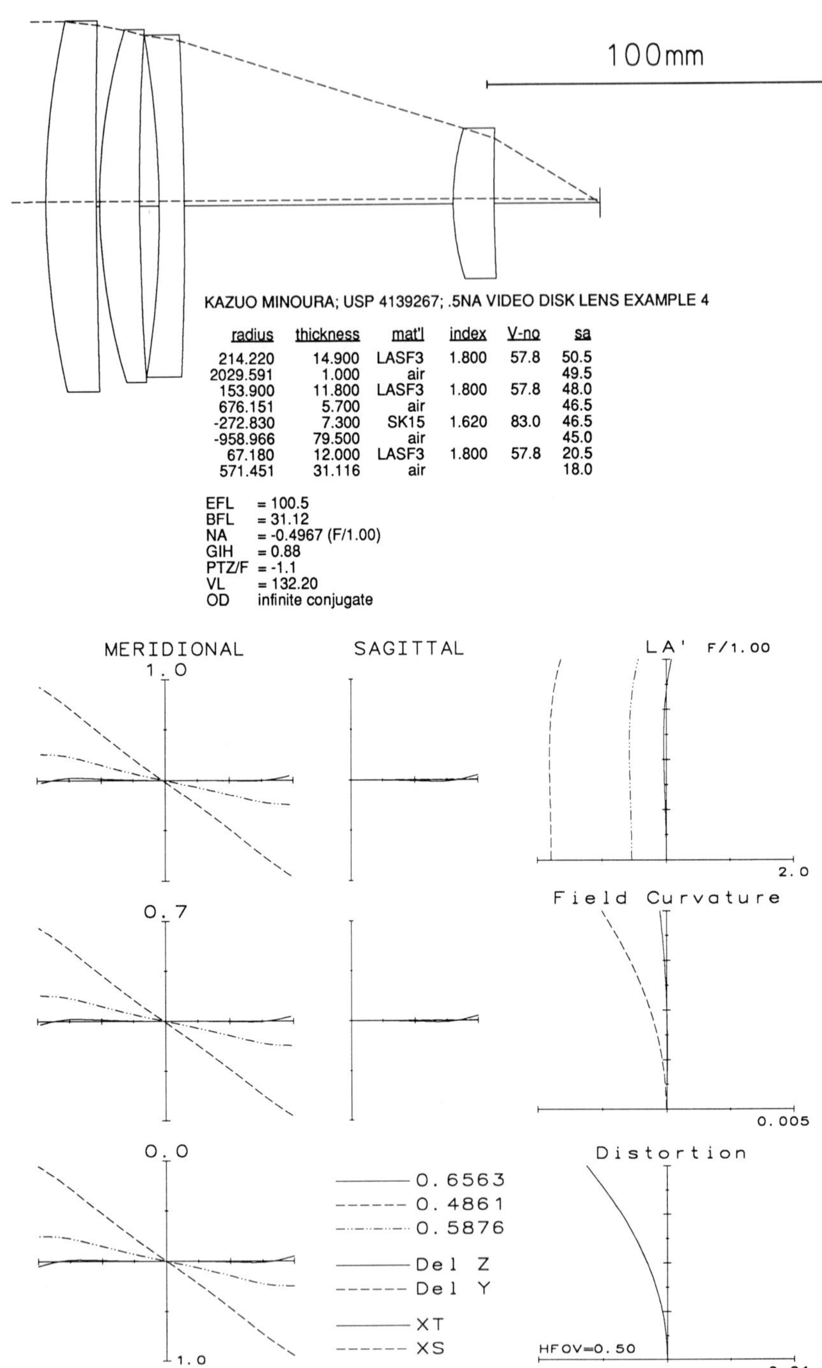

KAZUO MINOURA; USP 4139267; .5NA VIDEO DISK LENS EXAMPLE 4

radius	thickness	mat'l	index	V-no	sa
214.220	14.900	LASF3	1.800	57.8	50.5
2029.591	1.000	air			49.5
153.900	11.800	LASF3	1.800	57.8	48.0
676.151	5.700	air			46.5
-272.830	7.300	SK15	1.620	83.0	46.5
-958.966	79.500	air			45.0
67.180	12.000	LASF3	1.800	57.8	20.5
571.451	31.116	air			18.0

EFL = 100.5
BFL = 31.12
NA = -0.4967 (F/1.00)
GIH = 0.88
PTZ/F = -1.1
VL = 132.20
OD infinite conjugate

Figure 22.12

pletely flattened the tangential field and also almost completely eliminated the spherochromatism.

Figure 22.13 is an example of a diverging lens that can be used to produce a well-corrected point image or serve as the negative component for a galilean beam expander. With a well-corrected negative lens like this, a variety of independently corrected positive components can be used to produce several different powers of beam expander.

Figures 22.14, 22.15, and 22.16 illustrate, first, a laser collimating or focusing lens and, second, how the same general configuration was modified to produce a 2× zooming version for the same application.

The final design of this section, Fig. 22.17, is a very highly corrected finite conjugate (0.2×) telecentric objective for use in the near ultraviolet, using only BK7 (517-642) glass. It is a difficult lens to categorize.

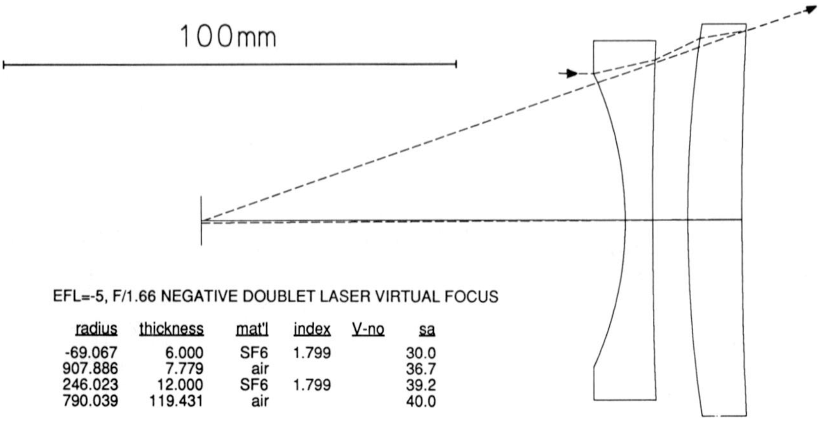

100mm

EFL=-5, F/1.66 NEGATIVE DOUBLET LASER VIRTUAL FOCUS

radius	thickness	mat'l	index	V-no	sa
-69.067	6.000	SF6	1.799		30.0
907.886	7.779	air			36.7
246.023	12.000	SF6	1.799		39.2
790.039	119.431	air			40.0

EFL = -100
BFL = -119.4
NA = 0.3064 (F/1.67)
GIH = -0.50
PTZ/F = -1.838
VL = 25.78
OD infinite conjugate

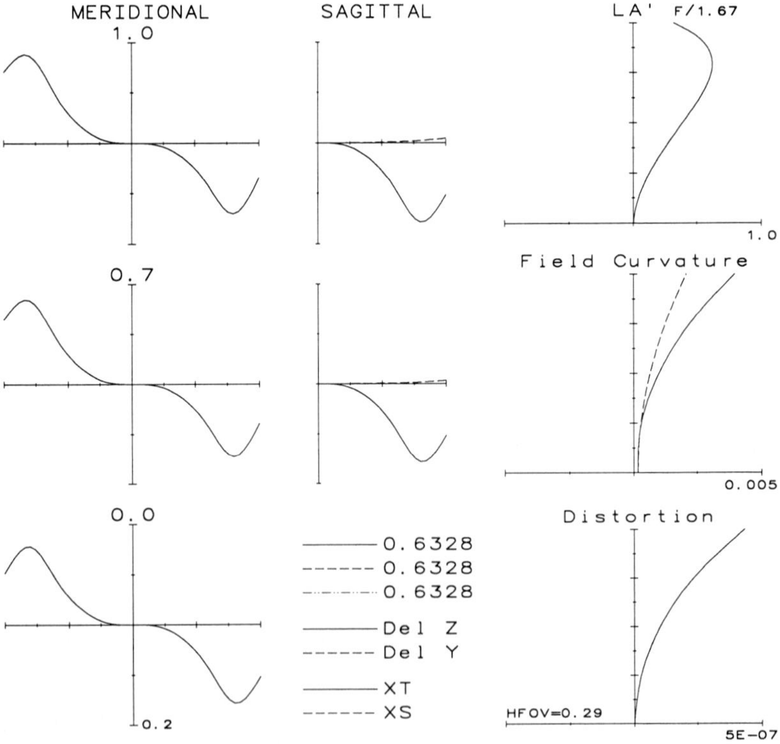

MERIDIONAL SAGITTAL LA' F/1.67
1.0
1.0
Field Curvature
0.7
0.005
0.0
Distortion
0.6328
0.6328
0.6328
Del Z
Del Y
XT
XS
0.2
HFOV=0.29
5E-07

Figure 22.13

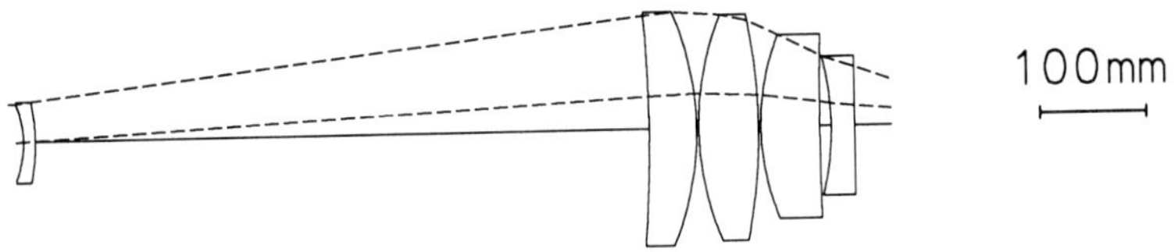

LASER COLLIMATING LENS, F/1.5, EFL=5 MM, WAVELENGTH=.4416

radius	thickness	mat'l	index	V-no	sa
-84.839	11.046	BK7	1.526		33.4
-171.722	579.255	air			35.4
-1150.060	44.186	BK7	1.526		99.0
-250.052	2.209	air			101.8
244.882	55.233	BK7	1.526		98.3
-453.795	2.209	air			93.4
198.752	55.233	BK7	1.526		80.0
2783.112	10.938	air			62.2
-230.604	22.093	BK7	1.526		60.6
	154.667	air			51.6

EFL = 100
BFL = 154.7
NA = -0.3183 (F/1.50)
GIH = 5.00
PTZ/F = 41.35
VL = 782.40
OD infinite conjugate

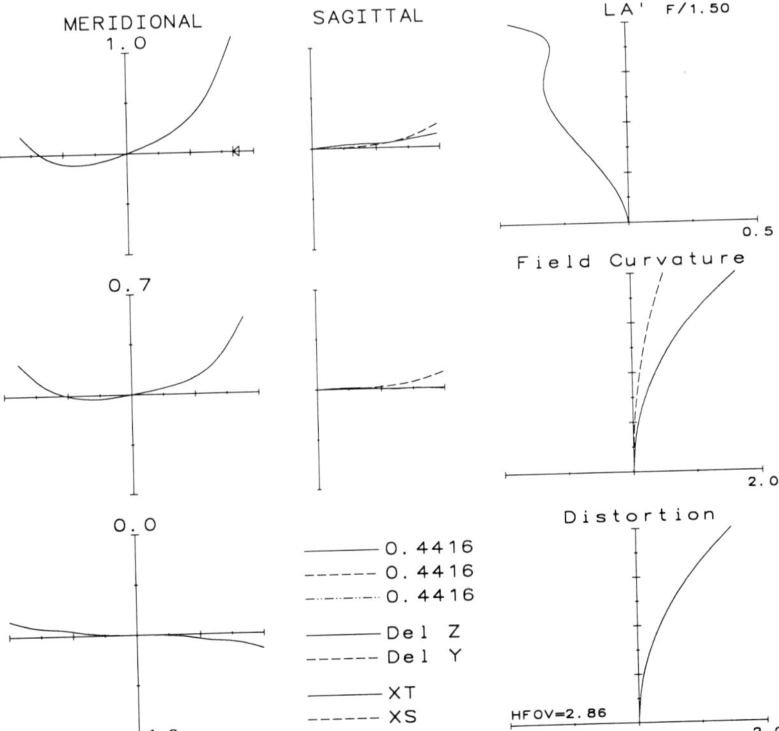

Figure 22.14

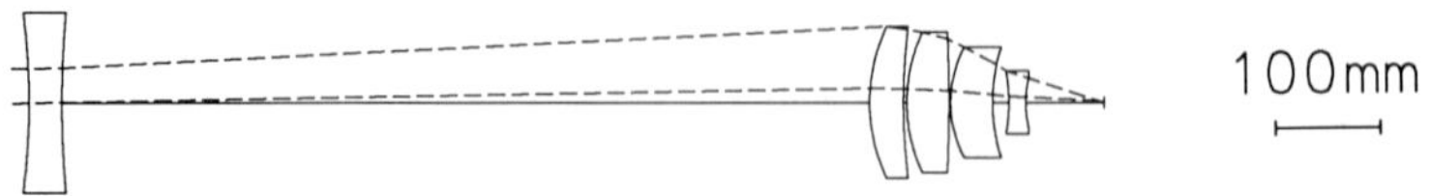

LASER ZOOM COLLIMATOR FEED LENS; EFL=4-8 MM - SHORT EFL

radius	thickness	mat'l	index	V-no	sa
-566.418	31.802	BK7	1.526		79.5
866.479	768.000	air			79.5
148.977	31.802	BK7	1.526		67.1
539.639	3.235	air			64.4
140.065	39.752	BK7	1.526		62.3
-1167.605	1.590	air			56.0
100.733	39.752	BK7	1.526		48.8
154.535	15.320	air			33.7
-135.259	15.901	BK7	1.526		27.9
88.765	74.032	air			21.7

EFL = 100
BFL = 74.03
NA = -0.2912 (F/1.67)
GIH = 1.58
PTZ/F = 5.19
VL = 947.15
OD infinite conjugate

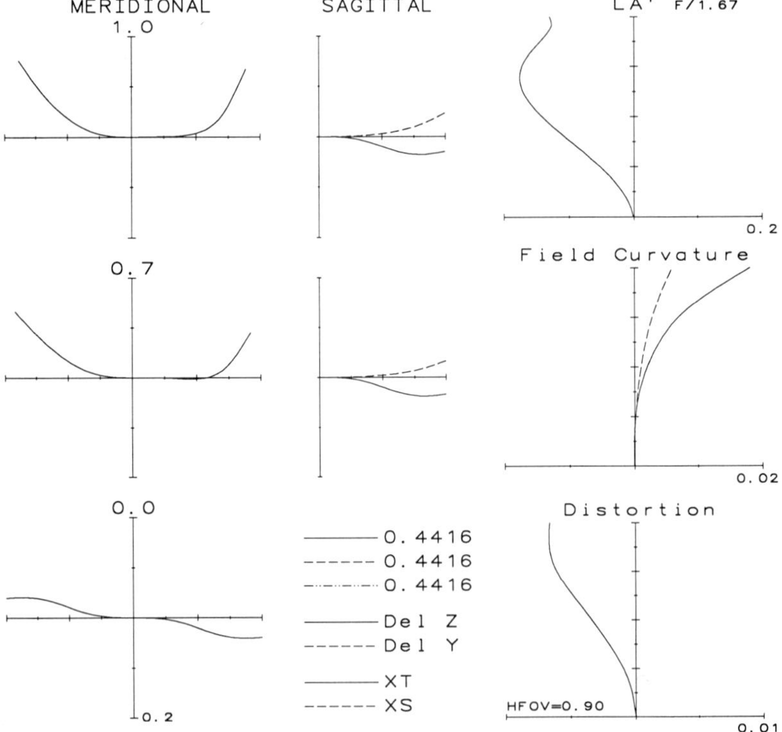

Figure 22.15

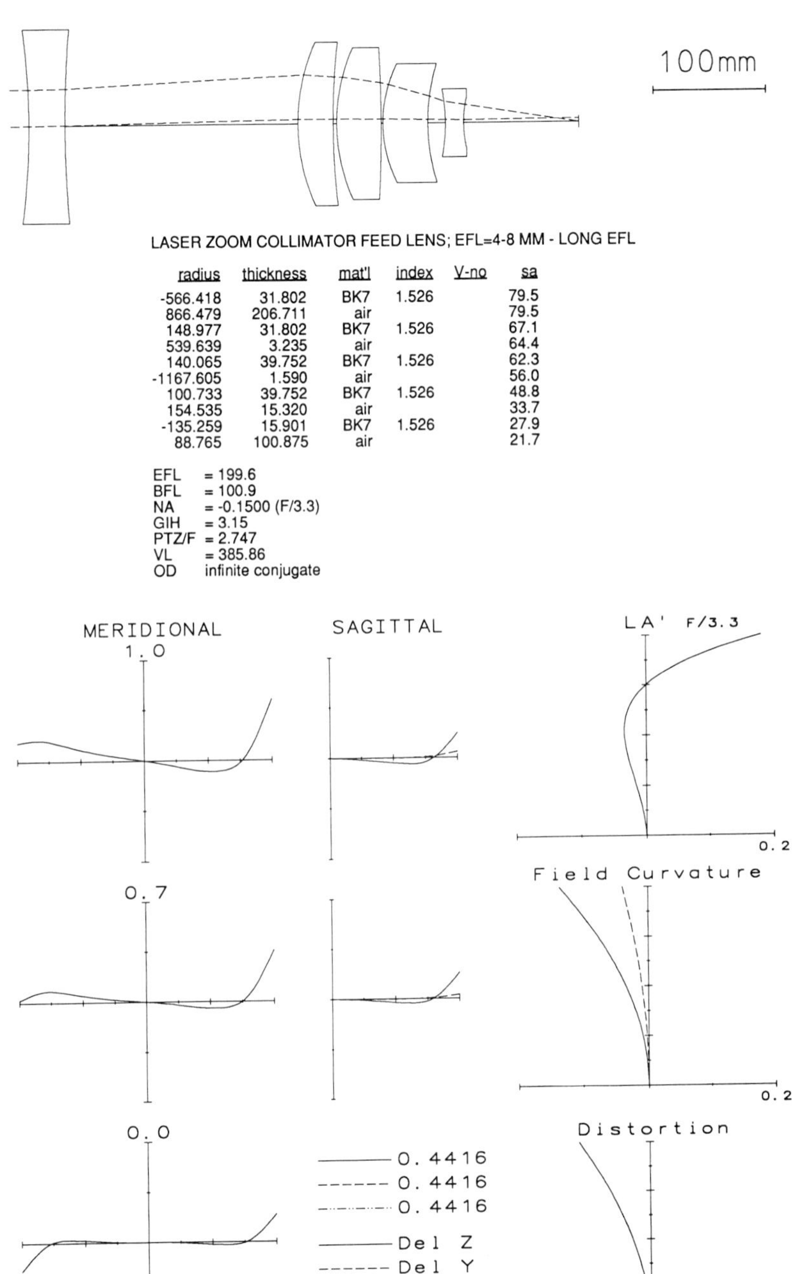

LASER ZOOM COLLIMATOR FEED LENS; EFL=4-8 MM - LONG EFL

radius	thickness	mat'l	index	V-no	sa
-566.418	31.802	BK7	1.526		79.5
866.479	206.711	air			79.5
148.977	31.802	BK7	1.526		67.1
539.639	3.235	air			64.4
140.065	39.752	BK7	1.526		62.3
-1167.605	1.590	air			56.0
100.733	39.752	BK7	1.526		48.8
154.535	15.320	air			33.7
-135.259	15.901	BK7	1.526		27.9
88.765	100.875	air			21.7

EFL = 199.6
BFL = 100.9
NA = -0.1500 (F/3.3)
GIH = 3.15
PTZ/F = 2.747
VL = 385.86
OD infinite conjugate

Figure 22.16

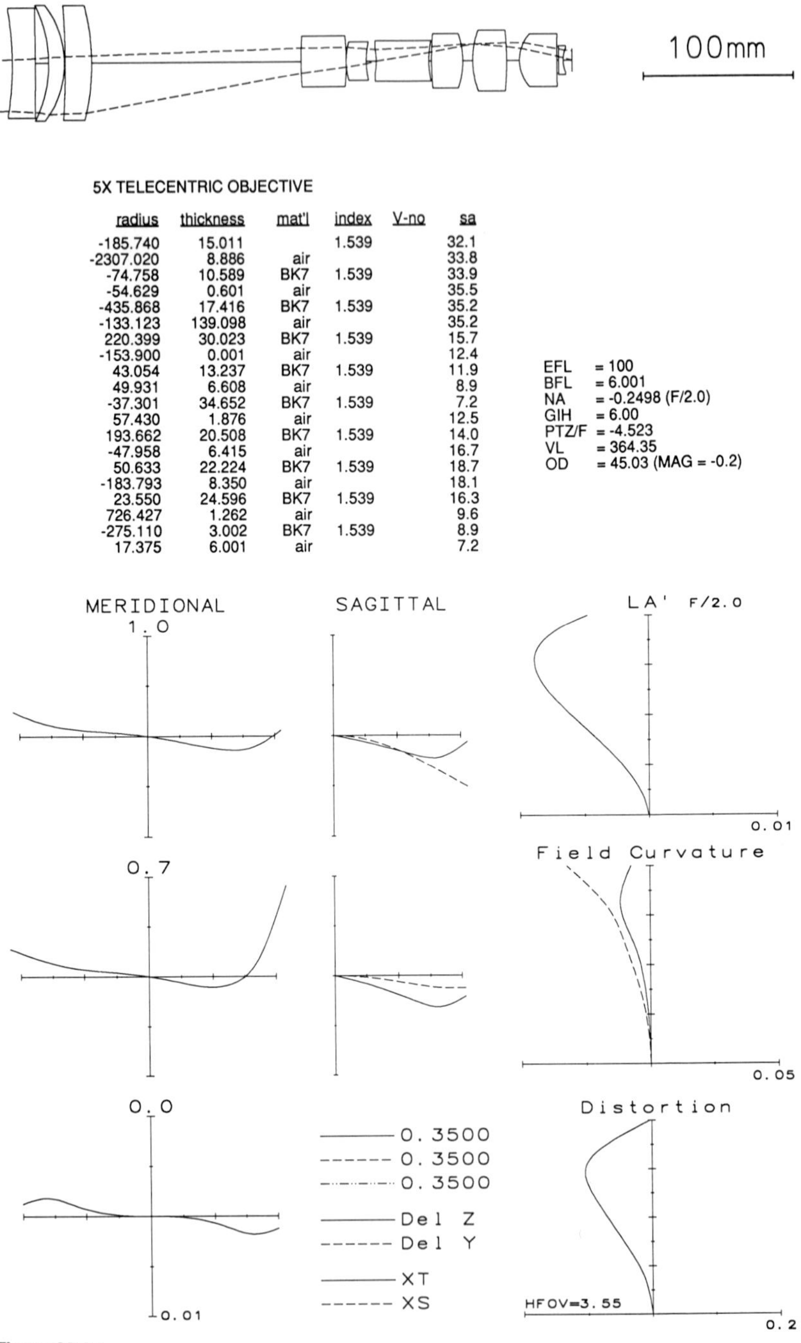

5X TELECENTRIC OBJECTIVE

radius	thickness	mat'l	index	V-no	sa
-185.740	15.011		1.539		32.1
-2307.020	8.886	air			33.8
-74.758	10.589	BK7	1.539		33.9
-54.629	0.601	air			35.5
-435.868	17.416	BK7	1.539		35.2
-133.123	139.098	air			35.2
220.399	30.023	BK7	1.539		15.7
-153.900	0.001	air			12.4
43.054	13.237	BK7	1.539		11.9
49.931	6.608	air			8.9
-37.301	34.652	BK7	1.539		7.2
57.430	1.876	air			12.5
193.662	20.508	BK7	1.539		14.0
-47.958	6.415	air			16.7
50.633	22.224	BK7	1.539		18.7
-183.793	8.350	air			18.1
23.550	24.596	BK7	1.539		16.3
726.427	1.262	air			9.6
-275.110	3.002	BK7	1.539		8.9
17.375	6.001	air			7.2

EFL = 100
BFL = 6.001
NA = -0.2498 (F/2.0)
GIH = 6.00
PTZ/F = -4.523
VL = 364.35
OD = 45.03 (MAG = -0.2)

Figure 22.17

Chapter

23

Tolerance Budgeting

23.1 The Tolerance Budget

A lens design is not truly complete until the designer has defined a set of tolerances which will assure that, when fabricated, the lens will perform as desired. Establishing a tolerance budget is almost as much an art as is lens design itself. The budget should take into account not only the characteristics of the optical system (and its application) but also the capabilities of the shop which is to build it, as well as the effect of the tolerances on the cost and delivery schedule.

The dimensions and characteristics for which tolerances should be defined may include the following:

Surface geometry

- Radius value
- Departure from nominal shape

Surface finish

- Quality (scratch and dig)
- Roughness, scatter, etc.

Surface separation

- Element thickness
- Spacing

Index of refraction

- At the central wavelength
- Total dispersion (*V* value)
- Partial dispersion
- Homogeneity

Alignment

- Surface tilt
- Element tilt and/or decentration
- Component tilt and/or decentration
- Prism or mirror angles and alignment

Transmission

- Optical material
- Filters
- Coatings

Physical characteristics

- Thermal expansion
- Stability
- Durability

The nominal surface geometry is specified by a spherical radius of curvature (in most cases), by a conic section, or as a general (power series) aspheric (e.g., Eq. 5.1). For a spherical surface, the specification is often given with respect to a specific test plate. The value of the radius is indicated by the departure of the lens surface from the test plate, given in units of interference fringes, or Newton's rings (at a specified wavelength). Each fringe indicates a change in the space between lens and test plate of one-half wave. The accuracy of the geometry of the surface is specified as a departure from the best-fit perfect sphere, and is often referred to as *asphericity* or *irregularity*. Because it is difficult to detect a small irregularity if there is a large number of fringes visible in a test, it is common practice to require that there be no more than four or five times as many fringes present as the irregularity specification allows. This requirement is equally valid when the irregularity is tested on any other type of interferometer. The same general approach is used to specify conics or general aspherics, i.e., a tolerance on the value of the basic surface curvature and a separate tolerance on the departure from the nominal geometry.

The difference in surface radius corresponding to N fringes departure from a test plate is given by

$$\Delta R = N\lambda\left(\frac{2R}{d}\right)^2 \tag{23.1}$$

where R is the nominal radius, λ is the test wavelength, and d is the diameter over which the N fringes are observed.

The effect of a surface *irregularity* on the shape of the system wavefront can be calculated from

$$\text{OPD} = 0.5(n' - n)(\text{number of fringes}) \tag{23.2}$$

where OPD is the wavefront deformation in wavelengths, $(n' - n)$ is the index change across the surface, and (number of fringes) is the height of the surface bump or irregularity in interference fringes. Note that, in most cases, irregularity takes the form of an astigmatic or toric surface, and that a compromise focus usually reduces its effect by a factor of 2. In an assembly, a random orientation of the cylinder axes can effectively cancel out much of the astigmatism. However, axially symmetrical irregularities of the type referred to by opticians as a *gull-wing* or *hair-pin* surface (after the shape of the test-plate fringes) are always directly additive in a worst-case sense, and can be a serious problem when the optics are fabricated "one up."

For *most* optical systems, scratches, digs, pits, bubbles, inclusions, and the like are purely cosmetic defects with little or no effect on the function of the system. Under such circumstances, their specification can best be regarded as an agreement between buyer and vendor as to an acceptable level of workmanship; this level usually depends on whether or not the optic can be seen. The functional effect is primarily scattering. If one expresses the area of the defect as a fraction of the area of the light beam, this will give the percentage of the light intercepted by the defect; most of this is scattered, some may be absorbed. The scattered light is probably distributed into 4π steradians; one can easily determine the resultant level of relative illumination in the image plane which is caused by the defect. It is usually totally negligible. If the defect is on a surface close to an image plane, the defect may then be visible; its visibility (contrast) and significance can be used to determine an acceptable size for such a defect.

There are, of course, exceptions to this. In laser systems, particularly those with high power levels, scattering resulting from surface roughness may be significant, and even small defects may become damage centers which can lead to the degradation or destruction of the system. Systems which are sensitive to low levels of scattered light, especially those where the defects may be brightly illuminated, constitute another class of exceptions.

Figure 23.1 is a table of typical tolerances which can be used as a

	Surface Quality	Diameter, mm	Deviation (concentricity), min	Thickness, mm	Radius	Regularity (asphericity)	Linear Dimension, mm	Angles
ow cost	120-80	± 0.2	> 10	± 0.5	Gage	Gage	± 0.5	Degrees
ommercial	80-50	± 0.07	3–10	± 0.25	10 Fr	3 Fr	± 0.25	± 15′
recision	60-40	± 0.02	1–3	± 0.1	5 Fr	1 Fr	± 0.1	± 5′–10′
xtraprecise	60-40	± 0.01	< 1	± 0.05	1 Fr	⅕ Fr	As req'd.	Seconds
astic	80-50		1	± 0.02	10 Fr	5 Fr	0.02	minutes

Figure 23.1 Tabulation of typical optical fabrication tolerances.

guide to tolerance sizes which are commonly considered appropriate. While these values are in fact fairly typical of ordinary optical shop practice, one should remember that there are many special cases which cannot be covered in a summary tabulation of this type.

The glass catalog tolerance on index is ± 0.001 for most glass types; tolerances of ± 0.0005 or ± 0.0002 can be had by selection, at a modest increase in cost (and sometimes delivery time). V-value tolerances are typically about ± 0.8 percent; again, tighter tolerances can be obtained. The catalog tolerance on index homogeneity is usually given as 0.0001. It is easy to calculate that, if the index varied this much within an element, the resultant wavefront deformation ($\Delta n \times$ thickness) would be overwhelming. This variation is, however, that which may occur within a *melt,* not a blank. The index variation within an element is 1 or 2 orders of magnitude less than this, but, in high-quality systems, even this can be important.

Unfortunately, index tolerances and the like for other (i.e., nonglass) optical materials are not well established. There is a tendency to regard the optical characteristics of crystals and other materials as if they were exact constants, despite the fact that significant variations often show up in measurements made on different material samples. This is an area which could profit from further investigation.

23.2 Additive Tolerances

In analyzing an optical system to determine the size of the tolerances to be applied to specific dimensions, one can readily calculate the partials of the system characteristics with respect to the dimensions under consideration. Thus one obtains the value of the partial derivative of the focal length (for example) with respect to each thickness, spacing, curvature, and index; likewise for the other characteristics, which may include back focus, magnification, and field coverage, as well as the aberrations or wavefront deformations. Then each dimensional tolerance, multiplied by the appropriate derivative, indicates the contribution of that tolerance to the variation of the characteristic. Now if it were necessary to be *absolutely* certain that (for example) the focal length did not vary more than a certain amount, one would be forced to establish the parameter tolerances so that the sum of the absolute values of the derivative-tolerance products did not exceed the allowable variance. Although this worst-case approach is occasionally necessary, one can frequently allow much larger tolerances by taking advantage of the laws of probability and statistical combination.

As a simple example, let us consider a stack of disks, each 0.1 in thick. We will assume that each disk is made to a tolerance of ± 0.005 in and that the probability of the thickness of the disk being any given value between 0.095 and 0.105 in is the same as the probability of its being any other value in this range. This situation is represented by the rectangular frequency distribution curve of Fig. 23.2*a*. Thus, for example, there is a 1 in 10 chance that any given disk will have a thickness between 0.095 and 0.096 in. Now if we stack two disks, we know that it is *possible* for their combined thicknesses to range from 0.190 to 0.210 in. However, the *probability* of the combination having either of these extreme thickness values is quite low. In a frequency distribution curve such as those shown in Fig. 23.2, the area under the curve between two abscissa values represents the (relative) number of pieces which will fall between the two abscissa values. Thus the probability of a characteristic falling between two values is the area under the curve between the two abscissas divided by the total area under the curve. Since the probability of either of the disks having a thickness between 0.095 and 0.096 is 1 in 10, if we randomly select two disks, the probability of *both* falling in this range is 1 in 100. Thus the probability of a pair of disks having a thickness between 0.190 and 0.192 is 1 in 100; similarly for a combined thickness of 0.208 to 0.210 in. The probability of a combined thickness of 0.190 to 0.191 (or 0.209 to 0.210) is much less; 1 in 400.

The frequency distribution curve representing this situation is

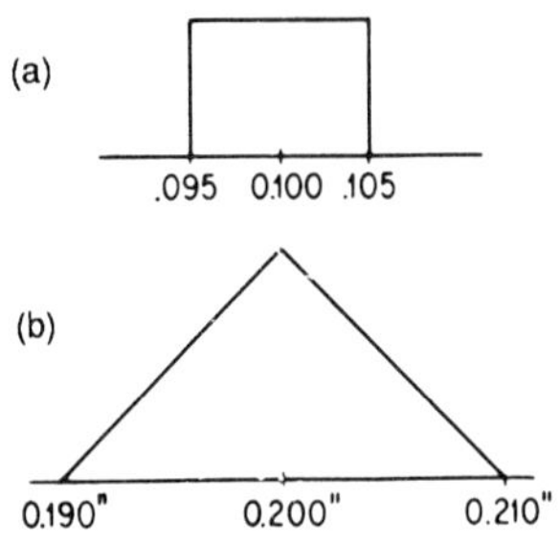

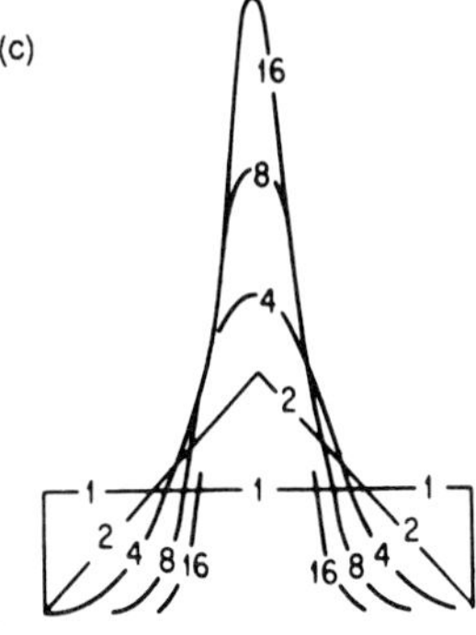

Figure 23.2 How additive tolerances combine in a random assembly. (*a*) Uniform probability distribution in a dimension of a single piece. (*b*) Resulting frequency distribution when two such pieces are combined. (*c*) Normalized curves for assemblies of 1, 2, 4, 8, and 16 pieces.

shown in Fig. 23.2*b* as a triangular distribution. Figure 23.2*c* shows frequency distribution curves for 1-, 2-, 4-, 8-, and 16-element assemblies. These curves have been normalized so that the area under each is the same and the extreme variations have been equalized. The important point here is that the probability of an assembly taking on an extreme value is tremendously reduced when the number of elements making up the assembly is increased. For example, in a stack of 16 disks with a nominal total thickness of 1.6 in and a possible variation in thickness of ±0.080 in, the probability of a random stack having a thickness less than 1.568 in or more than 1.632 in (i.e., ±0.032 in) is less than 1 in 100.

The importance of this in setting tolerances is immediately apparent. In the stacked-disks example, if the range of thicknesses represented by 1.568 to 1.632 in for 16 disks were the greatest variation that could be tolerated, we could be absolutely sure of meeting this requirement *only* by tolerancing each individual disk at ±0.002 in. However, if we were willing to accept a rejection rate of 1 percent in large-scale production, we could set the thickness tolerance at ±0.005 in. If the cost of the pieces made to the tighter tolerance exceeded the cost of the pieces made to the looser tolerance by as little as 1 percent (plus one sixteen-hundredth of the assembly, processing, and final inspection costs), the looser tolerance would result in a less costly product.

The peaking-up characteristic of multiple assemblies can also be represented by the two plots shown in Fig. 23.3. The graph on the left shows the percentage of assemblies which fall within a given central fraction of the total tolerance range as a function of that fraction. The number of elements per assembly is indicated on each curve. These curves were derived from Fig. 23.2*c*. The graph on the right in Fig. 23.3 is simply another way of presenting the same data. If one were interested in an assembly of 10 elements, the intersection of the abscissa corresponding to 10 and the appropriate curve would indicate

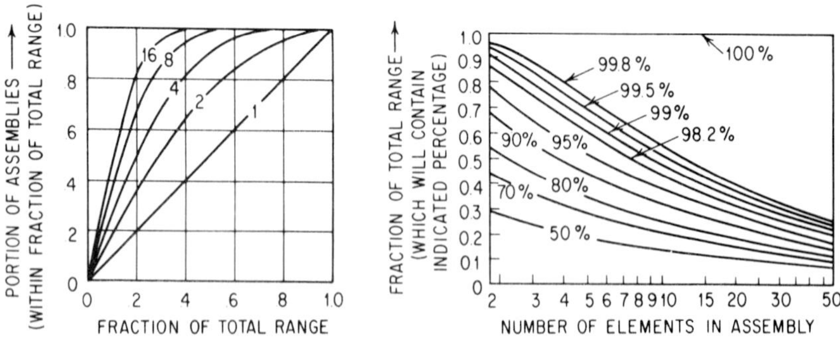

Figure 23.3 Probability distributions of additive tolerances in multiple assemblies, assuming a uniform distribution for a single part.

that all but 0.2 percent (using the 99.8 percent curve) of the assemblies would fall within 0.55 of the total tolerance range represented by the absolute sum of all 10 tolerances, and that over one-half the assemblies (using the 50 percent curve) would fall within 0.15 of the total possible range.

The preceding discussion has been based upon the simplified assumptions that (1) each individual piece had a rectangular frequency distribution, and (2) each tolerance was equal in effect. This is very rarely true in practice. The frequency distribution will, of course, depend on the techniques and controls used in fabricating the part, and the tolerance sizes may represent the partial derivative tolerance products from such diverse sources as tolerances on index, thickness, spacing, and curvature. Note, however, that in Fig. 23.2*c* the progression of curves may be started at any point. If, for example, the production methods produce a triangular distribution (such as that shown for an assembly of two elements), then the curve marked 4 (for *four elements*) will be the frequency distribution for two elements (of triangular distribution) and so on.

Note also that as more and more elements are included in the assembly, the curve becomes a closer and closer approximation to the normal distribution curve which is so useful in statistical analysis (except that the tolerance-type curves do not go to infinity as do normal curves). One useful property of the normal curve for an additive assembly is that its "peakedness" is proportional to the square root of the number of elements in assembly. Thus if 90 percent of the individual pieces are expected to fall within some given range, then for an assembly of 16 elements, 90 percent would be expected to fall within $\sqrt{1/16}$, or one-quarter of the total range. A brief examination will indicate that even the rectangular distribution assumed for Figs. 23.2 and 23.3 tends to follow this rule when there are more than a few elements in the assembly.

A rule frequently used to establish tolerances may be represented as follows:

$$T = \sqrt{\sum_{i=1}^{n} t_i^2} \tag{23.3}$$

This is frequently referred to as the RSS rule, shorthand for the square root of the sum of the squares. What the RSS rule means is this: if some percentage (say 99 percent) of the parts have tolerances which produce effects less than t_i (and which vary according to a normal, or gaussian, distribution), then the same percentage (i.e., 99 percent in our example) of the assemblies will show a total tolerance effect less than T.

While this section may seem to be a far cry from optical design, consider that a simple Cooke triplet has the following dimensions which affect its focal length and aberrations: six curvatures, three thicknesses, two spacings, three indices, and three V values. These total 14 for monochromatic characteristics and 17 for chromatic aberrations. Such a system is eminently qualified for statistical treatment. *Note that the validity of this approach does not depend on a large production quantity; it depends on a random combination of a certain number of tolerance effects.*

There are two obvious features of the RSS rule which are well worth noting. One is the square root effect: If you have n tolerance effects of a size $\pm x$, then the RSS rule says that a random combination will produce an effect equal to $\pm x$ times the square root of n. For example, given 16 tolerance effects of ± 1 mm, we should expect a variation of only ± 4 mm, not ± 16 mm. The other feature is that the larger effects dominate the combination. As an example, consider a case with nine tolerances of ± 1 mm and one tolerance of ± 10 mm. If we use the RSS rule on this, we get an expected variation equal to the square root of 109, or ± 10.44 mm. Compare this with the fact that the single ± 10-mm tolerance has an RSS of ± 10 mm. The addition of the nine ± 1-mm tolerances changes the expected variation by only 4.4 percent.

23.3 Establishing the Tolerance Budget

One possible way to establish a tolerance budget using this principle is as follows:

1. Calculate the partial derivatives of the aberrations (and other characteristics to be controlled) with respect to the fabrication tolerances (radius, asphericity, thickness and spacing, index, homogeneity, surface tilt, etc.). Express the aberrations as OPDs (wavefront deformation).
2. Select a preliminary tolerance budget. Figure 23.1 can be used as a rough guide to appropriate tolerance values.
3. Multiply the individual tolerances by the partial derivatives calculated in step 1.
4. Compute RSS from Eq. 23.3 for all the aberrations for each tolerance individually. This will indicate the relative sensitivity of each tolerance.
5. Compute RSS for all of the effects calculated in step 4.
6. Compare the results of step 5 with the performance required of the system. This can be done by computing RSS for the design OPD (as

indicated by its MTF or whatever measure is convenient) with the tolerance budget OPD and using the material of Chap. 4 to determine the resulting MTF or Strehl ratio.

7. Adjust the tolerance budget so that the result of step 6 is equal to the required performance. Since the larger effects dominate the RSS, if you are tightening the tolerances (as is quite likely on the first go-round), you should tighten the most sensitive ones as determined in step 4 (and possibly loosen the least sensitive). Note that there is no economic gain if you loosen tolerances beyond the level at which costs or prices cease to go down. Conversely, one should be sure that the tolerances are not tightened beyond a level at which fabrication becomes impossible—since cost rises asymptotically toward infinity as this level is approached.
8. After one or two adjustments (steps 2 through 7), the tolerance budget should converge to one which is reasonable economically and which will yield an acceptable product.

If the tolerances necessary to get acceptable performance are too tight to be fabricated economically, there are several ways which are commonly used to ease the situation (as described at greater length in Sec. 2.9).

1. A *test plate fit* is a redesign of the system using the measured values of the radii of existing test plates. This eliminates the radius tolerance (except for the variations due to the test glass fit in the shop and any error in the measurement of the radius).
2. A *melt fit* can effectively eliminate the effects of index and dispersion variation. Again, this is a redesign, using the measured index of the actual piece of glass to be used, instead of the catalog values.
3. A *thickness fit* uses the measured thicknesses of the actual fabricated elements; this amounts to an adjustment of the airspaces during the assembly process.

The redesigns called for in all three fitting operations above, while hardly trivial, are not major undertakings when an automatic lens design program is used.

While the above may tend to induce a desirable relaxation in tolerances, one or two words of caution are in order. As previously mentioned, the index of refraction distribution within a melt or lot of glass may or may not be centered about the nominal value. When it is centered about a nonnominal value, the preceding analysis is valid only with respect to the central value, not the nominal value. Further, in some optical shops, there is a tendency to make lens elements to the high side of the thickness tolerance; this allows scratched surfaces to be reprocessed and will, of course, upset the theoretical probabilities. Another tendency is for polishers to try for a *hollow* test glass fit, i.e., one in which there is a convex air lens between the test plate and the work. This is done because a block of lenses which is polished over is difficult to bring back. Surprisingly, these nonnormal distributions have very little effect on Eq. 23.3 (if there are enough elements in the assembly). Actually, Eq. 23.3 seems to be *quite* conservative in practice.

Thus the situation is seen to be a complex one, but, nonetheless, one in which a little careful thought in relaxing tolerances to the greatest allowable extent can pay handsome dividends. For those who wish to avoid the labor of a detailed analysis, the use of Eq. 23.3, or even the assumption that the tolerance buildup will not exceed one-half or one-third of the possible maximum variation, are fairly safe procedures in assemblies of more than a few elements. Above all, when cost is important, one should try to establish tolerances which are readily held by normal shop practices.

Formulary

This section is a condensed collection of formulas that are regularly used in optical design. It is assumed that the reader is generally familiar with the subject and that no detailed explanations are necessary. Thus this formulary is more in the style of a handbook than that of a textbook, since its purpose is that of a reference listing rather than a tutorial. However, each section is appropriately labeled, defined, and illustrated, so that one generally familiar with the subject can readily apply the formulas. If a more extensive explanation is desired, the reader is referred to W. J. Smith, *Modern Optical Engineering,* 2d ed., McGraw-Hill, New York, 1990.

F.1 Sign Conventions, Symbols, and Definitions

A primed symbol refers to a quantity after refraction (or reflection) by a surface or by a lens, or to a quantity associated with the image. The subscript k is used as a generalization for the last surface or lens of the system. Lowercase symbols are used for paraxial quantities; uppercase, for trigonometrically exact quantities. The data of a chief, or principal, ray is signified by the subscript p, e.g., u_p, y_p (in some works a bar over the symbol is used). Letter or numerical subscripts are used to identify surfaces or lenses in a sequence.

1. *Heights.* A height above the axis or above a reference point is positive, below is negative.

h, H — Object or image height; the intersection of a ray with the object or image surface

y, Y — Ray height; the height at which a ray intersects a surface or element

2. *Distances.* A distance to the right of a reference point is positive, to the left is negative. The reference point may be at a surface, a lens,

a principal point or a focal point. In ray tracing, the reference point moves sequentially from surface to surface (or from lens to lens).

s	Axial distance from a principal point
l	Axial distance from a surface
x	Axial distance from a focal point
d	Axial spacing between surfaces, elements, or principal planes

3. *Angles.* An angle or ray slope is positive if the ray is rotated clockwise to reach the axis or surface normal.

u, U	Ray slope angle
i, I	Angle of incidence or refraction

4. *Radius of curvature.* A radius or curvature is positive if the center of curvature is to the right of the surface.

r	Radius of curvature
c	Curvature, reciprocal of the radius, equal to $1/r$

5. *Focal length and power.* The focal length and power are positive if the lens or surface bends the ray toward the axis, i.e., if the lens converges light.

f	Effective focal length; the distance from the second principal point to the second focal point
ϕ	Power; for a lens, the reciprocal of the focal length; for a surface, $\phi = (n' - n)/r$

6. *Index of refraction.* The index is positive if light rays travel from left to right, negative if they travel right to left (as after a reflection).

n	Index of refraction, equal to the velocity of light in a vacuum (in practice, the velocity in air is usually used) divided by the velocity in the medium.

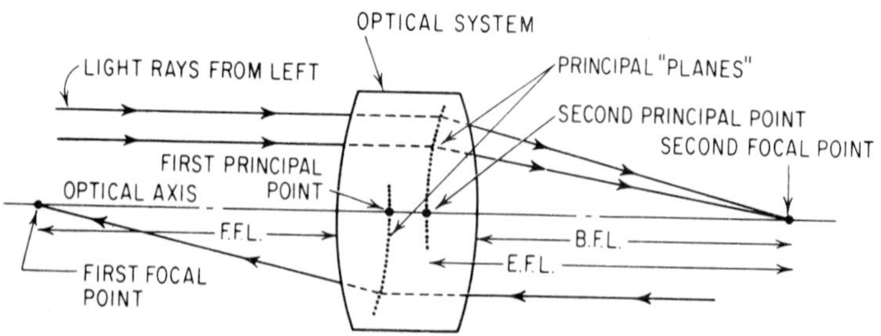

Figure F.1 The locations of the focal points and principal points of an optical system.

F.2 The Cardinal Points (Fig. F.1)

First/second focal point	The focus of rays emanating from an axial object point an infinite distance to the right/left
First/second principal point	The axial point corresponding to the locus at which the first/second focal-point-defining ray appears to be bent
First/second nodal points	A pair of axial points such that an oblique ray directed toward the first nodal point appears to emerge from the second, making the same angle to the optical axis. For a system in air, they coincide with the principal points.
efl = f	Effective focal length: the distance from the second principal point to the second focal point
bfl	Back focal length: the distance from the vertex of the last surface to the second focal point
ffl	Front focal length: the distance from the vertex of the first surface to the first focal point

To determine efl and bfl, trace a paraxial ray with $u_1 = 0$; then efl $= f = -y_1/u_k'$ and bfl $= -y_k/u_k'$.

F.3 Image Equations (Fig. F.2)

$$x' = \frac{-f^2}{x} \tag{F.3.1}$$

$$\frac{1}{s'} = \frac{1}{s} + \frac{1}{f} \tag{F.3.2}$$

$$s' = \frac{sf}{(s + f)} \tag{F.3.3}$$

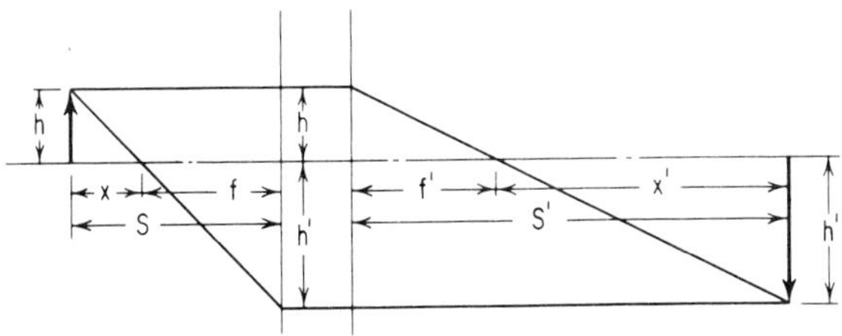

Figure F.2 The dimensions used in the imaging equations.

$$f = \frac{ss'}{(s - s')} \tag{F.3.4}$$

$$m = \frac{h'}{h} = \frac{s'}{s} = \frac{-x'}{f} = \frac{f}{x} \tag{F.3.5}$$

$$\overline{m} = \frac{s'_2 - s'_1}{s_2 - s_1} = \frac{s'_1/s_1}{s'_2/s_2} \tag{F.3.6}$$

$$= m_1 \times m_2 \qquad \text{(approaches } m^2 \text{ as } s_2 \text{ approaches } s_1\text{)}$$

F.4 Paraxial Ray Tracing (Surface by Surface) (Figs. F.3 and F.4)

$$u = \frac{-y}{l} \qquad l' = \frac{-y}{u'} \qquad y = -lu = -l'u' \tag{F.4.1}$$

$$n'u' = nu - \frac{y(n' - n)}{r} = nu - y(n' - n)c \tag{F.4.2}$$

$$y_{j+1} = y_j + du'_j \tag{F.4.3}$$

$$\frac{n'}{l'} = \frac{n}{l} + \frac{n' - n}{r} \tag{F.4.4}$$

$$i = cy + u \tag{F.4.5}$$

$$i' = \frac{ni}{n'} \tag{F.4.6}$$

$$u' = u - i + i' \tag{F.4.7}$$

$$m = \frac{h'}{h} = \frac{nu}{n'u'} = \frac{\text{object } nu}{\text{image } nu} \tag{F.4.8}$$

To get the data of a new (third) ray from the data of two previously traced rays:

$$y_{\text{new}} = Ay + By_p \tag{F.4.9}$$

$$u_{\text{new}} = Au + Bu_p \tag{F.4.10}$$

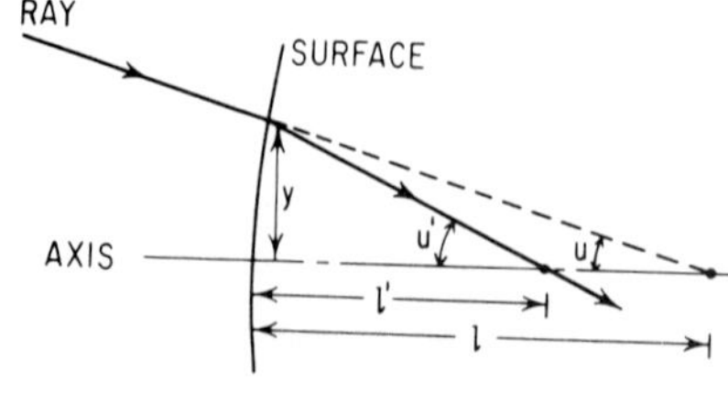

Figure F.3 The relationship $y = -lu = -l'u'$ for paraxial rays.

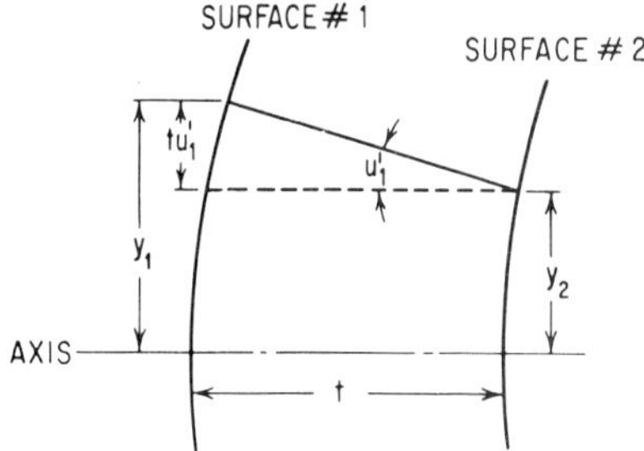

Figure F.4 The transfer of a paraxial ray from surface to surface by $y_{j+1} = y_j + tu_j'$.

F.5 Invariants

$$\text{Inv} = y_p nu - ynu_p = y_p n'u' - yn'u'_p \tag{F.5.1}$$

$$\text{Inv} = hnu = h'n'u' \tag{F.5.2}$$

$$Q = \frac{y_{p1} - y_{p2}}{y} \tag{F.5.3}$$

F.6 Paraxial Ray Tracing (Component by Component) (Fig. F.5)

$$u = \frac{-y}{s} \qquad u' = \frac{-y}{s'} \tag{F.6.1}$$

$$u' = u - y\phi \tag{F.6.2}$$

$$y_{j+1} \doteq y_j + du'_j \tag{F.6.3}$$

$$m = \frac{h'}{h} = \frac{u_1}{u'_k} \tag{F.6.4}$$

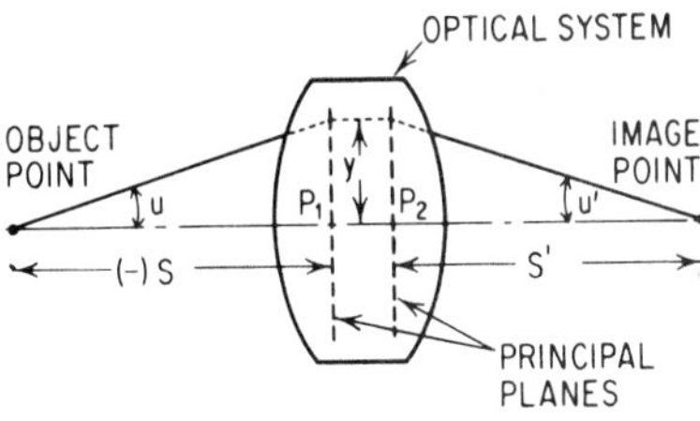

Figure F.5 The principal planes are *planes of unit magnification*, so that a ray appears to leave the second principal plane at the same height (y) that it appears to strike the first principal plane. Note that $y = -su = -s'u'$.

F.7 Two-Component Relationships

- Given: f_a, f_b, and d (or ϕ_a and ϕ_b) (Fig. F.6)

$$\phi_{ab} = \phi_a + \phi_b - d\phi_a \phi_b \tag{F.7.1}$$

$$f_{ab} = \frac{f_a f_b}{f_a + f_b - d} \tag{F.7.2}$$

$$B = \frac{f_{ab}(f_a - d)}{f_a} \tag{F.7.3}$$

- Given: f_{ab}, d, and B (Fig. F.6)

$$f_a = \frac{df_{ab}}{f_{ab} - B} \tag{F.7.4}$$

$$f_b = \frac{-dB}{f_{ab} - B - d} \tag{F.7.5}$$

- Given: s, s', d, and $m = h'/h = u/u'$ (Fig. F.7)

$$\phi_a = \frac{ms - md - s'}{msd} \tag{F.7.6}$$

$$\phi_b = \frac{d - ms + s'}{ds'} \tag{F.7.7}$$

- Given ϕ_a, ϕ_b, T, and $m = h'/h = u/u'$ (Fig. F.7)

Solve this quadratic for d:

$$0 = d^2 - dT + T(f_a + f_b) + \frac{(m-1)^2 f_a f_b}{m} \tag{F.7.8}$$

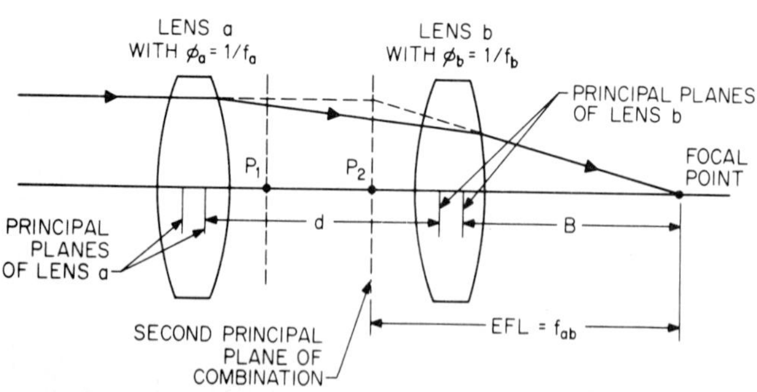

Figure F.6 Two separated components, object at infinity, showing the symbols used in Sec. F.7.

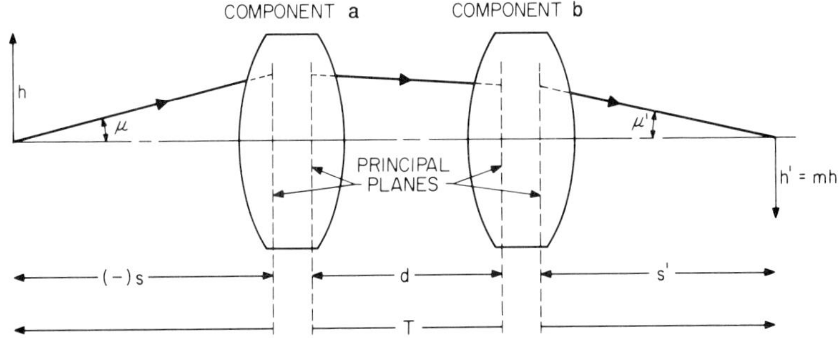

Figure F.7 Two separated components, object at a finite distance, showing the symbols used in Sec. F.7.

$$s = \frac{(m - 1)d + T}{(m - 1) - m\,d\phi_a} \tag{F.7.9}$$

$$s' = T + s - d \tag{F.7.10}$$

F.8 Third-Order Aberrations—Surface Contributions

Given: the paraxial ray data from an axial marginal ray (y, u, i) and a principal ray (y_p, u_p, i_p), calculated for each surface using Eqs. F.4.2, F.4.3, and F.4.5; h is the image height (i.e., y_p at the paraxial focus).

$$B = \frac{n(n' - n)y(u' + i)}{2n'hn_k'u_k'} \tag{F.8.1}$$

$$B_p = \frac{n(n' - n)y_p(u'_p + i_p)}{2n'hn_k'u_k'} \tag{F.8.2}$$

The following are the contributions of the surface to the transverse aberrations. The aberrations of the image are the sum of all the surface contributions:

Spherical:

$$\text{TSC} = Bi^2h \tag{F.8.3}$$

Sagittal coma:

$$\text{CC} = Bii_ph \tag{F.8.4}$$

Astigmatism:

$$\text{TAC} = Bi_p^2h \tag{F.8.5}$$

Petzval:

$$\text{TPC} = \frac{(n' - n)ch^2 n_k' u_k'}{2nn'} \tag{F.8.6}$$

Distortion:

$$\text{DC} = h\left[B_p i i_p + \frac{u'^2_p - u^2_p}{2}\right] \tag{F.8.7}$$

Axial chromatic:

$$\text{TAchC} = \frac{-yi(dn - dn'n/n')}{n_k' u_k'} \tag{F.8.8}$$

Lateral chromatic:

$$\text{TchC} = \frac{-yi_p(dn - dn'n/n')}{n_k' u_k'} \tag{F.8.9}$$

where

$$dn = n_F - n_C = \frac{n - 1}{V}$$

- Conversion from transverse to longitudinal aberration:

$$\text{Longitudinal} = \frac{\text{transverse}}{-u_k'} \tag{F.8.10}$$

$$\text{PC} = \frac{-\text{TPC}}{u_k'} \tag{F.8.11}$$

$$\text{AC} = \frac{-\text{TAC}}{u_k'} \tag{F.8.12}$$

$$x_S = \text{PC} + \text{AC} \tag{F.8.13}$$

$$x_T = \text{PC} + 3\text{AC} \tag{F.8.14}$$

$$\text{Petzval radius } \rho = \frac{h^2}{2\text{PC}} \tag{F.8.15}$$

In Eqs. F.8.13 to F.8.15, PC and AC are the sums of the contributions from all the surfaces.

- Conversion to Seidel coefficients

$$\text{S} = \text{transverse}(-2n_k' u_k') \tag{F.8.16}$$

F.9 Third-Order Aberrations—Thin Lens Contributions

Given: the paraxial ray data (y, u, y_p, u_p) for an axial marginal ray and a principal ray (calculated for each element using Eqs. F.6.2 and F.6.3), the element power ϕ, the element curvature $c = \phi/(n - 1)$, the curvature of the left surface c_1, and $v = u/y$.

$$G_1 = \frac{n^2(n - 1)}{2} \qquad G_2 = \frac{(2n + 1)(n - 1)}{2}$$

$$G_3 = \frac{(3n + 1)(n - 1)}{2} \qquad G_4 = \frac{(n + 2)(n - 1)}{2n} \tag{F.9.1}$$

$$G_5 = \frac{2(n + 1)(n - 1)}{n} \qquad G_6 = \frac{(3n + 2)(n - 1)}{2n}$$

$$G_7 = \frac{(2n + 1)(n - 1)}{2n} \qquad G_8 = \frac{n(n - 1)}{2}$$

With the stop at the element, the thin lens contributions are given by the following equations. Use the stop shift equations from Sec. F.10 with $Q = y_p/y$ to determine the contributions with the stop displaced from the lens.

Spherical:

$$\text{TSC} = \left(\frac{y^4}{u_k'}\right)(G_1c^3 - G_2c^2c_1 - G_3c^2v + G_4cc_1^2 + G_5cc_1v + G_6cv^2) \tag{F.9.2}$$

Sagittal coma:

$$\text{CC} = -hy^2(0.25G_5cc_1 + G_7cv - G_8c^2) \tag{F.9.3}$$

Astigmatism:

$$\text{TAC} = \frac{h^2\phi u_k'}{2} \tag{F.9.4}$$

Petzval:

$$\text{TPC} = \frac{h^2\phi u_k'}{2n} \tag{F.9.5}$$

Distortion:

$$\text{DC} = 0 \tag{F.9.6}$$

Axial chromatic:

$$\text{TAchC} = \frac{y^2\phi}{Vu_k'} \tag{F.9.7}$$

where

$$V = \frac{n_d - 1}{n_F - n_C}$$

Lateral chromatic:

$$\text{TchC} = 0 \tag{F.9.8}$$

Secondary chromatic:

$$\text{TSchC} = \frac{y^2\phi P}{Vu_k'} \tag{F.9.9}$$

where

$$P = \frac{n_d - n_C}{n_F - n_C}$$

F.10 Stop Shift Equations

The stop shift coefficient Q is defined by

$$Q = \frac{(y^*_p - y_p)}{y} \tag{F.10.1}$$

(The asterisk (*) indicates data with the stop in a new, shifted position.) Since Q is an invariant, it may be determined at any surface or element of a system, because it is everywhere the same. For this reason the following expressions may be applied to the third-order contributions of a single surface, or a group of surfaces, or a complete lens. Note that because the thin lens equations of Sec. F.9 apply with the stop in contact, they assume y_p is zero. Thus when the stop shift equations are used to determine the thin lens aberrations, Q reduces to $Q = (y^*_p - 0)/y$, or, when one has traced the principal ray, to

$$Q = \frac{y_p}{y} \tag{F.10.2}$$

$$\text{TSC}^* = \text{TSC} \tag{F.10.3}$$

$$\text{CC}^* = \text{CC} + Q \cdot \text{TSC} \tag{F.10.4}$$

$$\text{TAC}^* = \text{TAC} + 2Q \cdot \text{CC} + Q^2\text{TSC} \tag{F.10.5}$$

$$\text{TPC}^* = \text{TPC} \tag{F.10.6}$$

$$\text{DC}^* = \text{DC} + Q(\text{TPC} + 3\text{TAC}) + 3Q^2\text{CC} + Q^3\text{TSC} \tag{F.10.7}$$

$$\text{TAchC}^* = \text{TAchC} \tag{F.10.8}$$

$$\text{TchC}^* = \text{TchC} + Q \cdot \text{TAchC} \tag{F.10.9}$$

F.11 Third-Order Aberrations—Contributions from Aspheric Surfaces

Determine the contributions from the spherical surface or element, then add the following to determine the total contribution from the aspheric surface or element.

K = equivalent fourth-order deformation coefficient

$= \dfrac{\text{conic constant } \kappa}{8r^3}$ for pure conic sections

$= \dfrac{\kappa}{8r^3}$ plus the fourth-order deformation coefficient for conics with aspheric deformations

$W = \dfrac{4K(n' - n)}{hn_k'u_k'}$ ($hn_k'u_k'$ = the invariant)

$$\text{TSC}_a = Wy^4h \tag{F.11.1}$$

$$\text{CC}_a = Wy^3y_ph \tag{F.11.2}$$

$$\text{TAC}_a = Wy^2y_p{}^2h \tag{F.11.3}$$

$$\text{TPC}_a = 0 \tag{F.11.4}$$

$$\text{DC}_a = Wyy_p{}^3h \tag{F.11.5}$$

$$\text{TAchC}_a = 0 \tag{F.11.6}$$

$$\text{TchC}_a = 0 \tag{F.11.7}$$

F.12 Conversion of Aberrations to Wavefront Deformation (OPD, Optical Path Difference)

The following expressions convert aberrations to peak-to-peak or peak-to-valley wavefront deformations, when the reference point is chosen to minimize the OPD. Note that for low-order aberrations, rms OPD and peak-to-valley (P-V) OPD are related by

$$\text{rms OPD} = \frac{\text{P-V OPD}}{3.5} \qquad \text{(F.12.1)}$$

and the Strehl ratio is approximated by

$$\text{Strehl ratio} = (1 - 2\pi^2 w^2)^2 \qquad \text{(F.12.2)}$$

where w = rms OPD. In the following, NA = numerical aperture = $n_k' \sin u_k'$ and λ = wavelength.

Longitudinal defocus, field curvature:

$$\text{OPD} = \frac{(\text{defocus}) \cdot \text{NA}^2}{2\lambda n_k'} \text{ waves} \qquad \text{(F.12.3)}$$

Transverse third-order spherical (at best focus):

$$\text{OPD} = \frac{\text{TA}_m \cdot \text{NA}}{16\,\lambda} \text{ waves} \qquad \text{(F.12.4)}$$

Transverse zonal spherical ($\text{TA}_m = 0$) (at best focus):

$$\text{OPD} = \frac{\text{TA}_z \cdot \text{NA}}{16.8\lambda} \text{ waves} \qquad \text{(F.12.5)}$$

Tangential coma:

$$\text{OPD} = \frac{\text{coma}_t \cdot \text{NA}}{6\lambda} \text{ waves} \qquad \text{(F.12.6)}$$

Transverse axial chromatic:†

$$\text{OPD} = \frac{\text{TAch} \cdot \text{NA}}{4\lambda} \text{ waves} \qquad \text{(F.12.7)}$$

Lateral chromatic:†

$$\text{OPD} = \frac{\text{TchA} \cdot \text{NA}}{2\lambda} \text{ waves} \qquad \text{(F.12.8)}$$

†These chromatic expressions yield the OPD for the extreme wavelengths of the spectral band. For values correlating well with the other relationships given here as to the effect on image quality (e.g., MTF), divide the chromatic OPD from these expressions by 2.5 for ordinary chromatic, and by 4.5 for secondary spectrum.

Appendix

Lens Listings

TABLE A.1 Lens Listing in Order of Increasing Numerical Aperture

[Column A is the figure number, B the relative aperture (*F*-number), C the numerical aperture (NA), D the half field of view in degrees (HFOV), E the number of elements, and F the type.]

A Fig.	B F/#	C NA	D HFOV	E Elem	F Type
20.06	133.9	.0037	13.67	5	Zoom Scanner
20.06	94.6	.0053	13.67	5	Zoom Scanner
20.06	66.8	.0075	13.67	5	Zoom Scanner
22.02	50.0	.0100	25.50	3	Scanner Obj.
22.05	23.8	.0210	30.50	5	Scanner Obj.
22.03	19.7	.0254	30.10	4	Scanner Obj.
11.01	20.6	.0267	67.04	2	Double Meniscus
22.04	18.5	.0271	28.29	4	Scanner Obj.
7.12	14.7	.0340	17.22	4	Symmetrical EP
11.04	12.2	.0410	30.96	4	Double Meniscus
11.07	11.1	.0450	23.94	6	Double Meniscus (m=.999)
16.22	10.0	.0496	1.49	3	Catadioptric
16.13	8.0	.0594	1.75	4	Catadioptric
18.08	8.3	.0618	56.31	6	WA
11.05	8.0	.0622	26.57	6	Double Meniscus
8.18	8.0	.0624	14.00	3	Cooke Triplet
17.14	8.0	.0625	26.60	6	D-Gauss
9.17	8.0	.0626	85.40	9	Retrofocus
16.17	8.0	.0627	2.00	6	Catadioptric
9.18	7.9	.0632	80.00	4	Retrofocus
6.01	7.0	.0713	1.00	2	Tel. Obj. Gauss
6.08	7.0	.0714	1.00	2	Tel. Obj. Apochromat
6.02	7.0	.0714	1.00	2	Tel. Obj. Fraunhofer
6.03	7.0	.0714	1.00	2	Tel. Obj. Steinheil
6.12	7.0	.0714	1.00	3	Tel. Obj. Apochromat
16.14	7.0	.0714	2.50	5	Catadioptric
17.16	7.0	.0714	8.94	6	D-Gauss (m=.249)
6.10	7.0	.0723	1.00	3	Tel. Obj. Apochromat
6.11	7.0	.0723	1.00	3	Tel. Obj. Apochromat
11.02	6.3	.0787	50.00	4	Double Meniscus
12.01	6.3	.0793	30.11	4	Tessar
11.03	6.3	.0793	49.96	5	Double Meniscus
8.02	6.3	.0795	27.02	3	Cooke Triplet
8.16	6.3	.0797	30.11	3	Cooke Triplet
8.15	6.3	.0798	34.00	3	Cooke Triplet
8.17	6.3	.0800	27.02	3	Cooke Triplet
22.06	6.1	.0818	22.73	4	Telecentric Scan. Obj.
10.03	5.6	.0888	15.11	4	Telephoto
10.09	5.6	.0892	3.00	7	Telephoto
10.02	5.6	.0893	4.19	4	Telephoto
7.11	5.6	.0895	25.17	4	Symmetrical EP

TABLE A.1 Lens Listing in Order of Increasing Numerical Aperture (*Continued*)

A Fig.	B F/#	C NA	D HFOV	E Elem	F Type
10.05	5.6	.0898	5.00	4	Telephoto
18.06	5.6	.0901	37.55	6	WA
11.10	5.5	.0906	26.12	4	Celor (m=.251)
20.04	5.5	.0914	14.20	15	Zoom
11.09	5.4	.0925	27.66	4	Celor
12.04	5.4	.0931	22.04	4	Tessar
20.05	5.3	.0942	12.41	16	Zoom
7.03	5.1	.0968	15.00	2	Heugens EP
7.28	5.1	.0978	45.00	7	Eyepiece
7.05	5.1	.0982	18.00	3	Kellner EP
7.04	5.1	.0984	15.00	2	Ramsden EP
7.10	5.1	.0985	20.00	4	Symmetrical EP
7.27	5.1	.0985	29.84	6	Eyepiece
7.18	5.1	.0987	30.00	5	Erfle EP
7.09	5.1	.0988	30.00	4	Eyepiece
7.22	5.1	.0990	25.00	6	Eyepiece
7.25	5.1	.0990	30.11	6	Eyepiece
18.01	5.0	.0996	50.00	10	WA
16.06	5.0	.0998	.14	2	Cassegrain
7.19	5.0	.0999	34.99	5	Erfle EP
16.07	5.0	.1000	.29	2	Cassegrain
10.06	5.0	.1000	5.99	5	Telephoto
7.13	5.0	.1000	25.17	4	Symmetrical EP
7.08	5.0	.1001	20.00	4	Orthoscopic EP
7.14	5.0	.1001	24.99	4	Symmetrical EP
22.01	5.0	.1002	15.00	3	Scanner Obj.
7.20	5.0	.1002	32.50	5	Erfle EP
7.17	5.0	.1002	37.60	5	Eyepiece
7.21	5.0	.1003	34.99	6	Int. Foc. EP
18.04	5.0	.1004	40.03	8	WA
7.24	5.0	.1006	31.80	6	Eyepiece
7.16	5.0	.1007	34.99	5	Eyepiece
18.02	4.7	.1052	45.00	8	WA
8.14	4.7	.1054	13.60	3	Cooke Triplet (m=.052)
20.04	4.7	.1079	22.78	15	Zoom
10.11	4.5	.1099	3.09	5	Telephoto
12.03	4.5	.1103	27.92	4	Tessar
11.06	4.5	.1103	33.82	6	Double Meniscus
20.02	4.5	.1110	17.40	9	Zoom
10.04	4.5	.1111	15.11	4	Telephoto
12.02	4.5	.1111	27.92	4	Tessar
12.14	4.5	.1112	18.26	5	Heliar
11.08	4.5	.1113	25.17	4	Celor
17.19	4.5	.1114	15.50	6	D-Gauss
20.05	4.5	.1114	21.80	16	Zoom
12.13	4.5	.1114	30.11	5	Heliar
8.13	4.4	.1127	25.17	3	Cooke Triplet
10.12	4.4	.1139	3.39	5	Telephoto
12.19	4.4	.1144	27.92	5	Tessar
7.06	4.3	.1186	18.00	3	Eyepiece
20.05	4.2	.1186	28.12	16	Zoom (m=.181)
7.26	4.0	.1243	24.00	6	Eyepiece

TABLE A.1 Lens Listing in Order of Increasing Numerical Aperture (*Continued*)

A Fig.	B F/#	C NA	D HFOV	E Elem	F Type
8.03	4.0	.1247	22.98	3	Cooke Triplet
9.15	4.0	.1247	72.00	6	Retrofocus
17.32	4.0	.1250	23.00	7	D-Gauss
20.04	4.0	.1251	36.19	15	Zoom
7.29	4.0	.1252	45.00	7	Eyepiece
20.05	4.0	.1253	36.38	16	Zoom
9.09	4.0	.1255	40.00	8	Retrofocus
10.07	4.0	.1256	7.50	5	Telephoto
7.23	4.0	.1256	40.00	6	Eyepiece
18.07	4.0	.1259	31.50	4	WA
17.13	3.6	.1386	27.02	6	D-Gauss
12.06	3.6	.1391	27.92	4	Tessar
17.18	3.5	.1426	20.81	6	D-Gauss
17.20	3.5	.1427	6.78	6	D-Gauss
20.01	3.5	.1427	36.60	9	Zoom
12.15	3.5	.1428	25.17	5	Heliar
16.20	3.5	.1429	.50	3	Mirror
20.02	3.5	.1429	32.30	9	Zoom
12.05	3.5	.1430	21.80	4	Tessar
20.01	3.5	.1430	42.00	9	Zoom
20.01	3.5	.1433	31.30	9	Zoom
8.12	3.5	.1445	11.86	3	Cooke Triplet
18.03	3.4	.1466	40.03	8	WA
22.16	3.3	.1500	.90	5	Laser Zoom Colim.
7.07	3.3	.1508	25.17	4	Eyepiece
13.01	3.3	.1517	17.22	4	Petzval
7.15	3.3	.1529	25.17	4	Eyepiece
9.02	3.0	.1621	18.26	6	Retrofocus
12.18	3.0	.1659	23.27	5	Heliar
6.04	3.0	.1662	1.00	2	Tel. Obj. Cemented
12.16	3.0	.1663	23.27	5	Heliar
6.05	3.0	.1664	1.00	2	Tel. Obj.
22.07	3.0	.1667	9.55	6	Telecentric Scan. Obj.
8.04	3.0	.1667	18.98	3	Cooke Triplet
8.10	3.0	.1669	23.75	3	Cooke Triplet
12.07	3.0	.1682	28.00	4	Tessar
14.10	2.9	.1718	23.75	5	Sonnar
8.11	2.9	.1725	11.70	3	Cooke Triplet
12.09	2.8	.1746	21.00	4	Tessar
12.10	2.8	.1755	26.57	4	Tessar
8.09	2.8	.1771	23.50	3	Cooke Triplet
8.08	2.8	.1772	19.80	3	Cooke Triplet
12.12	2.8	.1774	25.17	4	Tessar
9.16	2.8	.1778	87.50	8	Retrofocus
12.08	2.8	.1782	20.30	4	Tessar
6.16	2.8	.1784	1.00	3	Tel. Obj.
6.18	2.8	.1784	1.00	3	Tel. Obj.
17.33	2.8	.1785	3.79	7	D-Gauss (m=.333)
9.08	2.8	.1785	37.00	5	Retrofocus
6.15	2.8	.1786	1.00	3	Tel. Obj.
6.17	2.8	.1786	1.00	3	Tel. Obj.
9.05	2.8	.1786	21.50	7	Retrofocus

TABLE A.1 Lens Listing in Order of Increasing Numerical Aperture (*Continued*)

A Fig.	B F/#	C NA	D HFOV	E Elem	F Type
6.06	2.8	.1787	1.00	2	Tel. Obj. High Index
20.03	2.8	.1787	6.17	12	Zoom
20.03	2.8	.1787	8.75	12	Zoom
20.03	2.8	.1787	15.11	12	Zoom
14.14	2.8	.1787	20.81	6	Ernostar
9.07	2.8	.1787	32.01	6	Retrofocus
6.14	2.8	.1788	1.00	3	Tel. Obj.
6.13	2.8	.1789	1.00	2	Tel. Obj.
6.07	2.8	.1789	1.00	2	Tel. Obj. Low Index
14.04	2.8	.1789	5.01	4	Split Triplet
10.08	2.8	.1790	7.00	5	Telephoto
12.11	2.8	.1790	23.27	4	Tessar
12.17	2.8	.1792	25.17	5	Heliar
18.05	2.8	.1795	38.00	7	WA
14.03	2.8	.1811	11.31	4	Split Triplet
14.11	2.7	.1830	19.80	4	Sonnar
12.20	2.7	.1847	16.17	5	Tessar
8.07	2.7	.1857	20.81	3	Cooke Triplet
17.17	2.6	.1913	11.68	6	D-Gauss (m=.114)
19.06	2.6	.1913	11.68	6	D-Gauss (m=.114)
8.05	2.5	.1979	16.01	3	Cooke Triplet
7.02	2.5	.1979	21.80	6	Magnifier (Brueke)
14.01	2.5	.1987	22.72	4	Split Triplet
14.05	2.5	.1995	17.00	4	Split Triplet
22.09	2.5	.1997	1.01	2	Laser Collim.
19.04	2.5	.1997	14.94	6	D-Gauss (m=.025)
17.40	2.5	.2000	25.17	7	D-Gauss
8.06	2.5	.2001	16.17	3	Cooke Triplet
19.05	2.5	.2002	14.75	6	D-Gauss
19.03	2.5	.2003	15.00	6	D-Gauss
17.39	2.5	.2005	27.92	7	D-Gaiss
9.11	2.5	.2027	45.00	8	Retrofocus
16.05	2.0	.2284	.50	2	Cassegrain
17.12	2.0	.2438	23.00	6	D-Gauss
16.03	2.0	.2462	.50	2	Cassegrain
16.08	2.0	.2462	.50	2	Gregorian
14.12	2.0	.2475	19.80	6	Sonnar
17.09	2.0	.2490	21.80	6	D-Gauss
9.10	2.0	.2494	42.00	11	Retrofocus
17.34	2.0	.2496	12.00	7	D-Gauss
17.38	2.0	.2496	12.41	7	D-Gauss
22.17	2.0	.2498	2.86	10	UV Telecentric (m=.20)
17.37	2.0	.2499	20.81	7	D-Gauss
16.09	2.0	.2500	.50	2	Gregorian
16.10	2.0	.2500	.50	2	Schwarzschild
17.11	2.0	.2500	27.92	6	D-Gauss
16.04	2.0	.2503	.50	2	Cassegrain
14.06	2.0	.2505	15.11	4	Split Triplet
17.08	2.0	.2505	21.80	6	D-Gauss
17.15	2.0	.2505	28.00	6	D-Gauss
17.07	2.0	.2506	15.02	6	D-Gauss
17.10	2.0	.2507	22.29	6	D-Gauss

TABLE A.1 Lens Listing in Order of Increasing Numerical Aperture (*Continued*)

A Fig.	B F/#	C NA	D HFOV	E Elem	F Type
14.07	2.0	.2507	22.50	4	Split Triplet
9.06	2.0	.2510	32.04	7	Retrofocus
21.06	2.0	.2512	7.00	4	IR Quad.
9.12	2.0	.2515	47.00	13	Retrofocus
7.01	2.0	.2549	17.43	2	Magnifier (E.P.)
16.18	1.90	.2600	2.00	7	Catadioptric
14.02	1.90	.2621	14.04	4	Split Triplet
14.08	1.90	.2632	17.22	4	Split Triplet
17.06	1.80	.2763	23.00	6	D-Gauss
10.10	1.80	.2779	9.20	7	Telephoto
12.21	1.80	.2804	18.26	6	Hektor
17.26	1.75	.2875	24.70	7	D-Gauss
22.15	1.70	.2912	.91	5	Laser Zoom Colim.
17.05	1.70	.2931	18.78	6	D-Gauss
17.25	1.70	.2942	15.11	7	D-Gauss
22.13	1.63	.3064	.29	2	Negative Doublet
13.03	1.60	.3095	9.09	4	Petzval
13.05	1.60	.3124	6.79	5	Petzval
13.02	1.60	.3128	6.79	4	Petzval
17.04	1.60	.3129	5.26	6	D-Gauss
16.23	1.60	.3130	1.00	3	Catadioptric
13.06	1.60	.3139	9.09	4	Petzval
22.14	1.55	.3183	2.86	5	Laser Collim.
21.07	1.55	.3255	8.79	4	IR Quad.
16.11	1.50	.3328	1.25	5	Catadioptric
21.08	1.50	.3330	2.55	4	IR Petzval
21.02	1.50	.3331	2.29	2	IR Tel. Obj.
21.01	1.50	.3331	2.29	2	IR Tel. Obj>
14.09	1.50	.3331	15.11	4	Split Triplet
21.03	1.50	.3334	2.29	2	IR Tel. Obj.
21.09	1.50	.3334	3.00	4	IR Petzval
13.04	1.50	.3335	10.20	4	Petzval
17.24	1.50	.3339	21.80	7	D-Gauss
17.47	1.50	.3342	21.80	7	D-Gauss
17.23	1.50	.3342	22.78	7	D-Gauss
17.28	1.50	.3347	11.86	7	D-Gauss
17.22	1.50	.3360	22.78	7	D-Gauss
16.16	1.48	.3374	3.20	5	Catadioptric
17.45	1.40	.3467	22.50	7	D-Gauss
17.46	1.40	.3476	32.00	7	D-Gauss
14.15	1.40	.3502	15.11	5	Ernostar
17.30	1.40	.3531	23.75	7	D-Gauss
17.35	1.40	.3559	15.11	7	D-Gauss
17.21	1.40	.3563	23.00	7	D-Gauss
13.09	1.40	.3571	12.41	6	Petzval
17.29	1.40	.3571	23.00	7	D-Gauss
17.31	1.40	.3573	23.00	7	D-Gauss
13.08	1.40	.3574	6.79	6	Petzval
13.07	1.40	.3576	6.79	6	Petzval
17.03	1.40	.3577	17.22	6	D-Gauss
17.36	1.40	.3579	12.41	7	D-Gauss
13.10	1.39	.3613	6.79	6	Petzval

TABLE A.1 Lens Listing in Order of Increasing Numerical Aperture (*Continued*)

A Fig.	B F/#	C NA	D HFOV	E Elem	F Type
17.02	1.35	.3676	12.95	6	D-Gauss
13.12	1.32	.3788	15.11	6	Petzval
17.01	1.25	.3992	12.41	6	D-Gauss
13.11	1.25	.4008	12.41	6	Petzval
17.44	1.20	.4151	2.29	8	D-Gauss
9.03	1.20	.4151	12.86	7	Retrofocus (m=.005)
17.27	1.20	.4163	15.64	7	D-Gauss
9.04	1.20	.4164	12.92	9	Retrofocus (m=.006)
16.15	1.20	.4166	5.14	5	Catadioptric
17.43	1.20	.4170	18.78	8	D-Gauss
14.13	1.19	.4185	18.00	5	Sonnar
22.11	1.18	.4216	1.44	4	Laser Disk Obj.
17.41	1.10	.4521	15.11	8	D-Gauss
19.01	1.09	.4614	20.69	3	TV Projection (m=.047)
13.14	1.01	.4929	15.00	6	Petzval
22.12	1.00	.4967	.50	4	Laser Disk Obj.
9.14	1.00	.5004	72.00	13	Retrofocus
22.10	0.99	.5028	.87	3	Laser Disk Obj.
17.42	0.99	.5042	21.80	8	D-Gauss
16.12	0.99	.5055	1.25	6	Catadioptric
13.13	0.91	.5493	6.84	5	Petzval
22.08	0.91	.5500	.57	1	Laser Disk Obj.
19.02	0.89	.5620	25.49	4	TV Projection (m=.024)
15.05	0.88	.5658	2.14	9	Microscope
15.06	0.77	.6521	4.87	4	Microscope (m=.025)
21.05	0.75	.6640	6.00	3	IR Triplet
21.04	0.75	.6647	6.00	3	IR Triplet
21.10	0.64	.7795	10.00	4	IR Quad.
15.07	0.62	.8021	1.39	12	Microscope (m=.025)
21.11	0.55	.9055	10.00	4	IR Quad.
15.09	0.53	.9396	11.22	11	Microscope (m=.009)
15.08	0.53	.9472	3.13	7	Microscope (m=.016)
15.10	0.61	1.2505	.05	10	Microscope (m=.015)
15.02	0.60	1.2886	5.52	6	Microscope (m=.016)
21.14	Afocal	16x		4	IR
21.12	Afocal	12x		5	IR Catadioptric
21.15	Afocal	7x		6	IR
16.21	Afocal	5x		3	Mirror
21.13	Afocal	5x		6	IR

TABLE A.2 Lens Listing in Order of Increasing Field of View

[Column A is the figure number, B the relative aperture (F-number), C the numerical aperture (NA), D the half field of view in degrees (HFOV), E the number of elements, and F the type.]

A Fig.	B F/#	C NA	D HFOV	E Elem	F Type
15.10	0.61	1.2505	.05	10	Microscope (m=.015)
16.06	5.0	.0998	.14	2	Cassegrain
16.07	5.0	.1000	.29	2	Cassegrain
22.13	1.63	.3064	.29	2	Negative Doublet
16.05	2.0	.2284	.50	2	Cassegrain
16.03	2.0	.2462	.50	2	Cassegrain
16.04	2.0	.2503	.50	2	Cassegrain
16.08	2.0	.2462	.50	2	Gregorian
16.09	2.0	.2500	.50	2	Gregorian
16.10	2.0	.2500	.50	2	Schwarzschild
16.20	3.5	.1429	.50	3	Mirror
22.12	1.00	.4967	.50	4	Laser Disk Obj.
22.08	0.91	.5500	.57	1	Laser Disk Obj.
22.10	0.99	.5028	.87	3	Laser Disk Obj.
22.16	3.3	.1500	.90	5	Laser Zoom Colim.
22.15	1.70	.2912	.91	5	Laser Zoom Colim.
6.05	3.0	.1664	1.00	2	Tel. Obj.
6.13	2.8	.1789	1.00	2	Tel. Obj.
6.08	7.0	.0714	1.00	2	Tel. Obj. Apochromat
6.04	3.0	.1662	1.00	2	Tel. Obj. Cemented
6.02	7.0	.0714	1.00	2	Tel. Obj. Fraunhofer
6.01	7.0	.0713	1.00	2	Tel. Obj. Gauss
6.06	2.8	.1787	1.00	2	Tel. Obj. High Index
6.07	2.8	.1789	1.00	2	Tel. Obj. Low Index
6.03	7.0	.0714	1.00	2	Tel. Obj. Steinheil
16.23	1.60	.3130	1.00	3	Catadioptric
6.16	2.8	.1784	1.00	3	Tel. Obj.
6.18	2.8	.1784	1.00	3	Tel. Obj.
6.15	2.8	.1786	1.00	3	Tel. Obj.
6.17	2.8	.1786	1.00	3	Tel. Obj.
6.14	2.8	.1788	1.00	3	Tel. Obj.
6.12	7.0	.0714	1.00	3	Tel. Obj. Apochromat
6.10	7.0	.0723	1.00	3	Tel. Obj. Apochromat
6.11	7.0	.0723	1.00	3	Tel. Obj. Apochromat
22.09	2.5	.1997	1.01	2	Laser Collim.
16.11	1.50	.3328	1.25	5	Catadioptric
16.12	0.99	.5055	1.25	6	Catadioptric
15.07	0.62	.8021	1.39	12	Microscope (m=.025)
22.11	1.18	.4216	1.44	4	Laser Disk Obj.
16.22	10.0	.0496	1.49	3	Catadioptric
16.13	8.0	.0594	1.75	4	Catadioptric

TABLE A.2 Lens Listing in Order of Increasing Field of View (*Continued*)

A Fig.	B F/#	C NA	D HFOV	E Elem	F Type
16.17	8.0	.0627	2.00	6	Catadioptric
16.18	1.90	.2600	2.00	7	catadioptric
15.05	0.88	.5658	2.14	9	Microscope
21.02	1.50	.3331	2.29	2	IR Tel. Obj.
21.03	1.50	.3334	2.29	2	IR Tel. Obj.
21.01	1.50	.3331	2.29	2	IR Tel. Obj>
17.44	1.20	.4151	2.29	8	D-Gauss
16.14	7.0	.0714	2.50	5	Catadioptric
21.08	1.50	.3330	2.55	4	IR Petzval
22.14	1.55	.3183	2.86	5	Laser Collim.
22.17	2.0	.2498	2.86	10	UV Telecentric (m=.20)
21.09	1.50	.3334	3.00	4	IR Petzval
10.09	5.6	.0892	3.00	7	Telephoto
10.11	4.5	.1099	3.09	5	Telephoto
15.08	0.53	.9472	3.13	7	Microscope (m=.016)
16.16	1.48	.3374	3.20	5	Catadioptric
10.12	4.4	.1139	3.39	5	Telephoto
17.33	2.8	.1785	3.79	7	D-Gauss (m=.333)
10.02	5.6	.0893	4.19	4	Telephoto
15.06	0.77	.6521	4.87	4	Microscope (m=.025)
10.05	5.6	.0898	5.00	4	Telephoto
14.04	2.8	.1789	5.01	4	Split Triplet
16.15	1.20	.4166	5.14	5	Catadioptric
17.04	1.60	.3129	5.26	6	D-Gauss
15.02	0.60	1.2886	5.52	6	Microscope (m=.016)
10.06	5.0	.1000	5.99	5	Telephoto
21.05	0.75	.6640	6.00	3	IR Triplet
21.04	0.75	.6647	6.00	3	IR Triplet
20.03	2.8	.1787	6.17	12	Zoom
17.20	3.5	.1427	6.78	6	D-Gauss
13.02	1.60	.3128	6.79	4	Petzval
13.05	1.60	.3124	6.79	5	Petzval
13.08	1.40	.3574	6.79	6	Petzval
13.07	1.40	.3576	6.79	6	Petzval
13.10	1.39	.3613	6.79	6	Petzval
13.13	0.91	.5493	6.84	5	Petzval
21.06	2.0	.2512	7.00	4	IR Quad.
10.08	2.8	.1790	7.00	5	Telephoto
10.07	4.0	.1256	7.50	5	Telephoto
20.03	2.8	.1787	8.75	12	Zoom
21.07	1.55	.3255	8.79	4	IR Quad.
17.16	7.0	.0714	8.94	6	D-Gauss (m=.249)
13.03	1.60	.3095	9.09	4	Petzval
13.06	1.60	.3139	9.09	4	Petzval
10.10	1.80	.2779	9.20	7	Telephoto
22.07	3.0	.1667	9.55	6	Telecentric Scan. Obj:
21.10	0.64	.7795	10.00	4	IR Quad.
21.11	0.55	.9055	10.00	4	IR Quad.
13.04	1.50	.3335	10.20	4	Petzval
15.09	0.53	.9396	11.22	11	Microscope (m=.009)
14.03	2.8	.1811	11.31	4	Split Triplet
17.17	2.6	.1913	11.68	6	D-Gauss (m=.114)

TABLE A.2 Lens Listing in Order of Increasing Field of View (*Continued*)

A Fig.	B F/#	C NA	D HFOV	E Elem	F Type
19.06	2.6	.1913	11.68	6	D-Gauss (m=.114)
8.11	2.9	.1725	11.70	3	Cooke Triplet
8.12	3.5	.1445	11.86	3	Cooke Triplet
17.28	1.50	.3347	11.86	7	D-Gauss
17.34	2.0	.2496	12.00	7	D-Gauss
17.01	1.25	.3992	12.41	6	D-Gauss
13.09	1.40	.3571	12.41	6	Petzval
13.11	1.25	.4008	12.41	6	Petzval
17.38	2.0	.2496	12.41	7	D-Gauss
17.36	1.40	.3579	12.41	7	D-Gauss
20.05	5.3	.0942	12.41	16	Zoom
9.03	1.20	.4151	12.86	7	Retrofocus (m=.005)
9.04	1.20	.4164	12.92	9	Retrofocus (m=.006)
17.02	1.35	.3676	12.95	6	D-Gauss
8.14	4.7	.1054	13.60	3	Cooke Triplet (m=.052)
20.06	133.9	.0037	13.67	5	Zoom Scanner
20.06	94.6	.0053	13.67	5	Zoom Scanner
20.06	66.8	.0075	13.67	5	Zoom Scanner
8.18	8.0	.0624	14.00	3	Cooke Triplet
14.02	1.90	.2621	14.04	4	Split Triplet
20.04	5.5	.0914	14.20	15	Zoom
19.05	2.5	.2002	14.75	6	D-Gauss
19.04	2.5	.1997	14.94	6	D-Gauss (m=.025)
7.03	5.1	.0968	15.00	2	Heugens EP
7.04	5.1	.0984	15.00	2	Ramsden EP
22.01	5.0	.1002	15.00	3	Scanner Obj.
19.03	2.5	.2003	15.00	6	D-Gauss
13.14	1.01	.4929	15.00	6	Petzval
17.07	2.0	.2506	15.02	6	D-Gauss
14.06	2.0	.2505	15.11	4	Split Triplet
14.09	1.50	.3331	15.11	4	Split Triplet
10.03	5.6	.0888	15.11	4	Telephoto
10.04	4.5	.1111	15.11	4	Telephoto
14.15	1.40	.3502	15.11	5	Ernostar
13.12	1.32	.3788	15.11	6	Petzval
17.25	1.70	.2942	15.11	7	D-Gauss
17.35	1.40	.3559	15.11	7	D-Gauss
17.41	1.10	.4521	15.11	8	D-Gauss
20.03	2.8	.1787	15.11	12	Zoom
17.19	4.5	.1114	15.50	6	D-Gauss
17.27	1.20	.4163	15.64	7	D-Gauss
8.05	2.5	.1979	16.01	3	Cooke Triplet
8.06	2.5	.2001	16.17	3	Cooke Triplet
12.20	2.7	.1847	16.17	5	Tessar
14.05	2.5	.1995	17.00	4	Split Triplet
13.01	3.3	.1517	17.22	4	Petzval
14.08	1.90	.2632	17.22	4	Split Triplet
7.12	14.7	.0340	17.22	4	Symmetrical EP
17.03	1.40	.3577	17.22	6	D-Gauss
20.02	4.5	.1110	17.40	9	Zoom
7.01	2.0	.2549	17.43	2	Magnifier (E.P.)
7.06	4.3	.1186	18.00	3	Eyepiece

TABLE A.2 Lens Listing in Order of Increasing Field of View (*Continued*)

A Fig.	B F/#	C NA	D HFOV	E Elem	F Type
7.05	5.1	.0982	18.00	3	Kellner EP
14.13	1.19	.4185	18.00	5	Sonnar
12.14	4.5	.1112	18.26	5	Heliar
12.21	1.80	.2804	18.26	6	Hektor
9.02	3.0	.1621	18.26	6	Retrofocus
17.05	1.70	.2931	18.78	6	D-Gauss
17.43	1.20	.4170	18.78	8	D-Gauss
8.04	3.0	.1667	18.98	3	Cooke Triplet
8.08	2.8	.1772	19.80	3	Cooke Triplet
14.11	2.7	.1830	19.80	4	Sonnar
14.12	2.0	.2475	19.80	6	Sonnar
7.08	5.0	.1001	20.00	4	Orthoscopic EP
7.10	5.1	.0985	20.00	4	Symmetrical EP
12.08	2.8	.1782	20.30	4	Tessar
19.01	1.09	.4614	20.69	3	TV Projection (m=.047)
8.07	2.7	.1857	20.81	3	Cooke Triplet
17.18	3.5	.1426	20.81	6	D-Gauss
14.14	2.8	.1787	20.81	6	Ernostar
17.37	2.0	.2499	20.81	7	D-Gauss
12.09	2.8	.1746	21.00	4	Tessar
9.05	2.8	.1786	21.50	7	Retrofocus
12.05	3.5	.1430	21.80	4	Tessar
17.09	2.0	.2490	21.80	6	D-Gauss
17.08	2.0	.2505	21.80	6	D-Gauss
7.02	2.5	.1979	21.80	6	Magnifier (Brueke)
17.24	1.50	.3339	21.80	7	D-Gauss
17.47	1.50	.3342	21.80	7	D-Gauss
17.42	0.99	.5042	21.80	8	D-Gauss
20.05	4.5	.1114	21.80	16	Zoom
12.04	5.4	.0931	22.04	4	Tessar
17.10	2.0	.2507	22.29	6	D-Gauss
14.07	2.0	.2507	22.50	4	Split Triplet
17.45	1.40	.3467	22.50	7	D-Gauss
14.01	2.5	.1987	22.72	4	Split Triplet
22.06	6.1	.0818	22.73	4	Telecentric Scan. Obj.
17.23	1.50	.3342	22.78	7	D-Gauss
17.22	1.50	.3360	22.78	7	D-Gauss
20.04	4.7	.1079	22.78	15	Zoom
8.03	4.0	.1247	22.98	3	Cooke Triplet
17.12	2.0	.2438	23.00	6	D-Gauss
17.06	1.80	.2763	23.00	6	D-Gauss
17.32	4.0	.1250	23.00	7	D-Gauss
17.21	1.40	.3563	23.00	7	D-Gauss
17.29	1.40	.3571	23.00	7	D-Gauss
17.31	1.40	.3573	23.00	7	D-Gauss
12.11	2.8	.1790	23.27	4	Tessar
12.18	3.0	.1659	23.27	5	Heliar
12.16	3.0	.1663	23.27	5	Heliar
8.09	2.8	.1771	23.50	3	Cooke Triplet
8.10	3.0	.1669	23.75	3	Cooke Triplet
14.10	2.9	.1718	23.75	5	Sonnar
17.30	1.40	.3531	23.75	7	D-Gauss

TABLE A.2 Lens Listing in Order of Increasing Field of View (*Continued*)

A Fig.	B F/#	C NA	D HFOV	E Elem	F Type
11.07	11.1	.0450	23.94	6	Double Meniscus (m=.999)
7.26	4.0	.1243	24.00	6	Eyepiece
17.26	1.75	.2875	24.70	7	D-Gauss
7.14	5.0	.1001	24.99	4	Symmetrical EP
7.22	5.1	.0990	25.00	6	Eyepiece
8.13	4.4	.1127	25.17	3	Cooke Triplet
11.08	4.5	.1113	25.17	4	Celor
7.07	3.3	.1508	25.17	4	Eyepiece
7.15	3.3	.1529	25.17	4	Eyepiece
7.11	5.6	.0895	25.17	4	Symmetrical EP
7.13	5.0	.1000	25.17	4	Symmetrical EP
12.12	2.8	.1774	25.17	4	Tessar
12.15	3.5	.1428	25.17	5	Heliar
12.17	2.8	.1792	25.17	5	Heliar
17.40	2.5	.2000	25.17	7	D-Gauss
19.02	0.89	.5620	25.49	4	TV Projection (m=.024)
22.02	50.0	.0100	25.50	3	Scanner Obj.
11.10	5.5	.0906	26.12	4	Celor (m=.251)
12.10	2.8	.1755	26.57	4	Tessar
11.05	8.0	.0622	26.57	6	Double Meniscus
17.14	8.0	.0625	26.60	6	D-Gauss
8.02	6.3	.0795	27.02	3	Cooke Triplet
8.17	6.3	.0800	27.02	3	Cooke Triplet
17.13	3.6	.1386	27.02	6	D-Gauss
11.09	5.4	.0925	27.66	4	Celor
12.03	4.5	.1103	27.92	4	Tessar
12.02	4.5	.1111	27.92	4	Tessar
12.06	3.6	.1391	27.92	4	Tessar
12.19	4.4	.1144	27.92	5	Tessar
17.11	2.0	.2500	27.92	6	D-Gauss
17.39	2.5	.2005	27.92	7	D-Gaiss
12.07	3.0	.1682	28.00	4	Tessar
17.15	2.0	.2505	28.00	6	D-Gauss
20.05	4.2	.1186	28.12	16	Zoom (m=.181)
22.04	18.5	.0271	28.29	4	Scanner Obj.
7.27	5.1	.0985	29.84	6	Eyepiece
7.09	5.1	.0988	30.00	4	Eyepiece
7.18	5.1	.0987	30.00	5	Erfle EP
22.03	19.7	.0254	30.10	4	Scanner Obj.
8.16	6.3	.0797	30.11	3	Cooke Triplet
12.01	6.3	.0793	30.11	4	Tessar
12.13	4.5	.1114	30.11	5	Heliar
7.25	5.1	.0990	30.11	6	Eyepiece
22.05	23.8	.0210	30.50	5	Scanner Obj.
11.04	12.2	.0410	30.96	4	Double Meniscus
20.01	3.5	.1433	31.30	9	Zoom
18.07	4.0	.1259	31.50	4	WA
7.24	5.0	.1006	31.80	6	Eyepiece
17.46	1.40	.3476	32.00	7	D-Gauss
9.07	2.8	.1787	32.01	6	Retrofocus
9.06	2.0	.2510	32.04	7	Retrofocus
20.02	3.5	.1429	32.30	9	Zoom

TABLE A.2 Lens Listing in Order of Increasing Field of View (*Continued*)

A Fig.	B F/#	C NA	D HFOV	E Elem	F Type
7.20	5.0	.1002	32.50	5	Erfle EP
11.06	4.5	.1103	33.82	6	Double Meniscus
8.15	6.3	.0798	34.00	3	Cooke Triplet
7.19	5.0	.0999	34.99	5	Erfle EP
7.16	5.0	.1007	34.99	5	Eyepiece
7.21	5.0	.1003	34.99	6	Int. Foc. EP
20.04	4.0	.1251	36.19	15	Zoom
20.05	4.0	.1253	36.38	16	Zoom
20.01	3.5	.1427	36.60	9	Zoom
9.08	2.8	.1785	37.00	5	Retrofocus
18.06	5.6	.0901	37.55	6	WA
7.17	5.0	.1002	37.60	5	Eyepiece
18.05	2.8	.1795	38.00	7	WA
7.23	4.0	.1256	40.00	6	Eyepiece
9.09	4.0	.1255	40.00	8	Retrofocus
18.04	5.0	.1004	40.03	8	WA
18.03	3.4	.1466	40.03	8	WA
20.01	3.5	.1430	42.00	9	Zoom
9.10	2.0	.2494	42.00	11	Retrofocus
7.28	5.1	.0978	45.00	7	Eyepiece
7.29	4.0	.1252	45.00	7	Eyepiece
9.11	2.5	.2027	45.00	8	Retrofocus
18.02	4.7	.1052	45.00	8	WA
9.12	2.0	.2515	47.00	13	Retrofocus
11.03	6.3	.0793	49.96	5	Double Meniscus
11.02	6.3	.0787	50.00	4	Double Meniscus
18.01	5.0	.0996	50.00	10	WA
18.08	8.3	.0618	56.31	6	WA
11.01	20.6	.0267	67.04	2	Double Meniscus
9.15	4.0	.1247	72.00	6	Retrofocus
9.14	1.00	.5004	72.00	13	Retrofocus
9.18	7.9	.0632	80.00	4	Retrofocus
9.17	8.0	.0626	85.40	9	Retrofocus
9.16	2.8	.1778	87.50	8	Retrofocus
21.14	Afocal	16x		4	IR
21.12	Afocal	12x		5	IR Catadioptric
21.15	Afocal	7x		6	IR
21.13	Afocal	5x		6	IR
16.21	Afocal	5x		3	Mirror

Index

About the Authors

WARREN J. SMITH is perhaps the world's leading authority in optical design and engineering. Currently chief scientist at Kaiser Electro-Optics, Inc. (Carlsbad, California), he is a fellow and past president of both the Optical Society of America and the International Society for Optical Engineering (SPIE). He is the author of McGraw-Hill's *Modern Optical Engineering*, now in a second edition, and a series editor of the *McGraw-Hill Optical and Electro-Optical Engineering Series*.

GENESEE OPTICS SOFTWARE, INC. (Rochester, New York) is the world's largest supplier of personal computer software for optical design, and the second largest supplier of optical design software in general.